Xinli Lu

April 20, 10

Analysis and Evaluation of Pumping Test Data

Second Edition (Completely Revised)

Analysis and Evaluation of Pumping Test Data

Second Edition (Completely Revised)

G.P. Kruseman
Senior hydrogeologist, TNO Institute of Applied Geoscience, Delft

N.A. de Ridder
Senior hydrogeologist, International Institute for Land Reclamation and Improvement, Wageningen
and
Professor in Hydrogeology, Free University, Amsterdam

With assistance from
J.M. Verweij
Freelance hydrogeologist

Publication 47

International Institute for Land Reclamation and Improvement,
P.O. Box 45, 6700 AA Wageningen, The Netherlands, 1991

The first edition of this book appeared as No. 11 in the series of Bulletins of the International Institute for Land Reclamation and Improvement/ILRI. Because the ILRI Bulletins have now been discontinued, this completely revised edition of the book appears as ILRI Publication 47.

The production of the book was made possible by cooperation between the following institutions:

International Institute for Land Reclamation and Improvement, Wageningen
TNO Institute of Applied Geoscience, Delft
Institute for Earth Sciences, Free University/VU, Amsterdam

First Edition 1970
Reprinted 1973
Reprinted 1976
Reprinted 1979
Reprinted 1983
Reprinted 1986
Reprinted 1989

Second Edition 1990
Reprinted 1991

ISBN 90 70754 207

Printed in The Netherlands

Preface

This is the second edition of *Analysis and Evaluation of Pumping Test Data.* Readers familiar with the first edition and its subsequent impressions will note a number of changes in the new edition. These changes involve the contents of the book, but not the philosophy behind it, which is to be a practical guide to all who are organizing, conducting, and interpreting pumping tests.

What changes have we made? In the first place, we have included the step-drawdown test, the slug test, and the oscillation test. We have also added three chapters on pumping tests in fractured rocks. This we have done because of comments from some of our reviewers, who regretted that the first edition contained nothing about tests in fractured rocks. It would be remiss of us, however, not to warn our readers that, in spite of the intense research that fractured rocks have undergone in the last two decades, the problem is still the subject of much debate. What we present are some of the common methods, but are aware that they are based on ideal conditions which are rarely met in nature. All the other methods, however, are so complex that one needs a computer to apply them.

We have also updated the book in the light of developments that have taken place since the first edition appeared some twenty years ago. We present, for instance, a more modern method of analyzing pumping tests in unconfined aquifers with delayed yield. We have also re-evaluated some of our earlier field examples and have added several new ones.

Another change is that, more than before, we emphasize the intricacy of analyzing field data, showing that the drawdown behaviour of totally different aquifer systems can be very similar.

It has become a common practice nowadays to use computers in the analysis of pumping tests. For this edition of our book, we seriously considered adding computer codes, but eventually decided not to because they would have made the book too voluminous and therefore too costly. Other reasons were the possible incompatibility of computer codes and, what is even worse, many of the codes are based on 'black box' methods which do not allow the quality of the field data to be checked. Interpreting a pumping test is not a matter of feeding a set of field data into a computer, tapping a few keys, and expecting the truth to appear. The only computer codes with merit are those that take over the tedious work of plotting the field data and the type curves, and display them on the screen. These computer techniques are advancing rapidly, but we have refrained from including them. Besides, the next ILRI Publication (No. 48, *SATEM: Selected Aquifer Test Evaluation Methods* by J. Boonstra) presents the most common well-flow equations in computerized form. As well, the International Ground-Water Modelling Centre in Indianapolis, U.S.A., or its branch office in Delft, The Netherlands, can provide all currently available information on computer codes.

Our wish to revise and update our book could never have been realized without the support and help of many people. We are grateful to Mr. F. Walter, Director of TNO Institute of Applied Geoscience, who made it possible for the first author and Ms

Hanneke Verwey to work on the book. We are also grateful to Brigadier (Retired) K.G. Ahmad, General Manager (Water) of the Water and Power Development Authority, Pakistan, for granting us permission to use pumping test data not officially published by his organization.

We also express our thanks to Dr J.A.H. Hendriks, Director of ILRI, who allowed the second author time to work on the book, and generously gave us the use of ILRI's facilities, including the services of Margaret Wiersma-Roche, who edited our manuscript and corrected our often wordy English. We are indebted to Betty van Aarst and Joop van Dijk for their meticulous drawings, and to Trudy Pleijsant-Paes for her patience and perseverance in processing the words and the equations of the book. Last, but by no means least, we thank ILRI's geohydrologist, Dr J. Boonstra, for his discussion of the three chapters on fractured rocks and his valuable contribution to their final draft.

We hope that this revised and updated edition of *Analysis and Evaluation of Pumping Test Data* will serve its readers as the first edition did. Any comments anyone would care to make will be received with great interest.

G.P. Kruseman
N.A. de Ridder

Contents

1 Basic concepts and definitions

When working on problems of groundwater flow, the geologist or engineer has to find reliable values for the hydraulic characteristics of the geological formations through which the groundwater is moving. Pumping tests have proved to be one of the most effective ways of obtaining such values.

Analyzing and evaluating pumping test data, however, is as much an art as a science. It is a science because it is based on theoretical models that the geologist or engineer must understand and on thorough investigations that he must conduct into the geological formations in the area of the test. It is an art because different types of aquifers can exhibit similar drawdown behaviours, which demand interpretational skills on the part of the geologist or engineer. We hope that this book will serve as a guide in both the science and the art.

The equations we present in this book are from well hydraulics. We have omitted any lengthy derivations of the equations because these can be found in the original publications listed in our References. With some exceptions, we present the equations in their final form, emphasizing the assumptions and conditions that underlie them, and outlining the procedures that are to be followed for their successful application.

'Hard rocks', both as potential sources of water and depositories for chemical or radioactive wastes, are receiving increasing attention in hydrogeology. We shall therefore be discussing some recent developments in the interpretation of pumping test data from such rocks.

This chapter summarizes the basic concepts and definitions of terms relevant to our subject. The next chapter describes how to conduct a pumping test. The remaining chapters all deal with the analysis and evaluation of pumping test data from a variety of aquifer types or aquifer systems, and from tests conducted under particular technical conditions.

1.1 Aquifer, aquitard, and aquiclude

An aquifer is defined as a saturated permeable geological unit that is permeable enough to yield economic quantities of water to wells. The most common aquifers are unconsolidated sand and gravels, but permeable sedimentary rocks such as sandstone and limestone, and heavily fractured or weathered volcanic and crystalline rocks can also be classified as aquifers.

An aquitard is a geological unit that is permeable enough to transmit water in significant quantities when viewed over large areas and long periods, but its permeability is not sufficient to justify production wells being placed in it. Clays, loams and shales are typical aquitards.

An aquiclude is an impermeable geological unit that does not transmit water at all. Dense unfractured igneous or metamorphic rocks are typical aquicludes. In nature, truly impermeable geological units seldom occur; all of them leak to some extent, and must therefore be classified as aquitards. In practice, however, geological units can be classified as aquicludes when their permeability is several orders of magnitude lower than that of an overlying or underlying aquifer.

The reader will note that the above definitions are relative ones; they are purposely imprecise with respect to permeability.

1.2 Aquifer types

There are three main types of aquifer: confined, unconfined, and leaky (Figure 1.1).

1.2.1 Confined aquifer

A confined aquifer (Figure 1.1A) is bounded above and below by an aquiclude. In a confined aquifer, the pressure of the water is usually higher than that of the atmosphere, so that if a well taps the aquifer, the water in it stands above the top of the aquifer, or even above the ground surface. We then speak of a free-flowing or artesian well.

1.2.2 Unconfined aquifer

An unconfined aquifer (Figure 1.1B), also known as a watertable aquifer, is bounded below by an aquiclude, but is not restricted by any confining layer above it. Its upper boundary is the watertable, which is free to rise and fall. Water in a well penetrating an unconfined aquifer is at atmospheric pressure and does not rise above the watertable.

1.2.3 Leaky aquifer

A leaky aquifer (Figure 1.1C and D), also known as a semi-confined aquifer, is an aquifer whose upper and lower boundaries are aquitards, or one boundary is an aquitard and the other is an aquiclude. Water is free to move through the aquitards, either upward or downward. If a leaky aquifer is in hydrological equilibrium, the water level in a well tapping it may coincide with the watertable. The water level may also stand above or below the watertable, depending on the recharge and discharge conditions.

In deep sedimentary basins, an interbedded system of permeable and less permeable layers that form a multi-layered aquifer system (Figure 1.1E), is very common. But such an aquifer system is more a succession of leaky aquifers, separated by aquitards, rather than a main aquifer type.

1.3 Anisotropy and heterogeneity

Most well hydraulics equations are based on the assumption that aquifers and aquitards are homogeneous and isotropic. This means that the hydraulic conductivity is the same throughout the geological formation and is the same in all directions (Figure

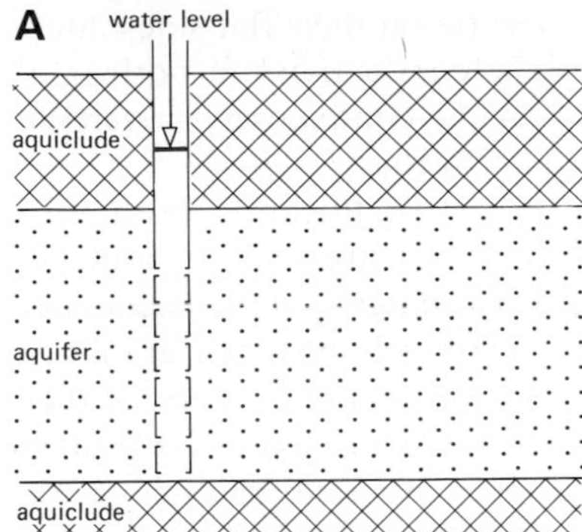

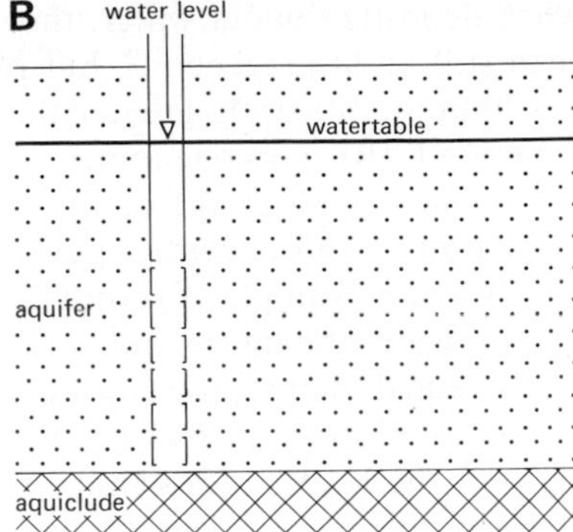

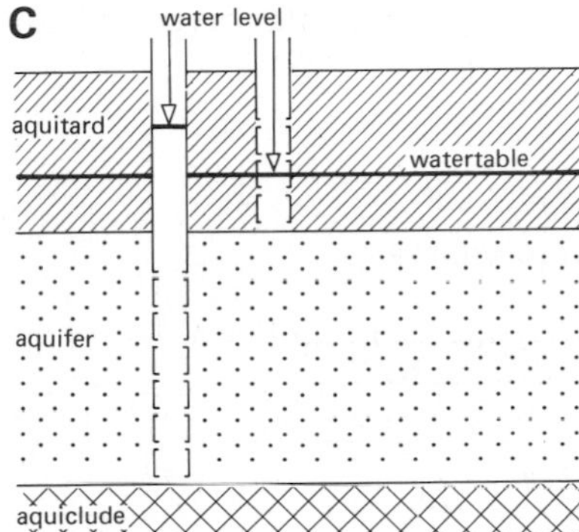

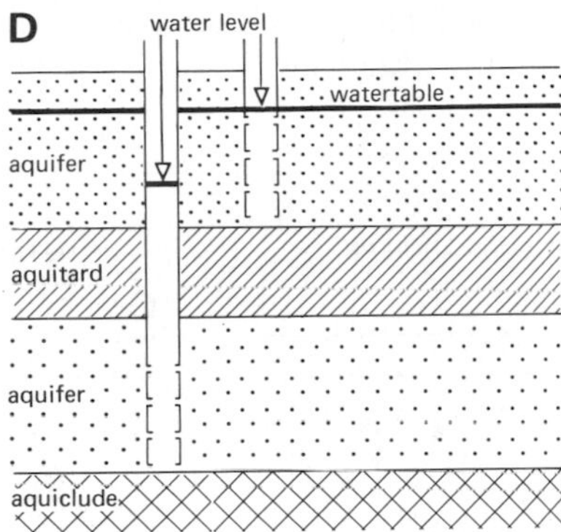

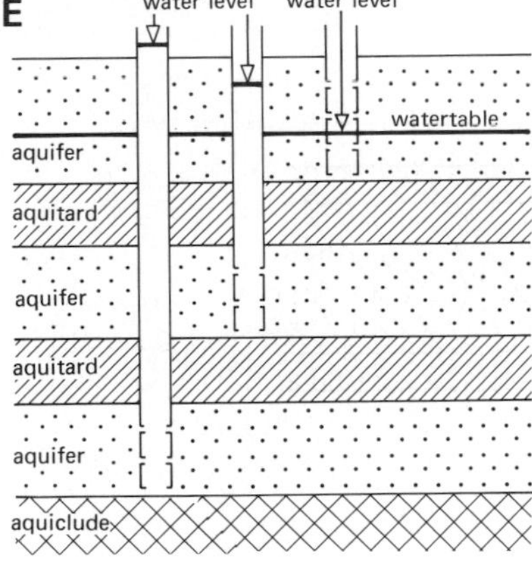

Figure 1.1 Different types of aquifers
A. Confined aquifer
B. Unconfined aquifer
C. and D. Leaky aquifers
E. Multi-layered leaky aquifer

1.2A). The individual particles of a geological formation, however, are seldom spherical so that, when deposited under water, they tend to settle on their flat sides. Such a formation can still be homogeneous, but its hydraulic conductivity in horizontal direction, K_h, will be significantly greater than its hydraulic conductivity in vertical direction, K_v (Figure 1.2B). This phenomenon is called anisotropy.

The lithology of most geological formations tends to vary significantly, both horizontally and vertically. Consequently, geological formations are seldom homogeneous. Figure 1.2C is an example of layered heterogeneity. Heterogeneity occurs not only in the way shown in the figure: individual layers may pinch out; their grain size may vary in horizontal direction; they may contain lenses of other grain sizes; or they may be discontinuous by faulting or scour-and-fill structures. In horizontally-stratified alluvial formations, the K_h/K_v ratios range from 2 to 10, but values as high as 100 can occur, especially where clay layers are present.

Anisotropy is a common property of fractured rocks (Figure 1.2D). The hydraulic conductivity in the direction of the main fractures is usually significantly greater than that normal to those fractures.

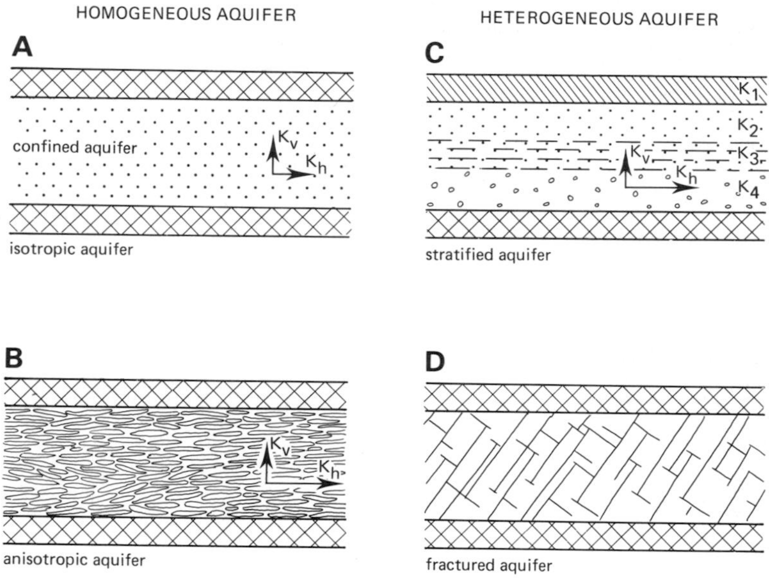

Figure 1.2 Homogeneous and heterogeneous aquifers, isotropic and anisotropic
A. Homogeneous aquifer, isotropic
B. Homogeneous aquifer, anisotropic
C. Heterogeneous aquifer, stratified
D. Heterogeneous aquifer, fractured

If the principal directions of anisotropy are known, one can transform an anisotropic system into an isotropic system by changing the coordinates. In the new coordinate system, the basic well-flow equation is again isotropic and the common equations can be used.

1.4 Bounded aquifers

Another common assumption in well hydraulics is that the pumped aquifer is horizontal and of infinite extent. But, viewed on a regional scale, some aquifers slope, and none of them extend to infinity because complex geological processes cause interfingering of layers and pinchouts of both aquifers and aquitards. At some places, aquifers and aquitards are cut by deeply incised channels, estuaries, or the ocean. In other words, aquifers and aquitards are laterally bounded in one way or another. Figure 1.3 shows some examples. The interpretation of pumping tests conducted in the vicinity of such boundaries requires special techniques, which we shall be discussing.

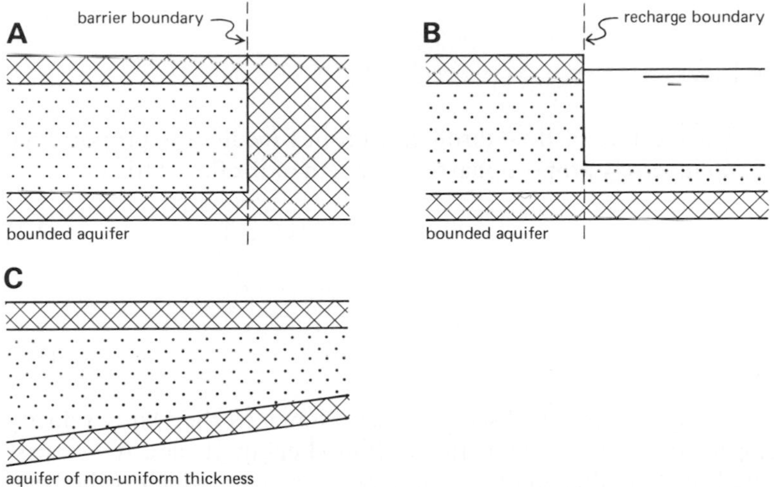

Figure 1.3 Bounded aquifers
A, B, and C

1.5 Steady and unsteady flow

There are two types of well-hydraulics equations: those that describe steady-state flow towards a pumped well and those that describe the unsteady-state flow.

Steady-state flow is independent of time. This means that the water level in the pumped well and in surrounding piezometers does not change with time. Steady-state flow occurs, for instance, when the pumped aquifer is recharged by an outside source, which may be rainfall, leakage through aquitards from overlying and/or underlying unpumped aquifers, or from a body of open water that is in direct hydraulic contact with the pumped aquifer. In practice, it is said that steady-state flow is attained if the changes in the water level in the well and piezometers have become so small with time that they can be neglected. As pumping continues, the water level may drop further, but the hydraulic gradient induced by the pumping will not change. In other words, the flow towards the well has attained a pseudo-steady-state.

In well hydraulics of fractured aquifers, the term pseudo-steady-state is used for the interporosity flow from the matrix blocks to the fractures. This flow occurs in response to the difference between the average hydraulic head in the matrix blocks and the average hydraulic head in the fractures. Spatial variation in hydraulic head gradients in the matrix blocks is ignored and the flow through the fractures to the well is radial and unsteady.

Unsteady-state flow occurs from the moment pumping starts until steady-state flow is reached. Consequently, if an infinite, horizontal, completely confined aquifer of constant thickness is pumped at a constant rate, there will always be unsteady-state flow. In practice, the flow is considered to be unsteady as long as the changes in water level in the well and piezometers are measurable or, in other words, as long as the hydraulic gradient is changing in a measurable way.

1.6 Darcy's law

Darcy's law states that the rate of flow through a porous medium is proportional to the loss of head, and inversely proportional to the length of the flow path, or

$$v = K\frac{\Delta h}{\Delta l} \tag{1.1}$$

or, in differential form

$$v = K\frac{dh}{dl} \tag{1.2}$$

where $v = Q/A$, which is the specific discharge, also known as the Darcy velocity or Darcy flux (Length/Time), Q = volume rate of flow (Length^3/Time), A = cross-sectional area normal to flow direction (Length^2), $\Delta h = h_2 - h_1$, which is the head loss, whereby h_1 and h_2 are the hydraulic heads measured at Points 1 and 2 (Length), Δl = the distance between Points 1 and 2 (Length), $dh/dl = i$, which is the hydraulic gradient (dimensionless), and K = constant of proportionality known as the hydraulic conductivity (Length/Time).

Alternatively, Darcy's law can be written as

$$Q = K\frac{dh}{dl}A \tag{1.3}$$

Note that the specific discharge v has the dimensions of a velocity, i.e. Length/Time. The concept specific discharge assumes that the water is moving through the entire porous medium, solid particles as well as pores, and is thus a macroscopic concept. The great advantage of this concept is that the specific discharge can be easily measured. It must, however, be clearly differentiated from the microscopic velocities, which are real velocities. Hence, if we are interested in real flow velocities, as in problems of groundwater pollution and solute transport, we must consider the actual paths of individual water particles as they find their way through the pores of the medium. In other words, we must consider the porosity of the transmitting medium and can write

$$v_a = \frac{v}{n} \text{ or } v_a = \frac{Q}{nA} \tag{1.4}$$

where v_a = real velocity of the flow, and n = porosity of the water-transmitting medium.

In using Darcy's law, one must know the range of its validity. After all, Darcy (1856) conducted his experiments on sand samples in the laboratory. So, Darcy's law is valid for laminar flow, but not for turbulent flow, as may happen in cavernous limestone or fractured basalt. In case of doubt, one can use the Reynolds number as a criterion to distinguish between laminar and turbulent flow. The Reynolds number is expressed as

$$N_R = \rho \frac{vd}{\mu} \tag{1.5}$$

where ρ is the fluid density, v is the specific discharge, μ is the viscosity of the fluid, and d is a representative length dimension of the porous medium, usually taken as a mean grain diameter or a mean pore diameter.

Experiments have shown that Darcy's law is valid for $N_R < 1$ and that no serious errors are created up to $N_R = 10$. This value thus represents an upper limit to the validity of Darcy's law. It should not be considered a unique limit, however, because turbulence occurs gradually. At full turbulence ($N_R < 100$), the head loss varies approximately with the second power of the velocity rather than linearly. Fortunately, most groundwater flow occurs with $N_R < 1$ so that Darcy's law applies. Only in exceptional situations, as in a rock with wide openings, or where steep hydraulic gradients exist, as in the near vicinity of a pumped well, will the criterion of laminar flow not be satisfied and Darcy's law will be invalid.

Darcy's law is also invalid at low hydraulic gradients, as may occur in compact clays, because, for low values of i, the relation between v and i is not linear. It is impossible to give a unique lower limit to the hydraulic gradients at which Darcy's law is still valid, because the values of i vary with the type and structure of the clay, while the mineral content of the water also plays a role (De Marsily 1986).

1.7 Physical properties

In the equations describing the flow to a pumped well, various physical properties and parameters of aquifers and aquitards appear. These will be discussed below.

1.7.1 Porosity (n)

The porosity of a rock is its property of containing pores or voids. If we divide the total unit volume V_T of an unconsolidated material into the volume of its solid portion

V_s and the volume of its voids V_v, we can define the porosity as $n = V_v/V_T$. Porosity is usually expressed as a decimal fraction or as a percentage.

With consolidated and hard rocks, a distinction is usually made between primary porosity, which is present when the rock is formed, and secondary porosity, which develops later as a result of solution or fracturing. As Figure 1.4 shows, fractures can be oriented in three main directions, which cut the rock into blocks. In theory, the primary porosity of a dense solid rock may be zero and the rock matrix will be impermeable. Such a rock can be regarded as a single-porosity system (Figure 1.4A). In some rocks, notably crystalline rocks, the main fractures are accompanied by a dense system of microfissures, which considerably increase the porosity of the rock matrix (Figure 1.4B). In contrast, the primary porosity of granular geological formations (e.g. sandstone) can be quite significant (Figure 1.4C). When such a formation is fractured, it can be regarded as a double-porosity system because the two types of porosities coexist: the primary or matrix porosity and the secondary or fracture porosity.

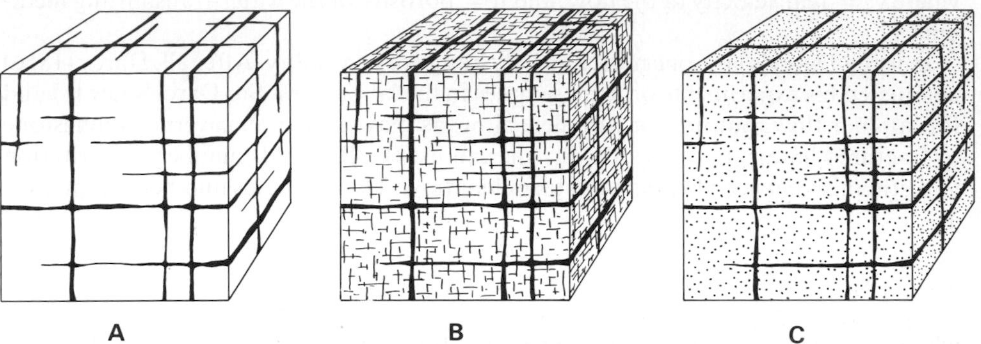

Figure 1.4 Porosity systems
A. Single porosity
B. Microfissures
C. Double porosity

Table 1.1 gives some porosity values for unconsolidated materials and rocks.

Table 1.1 Range of porosity values (n) in percentages

Rocks		Unconsolidated materials	
Sandstone	5–30	Gravel	25–40
Limestone	0–20	Sand	25–50
Karstic limestone	5–50	Silt	35–50
Shale	0–10	Clay	40–70
Basalt, fractured	5–50		
Crystalline rock	0–5		
Crystalline rock, fractured	0–10		

1.7.2 Hydraulic conductivity (K)

The hydraulic conductivity is the constant of proportionality in Darcy's law (Equation 1.3). It is defined as the volume of water that will move through a porous medium in unit time under a unit hydraulic gradient through a unit area measured at right angles to the direction of flow. Hydraulic conductivity can have any units of Length/Time, for example m/d.

The hydraulic conductivity of fractured rocks depends largely on the density of the fractures and the width of their apertures. Fractures can increase the hydraulic conductivity of solid rocks by several orders or magnitude.

The significant effect that fractures can have on the hydraulic conductivity of hard rocks has been treated by various authors. Maini and Hocking (1977), for example, as quoted by De Marsily (1986), give the equivalence between the hydraulic conductivity of a fractured rock and that of a porous (granular) aquifer. From their diagram, it follows that the flow through, say, a 100 m thick cross-section of a porous medium with a hydraulic conductivity of 10^{-12} m/d could, in a fractured medium with an impermeable rock matrix, also come from one single fracture only 0.2 mm wide.

For orders of magnitude of K for different materials, see Table 1.2.

Table 1.2 Order of magnitude of K for different kinds of rock (from Bouwer 1978)

	Geological classification	K (m/d)
Unconsolidated materials:		
	Clay	10^{-8} – 10^{-2}
	Fine sand	1 – 5
	Medium sand	5 – 2×10^{1}
	Coarse sand	2×10^{1} – 10^{2}
	Gravel	10^{2} – 10^{3}
	Sand and gravel mixes	5 – 10^{2}
	Clay, sand, gravel mixes (e.g. till)	10^{-3} – 10^{-1}
Rocks:		
	Sandstone	10^{-3} –1
	Carbonate rock with secondary porosity	10^{-2} –1
	Shale	10^{-7}
	Dense solid rock	$< 10^{-5}$
	Fractured or weathered rock	
	(Core samples)	Almost 0 – 3×10^{2}
	Volcanic rock	Almost 0 – 10^{3}

1.7.3 Interporosity flow coefficient (λ)

When a confined fractured aquifer of the double-porosity type is pumped, the interporosity flow coefficient controls the flow in the aquifer. It indicates how easily water can flow from the aquifer matrix blocks into the fractures, and is defined as

$$\lambda = \alpha r^2 \frac{K_m}{K_f} \qquad (1.6)$$

where α is a shape factor that reflects the geometry of the matrix blocks, r is the distance to the well, K is hydraulic conductivity, f is the fracture, and m is matrix block. The dimension of λ is reciprocal area.

1.7.4 Compressibility (α and β)

Compressibility is an important material and fluid property in the analysis of unsteady flow to wells. It describes the change in volume or the strain induced in an aquifer (or aquitard) under a given stress, or

$$\alpha = \frac{-dV_T/V_T}{d\sigma_e} \tag{1.7}$$

where V_T is the total volume of a given mass of material and $d\sigma_e$ is the change in effective stress. Compressibility is expressed in m^2/N or Pa^{-1}. Its value for clay ranges from 10^{-6} to 10^{-8}, for sand from 10^{-7} to 10^{-9}, for gravel and fractured rock from 10^{-8} to 10^{-10} m^2/N.

Similarly, the compressibility of water is defined as

$$\beta = \frac{-dV_w/V_w}{dp} \tag{1.8}$$

A change in the water pressure dp induces a change in the volume V_w of a given mass of water. The compressibility of groundwater under the range of temperatures that are usually encountered can be taken constant as 4.4×10^{-10} m^2/N (or Pa^{-1}).

1.7.5 Transmissivity (KD or T)

Transmissivity is the product of the average hydraulic conductivity K and the saturated thickness of the aquifer D. Consequently, transmissivity is the rate of flow under a unit hydraulic gradient through a cross-section of unit width over the whole saturated thickness of the aquifer. The effective transmissivity, as used for fractured media, is defined as

$$T = \sqrt{T_{f(x)} T_{f(y)}} \tag{1.9}$$

where f refers to the fractures and x and y to the principal axes of permeability. Transmissivity has the dimensions of $Length^3/Time \times Length$ or $Length^2/Time$ and is, for example, expressed in m^2/d or m^2/s.

1.7.6 Specific storage (S_s)

The specific storage of a saturated confined aquifer is the volume of water that a unit volume of aquifer releases from storage under a unit decline in hydraulic head. This release of water from storage under conditions of decreasing head h stems from the compaction of the aquifer due to increasing effective stress σ_e and the expansion of the water due to decreasing pressure p. Hence, the earlier-defined compressibilities of material and water play a role in these two mechanisms. The specific storage is defined as

$$S_s = \rho g(\alpha + n\beta) \qquad (1.10)$$

where ρ is the mass density of water (M/L^3), g is the acceleration due to gravity (N/L^3), and the other symbols are as defined earlier. The dimension of specific storage is $Length^{-1}$.

1.7.7 Storativity (S)

The storativity of a saturated confined aquifer of thickness D is the volume of water released from storage per unit surface area of the aquifer per unit decline in the component of hydraulic head normal to that surface. In a vertical column of unit area extending through the confined aquifer, the storativity S equals the volume of water released from the aquifer when the piezometric surface drops over a unit distance. Storativity is defined as

$$S = \rho g D(\alpha + n\beta) = S_s D \qquad (1.11)$$

As storativity involves a volume of water per volume of aquifer, it is a dimensionless quantity. Its values in confined aquifers range from 5×10^{-5} to 5×10^{-3}.

1.7.8 Storativity ratio (ω)

The storativity ratio is a parameter that controls the flow from the aquifer matrix blocks into the fractures of a confined fractured aquifer of the double-porosity type. (See also Sections 1.7.1 and 1.7.3.) It is defined as

$$\omega = \frac{S_f}{S_f + S_m} \qquad (1.12)$$

where S is the storativity and f is fracture and m is matrix block. Being a ratio, ω is dimensionless.

1.7.9 Specific yield (S_y)

The specific yield is the volume of water that an unconfined aquifer releases from storage per unit surface area of aquifer per unit decline of the watertable. The values of the specific yield range from 0.01 to 0.30 and are much higher than the storativities of confined aquifers. In unconfined aquifers, the effects of the elasticity of the aquifer matrix and of the water are generally negligible. Specific yield is sometimes called effective porosity, unconfined storativity, or drainable pore space. Small interstices do not contribute to the effective porosity because the retention forces in them are greater than the weight of water. Hence, no groundwater will be released from small interstices by gravity drainage.

It is obvious that water can only move through pores that are interconnected. Hard rocks may contain numerous unconnected pores in which the water is stagnant. The most common example is that of secondary dolomite. Dolomitization increases the porosity because the diagenetic transformation of calcite into dolomite is accompanied by a 13% reduction in volume of the rock (Matthess 1982). The porosity of secondary dolomite is high, 20 to 30%, but the effective porosity is low because the pores are seldom interconnected. Water in 'dead-end' pores is also almost stagnant, so such pores are excluded from the effective porosity. They do play a role, of course, when one is studying the mechanisms of compressibility and solute transport in porous media.

In fractured rocks, water only moves through the fractures, even if the unfractured matrix blocks are porous. This means that the effective porosity of the rock mass is linked to the volume of these fractures. A fractured granite, for example, has a matrix porosity of 1 to 2 %, but its effective porosity is less than 1% because the matrix itself has a very low permeability (De Marsily 1986).

Table 1.3 gives some representative values of specific yields for different materials.

Table 1.3 Representative values of specific yield (Johnson 1967)

Material	S_y	Material	S_y
Coarse gravel	23	Limestone	14
Medium gravel	24	Dune sand	38
Fine gravel	25	Loess	18
Coarse sand	27	Peat	44
Medium sand	28	Schist	26
Fine sand	23	Siltstone	12
Silt	8	Silty till	6
Clay	3	Sandy till	16
Fine-grained sandstone	21	Gravelly till	16
Medium-grained sandstone	27	Tuff	21

1.7.10 Diffusivity (KD/S)

The hydraulic diffusivity is the ratio of the transmissivity and the storativity of a saturated aquifer. It governs the propagation of changes in hydraulic head in the aquifer. Diffusivity has the dimension of $\text{Length}^2/\text{Time}$.

1.7.11 Hydraulic resistance (c)

The hydraulic resistance characterizes the resistance of an aquitard to vertical flow, either upward or downward. It is the reciprocal of the leakage or leakage coefficient K'/D' in Darcy's law when this law is used to characterize the amount of leakage through the aquitard; K' = the hydraulic conductivity of the aquitard for vertical flow, and D' = the thickness of the aquitard. The hydraulic resistance is thus defined as

$$c = \frac{D'}{K'} \tag{1.13}$$

and has the dimension of Time. It is often expressed in days. Values of c vary widely, from some hundreds of days to several ten thousand days; for aquicludes, c is infinite.

1.7.12 Leakage factor (L)

The leakage factor, or characteristic length, is a measure for the spatial distribution of the leakage through an aquitard into a leaky aquifer and vice versa. It is defined as

$$L = \sqrt{KDc} \tag{1.14}$$

Large values of L indicate a low leakage rate through the aquitard, whereas small values of L mean a high leakage rate. The leakage factor has the dimension of Length, expressed, for example, in metres.

2 Pumping tests

2.1 The principle

The principle of a pumping test is that if we pump water from a well and measure the discharge of the well and the drawdown in the well and in piezometers at known distances from the well, we can substitute these measurements into an appropriate well-flow equation and can calculate the hydraulic characteristics of the aquifer (Figure 2.1).

2.2 Preliminary studies

Before a pumping test is conducted, geological and hydrological information on the following should be collected:

- The geological characteristics of the subsurface (i.e. all those lithological, stratigraphic, and structural features that may influence the flow of groundwater);
- The type of aquifer and confining beds;
- The thickness and lateral extent of the aquifer and confining beds:
  - The aquifer may be bounded laterally by barrier boundaries of impermeable material (e.g. the bedrock sides of a buried valley, a fault, or simply lateral changes in the lithology of the aquifer material);
  - Of equal importance are any lateral recharge boundaries (e.g. where the aquifer is in direct hydraulic contact with a deeply incised perennial river or canal, a lake,

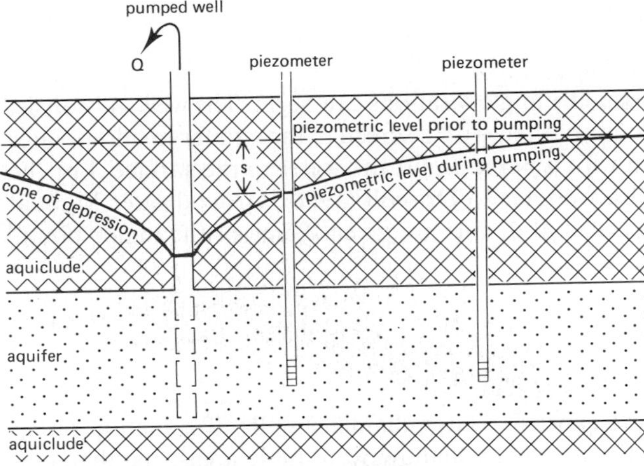

Figure 2.1 Drawdown in a pumped aquifer

or the ocean) or any horizontal recharge boundaries (e.g. where percolating rain or irrigation water causes the watertable of an unconfined aquifer to rise, or where an aquitard leaks and recharges the aquifer);
- Data on the groundwater-flow system: horizontal or vertical flow of groundwater, watertable gradients, and regional trends in groundwater levels;
- Any existing wells in the area. From the logs of these wells, it may be possible to derive approximate values of the aquifer's transmissivity and storativity and their spatial variation. It may even be possible to use one of those wells for the test, thereby reducing the cost of field work. Sometimes, however, such a well may produce uncertain results because details of its construction and condition are not available.

2.3 Selecting the site for the well

When an existing well is to be used for the test or when the hydraulic characteristics of a specific location are required, the well site is predetermined and one cannot move to another, possibly more suitable site. When one has the freedom to choose, however, the following points should be kept in mind:
- The hydrogeological conditions should not change over short distances and should be representative of the area under consideration, or at least a large part of it;
- The site should not be near railways or motorways where passing trains or heavy traffic might produce measurable fluctuations in the hydraulic head of a confined aquifer;
- The site should not be in the vicinity of existing discharging wells;
- The pumped water should be discharged in a way that prevents its return to the aquifer;
- The gradient of the watertable or piezometric surface should be low;
- Manpower and equipment must be able to reach the site easily.

2.4 The well

After the well site has been chosen, drilling operations can begin. The well will consist of an open-ended pipe, perforated or fitted with a screen in the aquifer to allow water to enter the pipe, and equipped with a pump to lift the water to the surface. For the design and construction of wells, we refer to Driscoll (1986), Groundwater Manual (1981), and Genetier (1984), where full details are given. Some of the major points are summarized below.

2.4.1 Well diameter

A pumping test does not require expensive large-diameter wells. If a suction pump placed on the ground surface is used, as in shallow watertable areas, the diameter of the well can be small. A submersible pump requires a well diameter large enough to accommodate the pump.

The diameter of the well can be varied without greatly affecting the yield of the well. Doubling the diameter would only increase the yield by about 10 per cent, other things being equal.

2.4.2 Well depth

The depth of the well will usually be determined from the log of an exploratory bore hole or from the logs of nearby existing wells, if any. The well should be drilled to the bottom of the aquifer, if possible, because this has various advantages, one of which is that it allows a longer well screen to be placed, which will result in a higher well yield.

During drilling operations, samples of the geological formations that are pierced should be collected and described lithologically. Records should be kept of these lithological descriptions, and the samples themselves should be stored for possible future reference.

2.4.3 Well screen

The length of the well screen and the depth at which it is placed will largely be decided by the depth at which the coarsest materials are found. In the lithological descriptions, therefore, special attention should be given to the grain size of the various materials. If geophysical well logs are run immediately after the completion of drilling, a preliminary interpretation of those logs will help greatly in determining the proper depth at which to place the screen.

If the aquifer consists of coarse gravel, the screen can be made locally by sawing, drilling, punching, or cutting openings in the pipe. In finer formations, finer openings are needed. These may vary in size from some tenths of a millimetre to several millimetres. Such precision-made openings can only be obtained in factory-made screens. To prevent the blocking of well screen openings by spherical grains, long narrow slits are preferable. The slots should retain 30 to 50 per cent of the aquifer material, depending on the uniformity coefficient of the aquifer sample. (For details, see Driscoll 1986; Huisman 1972.)

The well screen should be slotted or perforated over no more than 30 to 40 per cent of its circumference to keep the entrance velocity low, say less than about 3 cm/s. At this velocity, the friction losses in the screen openings are small and may even be negligible.

A general rule is to screen the well over at least 80 per cent of the aquifer thickness because this makes it possible to obtain about 90 per cent or more of the maximum yield that could be obtained if the entire aquifer were screened. Another even more important advantage of this screen length is that the groundwater flow towards the well can be assumed to be horizontal, an assumption that underlies almost all well-flow equations (Figure 2.2A).

There are some exceptions to the general rule:

- In unconfined aquifers, it is common practice to screen only the lower half or lower one-third of the aquifer because, if appreciable drawdowns occur, the upper part

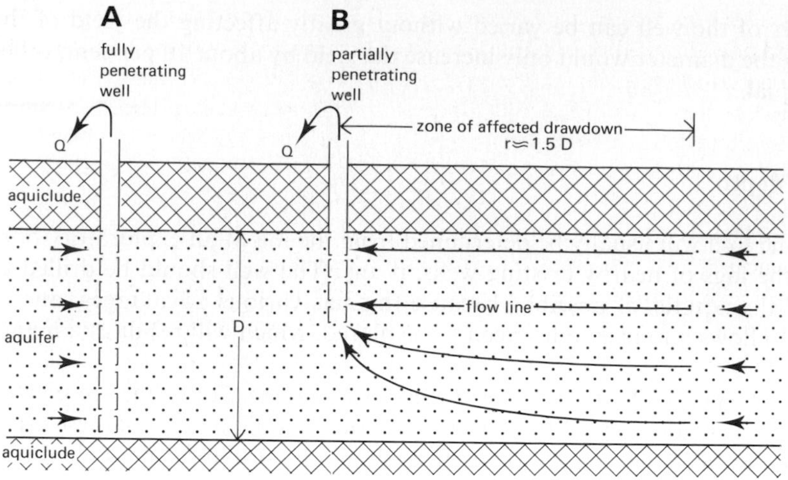

Figure 2.2 A) A fully penetrating well;
B) A partially penetrating well

of a longer well screen would fall dry;

- In a very thick aquifer, it will be obvious that the length of the screen will have to be less than 80 per cent, simply for reasons of economy. Such a well is said to be a partially penetrating well. It induces vertical-flow components, which can extend outwards from the well to distances roughly equal to 1.5 times the thickness of the aquifer (Figure 2.2B). Within this radius, the measured drawdowns have to be corrected before they can be used in calculating the aquifer characteristics;
- Wells in consolidated aquifers do not need a well screen because the material around the well is stable.

2.4.4 Gravel pack

It is easier for water to enter the well if the aquifer material immediately surrounding the screen is removed and replaced by artificially-graded coarser material. This is known as a gravel pack. When the well is pumped, the gravel pack will retain much of the aquifer material that would otherwise enter the well. With a gravel pack, larger slot sizes can be selected for the screen. The thickness of the pack should be in the range of 8 to 15 cm. Gravel pack material should be clean, smoothly-rounded grains. Details on the gravel sizes to be used in gravel packs are given by Driscoll (1986) and Huisman (1972).

2.4.5 The pump

After the well has been drilled, screened, and gravel-packed, as necessary, a pump has to be installed to lift the water. It is beyond the scope of this book to discuss the many kinds of pumps that might be used, so some general remarks must suffice:

- The pump and power unit should be capable of operating continuously at a constant discharge for a period of at least a few days. An even longer period may be required for unconfined or leaky aquifers, and especially for fractured aquifers. The same applies if drawdown data from piezometers at great distances from the well are to be analyzed. In such cases, pumping should continue for several days more;
- The capacity of the pump and the rate of discharge should be high enough to produce good measurable drawdowns in piezometers as far away as, say, 100 or 200 m from the well, depending on the aquifer conditions.

After the pump has been installed, the well should be developed by being pumped at a low discharge rate. When the initially cloudy water becomes clear, the discharge rate should be increased and pumping continued until the water clears again. This procedure should be repeated until the desired discharge rate for the test is reached or exceeded.

2.4.6 Discharging the pumped water

The water delivered by the well should be prevented from returning to the aquifer. This can be done by conveying the water through a large-diameter pipe, say over a distance of 100 or 200 m, and then discharging it into a canal or natural channel. The water can also be conveyed through a shallow ditch, but the bottom of the ditch should be sealed with clay or plastic sheets to prevent leakage. Piezometers can be used to check whether any water is lost through the bottom of the ditch.

2.5 Piezometers

A piezometer (Figure 2.3) is an open-ended pipe, placed in a borehole that has been drilled to the desired depth in the ground. The bottom tip of the piezometer is fitted with a perforated or slotted screen, 0.5 to 1 m long, to allow the inflow of water. A plug at the bottom and jute or cotton wrapped around the screen will prevent the entry of fine aquifer material.

The annular space around the screen should be filled with a gravel pack or uniform coarse sand to facilitate the inflow of water. The rest of the annular space can be filled with any material available, except where the presence of aquitards requires a seal of bentonite clay or cement grouting to prevent leakage along the pipe. Experience has taught us that very fine clayey sand provides almost as good a seal as bentonite. It produces an error of less than 0.03 m, even when the difference in head between the aquifers is more than 30 m.

The water levels measured in piezometers represent the average head at the screen of the piezometers. Rapid and accurate measurements can best be made in small-diameter piezometers. If their diameter is large, the volume of water contained in them may cause a time lag in changes in drawdown. When the depth to water is to be measured manually, the diameter of the piezometers need not be larger than 5 cm. If automatic water-level recorders or electronic water pressure transducers are used, larger-diameter piezometers will be needed. In a heterogeneous aquifer with intercalated

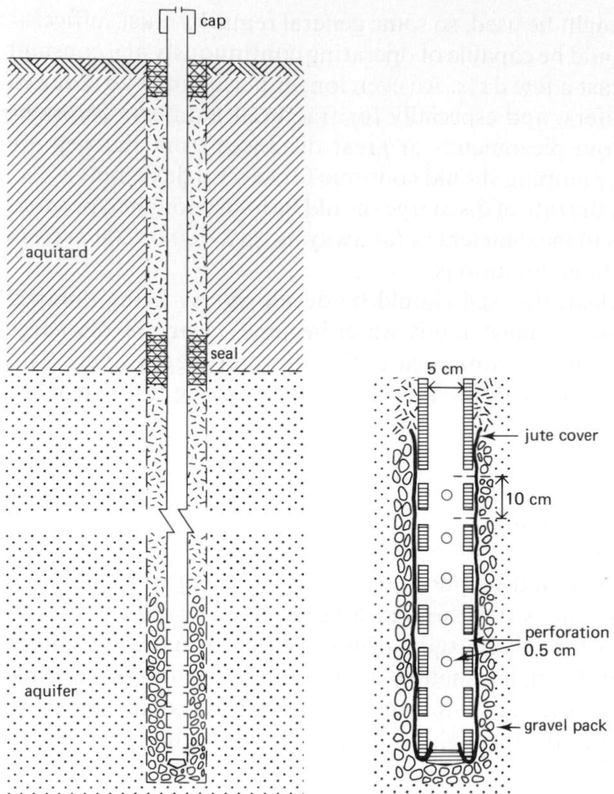

Figure 2.3 A piezometer

aquitards, the diameter of the bore holes should be large enough to allow a cluster of piezometers to be placed at different depths (Figure 2.4).

After the piezometers have been installed, it is advisable to pump or flush them for a short time to remove silt and clay particles. This will ensure that they function properly during the test.

After the well has been completed and its information analyzed, one has to decide how many piezometers to place, at what depths, and at what distances from the well.

2.5.1 The number of piezometers

The question of how many piezometers to place depends on the amount of information needed, and especially on its required degree of accuracy, but also on the funds available for the test.

Although it will be shown in later chapters that drawdown data from the well itself or from one single piezometer often permit the calculation of an aquifer's hydraulic characteristics, it is nevertheless always best to have as many piezometers as conditions permit. Three, at least, are recommended. The advantage of having more than one piezometer is that the drawdowns measured in them can be analyzed in two ways: by the time-drawdown relationship and by the distance-drawdown relationship. Obviously, the results of such analyses will be more accurate and will be representative of a larger volume of the aquifer.

2.5.2 Their distance from the well

Piezometers should be placed not too near the well, but not too far from it either. This rather vague statement needs some explanation. So, as will be outlined below, the distances at which piezometers should be placed depends on the type of aquifer, its transmissivity, the duration of pumping, the discharge rate, the length of the well screen, and whether the aquifer is stratified or fractured.

The type of aquifer

When a confined aquifer is pumped, the loss of hydraulic head propagates rapidly because the release of water from storage is entirely due to the compressibility of the aquifer material and that of the water. The drawdown will be measurable at great distances from the well, say several hundred metres or more.

In unconfined aquifers, the loss of head propagates slowly. Here, the release of water from storage is mostly due to the dewatering of the zone through which the

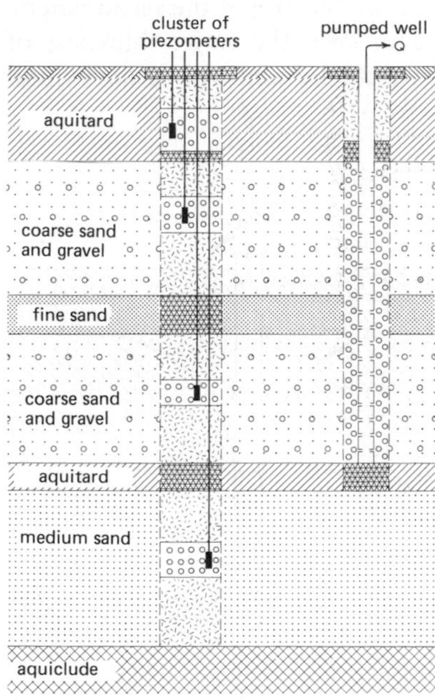

Figure 2.4 Cluster of piezometers in a heterogeneous aquifer intercalated with aquitards

water is moving, and only partially due to the compressibility of the water and aquifer material. Unless pumping continues for several days, the drawdown will only be measurable fairly close to the well, say not much more than about 100 m.

A leaky aquifer occupies an intermediate position. Depending on the hydraulic resistance of its confining aquitard (or aquitards), a leaky aquifer will resemble either a confined or an unconfined aquifer.

Transmissivity
When the transmissivity of the aquifer is high, the cone of depression induced by pumping will be wide and flat (Figure 2.5A). When the transmissivity is low, the cone will be steep and narrow (Figure 2.5B). In the first case, piezometers can be placed farther from the well than they can in the second.

The duration of the test
Theoretically, in an extensive aquifer, as long as the flow to the well is unsteady, the cone of depression will continue to expand as pumping continues. Therefore, for tests of long duration, piezometers can be placed at greater distances from the well than for tests of short duration.

The discharge rate
If the discharge rate is high, the cone of depression will be wider and deeper than if the discharge rate is low. With a high discharge rate, therefore, the piezometers can be placed at greater distances from the well.

The length of the well screen
The length of the well screen has a strong bearing on the placing of the piezometers. If the well is a fully penetrating one, i.e. it is screened over the entire thickness of the aquifer or at least 80 per cent of it, the flow towards the well will be horizontal

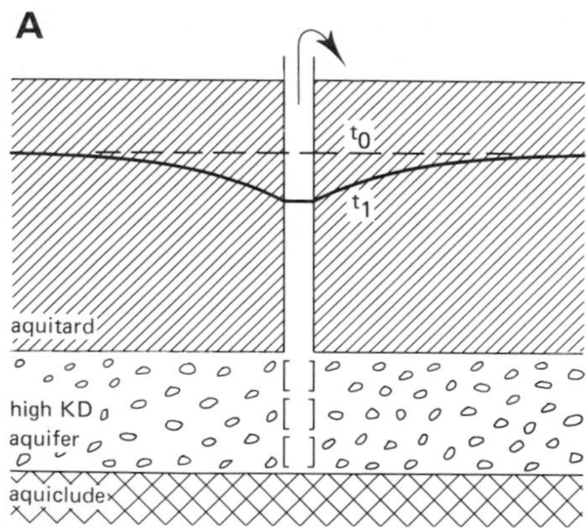

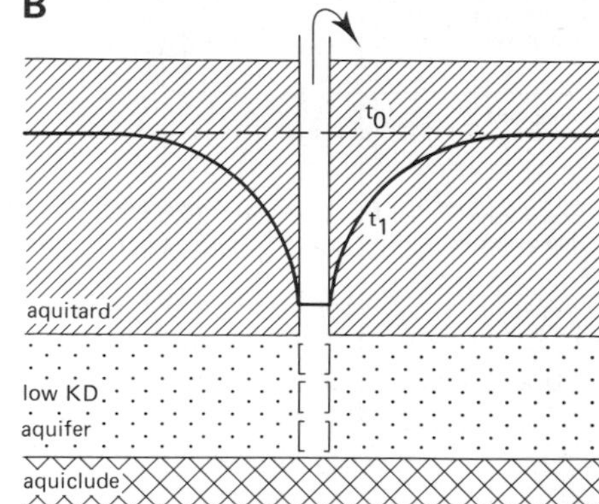

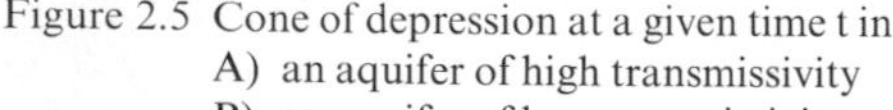
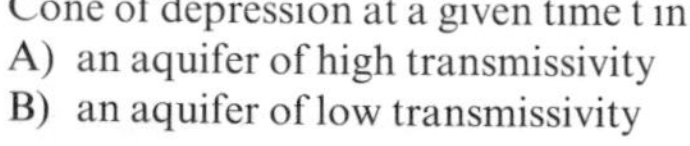
Figure 2.5 Cone of depression at a given time t in:
A) an aquifer of high transmissivity
B) an aquifer of low transmissivity

and piezometers can be placed close to the well. Obviously, if the aquifer is not very thick, it is always best to employ a fully penetrating well.

If the well is only partially penetrating, the relatively short length of well screen will induce vertical flow components, which are most noticeable near the well. If piezometers are placed near the well, their water-level readings will have to be corrected before being used in the analysis. These rather complicated corrections can be avoided if the piezometers are placed farther from the well, say at distances which are at least equal to 1.5 times the thickness of the aquifer. At such distances, it can be assumed that the flow is horizontal (see Figure 2.2).

Stratification

Homogeneous aquifers seldom occur in nature, most aquifers being stratified to some degree. Stratification causes differences in horizontal and vertical hydraulic conductivity, so that the drawdown observed at a certain distance from the well may differ at different depths within the aquifer. As pumping continues, these differences in drawdown diminish. Moreover, the greater the distance from the well, the less effect stratification has upon the drawdowns.

Fractured rock

Deciding on the number and location of piezometers in fractured rock poses a special problem, although the rock can be so densely fractured that its drawdown response to pumping resembles that of an unconsolidated homogeneous aquifer; if so, the number and location of the piezometers can be chosen in the same way as for such an aquifer.

If the fracture is a single vertical fracture, however, matters become more complicated. The number and location of piezometers will then depend on the orientation of the fracture (which may or may not be known) and on the transmissivity of the rock on opposite sides of the fracture (which may be the same or, as so often happens, is not the same). Further, the fracture may be open or closed. If it is open, its hydraulic conductivity can be regarded as infinite, and it will resemble a canal whose water level is suddenly lowered. There will then be no hydraulic gradient inside the fracture, so that it can be regarded as an 'extended well', or as a drain that receives water from the adjacent rock through parallel flow. This situation requires that piezometers be placed along a line perpendicular to the fracture. To check whether the fracture can indeed be regarded as an 'extended well', a few piezometers should be placed in the fracture itself.

If the hydraulic conductivity of the fracture is severely reduced by weathering or by the deposition of minerals on the fracture plane, pumping will cause hydraulic gradients to develop in the fracture and in the adjacent rock. This situation requires piezometers in the fracture and in the adjacent rock.

If the fracture is a single vertical open fracture of infinite hydraulic conductivity and known orientation, and if the transmissivity of the rock is the same on both sides of the fracture, two piezometers on the same side of the fracture are required to determine the perpendicular distances between the piezometers and the fracture (Figure 2.6A). In this figure, the piezometer closest to the pumped well is not the piezometer closest to the fracture. Regardless of the distances r_1 and r_2, the drawdown will be greatest in the piezometer closest to the fracture. To analyze pumping test data from such a fracture, we must know the distances between the piezometers and the fracture, x_1 and x_2, which we can calculate from r_1 and r_2, measured in the field, and the angles Θ_1 and Θ_2.

If the precise orientation of the fracture is not known, more than two piezometers will be needed. As can be seen in Figure 2.6B, if x_1 is small relative to x_2, two orientations are possible because x_1 may be on either side of the fracture. More piezometers must then be placed to find the orientation.

More piezometers are also required if there is geological evidence that the transmissivity of the rock on opposite sides of the fracture is significantly different.

Summarizing

As is obvious from the above, there are many factors to be taken into account in deciding how far from the well the piezometers should be placed. Nevertheless, if one has a proper knowledge of the test site (especially of the type of aquifer, its thickness, stratification or fracturing, and expected transmissivity), it will be easier to make the right decisions.

Although no fixed rule can be given and the ultimate choice depends entirely on local conditions, placing piezometers between 10 and 100 m from the well will give reliable data in most cases. For thick aquifers or stratified confined ones, the distances should be greater, say between 100 and 250 m or more from the well.

One or more piezometers should also be placed outside the area affected by the pumping so that the natural behaviour of the hydraulic head in the aquifer can be

A

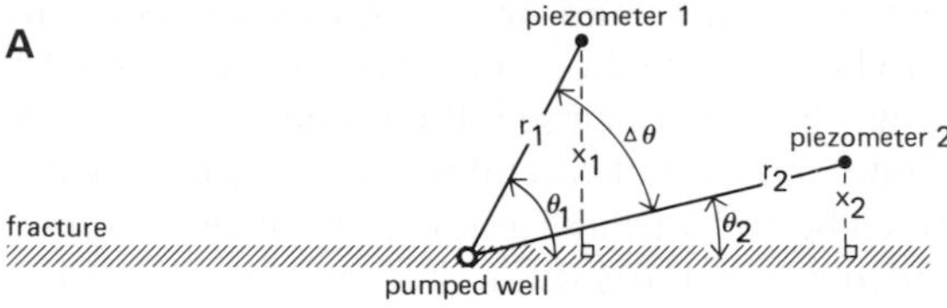

B

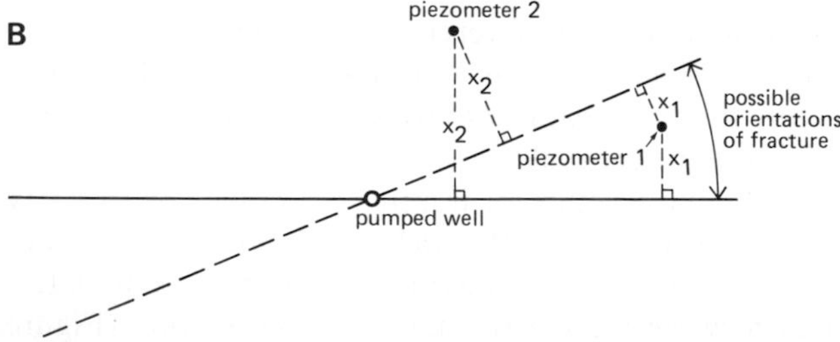

Figure 2.6 Piezometer arrangement near a fracture:
A) of known orientation
B) of unknown orientation

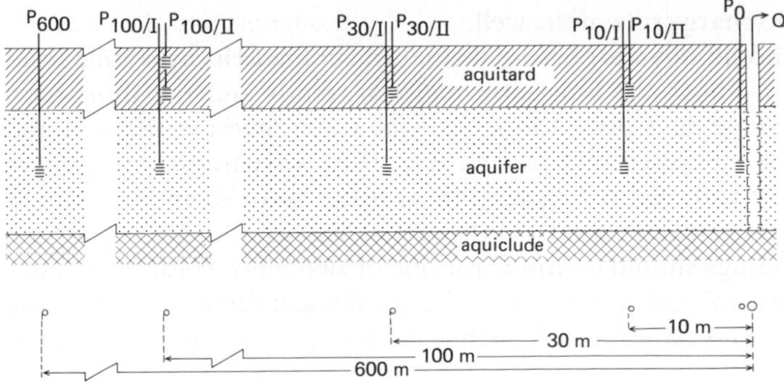

Figure 2.7 Example of a piezometer arrangement

measured. These piezometers should be several hundred metres away from the well, or in the case of truly confined aquifers, as far away as one kilometre or more. If the readings from these piezometers show water-level changes during the test (e.g. changes caused by natural discharge or recharge), these data will be needed to correct the drawdowns induced by the pumping.

An example of a piezometer arrangement in an unconsolidated leaky aquifer is shown in Figure 2.7.

2.5.3 Depth of the piezometers

The depth of the piezometers is at least as important as their distance from the well. In an isotropic and homogeneous aquifer, the piezometers should be placed at a depth that coincides with that of half the length of the well screen. For example, if the well is fully penetrating and its screen is between 10 and 20 m below the ground surface, the piezometers should be placed at a depth of about 15 m.

For heterogeneous aquifers made up of sandy deposits intercalated with aquitards, it is recommended that a cluster of piezometers be placed, i.e. one piezometer in each sandy layer (see Figure 2.4). The holes in the aquitards should be sealed to prevent leakage along the tubes. Despite these precautions, some leakage may still occur, so it is recommended that the screens be placed a few metres away from the upper and lower boundaries of the aquitards where the effect of this leakage is small.

If an aquifer is overlain by a partly saturated aquitard, piezometers should also be placed in the aquitard to check whether its watertable is affected when the underlying aquifer is pumped. This information is needed for the analysis of tests in leaky aquifers.

2.6 The measurements to be taken

The measurements to be taken during a pumping test are of two kinds:
- Measurements of the water levels in the well and the piezometers;

– Measurements of the discharge rate of the well.

Ideally, a pumping test should not start before the natural changes in hydraulic head in the aquifer are known – both the long-term regional trends and the short-term local variations. So, for some days prior to the test, the water levels in the well and the piezometers should be measured, say twice a day. If a hydrograph (i.e. a curve of time versus water level) is drawn for each of these observation points, the trend and rate of water-level change can be read. At the end of the test (i.e. after complete recovery), water-level readings should continue for one or two days. With these data, the hydrographs can be completed and the rate of natural water-level change during the test can be determined. This information can then be used to correct the drawdowns observed during the test.

Special problems arise in coastal aquifers whose hydraulic head is affected by tidal movements. Prior to the test, a complete picture of the changes in head should be obtained, including maximum and minimum water levels in each piezometer and their time of occurrence.

When a test is expected to last one or more days, measurements should also be made of the atmospheric pressure, the levels of nearby surface waters, if present, and any precipitation.

In areas where production wells are operating, the pumping test has to be conducted under less than ideal conditions. Nevertheless, the possibly significant effects of these interfering wells can be eliminated from the test data if their on-off times and discharge rates are monitored, both before and during the test. Even so, it is best to avoid the disturbing influence of such wells if at all possible.

2.6.1 Water-level measurements

The water levels in the well and the piezometers must be measured many times during a test, and with as much accuracy as possible. Because water levels are dropping fast during the first one or two hours of the test, the readings in this period should be made at brief intervals. As pumping continues, the intervals can be gradually lengthened. Table 2.1 gives a range of intervals for readings in the well. For single well tests (i.e. tests without the use of piezometers), the intervals in the first 5 to 10 minutes of the test should be shorter because these early-time drawdown data may reveal well-bore storage effects.

Table 2.1 Range of intervals between water-level measurements in well

Time since start of pumping	Time intervals
0– 5 minutes	0.5 minutes
5– 60 minutes	5 minutes
60–120 minutes	20 minutes
120–shutdown of the pump	60 minutes

Similarly, in the piezometers, water-level measurements should be taken at brief intervals during the first hours of the test, and at longer intervals as the test continues. Table 2.2 gives a range of intervals for measurements in those piezometers placed in the aquifer and located relatively close to the well; here, the water levels are immediately affected by the pumping. For piezometers farther from the well and for those in confining layers above or below the aquifer, the intervals in the first minutes of the test need not be so brief.

Table 2.2 Range of intervals between water-level measurements in piezometers

Time since start of pumping	Time intervals
0 – 2 minutes	approx. 10 seconds
2 – 5 minutes	30 seconds
5 – 15 minutes	1 minute
15 – 50 minutes	5 minutes
50 – 100 minutes	10 minutes
100 minutes – 5 hours	30 minutes
5 hours – 48 hours	60 minutes
48 hours – 6 days	3 times a day
6 days – shutdown of the pump	1 time a day

The suggested intervals need not be adhered to too rigidly as they should be adapted to local conditions, available personnel, etc. All the same, readings should be frequent in the first hours of the test because, in the analysis of the test data, time generally enters in a logarithmic form.

All manual measurements of water levels and times should preferably be noted on standard, pre-printed forms, with space available for all relevant field data. An example is shown in Figure 2.8. The completed forms should be kept on file.

After some hours of pumping, sufficient time will become available in the field to draw the time-drawdown curves for the well and for each piezometer. Log-log and semi-log paper should be used for this purpose, with the time in minutes on a logarithmic scale. These graphs can be helpful in checking whether the test is running well and in deciding on the time to shut down the pump.

After the pump has been shut down, the water levels in the well and the piezometers will start to rise – rapidly in the first hour, but more slowly afterwards. These rises can be measured in what is known as a recovery test. If the discharge rate of the well was not constant throughout the pumping test, recovery-test data are more reliable than the drawdown data because the watertable recovers at a constant rate, which is the average of the pumping rate. The data from a recovery test can also be used to check the calculations made on the basis of the drawdown data. The schedule for recovery measurements should be the same as that adhered to during the pumping test.

Pumping test VENNEBULTEN

OPSERVATIONS during PUMPING/RECOVERY

Piezometer W II/90; Depth 15m-m.s.; Distance 90 m

Pumping test by ICW; Directed by H. WITT

For project ACHTERHOEK

Location VENNEBULTEN

Start 28-10-65 10h27; Stop 29-10-65 11h59

Initial water level 1-385 m

Final water level 1-589 m

Reference level TOP OF PIEZOMETER = 22.322 +m.s.l.*

Remark TIME IN MINUTES

WATER LEVEL AND DRAWDOWN IN CM

DISCHARGE RATE m^3/h

time	water level	draw-down	time	water level	draw-down	discharge rate		
						time	flow-meter	discharge rate
0	138.5					0	183.54	0
1.17	138.9	0.4				60	219.91	36.37
1.34	139.4	0.9						

* mean sea level

Figure 2.8 Example of a pre-printed pumping-test form

2.6.1.1 Water-level-measuring devices

The most accurate recordings of water-level changes are made with fully-automatic microcomputer-controlled systems, as developed, for instance, by the TNO Institute of Applied Geoscience, The Netherlands (Figure 2.9). This system uses pressure transducers or acoustic transducers for continuous water-level recordings, which are stored on magnetic tape (see also Kohlmeier et al. 1983).

A good alternative is the conventional automatic recorder, which also produces a continuous record of water-level changes. Such recorders, however, require large-diameter piezometers.

Fairly accurate measurements can be taken by hand, but then the instant of each reading must be recorded with a chronometer. Experience has shown that it is possible to measure water levels to within 1 or 2 mm with one of the following:

- A floating steel tape and standard with pointer;
- An electrical sounder;
- The wetted-tape method.

For piezometers close to the well where water levels are changing rapidly during the first hours of the test, the most convenient device is the floating steel tape with pointer because it permits direct readings. For piezometers far from the well, conventional automatic recorders are the most suitable devices because only slow water-level changes can be interpreted from their graphs. For piezometers at intermediate distances, either floating or hand-operated water-level indicators can be used, but even when water levels are changing rapidly, accurate observations can be made with a recorder, provided a chronometer is used and the time of each reading is marked manually on the graph.

For detailed descriptions of automatic recorders, mechanical and electrical sounders, and other equipment for measuring water levels in wells, we refer to handbooks (e.g. Driscoll 1986; Genetier 1984; Groundwater Manual 1981).

2.6.2 Discharge-rate measurements

Amongst the arrangements to be made for a pumping test is a proper control of the discharge rate. This should preferably be kept constant throughout the test. During pumping, the discharge should be measured at least once every hour, and any necessary adjustments made to keep it constant.

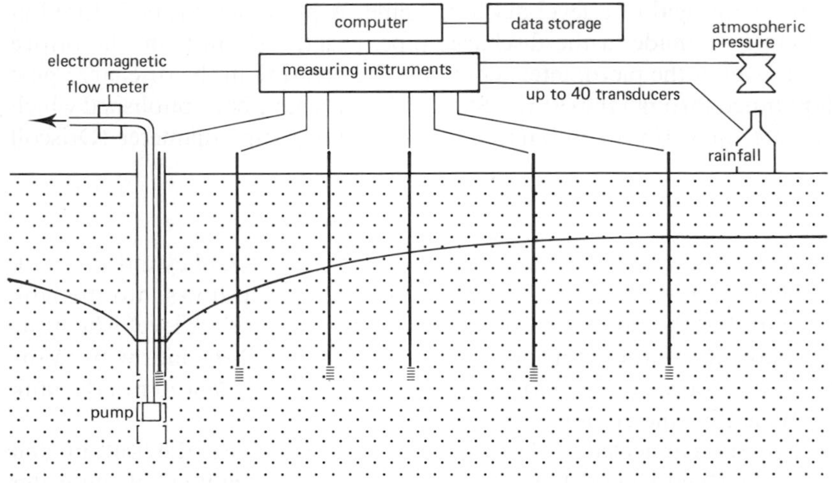

Figure 2.9 A fully-automated micro-computer-controlled recorder

The discharge can be kept constant by a valve in the discharge pipe. This is a more accurate method of control than changing the speed of the pump.

The fully-automatic computer-controlled system shown earlier in Figure 2.9 includes a magnetic flow meter for discharge measurements as part of a discharge-correction scheme to maintain a constant discharge.

A constant discharge rate, however, is not a prerequisite for the analysis of a pumping test. There are methods available that take variable discharge into account, whether it be due to natural causes or is deliberately provoked.

2.6.2.1 Discharge-measuring devices

To measure the discharge rate, a commercial water meter of appropriate capacity can be used. The meter should be connected to the discharge pipe in a way that ensures accurate readings being made: at the bottom of a U-bend, for instance, so that the pipe is running full. If the water is being discharged through a small ditch, a flume can be used to measure the discharge.

If no appropriate water meter or flume is available, there are other methods of measuring or estimating the discharge.

Container

A very simple and fairly accurate method is to measure the time it takes to fill a container of known capacity (e.g. an oil drum). This method can only be used if the discharge rate is low.

Orifice weir

The circular orifice weir is commonly used to measure the discharge from a turbine or centrifugal pump. It does not work when a piston pump is used because the flow from such a pump pulsates too much.

The orifice is a perfectly round hole in the centre of a circular steel plate which is fastened to the outer end of a level discharge pipe. A piezometer tube is fitted in a 0.32 or 0.64 cm hole made in the discharge pipe, exactly 61 cm from the orifice plate. The water level in the piezometer represents the pressure in the discharge pipe when water is pumped through the orifice. Standard tables have been published which show the flow rate for various combinations of orifice and pipe diameter (Driscoll 1986).

Orifice bucket

The orifice bucket was developed in the U.S.A. It consists of a small cylindrical tank with circular openings in the bottom. The water from the pump flows into the tank and discharges through the openings. The tank fills with water to a level where the pressure head causes the outflow through the openings to equal the inflow from the pump. If the tank overflows, one or more orifices are opened. If the water in the tank does not rise sufficiently, one or more orifices are closed with plugs.

A piezometer tube is connected to the outer wall of the tank near the bottom, and a vertical scale is fastened behind the tube to allow accurate readings of the water level in the tank. A calibration curve is required, showing the rate of discharge through a single orifice of a given size for various values of the pressure head. The discharge rate taken from this curve, multiplied by the number of orifices through which the water is being discharged, gives the total rate of discharge for any given water-level reading. If the orifice bucket is provided with many openings, a considerable range of pumping rates can be measured. A further advantage of the orifice bucket is that it tends to smooth out any pulsating flow from the pump, thus permitting the average pumping rate to be determined with fair accuracy.

Jet-stream method

If none of the above-mentioned methods can be applied, the jet-stream method (or open-pipe-flow method) can be used. By measuring the dimensions of a stream flowing either vertically or horizontally from an open pipe, one can roughly estimate the discharge.

If the water is discharged through a vertical pipe, estimates of the discharge can be made from the diameter of the pipe and the height to which the water rises above the top of the pipe. Driscoll (1986) has published a table showing the discharge rates for different pipe diameters and various heights of the crest of the stream above the top of the pipe.

If the water is discharged through a horizontal pipe, flowing full and with a free fall from the discharge opening, estimates of the discharge can be made from the horizontal and vertical distances from the end of the pipe to a point in the flowing stream of water. The point can be chosen at the outer surface of the stream or in its centre. Another table by Driscoll (1986) shows the discharge rates for different pipe diameters and for various horizontal distances of the stream of water.

2.7 Duration of the pumping test

The question of how many hours to pump the well in a pumping test is difficult to answer because the period of pumping depends on the type of aquifer and the degree of accuracy desired in establishing its hydraulic characteristics. Economizing on the period of pumping is not recommended because the cost of running the pump a few extra hours is low compared with the total costs of the test. Besides, better and more reliable data are obtained if pumping continues until steady or pseudo-steady flow has been attained. At the beginning of the test, the cone of depression develops rapidly because the pumped water is initially derived from the aquifer storage immediately around the well. But as pumping continues, the cone expands and deepens more slowly because, with each additional metre of horizontal expansion, a larger volume of stored water becomes available. This apparent stabilization of the cone often leads inexperienced observers to conclude that steady state has been reached. Inaccurate measurements of the drawdowns in the piezometers – drawdowns that are becoming smaller and smaller as pumping continues – can lead to the same wrong conclusion. In reality, the cone of depression will continue to expand until the recharge of the aquifer equals the pumping rate.

In some tests, steady-state or equilibrium conditions occur a few hours after the start of pumping; in others, they occur within a few days or weeks; in yet others, they never occur, even though pumping continues for years. It is our experience that, under average conditions, a steady state is reached in leaky aquifers after 15 to 20 hours of pumping; in a confined aquifer, it is good practice to pump for 24 hours; in an unconfined aquifer, because the cone of depression expands slowly, a longer period is required, say 3 days.

As will be demonstrated in later chapters, it is not absolutely necessary to continue pumping until a steady state has been reached, because methods are available to analyze unsteady-state data. Nevertheless, it is good practice to strive for a steady state, especially when accurate information on the aquifer characteristics is desired, say as a basis for the construction of a pumping station for domestic water supplies or other expensive works. If a steady state has been reached, simple equations can be used to analyze the data and reliable results will be obtained. Besides, the longer period of pumping required to reach steady state may reveal the presence of boundary conditions previously unknown, or in cases of fractured formations, will reveal the specific flows that develop during the test.

Preliminary plotting of drawdown data during the test will often show what is happening and may indicate how much longer the test should continue.

2.8 Processing the data

2.8.1 Conversion of the data

The water-level data collected before, during, and after the test should first be expressed in appropriate units. The measurement units of the International System are recommended (Annex 2.1), but there is no fixed rule for the units in which the field data and hydraulic characteristics should be expressed. Transmissivity, for instance, can be expressed in m^2/s or m^2/d. Field data are often expressed in units other than those in which the final results are presented. Time data, for instance, might be expressed in seconds during the first minutes of the test, minutes during the following hours, and actual time later on, while water-level data might be expressed in different units of length appropriate to the timing of the observations.

It will be clear that before the field data can be analyzed, they should first be converted: the time data into a single set of time units (e.g. minutes) and the drawdown data into a single set of length units (e.g. metres), or any other unit of length that is suitable (Annex 2.2).

2.8.2 Correction of the data

Before being used in the analysis, the observed water levels may have to be corrected for external influences (i.e. those not related to the pumping). To find out whether this is necessary, one has to analyze the local trend in the hydraulic head or watertable. The most suitable data for this purpose are the water-level measurements taken in a 'distant' piezometer during the test, but measurements taken at the test site for some days before and after the test can also be used.

If, after the recovery period, the same constant water level is observed as during the pre-testing period, it can safely be assumed that no external events influenced the hydraulic head during the test. If, however, the water level is subject to unidirectional or rhythmic changes, it will have to be corrected.

2.8.2.1 Unidirectional variation

The aquifer may be influenced by natural recharge or discharge, which will result in a rise or a fall in the hydraulic head. By interpolation from the hydrographs of the well and the piezometers, this natural rise or fall can be determined for the pumping and recovery periods. This information is then used to correct the observed water levels.

Example 2.1
Suppose that the hydraulic head in an aquifer is subject to unidirectional variation, and that the water level in a piezometer at the moment t_o (start of the pumping test) is h_o. From the interpolated hydrograph of natural variation, it can be read that, at a moment t_l, the water level would have been h_l if no pumping had occurred. The absolute value of water-level change due to natural variation at t_l is then: $h_o - h_l = \Delta h_l$. If the observed drawdown at t_l is s_l, where the observed drawdown is defined as the lowering of the water level with respect to the water level at $t = t_o$, the drawdown due to pumping is:
- With natural discharge: $s_l' = s_l - \Delta h_l$;
- With natural recharge: $s_l' = s_l + \Delta h_l$.

2.8.2.2 Rhythmic fluctuations

In confined and leaky aquifers, rhythmic fluctuations of the hydraulic head may be due to the influence of tides or river-level fluctuations, or to rhythmic variations in atmospheric pressure. In unconfined aquifers whose watertables are close to the ground surface, diurnal fluctuations of the watertable can be significant because of the great difference between day and night evapotranspiration. The watertable drops during the day because of the consumptive use by the vegetation and recovers during the night when the plant stomata are closed.

Hydrographs of the well and the piezometers, covering sufficiently long pre-test and post-recovery periods, will yield the information required to correct the water levels observed during the test.

Example 2.2
For this example, data from the pumping test 'Dalem' (see Chapter 4 and Figure 4.2) will be corrected for the piezometer at 400 m from the well. The piezometer was located 1900 m from the River Waal, which is under the influence of the tide in the North Sea. The Waal is hydraulically connected with the aquifer; hence the rise and fall of the river level affected the water levels in the piezometers. Piezometer readings covering a few days both prior to pumping and after complete recovery made it possible to interpolate the groundwater time-versus-tide curve for the pumping and recovery periods.

Figure 2.10A shows the curve of the groundwater tide with respect to a reference level, which was selected as the water level at the moment pumping started (08.04 hours). At 10.20 hours, it was low tide and the water levels had fallen 5 mm, independently of pumping. This meant that the water level observed at that moment was 5 mm lower than it would have been if there had been no tidal influence. The drawdown therefore has to be corrected accordingly. The correction term applied is read on the vertical axis of the time-tide curve.

Figure 2.10B shows the uncorrected time-drawdown curve and the same curve after being corrected. It will be noted that different vertical scales have been used in Parts A and B of Figure 2.10.

The same procedure is followed to correct the data from the other piezometers. For each, a time-tide curve, corresponding to the distance between the piezometer and the river, is used. Obviously, the closer a piezometer is to the river, the greater is the influence of the tide on its water levels.

2.8.2.3 Non-rhythmic regular fluctuations

Non-rhythmic regular fluctuations, due, for example, to changes in atmospheric pressure, can be detected on a hydrograph covering the pre-test period. In wells or piezometers tapping confined and leaky aquifers, the water levels are continuously changing as the atmospheric pressure changes. When the atmospheric pressure decreases, the water levels rise in compensation, and vice versa (Figure 2.11). By comparing the atmospheric changes, expressed in terms of a column of water, with the actual changes in water levels observed during the pre-test period, one can determine the barometric efficiency of the aquifer. The barometric efficiency (BE) is defined as the ratio of

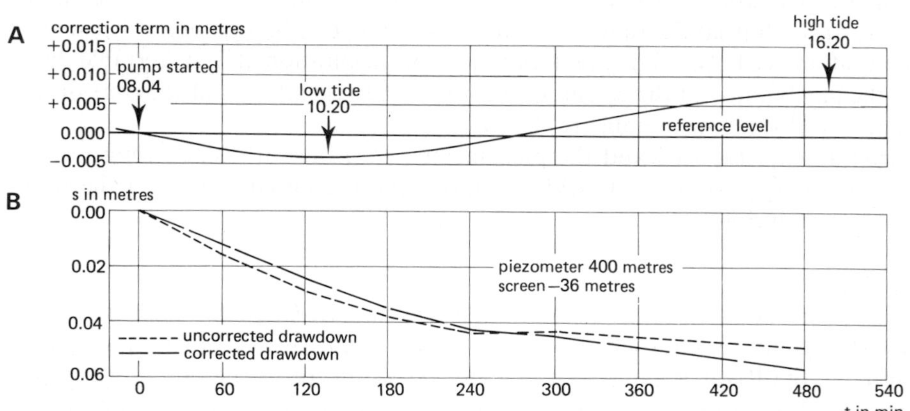

Figure 2.10 Correction of data for tidal influence
A) The curve of the groundwater tide under non-pumped conditions
B) Corrected and uncorrected drawdowns
Note: Vertical scales in upper and lower part of figure are different

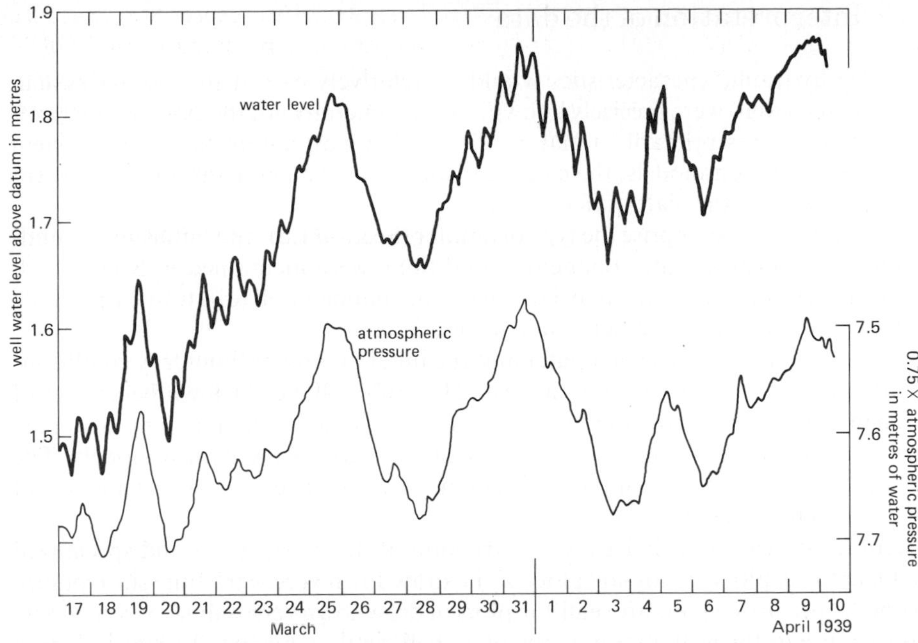

Figure 2.11 Response of water level in a well penetrating a confined aquifer to changes in atmospheric pressure, showing a barometric efficiency of 75 per cent (Robinson 1939)

change in water level (Δh) in a well to the corresponding change in atmospheric pressure (Δp), or $BE = \gamma \Delta h/\Delta p$, in which γ is the specific weight of water. BE usually ranges from 0.20 to 0.75.

From the changes in atmospheric pressure observed during a test, and the known relationship between Δp and Δh, the water-level changes due to changes in atmospheric pressure alone (Δh_p) can be calculated for the test period for the well and each piezometer. Subsequently, the actual drawdown during the test can be corrected for the water-level changes due to atmospheric pressure:

- For falling atmospheric pressures: $s' = s + \Delta h_p$;
- For rising atmospheric pressures: $s' = s - \Delta h_p$.

2.8.2.4 Unique fluctuations

In general, the water levels measured during a pumping test cannot be corrected for unique fluctuations due, say, to heavy rain or the sudden rise or fall of a nearby river or canal that is in hydraulic connection with the aquifer. In certain favourable circumstances, allowance can be made for such fluctuations by extrapolating the data from a control piezometer outside the zone of influence of the well. But, in general, the data of the test become worthless and the test has to be repeated when the situation has returned to normal.

2.9 Interpretation of the data

Calculating hydraulic characteristics would be relatively easy if the aquifer system (i.e. aquifer plus well) were precisely known. This is generally not the case, so interpreting a pumping test is primarily a matter of identifying an unknown system. System identification relies on models, the characteristics of which are assumed to represent the characteristics of the real aquifer system.

Theoretical models comprise the type of aquifer (Section 1.2), and initial and boundary conditions. Typical outer boundary conditions were mentioned in Section 1.4. Inner boundary conditions are associated with the pumped well (e.g. fully or partially penetrating, small or large diameter, well losses).

In a pumping test, the type of aquifer and the inner and outer boundary conditions dominate at different times during the test. They affect the drawdown behaviour of the system in their own individual ways. So, to identify an aquifer system, one must compare its drawdown behaviour with that of the various theoretical models. The model that compares best with the real system is then selected for the calculation of the hydraulic characteristics.

System identification includes the construction of diagnostic plots and specialized plots. Diagnostic plots are log-log plots of the drawdown versus the time since pumping started. Specialized plots are semi-log plots of drawdown versus time, or drawdown versus distance to the well; they are specific to a given flow regime. A diagnostic plot allows the dominating flow regimes to be identified; these yield straight lines on specialized plots. The characteristic shapes of the curves can help in selecting the appropriate model.

In a number of cases, a semi-log plot of drawdown versus time has more diagnostic value than a log-log plot. We therefore recommend that both types of graphs be constructed.

The choice of theoretical model is a crucial step in the interpretation of pumping tests. If the wrong model is chosen, the hydraulic characteristics calculated for the real aquifer will not be correct. A troublesome fact is that theoretical solutions to well-flow problems are usually not unique. Some models, developed for different aquifer systems, yield similar responses to a given stress exerted on them. This makes system identification and model selection a difficult affair. One can reduce the number of alternatives by conducting more field work, but that could make the total costs of the test prohibitive. In many cases, uncertainty as to which model to select will remain. We shall discuss this problem briefly below. The examples we give will illustrate that analyzing a pumping test is not merely a matter of opening a particular page of this book and applying the method described there.

2.9.1 Aquifer categories

Aquifers fall into two broad categories: unconsolidated aquifers and consolidated fractured aquifers. Within both categories, the aquifers may be confined, unconfined, or leaky (Section 1.2, Figure 1.1). We shall first consider all three types of unconsolidated aquifer, and then the consolidated aquifer, but only the confined type.

Figure 2.12 shows log-log and semi-log plots of the theoretical time-drawdown relationships for confined, unconfined, and leaky unconsolidated aquifers. We present these graphs in pairs because, although log-log plots are diagnostic, as the oil industry states, we believe that semi-log plots can sometimes be even more diagnostic. This becomes clear if we look at Parts A and A′ of Figure 2.12. These refer to an ideal, confined, unconsolidated aquifer, homogeneous and isotropic, and pumped at a constant rate by a fully penetrating well of very small diameter. From the semi-log plot (Part A′), we can see that the time-drawdown relationship at early pumping times is not linear, but at later times it is. If a linear relationship like this is found, it should be used to calculate the hydraulic characteristics because the results will be much more accurate than those obtained by matching field data plots with the curve of Part A. (We return to this subject in Sections 3.2.1 and 3.2.2.)

Parts B and B′ of Figure 2.12 show the curves for an unconfined, homogeneous, isotropic aquifer of infinite lateral extent and with a delayed yield. These two curves are characteristic. At early pumping times, the curve of the log-log plot (Part B) follows the curve for the confined aquifer shown in Part A. Then, at medium pumping times, it shows a flat segment. This reflects the recharge from the overlying, less permeable aquifer, which stabilizes the drawdown. At late times, the curve again follows a portion of the curve of Part A. The semi-log plot is even more characteristic: it shows two parallel straight-line segments at early and late pumping times. (We return to this subject in Section 5.1.1.)

Parts C and C′ of Figure 2.12 refer to a leaky aquifer. At early pumping times, the curves follow those of Parts A and A′. At medium pumping times, more and more water from the aquitard (or aquitards) is reaching the aquifer. Eventually, at late pumping times, all the water pumped is from leakage through the aquitard(s), and the flow towards the well has reached a steady state. This means that the drawdown

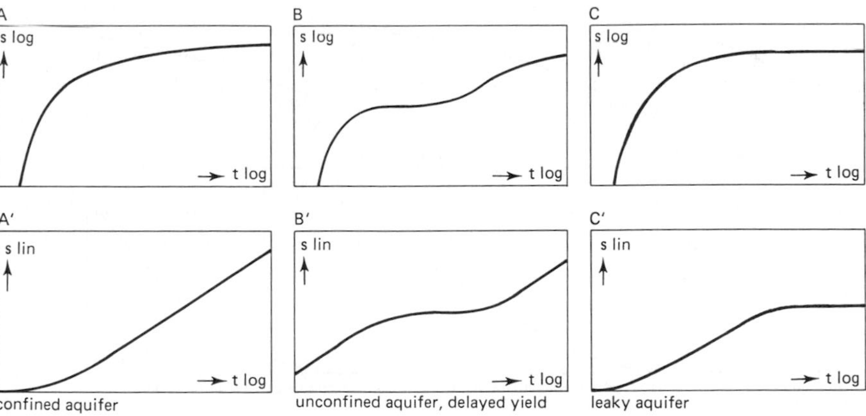

Figure 2.12 Log-log and semi-log plots of the theoretical time-drawdown relationships of unconsolidated aquifers:
Parts A and A′: Confined aquifer
Parts B and B′: Unconfined aquifer
Parts C and C′: Leaky aquifer

in the aquifer stabilizes, as is clearly reflected in both graphs. (We return to this subject in Sections 4.1.1 and 4.1.2.)

We shall now consider the category of confined, consolidated fractured aquifers, some examples of which are shown in Figure 2.13. Parts A and A′ of this figure refer to a confined, densely fractured, consolidated aquifer of the double-porosity type. In an aquifer like this, we recognize two systems: the fractures of high permeability and low storage capacity, and the matrix blocks of low permeability and high storage capacity. The flow towards the well in such a system is entirely through the fractures and is radial and in an unsteady state. The flow from the matrix blocks into the fractures is assumed to be in a pseudo-steady state. Characteristic of the flow in such a system is that three time periods can be recognized:

- Early pumping time, when all the flow comes from storage in the fractures;
- Medium pumping time, a transition period during which the matrix blocks feed their water at an increasing rate to the fractures, resulting in a (partly) stabilizing drawdown;
- Late pumping time, when the pumped water comes from storage in both the fractures and the matrix blocks.

(We return to this subject in Chapter 17.)

The shapes of the curves in Parts A and A′ of Figure 2.13 resemble those of Parts B and B′ of Figure 2.12, which refer to an unconfined, unconsolidated aquifer with delayed yield.

Parts B and B′ of Figure 2.13 present the curves for a well that pumps a single plane vertical fracture in a confined, homogeneous, and isotropic aquifer of low permeability. The fracture has a finite length and a high hydraulic conductivity. Characteristic of this system is that a log-log plot of early pumping time shows a straight-line segment of slope 0.5. This segment reflects the dominant flow regime in that period:

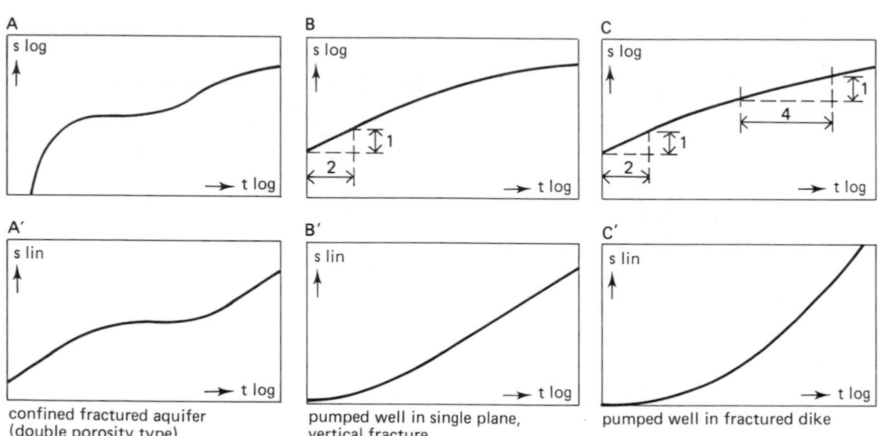

Figure 2.13 Log-log and semi-log plots of the theoretical time-drawdown relationships of consolidated, fractured aquifers:
Parts A and A′: Confined fractured aquifer, double porosity type
Parts B and B′: A single plane vertical fracture
Parts C and C′: A permeable dike in an otherwise poorly permeable aquifer

it is horizontal, parallel, and perpendicular to the fracture. This flow regime gradually changes, until, at late time, it becomes pseudo-radial. The shapes of the curves at late time resemble those of Parts A and A′ of Figure 2.12. (We return to this subject in Section 18.3.)

Parts C and C′ of Figure 2.13 refer to a well in a densely fractured, highly permeable dike of infinite length and finite width in an otherwise confined, homogeneous, isotropic, consolidated aquifer of low hydraulic conductivity and high storage capacity. Characteristic of such a system are the two straight-line segments in a log-log plot of early and medium pumping times. The first segment has a slope of 0.5 and thus resembles that of the well in the single, vertical, plane fracture shown in Part B of Figure 2.13. At early time, the flow towards the well is exclusively through the dike, and this flow is parallel. At medium time, the adjacent aquifer starts yielding water to the dike. The dominant flow regime in the aquifer is then near-parallel to parallel, but oblique to the dike. In a log-log plot, this flow regime is reflected by a one-fourth slope straight-line segment. At late time, the dominant flow regime is pseudo-radial, which, in a semi-log plot, is reflected by a straight line.

The one-fourth slope straight-line segment does not always appear in a log-log plot; whether it does or not depends on the hydraulic diffusity ratio between the dike and the adjacent aquifer. (We return to this subject in Section 19.3.)

2.9.2 Specific boundary conditions

When field data curves of drawdown versus time deviate from the theoretical curves of the main types of aquifer, the deviation is usually due to specific boundary conditions (e.g. partial penetration of the well, well-bore storage, recharge boundaries, or impermeable boundaries). Specific boundary conditions can occur individually (e.g. a partially penetrating well in an otherwise homogeneous, isotropic aquifer of infinite extent), but they often occur in combination (e.g. a partially penetrating well near a deeply incised river or canal). Obviously, specific boundary conditions can occur in all types of aquifers, but the examples we give below refer only to unconsolidated, confined aquifers.

Partial penetration of the well

Theoretical models usually assume that the pumped well fully penetrates the aquifer, so that the flow towards the well is horizontal. With a partially penetrating well, the condition of horizontal flow is not satisfied, at least not in the vicinity of the well. Vertical flow components are thus induced in the aquifer, and these are accompanied by extra head losses in and near the well. Figure 2.14 shows the effect of partial penetration. The extra head losses it induces are clearly reflected. (We return to this subject in Chapter 10.)

Well-bore storage

All theoretical models assume a line source or sink, which means that well-bore storage effects can be neglected. But all wells have a certain dimension and thus store some water, which must first be removed when pumping begins. The larger the diameter of the well, the more water it will store, and the less the condition of line source or

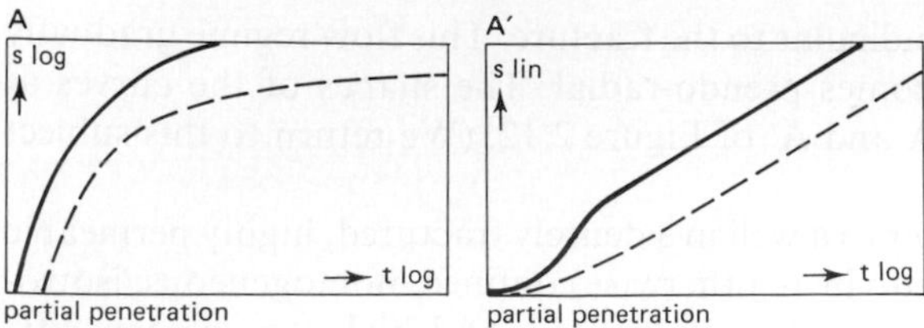

Figure 2.14 The effect of the well's partial penetration on the time-drawdown relationship in an unconsolidated, confined aquifer. The dashed curves are those of Parts A and A′ of Figure 2.12

sink will be satisfied. Obviously, the effects of well-bore storage will appear at early pumping times, and may last from a few minutes to many minutes, depending on the storage capacity of the well. In a log-log plot of drawdown versus time, the effect of well-bore storage is reflected by a straight-line segment with a slope of unity. (We return to this subject in Section 15.1.1.)

If a pumping test is conducted in a large-diameter well and drawdown data from observation wells or piezometers are used in the analysis, it should not be forgotten that those data will also be affected by the well-bore storage in the pumped well. At early pumping time, the data will deviate from the theoretical curve, although, in a log-log plot, no early-time straight-line segment of slope unity will appear. Figure 2.15 shows the effect of well-bore storage on time-drawdown plots of observation wells or piezometers. (We return to this subject in Section 11.1.)

Recharge or impermeable boundaries

The theoretical curves of all the main aquifer types can also be affected by recharge or impermeable boundaries. This effect is shown in Figure 2.16. Parts A and A′ of that figure show a situation where the cone of depression reaches a recharge boundary. When this happens, the drawdown in the well stabilizes. The field data curve then begins to deviate more and more from the theoretical curve, which is shown in the dashed segment of the curve. Impermeable (no-flow) boundaries have the opposite effect on the drawdown. If the cone of depression reaches such a boundary, the drawdown will double. The field data curve will then steepen, deviating upward from the theoretical curve. This is shown in Parts B and B′ of Figure 2.16. (We return to this subject in Chapter 6.)

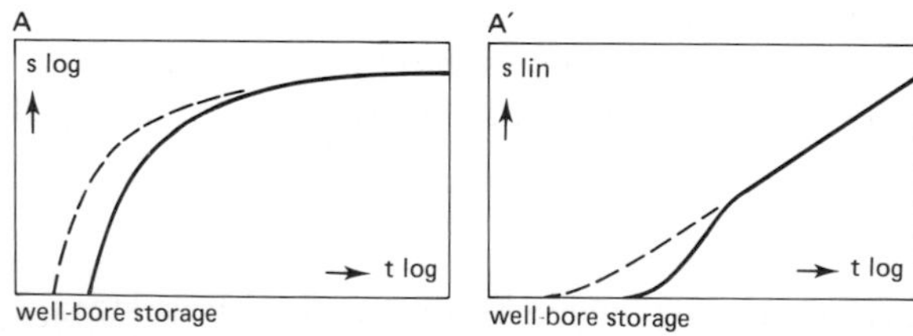

Figure 2.15 The effect of well-bore storage in the pumped well on the theoretical time-drawdown plots of observation wells or piezometers. The dashed curves are those of Parts A and A′ of Figure 2.12

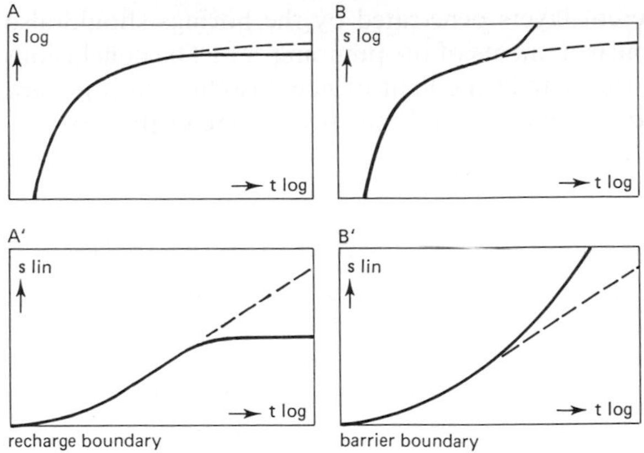

Figure 2.16 The effect of a recharge boundary (Parts A and A′) and an impermeable boundary Parts B and B′) on the theoretical time-drawdown relationship in a confined unconsolidated aquifer. The dashed curves are those of Parts A and A′ of Figure 2.12

2.10 Reporting and filing of data

2.10.1 Reporting

When the evaluation of the test data has been completed, a report should be written about the results. It is beyond the scope of this book to say what this report should contain, but it should at least include the following items:

- A map, showing the location of the test site, the well and the piezometers, and recharge and barrier boundaries, if any;
- A lithological cross-section of the test site, based on the data obtained from the bore holes, and showing the depth of the well screen and the number, depth, and distances of the piezometers;
- Tables of the field measurements made of the well discharge and the water levels in the well and the piezometers;
- Hydrographs, illustrating the corrections applied to the observed data, if applicable;
- Time-drawdown curves and distance-drawdown curves;
- The considerations that led to the selection of the theoretical model used for the analysis;
- The calculations in an abbreviated form, including the values obtained for the aquifer characteristics and a discussion of their accuracy;
- Recommendations for further investigations, if applicable;
- A summary of the main results.

2.10.2 Filing of data

A copy of the report should be kept on file for further reference and for use in any later studies. Samples of the different layers penetrated by the borings should also be filed, as should the basic field measurements of the pumping test. The conclusions drawn from the test may become obsolete in the light of new insights, but the hard facts, carefully collected in the field, remain facts and can always be re-evaluated.

3 Confined aquifers

When a fully penetrating well pumps a confined aquifer (Figure 3.1), the influence of the pumping extends radially outwards from the well with time, and the pumped water is withdrawn entirely from the storage within the aquifer. In theory, because the pumped water must come from a reduction of storage within the aquifer, only unsteady-state flow can exist. In practice, however, the flow to the well is considered to be in a steady state if the change in drawdown has become negligibly small with time.

Methods for evaluating pumping tests in confined aquifers are available for both steady-state flow (Section 3.1) and unsteady-state flow (Section 3.2).

The assumptions and conditions underlying the methods in this chapter are:

1) The aquifer is confined;
2) The aquifer has a seemingly infinite areal extent;
3) The aquifer is homogeneous, isotropic, and of uniform thickness over the area influenced by the test;
4) Prior to pumping, the piezometric surface is horizontal (or nearly so) over the area that will be influenced by the test;
5) The aquifer is pumped at a constant discharge rate;
6) The well penetrates the entire thickness of the aquifer and thus receives water by horizontal flow.

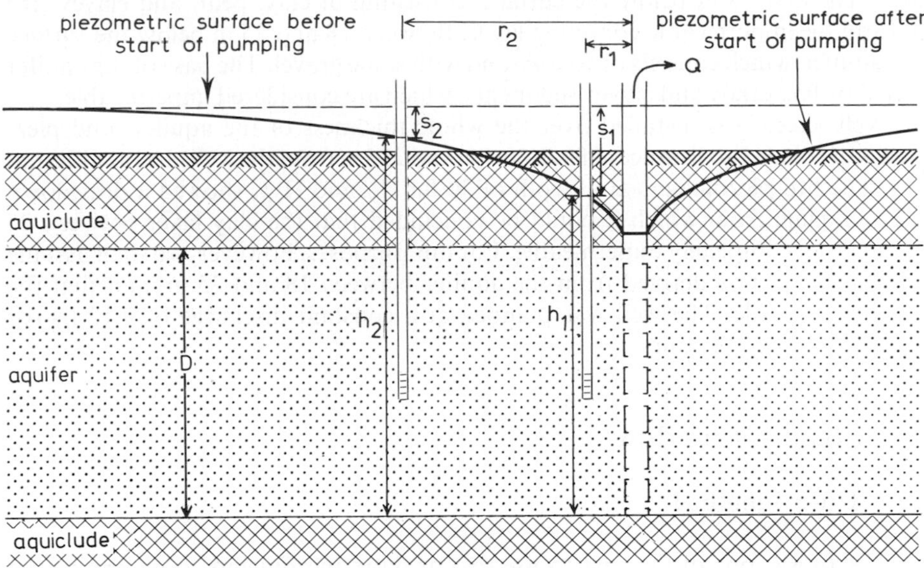

Figure 3.1 Cross-section of a pumped confined aquifer

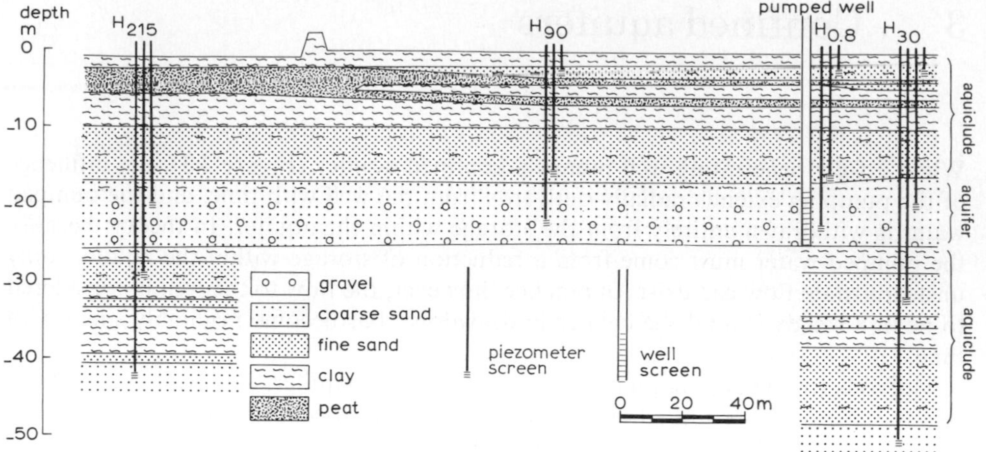

Figure 3.2 Lithological cross-section of the pumping-test site 'Oude Korendijk', The Netherlands (after Wit 1963)

And, in addition, for unsteady-state methods:

7) The water removed from storage is discharged instantaneously with decline of head;
8) The diameter of the well is small, i.e. the storage in the well can be neglected.

The methods described in this chapter will be illustrated with data from a pumping test conducted in the polder 'Oude Korendijk', south of Rotterdam, The Netherlands (Wit 1963).

Figure 3.2 shows a lithological cross-section of the test site as derived from the borings. The first 18 m below the surface, consisting of clay, peat, and clayey fine sand, form the impermeable confining layer. Between 18 and 25 m below the surface lies the aquifer, which consists of coarse sand with some gravel. The base of the aquifer is formed by fine sandy and clayey sediments, which are considered impermeable.

The well screen was installed over the whole thickness of the aquifer, and piezometers were placed at distances of 0.8, 30, 90, and 215 m from the well, and at different depths. The two piezometers at a depth of 30 m, H_{30} and H_{215}, showed a drawdown during pumping, from which it could be concluded that the clay layer between 25 and 27 m is not completely impermeable. For our purposes, however, we shall assume that all the water was derived from the aquifer between 18 and 25 m, and that the base is impermeable. The well was pumped at a constant discharge of 9.12 l/s (or 788 m^3/d) for nearly 14 hours.

3.1 Steady-state flow

3.1.1 Thiem's method

Thiem (1906) was one of the first to use two or more piezometers to determine the transmissivity of an aquifer. He showed that the well discharge can be expressed as

$$Q = \frac{2\pi KD(h_2 - h_1)}{\ln(r_2/r_1)} = \frac{2\pi KD(h_2 - h_1)}{2.30 \log (r_2/r_1)} \quad (3.1)$$

where

Q = the well discharge in m^3/d

KD = the transmissivity of the aquifer in m^2/d

r_1 and r_2 = the respective distances of the piezometers from the well in m

h_1 and h_2 = the respective steady-state elevations of the water levels in the piezometers in m.

For practical purposes, Equation 3.1 is commonly written as

$$Q = \frac{2\pi KD(s_{m1} - s_{m2})}{2.30 \log (r_2/r_1)} \quad (3.2)$$

where s_{m1} and s_{m2} are the respective steady-state drawdowns in the piezometers in m.

In cases where only one piezometer at a distance r_1 from the well is available

$$Q = \frac{2\pi KD(s_{mw} - s_{m1})}{2.30 \log (r_1/r_w)} \quad (3.3)$$

where s_{mw} is the steady-state drawdown in the well, and r_w is the radius of the well.

Equation 3.3 is of limited use because local hydraulic conditions in and near the well strongly influence the drawdown in the well (e.g. s_w is influenced by well losses caused by the flow through the well screen and the flow inside the well to the pump intake). Equation 3.3 should therefore be used with caution and only when other methods cannot be applied. Preferably, two or more piezometers should be used, located close enough to the well that their drawdowns are appreciable and can readily be measured.

With the Thiem (or equilibrium) equation, two procedures can be followed to determine the transmissivity of a confined aquifer. The following assumptions and conditions should be satisfied:

- The assumptions listed at the beginning of this chapter;
- The flow to the well is in steady state.

Procedure 3.1

- Plot the observed drawdowns in each piezometer against the corresponding time on a sheet of semi-log paper: the drawdowns on the vertical axis on a linear scale and the time on the horizontal axis on a logarithmic scale;
- Construct the time-drawdown curve for each piezometer; this is the curve that fits best through the points.
  It will be seen that for the late-time data the curves of the different piezometers run parallel. This means that the hydraulic gradient is constant and that the flow in the aquifer can be considered to be in a steady state;
- Read for each piezometer the value of the steady-state drawdown s_m;
- Substitute the values of the steady-state drawdown s_{m1} and s_{m2} for two piezometers into Equation 3.2, together with the corresponding values of r and the known value of Q, and solve for KD;

- Repeat this procedure for all possible combinations of piezometers. Theoretically, the results should show a close agreement; in practice, however, the calculations may give more or less different values of KD, e.g. because the condition of homogeneity of the aquifer was not satisfied. The mean is used as the final result.

Example 3.1
We shall illustrate Procedure 3.1 of the Thiem method with data from the pumping test 'Oude Korendijk'. On semi-log paper and using Table 3.1, we plot the drawdown versus time for all the piezometers, and draw the curves through the plotted points (Figure 3.3). As can be seen from this figure, the water levels in the piezometers at the end of the test (after 830 minutes of pumping) had not yet stabilized. In other words, steady-state flow had not been reached.

From Figure 3.3, however, it can also be seen that the curves of the piezometers H_{30} and H_{90} start to run parallel approximately 10 minutes after pumping began. This means that the drawdown difference between these piezometers after t = 10 minutes remained constant, i.e. the hydraulic gradient between these piezometers remained constant. This is the primary condition for which Thiem's equation is valid.

The reader will note that during the whole pumping period the cone of depression deepened and expanded. Even at late pumping times, the water levels in the piezometers continued to drop: a clear example of unsteady-state flow! Although the cone of depression deepened during the whole pumping period, after 10 minutes of pumping it deepened uniformly between the two piezometers under consideration: a typical case of what is sometimes called transient steady-state flow!

Wenzel (1942) was probably the first who proved the transient nature of the Thiem equation, but this important work has received little attention in the literature, until recently when Butler (1988) discussed the matter in detail.

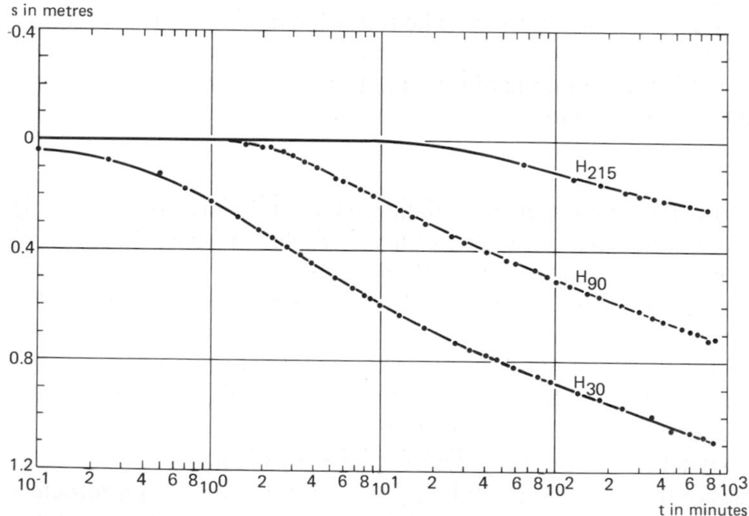

Figure 3.3 Time-drawdown plot of the piezometers H_{30}, H_{90} and H_{215}, pumping test 'Oude Korendijk

Table 3.1 Data pumping test 'Oude Korendijk' (after Wit 1963)

Piezometer H_{30} Screen depth 20 m

t (min)	s (m)	t/r^2(min/m^2)	t (min)	s (m)	t/r^2(min/m^2)
0	0	0	18	0.680	2.00×10^{-2}
0.1	0.04	1.11×10^{-4}	27	0.742	3.00
0.25	0.08	2.78	33	0.753	3.67
0.50	0.13	5.56	41	0.779	4.56
0.70	0.18	7.78×10^{-4}	48	0.793	5.33
1.0	0.23	1.11×10^{-3}	59	0.819	6.56
1.40	0.28	1.56	80	0.855	8.89×10^{-2}
1.90	0.33	2.11	95	0.873	1.06×10^{-1}
2.33	0.36	2.59	139	0.915	1.54
2.80	0.39	3.12	181	0.935	2.01
3.36	0.42	3.73	245	0.966	2.72
4.00	0.45	4.44	300	0.990	3.33
5.35	0.50	5.94	360	1.007	4.00
6.80	0.54	7.56	480	1.050	5.33
8.3	0.57	9.22	600	1.053	6.67
8.7	0.58	9.67×10^{-3}	728	1.072	8.09
10.0	0.60	1.11×10^{-2}	830	1.088	9.22×10^{-1}
13.1	0.64	1.46×10^{-2}			

Piezometer H_{90} Screen depth 24 m

t (min)	s (m)	t/r^2(min/m^2)	t (min)	s (m)	t/r^2(min/m^2)
0	0	0	40	0.404	4.94×10^{-3}
1.5	0.015	1.85×10^{-4}	53	0.429	6.54
2.0	0.021	2.47	60	0.444	7.41
2.16	0.023	2.67	75	0.467	9.26×10^{-3}
2.66	0.044	3.28	90	0.494	1.11×10^{-2}
3	0.054	3.70	105	0.507	1.30
3.5	0.075	4.32	120	0.528	1.48
4	0.090	4.94	150	0.550	1.85
4.33	0.104	5.35	180	0.569	2.22
5.5	0.133	6.79	248	0.593	3.06
6	0.153	7.41	301	0.614	3.72
7.5	0.178	9.26×10^{-4}	363	0.636	4.48
9	0.206	1.11×10^{-3}	422	0.657	5.21
13	0.250	1.60	542	0.679	6.69
15	0.275	1.85	602	0.688	7.43
18	0.305	2.22	680	0.701	8.40
25	0.348	3.08	785	0.718	9.69×10^{-2}
30	0.364	3.70×10^{-3}	845	0.716	1.04×10^{-1}

Piezometer H_{215} Screen depth 20 m

t (min)	s (m)	t/r^2(min/m^2)	t (min)	s (m)	t/r^2(min/m^2)
0	0	0	305	0.196	6.60×10^{-3}
66	0.089	1.43×10^{-3}	366	0.207	7.92×10^{-3}
127	0.138	2.75×10^{-3}	430	0.214	9.30×10^{-3}
185	0.165	4.00×10^{-3}	606	0.227	1.31×10^{-2}
251	0.186	5.43×10^{-3}	780	0.250	1.68×10^{-2}

From Figure 3.3, the reader will also note that the time-drawdown curve of piezometer H_{215} does not run parallel to that of the other piezometers, not even at very late pumping times. In applying Procedure 3.1 of the Thiem method, therefore, we shall disregard the data of this piezometer and shall use only the data from the piezometers H_{30} and H_{90} for $t > 10$ minutes. In doing so, and using Equation 3.2 after rearranging, we find

$$KD = \frac{788 \times 2.30}{2 \times 3.14\,(1.088 - 0.716)} \log \frac{90}{30} = 370\ m^2/d$$

Similar calculations were made for combinations of these piezometers with the piezometer $H_{0.80}$. The results are given in Table 3.2. The table shows only minor differences in the results. Our conclusion is that the transmissivity of the tested aquifer is approximately 385 m²/d.

Table 3.2 Results of the application of Thiem's method, Procedure 3.1, to data from the pumping test 'Oude Korendijk'

r_1 (m)	r_2 (m)	s_{m1} (m)	s_{m2} (m)	KD (m^2/d)
30	90	1.088	0.716	370
0.8	30	2.236	1.088	396
0.8	90	2.236	0.716	389
			Mean	385

Procedure 3.2

- Plot on semi-log paper the observed transient steady-state drawdown s_m of each piezometer against the distance r between the well and the piezometer (Figure 3.4);
- Draw the best-fitting straight line through the plotted points; this is the distance-drawdown graph;
- Determine the slope of this line Δs_m, i.e. the difference of drawdown per log cycle of r, giving $r_2/r_1 = 10$ or $\log r_2/r_1 = 1$. In doing so Equation 3.2 reduces to

$$Q = \frac{2\pi KD}{2.30} \Delta s_m \qquad (3.4)$$

- Substitute the numerical values of Q and Δs_m into Equation 3.4 and solve for KD.

Example 3.2

Using Procedure 3.2 of the Thiem method, we plot the values of s_m and r on semi-log paper (Figure 3.4). We then draw a straight line through the plotted points. Note that the plot of piezometer H_{215} falls below the straight line and is therefore discarded. The slope of the straight line is equal to a drawdown difference of 0.74 m per log cycle of r. Introducing this value and the value of Q into Equation 3.4 yields

$$KD = \frac{2.30Q}{2\pi\Delta s} = \frac{2.30 \times 788}{2 \times 3.14 \times 0.74} = 390\ m^2/d$$

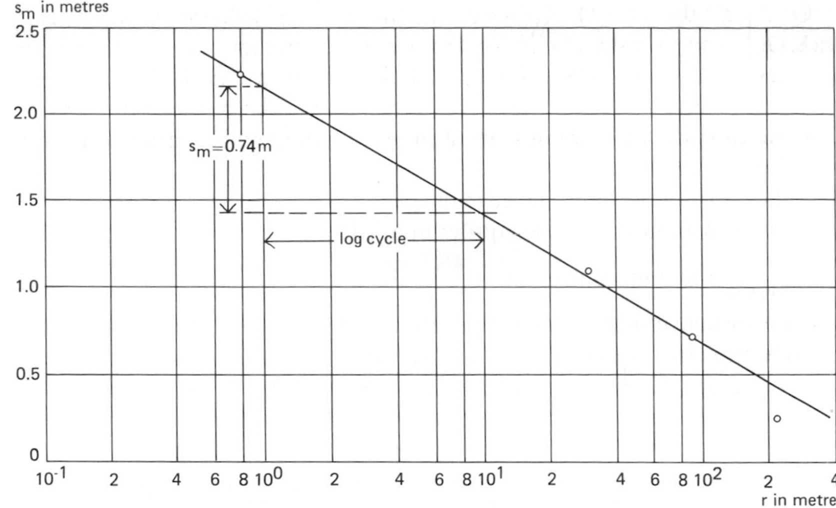

Figure 3.4 Analysis of data from pumping test 'Oude Korendijk' with the Thiem method, Procedure 3.2

This result agrees very well with the average value obtained with the Thiem method, Procedure 3.1.

Remarks

- Steady-state has been defined here as the situation where variations of the drawdown with time are negligible, or where the hydraulic gradient has become constant. The reader will know, however, that true steady state, i.e. drawdown variations are zero, is impossible in a confined aquifer;
- Field conditions may be such that considerable time is required to reach steady-state flow. Such long pumping times are not always required, however, because transient steady-state flow, i.e. flow under a constant hydraulic gradient, may be reached much earlier as we have shown in Example 3.1.

3.2 Unsteady-state flow

3.2.1 Theis's method

Theis (1935) was the first to develop a formula for unsteady-state flow that introduces the time factor and the storativity. He noted that when a well penetrating an extensive confined aquifer is pumped at a constant rate, the influence of the discharge extends outward with time. The rate of decline of head, multiplied by the storativity and summed over the area of influence, equals the discharge.

The unsteady-state (or Theis) equation, which was derived from the analogy between the flow of groundwater and the conduction of heat, is written as

$$s = \frac{Q}{4\pi KD} \int_u^\infty \frac{e^{-y}\,dy}{y} = \frac{Q}{4\pi KD} W(u) \qquad (3.5)$$

where

s = the drawdown in m measured in a piezometer at a distance r in m from the well

Q = the constant well discharge in m^3/d

KD = the transmissivity of the aquifer in m^2/d

$$u = \frac{r^2 S}{4KDt} \text{ and consequently } S = \frac{4KDtu}{r^2} \qquad (3.6)$$

S = the dimensionless storativity of the aquifer

t = the time in days since pumping started

$$W(u) = -0.5772 - \ln u + u - \frac{u^2}{2.2!} + \frac{u^3}{3.3!} - \frac{u^4}{4.4!} + \ldots$$

The exponential integral is written symbolically as W(u), which in this usage is generally read 'well function of u' or 'Theis well function'. It is sometimes found under the symbol -Ei(-u) (Jahnke and Embde 1945). A well function like W(u) and its argument u are also indicated as 'dimensionless drawdown' and 'dimensionless time', respectively. The values for W(u) as u varies are given in Annex 3.1.
From Equation 3.5, it will be seen that, if s can be measured for one or more values of r and for several values of t, and if the well discharge Q is known, S and KD can be determined. The presence of the two unknowns and the nature of the exponential integral make it impossible to effect an explicit solution.

Using Equations 3.5 and 3.6, Theis devised the 'curve-fitting method' (Jacob 1940) to determine S and KD. Equation 3.5 can also be written as

$$\log s = \log(Q/4\pi KD) + \log(W(u))$$

and Equation 3.6 as

$$\log(r^2/t) = \log(4KD/S) + \log(u)$$

Since $Q/4\pi KD$ and $4KD/S$ are constant, the relation between log s and log (r^2/t) must be similar to the relation between log W(u) and log (u). Theis's curve-fitting method is based on the fact that if s is plotted against r^2/t and W(u) against u on the same log-log paper, the resulting curves (the data curve and the type curve, respectively) will be of the same shape, but will be horizontally and vertically offset by the constants $Q/4\pi KD$ and $4KD/S$. The two curves can be made to match. The coordinates of an arbitrary matching point are the related values of s, r^2/t, u, and W(u), which can be used to calculate KD and S with Equations 3.5 and 3.6.

Instead of using a plot of W(u) versus (u) (normal type curve) in combination with a data plot of s versus r^2/t, it is frequently more convenient to use a plot of W(u) versus 1/u (reversed type curve) and a plot of s versus t/r^2 (Figure 3.5).

Theis's curve-fitting method is based on the assumptions listed at the beginning of this chapter and on the following limiting condition:
- The flow to the well is in unsteady state, i.e. the drawdown differences with time are not negligible, nor is the hydraulic gradient constant with time.

Procedure 3.3

– Prepare a type curve of the Theis well function on log-log paper by plotting values of W(u) against the arguments 1/u, using Annex 3.1 (Figure 3.5);

– Plot the observed data curve s versus t/r^2 on another sheet of log-log paper of the same scale;

– Superimpose the data curve on the type curve and, keeping the coordinate axes parallel, adjust until a position is found where most of the plotted points of the data curve fall on the type curve (Figure 3.6);

– Select an arbitrary match point A on the overlapping portion of the two sheets and read its coordinates W(u), 1/u, s, and t/r^2. Note that it is not necessary for the match point to be located along the type curve. In fact, calculations are greatly simplified if the point is selected where the coordinates of the type curve are W(u) = 1 and 1/u = 10;

– Substitute the values of W(u), s, and Q into Equation 3.5 and solve for KD;

– Calculate S by substituting the values of KD, t/r^2, and u into Equation 3.6.

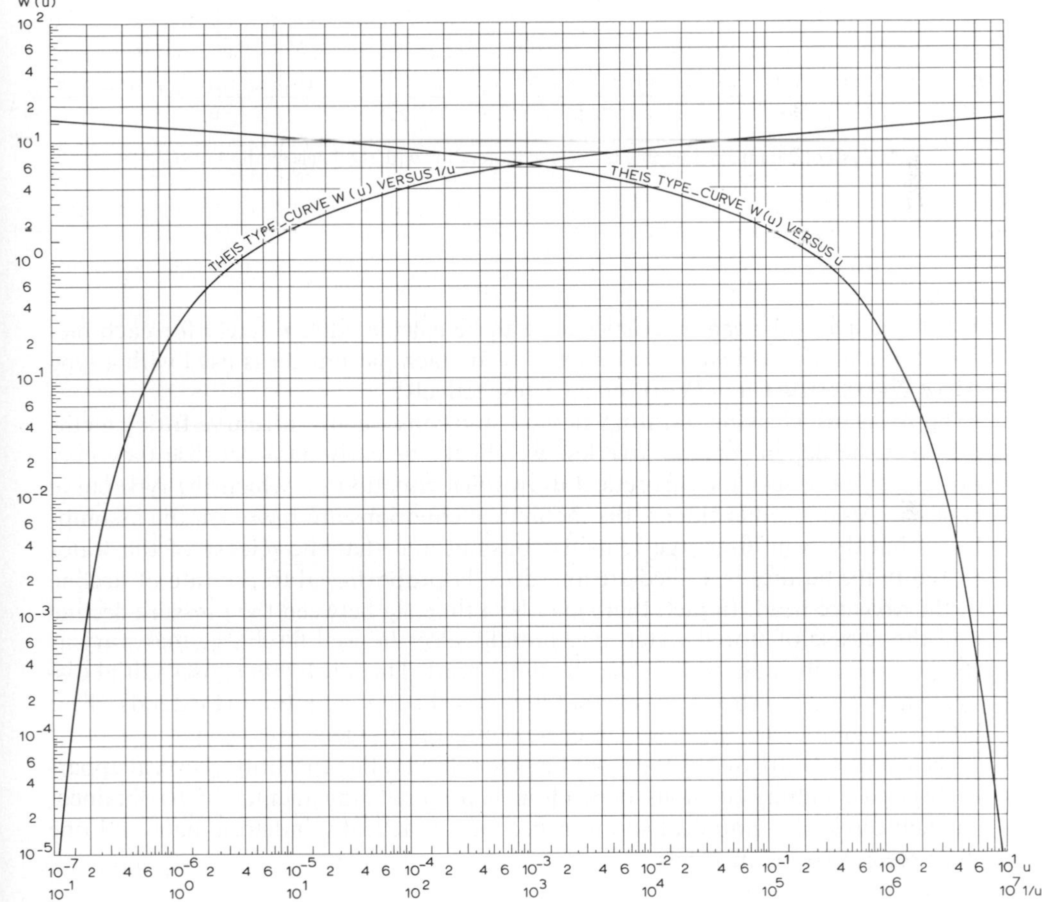

Figure 3.5 Theis type curve for W(u) versus u and W(u) versus l/u

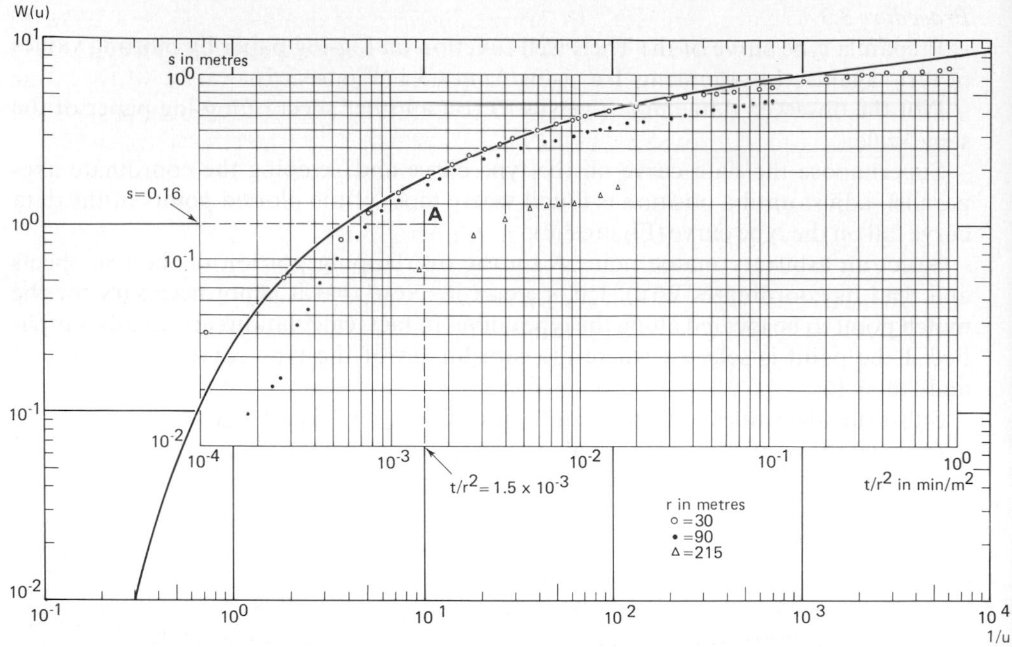

Figure 3.6 Analysis of data from pumping test 'Oude Korendijk' with the Theis method, Procedure 3.3

Remarks

- When the hydraulic characteristics have to be calculated separately for each piezometer, a plot of s versus t or s versus 1/t for each piezometer is used with a type curve W(u) versus 1/u or W(u) versus u, respectively;
- In applying the Theis curve-fitting method, and consequently all curve-fitting methods, one should, in general, give less weight to the early data because they may not closely represent the theoretical drawdown equation on which the type curve is based. Among other things, the theoretical equations are based on the assumptions that the well discharge remains constant and that the release of the water stored in the aquifer is immediate and directly proportional to the rate of decline of the pressure head. In fact, there may be a time lag between the pressure decline and the release of stored water, and initially also the well discharge may vary as the pump is adjusting itself to the changing head. This probably causes initial disagreement between theory and actual flow. As the time of pumping extends, these effects are minimized and closer agreement may be attained;
- If the observed data on the logarithmic plot exhibit a flat curvature, several apparently good matching positions, depending on personal judgement, may be obtained. In such cases, the graphical solution becomes practically indeterminate and one must resort to other methods.

Example 3.3

The Theis method will be applied to the unsteady-state data from the pumping test

'Oude Korendijk' listed in Table 3.1. Figure 3.6 shows a plot of the values of s versus t/r^2 for the piezometers H_{30}, H_{90} and H_{215} matched with the Theis type-curve, W(u) versus 1/u. The reader will note that for late pumping times the points do not fall exactly on the type curve. This may be due to leakage effects because the aquifer was not perfectly confined. Note the anomalous drawdown behaviour of piezometer H_{215} already noticed in Example 3.2. In the matching procedure, we have discarded the data of this piezometer. The match point A has been so chosen that the value of W(u) = 1 and the value of 1/u = 10. On the sheet with the observed data, the match point A has the coordinates $s_A = 0.16$ m and $(t/r^2)_A = 1.5 \times 10^{-3}$ min/m² $= 1.5 \times 10^{-3}/1440$ d/m². Introducing these values and the value of Q = 788 m³/d into Equations 3.5 and 3.6 yields

$$KD = \frac{Q}{4\pi S_A} W(u) = \frac{788}{4 \times 3.14 \times 0.16} \times 1 = 392 \text{ m}^2/\text{d}$$

and

$$S = \frac{4KD(t/r^2)_A}{1/u} = 4 \times 392 \times \frac{1.5 \times 10^{-3}}{1440} \times \frac{1}{10} = 1.6 \times 10^{-4}$$

3.2.2 Jacob's method

The Jacob method (Cooper and Jacob 1946) is based on the Theis formula, Equation 3.5

$$s = \frac{Q}{4\pi KD} W(u) = \frac{Q}{4\pi KD}(-0.5772 - \ln u + u - \frac{u^2}{2.2!} + \frac{u^3}{3.3!} - ....)$$

From $u = r^2S/4KDt$, it will be seen that u decreases as the time of pumping t increases and the distance from the well r decreases. Accordingly, for drawdown observations made in the near vicinity of the well after a sufficiently long pumping time, the terms beyond ln u in the series become so small that they can be neglected. So for small values of u ($u < 0.01$), the drawdown can be approximated by

$$s = \frac{Q}{4\pi KD}(-0.5772 - \ln \frac{r^2S}{4KDt})$$

with

an error less than	1%	2%	5%	10%
for u smaller than	0.03	0.05	0.1	0.15

After being rewritten and changed into decimal logarithms, this equation reduces to

$$s = \frac{2.30Q}{4\pi KD} \log \frac{2.25KDt}{r^2S} \tag{3.7}$$

Because Q, KD, and S are constant, if we use drawdown observations at a short distance r from the well, a plot of drawdown s versus the logarithm of t forms a straight line (Figure 3.7). If this line is extended until it intercepts the time-axis where s = 0, the interception point has the coordinates s = 0 and $t = t_0$. Substituting these values into Equation 3.7 gives

$$0 = \frac{2.30Q}{4\pi KD} \log \frac{2.25KDt_0}{r^2S}$$

and because $\frac{2.30Q}{4\pi KD} \neq 0$, it follows that $\frac{2.25KDt_0}{r^2S} = 1$

or

$$S = \frac{2.25KDt_0}{r^2} \tag{3.8}$$

The slope of the straight line (Figure 3.7), i.e. the drawdown difference Δs per log cycle of time $\log t/t_0 = 1$, is equal to $2.30Q/4\pi KD$. Hence

$$KD = \frac{2.30Q}{4\pi\Delta s} \tag{3.9}$$

Similarly, it can be shown that, for a fixed time t, a plot of s versus r on semi-log paper forms a straight line and the following equations can be derived

$$S = \frac{2.25KDt}{r_0^2} \tag{3.10}$$

and

$$KD = \frac{2.30Q}{2\pi\Delta s} \tag{3.11}$$

If all the drawdown data of all piezometers are used, the values of s versus t/r^2 can be plotted on semi-log paper. Subsequently, a straight line can be drawn through the

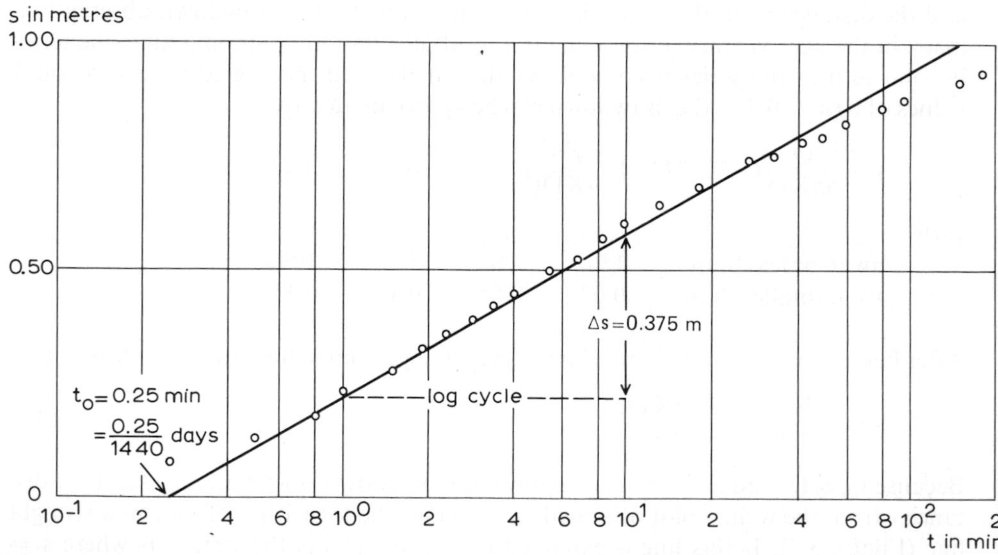

Figure 3.7 Analysis of data from pumping test 'Oude Korendijk' (r = 30 m) with the Jacob method, Procedure 3.4

plotted points. Continuing with the same line of reasoning as above, we derive the following formulas

$$S = 2.25KD(t/r^2)_0 \tag{3.12}$$

and

$$KD = \frac{2.30Q}{4\pi\Delta s} \tag{3.13}$$

Jacob's straight-line method can be applied in each of the three situations outlined above. (See Procedure 3.4 for r = constant, Procedure 3.5 for t = constant, and Procedure 3.6 when values of t/r^2 are used in the data plot.)
The following assumptions and conditions should be satisfied:
- The assumptions listed at the beginning of this chapter;
- The flow to the well is in unsteady state;
- The values of u are small ($u < 0.01$), i.e. r is small and t is sufficiently large.

The condition that u be small in confined aquifers is usually satisfied at moderate distances from the well within an hour or less. The condition $u < 0.01$ is rather rigid. For a five or even ten times higher value ($u < 0.05$ and $u < 0.10$), the error introduced in the result is less than 2 and 5%, respectively. Further, a visual inspection of the graph in the range $u < 0.01$ and $u < 0.1$ shows that it is difficult, if not impossible, to indicate precisely where the field data start to deviate from the straight-line relationship. For all practical purposes, therefore, we suggest using $u < 0.1$ as a condition for Jacob's method.

The reader will note that the use of Equation 3.7 for the determination of the difference in drawdown $s_1 - s_2$ between two piezometers at distances r_1 and r_2 from the well leads to an expression that is identical to the Thiem formula (Equation 3.2).

Procedure 3.4 (for r is constant)
- For one of the piezometers, plot the values of s versus the corresponding time t on semi-log paper (t on logarithmic scale), and draw a straight line through the plotted points (Figure 3.7);
- Extend the straight line until it intercepts the time axis where s = 0, and read the value of t_0;
- Determine the slope of the straight line, i.e. the drawdown difference Δs per log cycle of time;
- Substitute the values of Q and Δs into Equation 3.9 and solve for KD. With the known values of KD and t_0, calculate S from Equation 3.8.

Remarks
- Procedure 3.4 should be repeated for other piezometers at moderate distances from the well. There should be a close agreement between the calculated KD values, as well as between those of S;
- When the values of KD and S are determined, they are introduced into the equation $u = r^2S/4KDt$ to check whether $u < 0.1$, which is a practical condition for the applicability of the Jacob method.

Example 3.4
For this example, we use the drawdown data of the piezometer H_{30} in 'Oude Korendijk' (Table 3.1). We plot these data against the corresponding time data on semi-log paper (Figure 3.7), and fit a straight line through the plotted points. The slope of this straight line is measured on the vertical axis $\Delta s = 0.375$ m per log cycle of time. The intercept of the fitted straight line with the absciss (zero-drawdown axis) is $t_0 = 0.25$ min $= 0.25/1440$ d. The discharge rate $Q = 788$ m^3/d. Substitution of these values into Equation 3.9 yields

$$KD = \frac{2.30Q}{4\pi\Delta s} = \frac{2.30 \times 788}{4 \times 3.14 \times 0.375} = 385 \text{ m}^2/\text{d}$$

and into Equation 3.8

$$S = \frac{2.25KDt_0}{r^2} = \frac{2.25 \times 385}{30^2} \times \frac{0.25}{1440} = 1.7 \times 10^{-4}$$

Substitution of the values of KD, S, and r into $u = r^2S/4KDt$ shows that, for $t > 0.001$ d or $t > 1.4$ min, $u < 0.1$, as is required. The departure of the time-drawdown curve from the theoretical straight line is probably due to leakage through one of the assumed 'impermeable' layers.

The same method applied to the data collected in the piezometer at 90 m gives: $KD = 450$ m^2/d and $S = 1.7 \times 10^{-4}$ with $u < 0.1$ for $t > 11$ min. This result is less reliable because few points are available between $t = 11$ min. and the time that leakage probably starts to influence the drawdown data.

Procedure 3.5 (t is constant)
- Plot for a particular time t the values of s versus r on semi-log paper (r on logarithmic scale), and draw a straight line through the plotted points (Figure 3.8);
- Extend the straight line until it intercepts the r axis where $s = 0$, and read the value of r_0;
- Determine the slope of the straight line, i.e. the drawdown difference Δs per log cycle of r;
- Substitute the values of Q and Δs into Equation 3.11 and solve for KD. With the known values of KD and r_0, calculate S from Equation 3.10.

Remarks
- Note the difference in the denominator of Equations 3.9 and 3.11;
- The data of at least three piezometers are needed for reliable results;
- If the drawdown in the different piezometers is not measured at the same time, the drawdown at the chosen moment t has to be interpolated from the time-drawdown curve of each piezometer used in Procedure 3.4;
- Procedure 3.5 should be repeated for several values of t. The values of KD thus obtained should agree closely, and the same holds true for values of S.

Example 3.5
Here, we plot the (interpolated) drawdown data from the piezometers of 'Oude Korendijk' for $t = 140$ min ≈ 0.1 d against the distances between the piezometers and the well (Figure 3.8). In the previous examples, we explained why we discarded the point

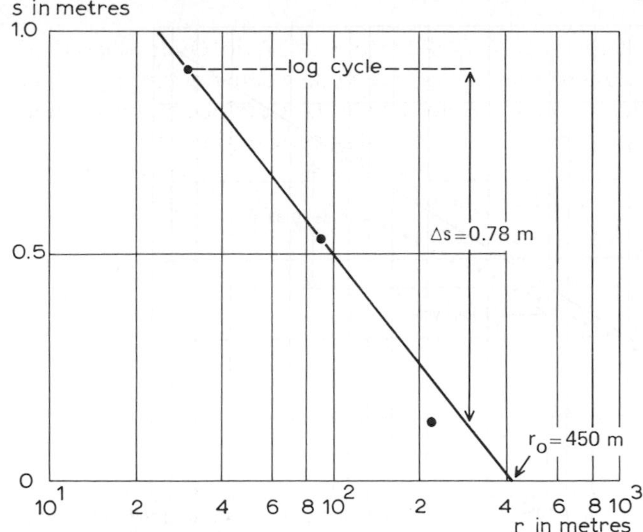

Figure 3.8 Analysis of data from pumping test 'Oude Korendijk' (t = 140 min) with the Jacob method, Procedure 3.5

of piezometer H_{215}. The slope of the straight line $\Delta s = 0.78$ m and the intercept with the absciss $r_0 = 450$ m. The discharge rate $Q = 788$ m³/d. Substitution of these values into Equation 3.11 yields

$$KD = \frac{2.30Q}{2\pi\Delta s} = \frac{2.30 \times 788}{2 \times 3.14 \times 0.78} = 370 \text{ m}^2/\text{d}$$

and into Equation 3.10

$$S = \frac{2.25KDt}{r_0^2} = \frac{2.25 \times 370 \times 0.1}{450^2} = 4.1 \times 10^{-4}$$

Procedure 3.6 (based on s versus t/r^2 data plot)

- Plot the values of s versus t/r^2 on semi-log paper (t/r^2 on the logarithmic axis), and draw a straight line through the plotted points (Figure 3.9);
- Extend the straight line until it intercepts the t/r^2 axis where s = 0, and read the value of $(t/r^2)_0$;
- Determine the slope of the straight line, i.e. the drawdown difference Δs per log cycle of t/r^2;
- Substitute the values of Q and Δs into Equation 3.13 and solve for KD. Knowing the values of KD and $(t/r^2)_0$, calculate S from Equation 3.12.

Example 3.6

As an example of the Jacob method, Procedure 3.6, we use the values of t/r^2 for all the piezometers of 'Oude Korendijk' (Table 3.1). In Figure 3.9, the values of s are plotted on semi-log paper against the corresponding values of t/r^2. Through those points, and neglecting the points for H_{215}, we draw a straight line, which intercepts

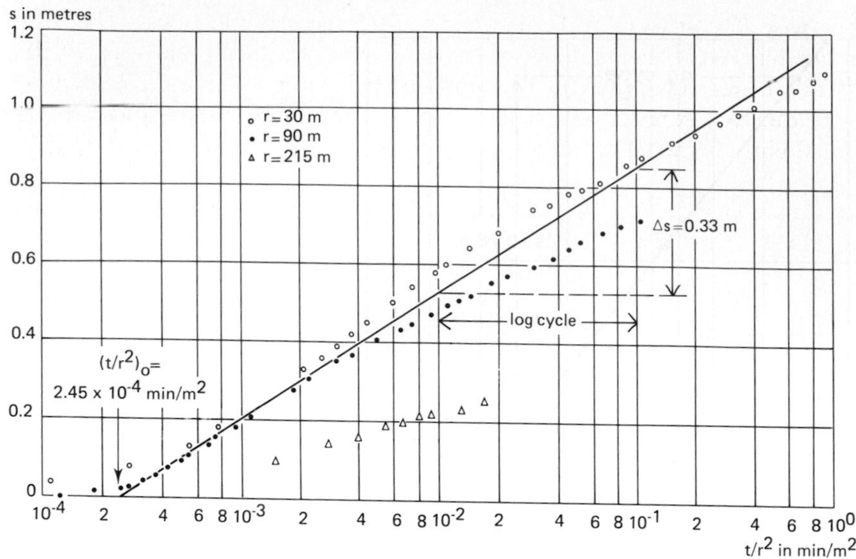

Figure 3.9 Analysis of data from pumping test 'Oude Korendijk' with the Jacob method, Procedure 3.6

the $s = 0$ axis (absciss) in $(t/r^2)_0 = 2.45 \times 10^{-4}$ min/m² or $(2.45/1440) \times 10^{-4}$ d/m². On the vertical axis, we measure the drawdown difference per log cycle of t/r^2 as $\Delta s = 0.33$ m. The discharge rate $Q = 788$ m³/d.
Introducing these values into Equation 3.13 gives

$$KD = \frac{2.30Q}{4\pi\Delta s} = \frac{2.30 \times 788}{4 \times 3.14 \times 0.33} = 437 \text{ m}^2/\text{d}$$

and into Equation 3.12

$$S = 2.25KD(t/r^2)_0 = 2.25 \times 437 \times \frac{2.45}{1440} \times 10^{-4} = 1.7 \times 10^{-4}$$

3.3 Summary

Using data from the pumping test 'Oude Korendijk' (Figure 3.2 and Table 3.1), we have illustrated the methods of analyzing (transient) steady and unsteady flow to a well in a confined aquifer. Table 3.3 summarizes the values we obtained for the aquifer's hydraulic characteristics.
When we compare the results of Table 3.3, we can conclude that the values of KD and S agree very well, except for those of the last two methods. The differences in the results are due to the fact that the late-time data have probably been influenced by leakage and that graphical methods of analysis are never accurate. Minor shifts of the data plot are often possible, giving an equally good match with a type curve, but yielding different values for the aquifer characteristics. The same is true for a semi-log plot whose points do not always fit on a straight line because of measuring

errors or otherwise. The analysis of the Jacob 2 method, for example, is weak, because the straight line has been fitted through only two points, the third point, that of the piezometer H_{215}, being unreliable. The anomalous behaviour of this far-field piezometer may be due to leakage effects, heterogeneity of the aquifer (the transmissivity at H_{215} being slightly higher than closer to the well), or faulty construction (partly clogged).

We could thus conclude that the aquifer at 'Oude Korendijk' has the following parameters: $KD = 390\ m^2/d$ and $S = 1.7 \times 10^{-4}$.

Table 3.3 Hydraulic characteristics of the confined aquifer at 'Oude Korendijk', obtained by the different methods

Method	KD (m^2/d)	S (−)
Thiem 1	385	–
Thiem 2	390	–
Theis	392	1.6×10^{-4}
Jacob 1	385	1.7×10^{-4}
Jacob 2	370	4.1×10^{-4}
Jacob 3	437	1.7×10^{-4}

4 Leaky aquifers

In nature, leaky aquifers occur far more frequently than the perfectly confined aquifers discussed in the previous chapter. Confining layers overlying or underlying an aquifer are seldom completely impermeable; instead, most of them leak to some extent. When a well in a leaky aquifer is pumped, water is withdrawn not only from the aquifer, but also from the overlying and underlying layers. In deep sedimentary basins, it is common for a leaky aquifer to be just one part of a multi-layered aquifer system as was shown in Figure 1.1E.

For the purpose of this chapter, we shall consider the three-layered system shown in Figure 4.1. The system consists of two aquifers, separated by an aquitard. The lower aquifer rests on an aquiclude. A well fully penetrates the lower aquifer and is screened over the total thickness of the aquifer. The well is not screened in the upper unconfined aquifer. Before the start of pumping, the system is at rest, i.e. the piezometric surface of the lower aquifer coincides with the watertable in the upper aquifer.

When the well is pumped, the hydraulic head in the lower aquifer will drop, thereby creating a hydraulic gradient not only in the aquifer itself, but also in the aquitard. The flow induced by the pumping is assumed to be vertical in the aquitard and horizontal in the aquifer. The error introduced by this assumption is usually less than 5 per cent if the hydraulic conductivity of the aquifer is two or more orders of magnitude greater than that of the aquitard (Neuman and Witherspoon 1969a).

The water that the pumped aquifer contributes to the well discharge comes from storage within that aquifer. The water contributed by the aquitard comes from storage within the aquitard and leakage through it from the overlying unpumped aquifer.

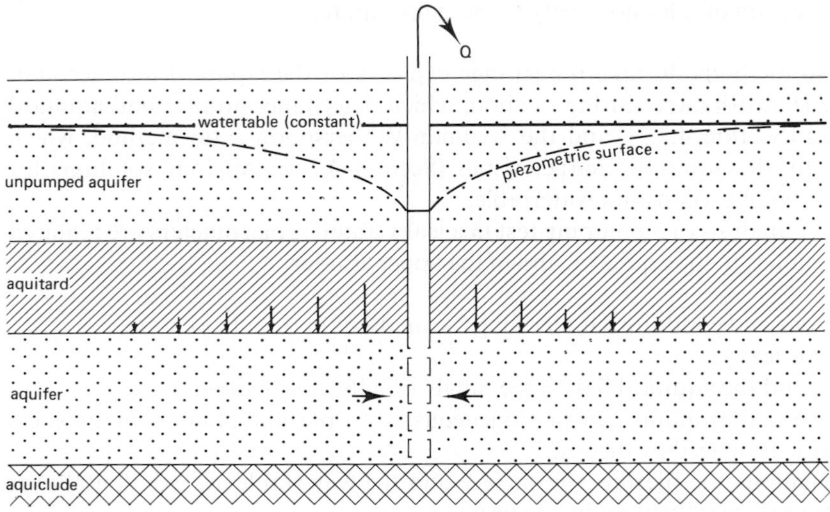

Figure 4.1 Cross-section of a pumped leaky aquifer

As pumping continues, more of the water comes from leakage from the unpumped aquifer and relatively less from aquitard storage. After a certain time, the well discharge comes into equilibrium with the leakage through the aquitard and a steady-state flow is attained. Under such conditions, the aquitard serves merely as a water-transmitting medium, and the water contributed from its storage can be neglected.

Solutions to the steady-state flow problem (Section 4.1) have been found on the basis of two very restrictive assumptions. The first is that, during pumping, the watertable in the upper aquifer remains constant; the second is that the rate of leakage into the leaky aquifer is proportional to the hydraulic gradient across the aquitard. But, as pumping continues, the watertable in the upper aquifer will drop because more and more of its water will be leaking through the aquitard into the pumped aquifer. The assumption of a constant watertable will only be satisfied if the upper aquifer is replenished by an outside source, say from surface water distributed over the aquifer via a system of narrowly spaced ditches. If the watertable can thus be kept constant as pumping continues, the well discharge will eventually be supplied entirely from the upper aquifer and steady-state flow will be attained. If the watertable cannot be controlled and does not remain constant and if pumping times are long, neglecting the drawdown in the upper aquifer can lead to considerable errors, unless its transmissivity is significantly greater than that of the pumped aquifer (Neuman and Witherspoon 1969b).

The second assumption completely ignores the storage capacity of the aquitard. This is justified when the flow to the well has become steady and the amount of water supplied from storage in the aquitard has become negligibly small (Section 4.1).

As long as the flow is unsteady, the effects of aquitard storage cannot be neglected. Yet, two of the solutions for unsteady flow (Sections 4.2.1 and 4.2.2) do neglect these effects, although, as pointed out by Neuman and Witherspoon (1972), this can result in:

- An overestimation of the hydraulic conductivity of the leaky aquifer;
- An underestimation of the hydraulic conductivity of the aquitard;
- A false impression of inhomogeneity in the leaky aquifer.

The other two methods do take the storage capacity of the aquitard into account. They are the Hantush curve-fitting method, which determines aquifer and aquitard characteristics (Section 4.2.3), and the Neuman-Witherspoon ratio method, which determines only the aquitard characteristics (Section 4.2.4). All four solutions for unsteady flow assume a constant watertable.

For a proper analysis of a pumping test in a leaky aquifer, piezometers are required in the leaky aquifer, in the aquitard, and in the upper aquifer.

The assumptions and conditions underlying the methods in this chapter are:

- The aquifer is leaky;
- The aquifer and the aquitard have a seemingly infinite areal extent;
- The aquifer and the aquitard are homogeneous, isotropic, and of uniform thickness over the area influenced by the test;
- Prior to pumping, the piezometric surface and the watertable are horizontal over the area that will be influenced by the test;
- The aquifer is pumped at a constant discharge rate;

- The well penetrates the entire thickness of the aquifer and thus receives water by horizontal flow;
- The flow in the aquitard is vertical;
- The drawdown in the unpumped aquifer (or in the aquitard, if there is no unpumped aquifer) is negligible.

And for unsteady-state conditions:
- The water removed from storage in the aquifer and the water supplied by leakage from the aquitard is discharged instantaneously with decline of head;
- The diameter of the well is very small, i.e. the storage in the well can be neglected.

The methods will be illustrated with data from the pumping test 'Dalem', The Netherlands (De Ridder 1961). Figure 4.2 shows a lithostratigraphical section of the test site as derived from the drilling data. The Kedichem Formation is regarded as the aquiclude. The Holocene layers form the aquitard overlying the leaky aquifer. The reader will note that there is no aquifer overlying the aquitard as in Figure 4.1. Instead, the aquitard extends to the surface where a system of narrowly spaced drainage ditches ensured a relatively constant watertable in the aquitard during the test.

The site lies about 1500 m north of the River Waal. The level of this river is affected by the tide and so too is the piezometric surface of the aquifer because it is in hydraulic connection with the river. The well was fitted with two screens. During the test, the lower screen was sealed and the entry of water was restricted to the upper screen, placed from 11 to 19 m below the surface. For 24 hours prior to pumping, the water levels in the piezometers were observed to determine the effect of the tide on the hydraulic head in the aquifer. By extrapolation of these data, time-tide curves for the

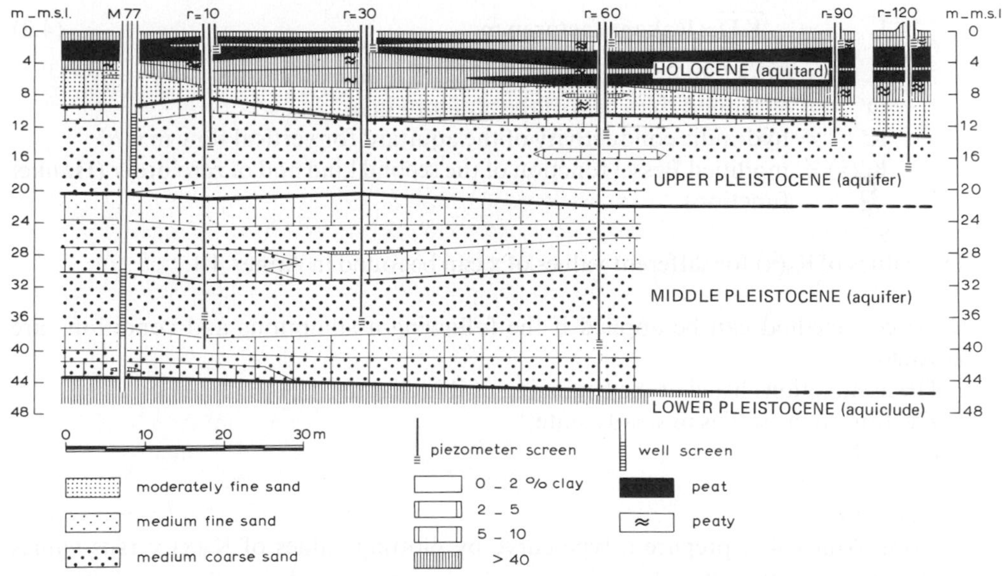

Figure 4.2 Lithostratigraphical cross-section of the pumping-test site 'Dalem', The Netherlands (after De Ridder 1961)

pumping period were established to allow a correction of the measured drawdowns (see Example 2.2). The data from the piezometers near the well were influenced by the effects of the well's partial penetration, for which allowance also had to be made (Example 10.1). The aquifer was pumped for 8 hours at a constant discharge of Q = 31.70 m^3/hr (or 761 m^3/d). The steady-state drawdown, which had not yet been reached, could be extrapolated from the time-drawdown curves.

4.1 Steady-state flow

The two methods presented below, both of which use steady-state drawdown data, allow the characteristics of the aquifer and the aquitard to be determined.

4.1.1 De Glee's method

For the steady-state drawdown in an aquifer with leakage from an aquitard proportional to the hydraulic gradient across the aquitard, De Glee (1930, 1951; see also Anonymous 1964, pp 35-41) derived the following formula

$$s_m = \frac{Q}{2\pi KD} K_0(\frac{r}{L}) \tag{4.1}$$

where

- s_m = steady-state (stabilized) drawdown in m in a piezometer at distance r in m from the well
- Q = discharge of the well in m^3/d
- L = $\sqrt{KDc}$: leakage factor in m (4.2)
- c = D'/K': hydraulic resistance of the aquitard in d
- D' = saturated thickness of the aquitard in m
- K' = hydraulic conductivity of the aquitard for vertical flow in m/d
- $K_0(x)$ = modified Bessel function of the second kind and of zero order (Hankel function)

The values of $K_0(x)$ for different values of x can be found in Annex 4.1.

De Glee's method can be applied if the following assumptions and conditions are satisfied:

- The assumptions listed at the beginning of this chapter;
- The flow to the well is in steady state;
- L > 3D.

Procedure 4.1

- Using Annex 4.1, prepare a type curve by plotting values of $K_0(x)$ versus values of x on log-log paper;
- On another sheet of log-log paper of the same scale, plot the steady-state (stabilized) drawdown in each piezometer s_m versus its corresponding value of r;

- Match the data plot with the type curve;
- Select an arbitrary point A on the overlapping portion of the sheets and note for A the values of s, r, $K_0(r/L)$, and $r/L(=x)$. It is convenient to select as point A the point where $K_0(r/L) = 1$ and $r/L = 1$;
- Calculate KD by substituting the known value of Q and the values of s_m and $K_0(r/L)$ into Equation 4.1;
- Calculate c by substituting the calculated value of KD and the values of r and r/L into Equation 4.2, written as

$$c = \frac{L^2}{KD} = \frac{1}{(r/L)^2} \times \frac{r^2}{KD}$$

Example 4.1

When the pump at 'Dalem' was shut down, steady-state drawdown had not yet been fully reached, but could be extrapolated from the time-drawdown curves. Table 4.1 gives the extrapolated steady-state drawdowns in the piezometers that had screens at a depth of 14 m (unless otherwise stated), corrected for the effects of the tide in the river and for partial penetration.

Table 4.1 Corrected extrapolated steady-state drawdowns of pumping test 'Dalem' (after De Ridder 1961)

Piezometer	P_{10}	P_{10}*	P_{30}	P_{30}*	P_{60}	P_{90}	P_{120}	P_{400}*
Drawdown in m	0.310	0.252	0.235	0.213	0.170	0.147	0.132	0.059

* screen depth 36 m

For this example, we first plot the drawdowns listed in Table 4.1 versus the corresponding distances, which we then fit with De Glee's type curve $K_0(x)$ versus x (Figure 4.3). As match point A, we choose the point where $K_0(r/L) = 1$ and $r/L = 1$. On the observed data sheet, point A has the coordinates $s_m = 0.057$ m and $r = 1100$ m. Substituting these values and the known value of $Q = 761\ m^3/d$ into Equation 4.1, we obtain

$$KD = \frac{Q}{2\pi s_m} K_0\left(\frac{r}{L}\right) = \frac{761}{2 \times 3.14 \times 0.057} \times 1 = 2126\ m^2/d$$

Further, $r/L = 1$, $L = r = 1100$ m. Hence

$$c = \frac{L^2}{KD} = \frac{(1100)^2}{2126} = 569\ d$$

4.1.2 Hantush-Jacob's method

Unaware of the work done many years earlier by De Glee, Hantush and Jacob (1955) also derived Equation 4.1. Hantush (1956, 1964) noted that if r/L is small (r/L ≤

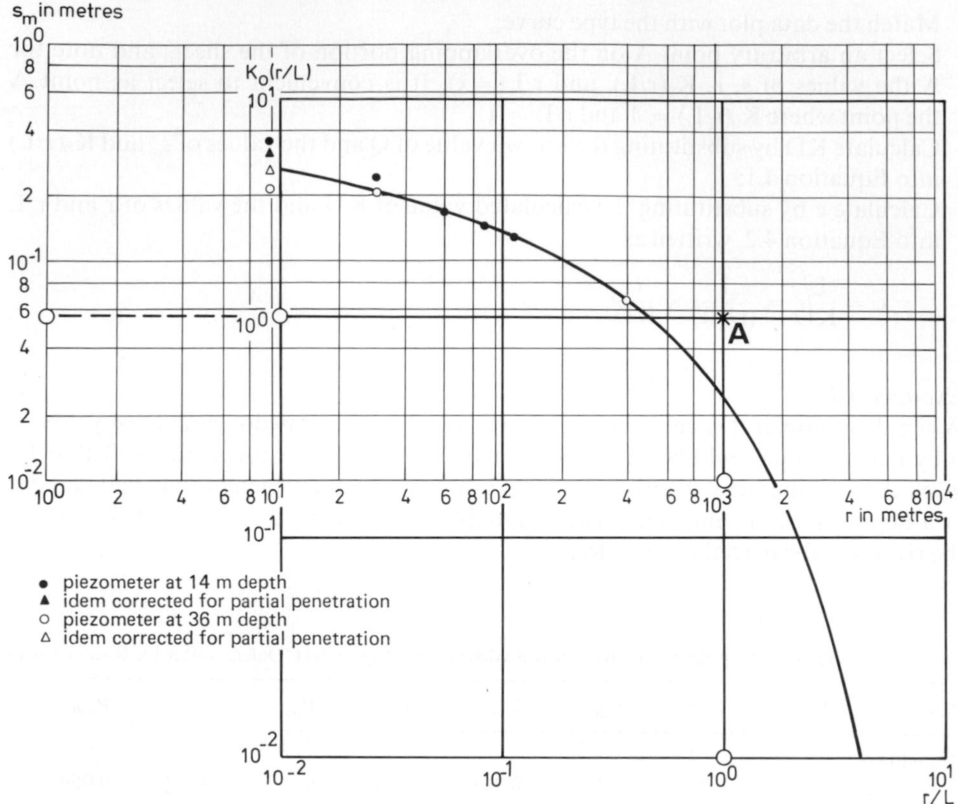

Figure 4.3 Analysis of data from pumping test 'Dalem' with the De Glee method

0.05), Equation 4.1 can, for practical purposes, be approximated by

$$s_m \approx \frac{2.30Q}{2\pi KD}\left(\log 1.12\frac{L}{r}\right) \tag{4.3}$$

For $r/L < 0.16, 0.22, 0.33$, and 0.45, the errors in using this equation instead of Equation 4.1 are less than 1, 2, 5, and 10 per cent, respectively (Huisman 1972). A plot of s_m against r on semi-log paper, with r on the logarithmic scale, will show a straight-line relationship in the range where r/L is small (Figure 4.4). In the range where r/L is large, the points fall on a curve that approaches the zero-drawdown axis asymptotically.

The slope of the straight portion of the curve, i.e. the drawdown difference Δs_m per log cycle of r, is expressed by

$$\Delta s_m = \frac{2.30Q}{2\pi KD} \tag{4.4}$$

The extended straight-line portion of the curve intercepts the r axis where the drawdown is zero. At the interception point, $s_m = 0$ and $r = r_0$ and thus Equation 4.3 reduces to

$$0 = \frac{2.30Q}{2\pi KD}\left(\log 1.12\frac{L}{r_0}\right)$$

from which it follows that

$$1.12\frac{L}{r_0} = \frac{1.12}{r_0}\sqrt{KDc} = 1$$

and hence

$$c = \frac{(r_0/1.12)^2}{KD} \tag{4.5}$$

The Hantush-Jacob method can be used if the following assumptions and conditions are satisfied:

- The assumptions listed at the beginning of this chapter;
- The flow to the well is in steady state;
- $L > 3D$;
- $r/L \leq 0.05$.

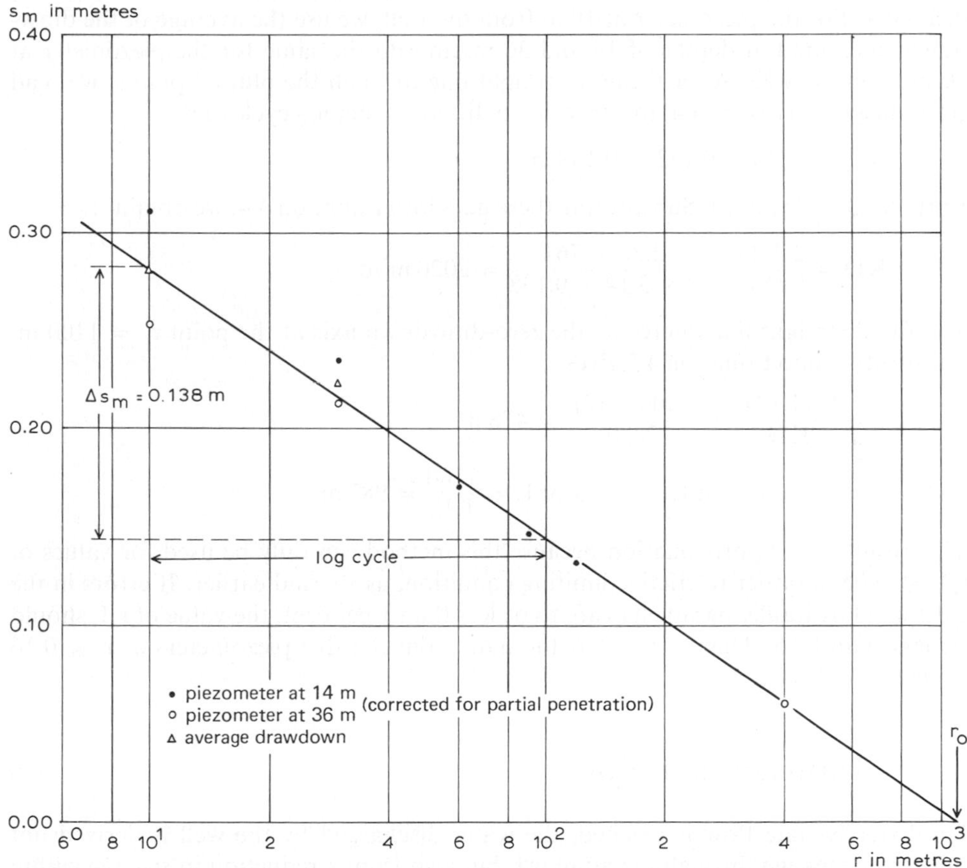

Figure 4.4 Analysis of data from pumping test 'Dalem' with the Hantush-Jacob method

Procedure 4.2

- On semi-log paper, plot s_m versus r (r on logarithmic scale);
- Draw the best-fit straight line through the points;
- Determine the slope of the straight line (Figure 4.4);
- Substitute the value of Δs_m and the known value of Q into Equation 4.4 and solve for KD;
- Extend the straight line until it intercepts the r axis and read the value of r_0;
- Calculate the hydraulic resistance of the aquitard c by substituting the values of r_0 and KD into Equation 4.5.

Another way to calculate c is:

- Select any point on the straight line and note its coordinates s_m and r;
- Substitute these values, together with the known values of Q and KD into Equation 4.3 and solve for L;
- Since $L = \sqrt{KDc}$, calculate c.

Example 4.2

For this example, using data from the pumping test 'Dalem', we first plot the steady-state drawdown data listed in Table 4.1 on semi-log paper versus the corresponding distances. For the piezometer at 10 m from the well, we use the average of the drawdowns measured at depths of 14 and 36 m, and do the same for the piezometer at 30 m from the well. After fitting a straight line through the plotted points, we read from the graph (Figure 4.4) the drawdown difference per log cycle of r

$$\Delta s_m = 0.281 - 0.143 = 0.138 \text{ m}$$

Further, $Q = 761 \text{ m}^3/\text{d}$. Substituting these data into Equation 4.4, we obtain

$$KD = \frac{2.30Q}{2\pi\Delta s_m} = \frac{2.30 \times 761}{2 \times 3.14 \times 0.138} = 2020 \text{ m}^2/\text{d}$$

The fitted straight line intercepts the zero-drawdown axis at the point $r_0 = 1100$ m. Substitution into Equation 4.5 gives

$$c = \frac{(r_o/1.12)^2}{KD} = \frac{(1100/1.12)^2}{2020} = 478 \text{ d}$$

and L is calculated from $1.12\frac{L}{r_0} = 1$ or $L = \frac{1100}{1.12} = 982$ m.

This result is an approximation because this method can only be used for values of $r/L \leq 0.05$, a rather restrictive limiting condition, as we said earlier. If errors in the calculated hydraulic parameters are to be less than 1 per cent, the value of r/L should be less than 0.16. This means that the data from the five piezometers at $r \leq 0.16 \times 982 = 157$ m can be used.

4.2 Unsteady-state flow

Until steady-state flow is reached, the water discharged by the well is derived not only from leakage through the aquitard, but also from a reduction in storage within both the aquitard and the pumped aquifer.

The methods available for analyzing data of unsteady-state flow are the Walton curve-fitting method, the Hantush inflection-point method (both of which, however, neglect the aquitard storage), the Hantush curve-fitting method, and the Neuman and Witherspoon ratio method (both of which do take aquitard storage into account).

4.2.1 Walton's method

With the effects of aquitard storage considered negligible, the drawdown due to pumping in a leaky aquifer is described by the following formula (Hantush and Jacob 1955)

$$s = \frac{Q}{4\pi KD} \int_u^\infty \frac{1}{y} \exp\left(-y - \frac{r^2}{4L^2y}\right) dy$$

or

$$s = \frac{Q}{4\pi KD} W(u,r/L) \tag{4.6}$$

where

$$u = \frac{r^2S}{4KDt} \tag{4.7}$$

Equation 4.6 has the same form as the Theis well function (Equation 3.5), but there are two parameters in the integral: u and r/L. Equation 4.6 approaches the Theis well function for large values of L, when the exponential term $r^2/4L^2y$ approaches zero.

On the basis of Equation 4.6, Walton (1962) developed a modification of the Theis curve-fitting method, but instead of using one type curve, Walton uses a type curve for each value of r/L. This family of type curves (Figure 4.5) can be drawn from the tables of values for the function W(u,r/L) as published by Hantush (1956) and presented in Annex 4.2.

Walton's method can be applied if the following assumptions and conditions are satisfied:

- The assumptions listed at the beginning of this chapter;
- The aquitard is incompressible, i.e. the changes in aquitard storage are negligible;
- The flow to the well is in unsteady state.

Procedure 4.3

- Using Annex 4.2, plot on log-log paper W(u,r/L) versus 1/u for different values of r/L; this gives a family of type curves (Figure 4.5);
- Plot for one of the piezometers the drawdown s versus the corresponding time t on another sheet of log-log paper of the same scale; this gives the observed time-drawdown data curve;
- Match the observed data curve with one of the type curves (Figure 4.6);
- Select a match point A and note for A the values of W(u,r/L), 1/u, s, and t;
- Substitute the values of W(u,r/L) and s and the known value of Q into Equation 4.6 and calculate KD;
- Substitute the value of KD, the reciprocal value of 1/u, and the values of t and r into Equation 4.7 and solve for S;

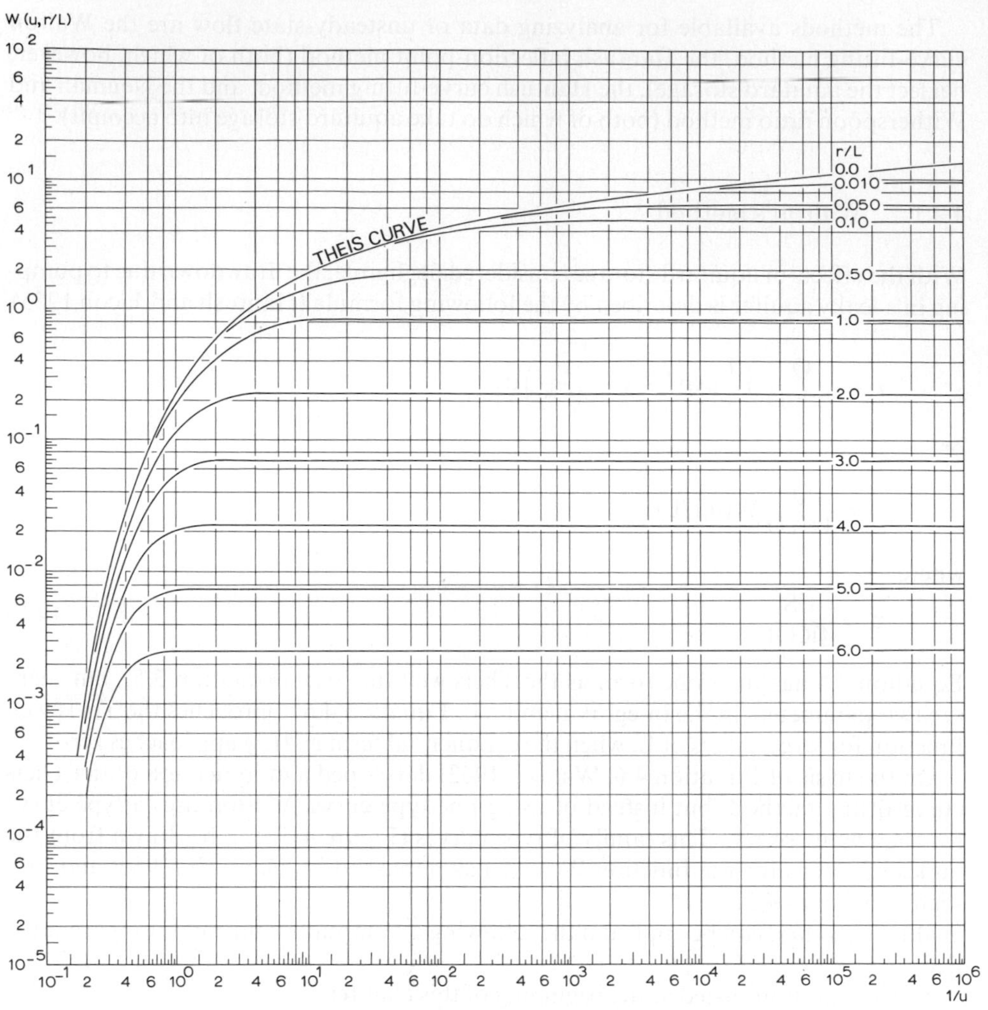

Figure 4.5 Family of Walton's type curves W(u,r/L) versus 1/u for different values of r/L

- From the type curve that best fits the observed data curve, take the numerical value of r/L and calculate L. Then, because $L = \sqrt{KDc}$, calculate c;
- Repeat the procedure for all piezometers. The calculated values of KD, S, and c should show reasonable agreement.

Remark

- To obtain the unique fitting position of the data plot with one of the type curves, enough of the observed data should fall within the period when leakage effects are negligible, or r/L should be rather large.

Example 4.3

Compiled from the pumping test 'Dalem', Table 4.2 presents the corrected drawdown data of the piezometers at 30, 60, 90, and 120 m from the well. Using the data from the piezometer at 90 m, we plot the drawdown data against the corresponding values of t on log-log paper. A comparison with the Walton family of type curves shows that the plotted points fall along the curve for r/L = 0.1 (Figure 4.6). The point where W(u,r/L) = 1 and 1/u = 10^2 is chosen as match point A_{90}. On the observed data sheet, this point has the coordinates s = 0.035 m and t = 0.22 d. Introducing the appropriate numerical values into Equations 4.6 and 4.7 yields

$$KD = \frac{Q}{4\pi s} W(u,r/L) = \frac{761}{4 \times 3.14 \times 0.035} \times 1 = 1731 \text{ m}^2/\text{d}$$

and

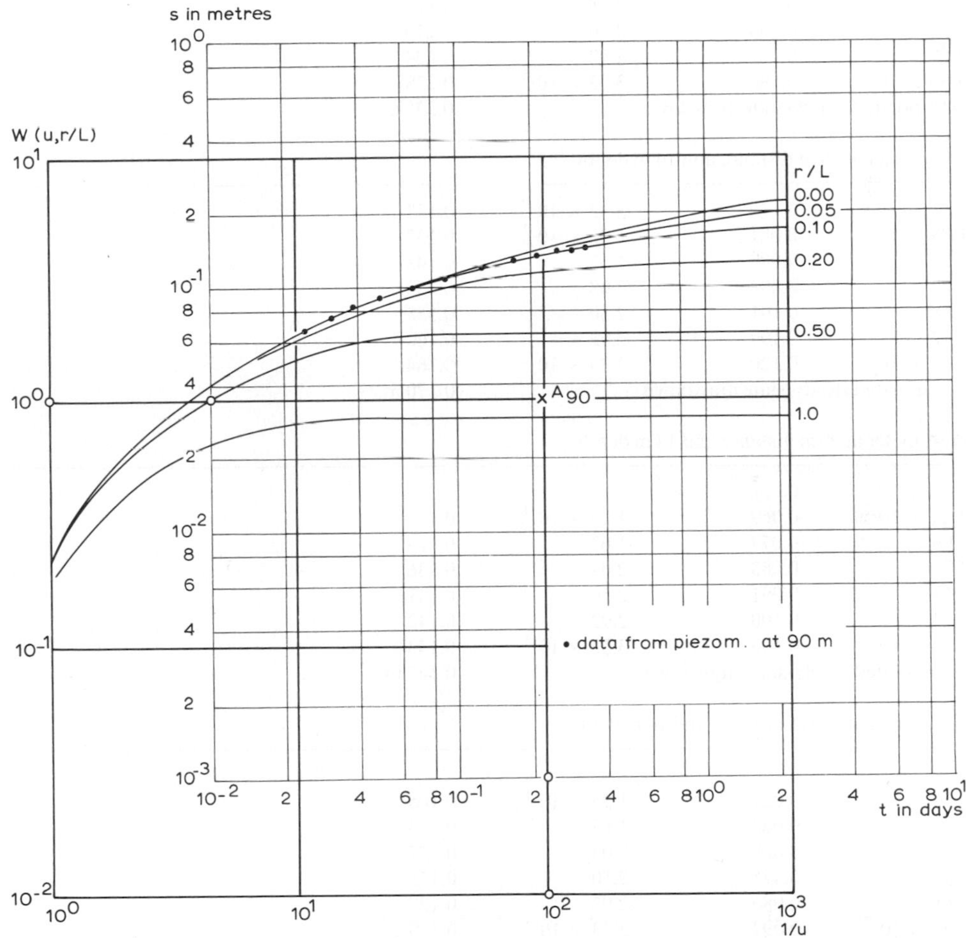

Figure 4.6 Analysis of data from pumping test 'Dalem' (r = 90 m) with the Walton method

$$S = \frac{4KDt}{r^2} u = \frac{4 \times 1731 \times 0.22}{90^2} \times \frac{1}{10^2} = 1.9 \times 10^{-3}$$

Further, because r = 90 m and r/L = 0.1, it follows that L = 900 m and hence $c = L^2/KD = (900)^2/1731 = 468$ d.

Table 4.2 Drawdown data from pumping test 'Dalem', The Netherlands (after De Ridder 1961)

Time (d)	Drawdown (m)	Time (d)	Drawdown (m)
Piezometer at 30 m distance and 14 m depth			
0	0		
1.53×10^{-2}	0.138	8.68×10^{-2}	0.190
1.81	0.141	1.25×10^{-1}	0.201
2.29	0.150	1.67	0.210
2.92	0.156	2.08	0.217
3.61	0.163	2.50	0.220
4.58	0.171	2.92	0.224
6.60×10^{-2}	0.180	3.33×10^{-1}	0.228
extrapolated steady-state drawdown			0.235 m
Piezometer at 60 m distance and 14 m depth			
0	0	8.82×10^{-2}	0.127
1.88×10^{-2}	0.081	1.25×10^{-1}	0.137
2.36	0.089	1.67	0.148
2.99	0.094	2.08	0.155
3.68	0.101	2.50	0.158
4.72	0.109	2.92	0.160
6.67×10^{-2}	0.120	3.33×10^{-1}	0.164
extrapolated steady-state drawdown			0.170 m
Piezometer at 90 m distance and 14 m depth			
0	0		
2.43×10^{-2}	0.069	1.25×10^{-1}	0.120
3.06	0.077	1.67	0.129
3.75	0.083	2.08	0.136
4.68	0.091	2.50	0.141
6.74	0.100	2.92	0.142
8.96×10^{-2}	0.109	3.33×10^{-1}	0.143
extrapolated steady-state drawdown			0.147 m
Piezometer at 120 m distance and 14 m depth			
0	0		
2.50×10^{-2}	0.057	1.25×10^{-1}	0.105
3.13	0.063	1.67	0.113
3.82	0.068	2.08	0.122
5.00	0.075	2.50	0.125
6.81	0.086	2.92	0.127
9.03×10^{-2}	0.092	3.33×10^{-1}	0.129
Extrapolated steady-state drawdown			0.132 m

4.2.2 Hantush's inflection-point method

Hantush (1956) developed several procedures for the analysis of pumping test data in leaky aquifers, all of them based on Equation 4.6

$$s = \frac{Q}{4\pi KD} W(u,r/L)$$

One of these procedures (Procedure 4.4) uses the drawdown data from a single piezometer; the other (Procedure 4.5) uses the data from at least two piezometers. To determine the inflection point P (which will be discussed further below), the steady-state drawdown s_m should be known, either from direct observations or from extrapolation. The curve of s versus t on semi-log paper has an inflection point P where the following relations hold

$$s_p = 0.5\, s_m = \frac{Q}{4\pi KD} K_0\left(\frac{r}{L}\right) \tag{4.8}$$

where K_0 is the modified Bessel function of the second kind and zero order

$$u_p = \frac{r^2 S}{4KDt_p} = \frac{r}{2L} \tag{4.9}$$

The slope of the curve at the inflection point Δs_p is given by

$$\Delta s_p = \frac{2.30Q}{4\pi KD} e^{-r/L} \tag{4.10}$$

or

$$r = 2.30L\left(\log\frac{2.30Q}{4\pi KD} - \log \Delta s_p\right) \tag{4.11}$$

At the inflection point, the relation between the drawdown and the slope of the curve is given by

$$2.30\frac{s_p}{\Delta s_p} = e^{r/L} K_0(r/L) \tag{4.12}$$

In Equations 4.8 to 4.12, the index p means 'at the inflection point'. Further, Δs stands for the slope of a straight line.

Either of Hantush's procedures of the inflection-point method can be used if the following assumptions and conditions are satisfied:
- The assumptions listed at the beginning of this chapter;
- The aquitard is incompressible, i.e. changes in aquitard storage are negligible;
- The flow to the well is in unsteady state;
- It must be possible to extrapolate the steady-state drawdown for each piezometer.

Procedure 4.4
- For one of the piezometers, plot s versus t on semi-log paper (t on logarithmic scale) and draw the curve that best fits through the plotted points (Figure 4.7);

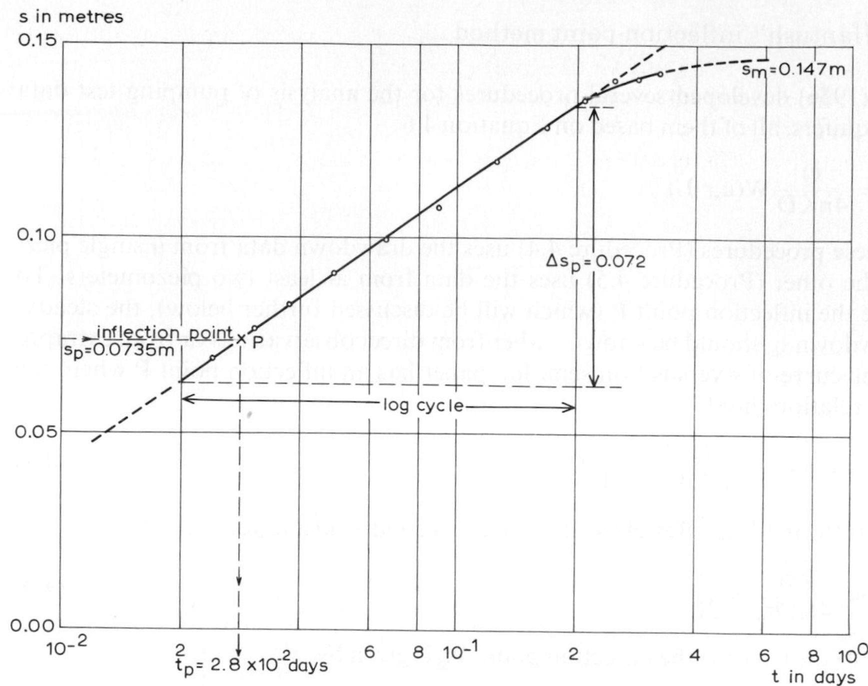

Figure 4.7 Analysis of data from pumping test 'Dalem' (r = 90 m) with Procedure 4.4 of the Hantush inflection-point method

- Determine the value of the maximum drawdown s_m by extrapolation. This is only possible if the period of the test was long enough;
- Calculate s_p with Equation 4.8: $s_p = (0.5)s_m$. The value of s_p on the curve locates the inflection point P;
- Read the value of t_p at the inflection point from the time-axis;
- Determine the slope Δs_p of the curve at the inflection point. This can be closely approximated by reading the drawdown difference per log cycle of time over the straight portion of the curve on which the inflection point lies, or over the tangent to the curve at the inflection point;
- Substitute the values of s_p and Δs_p into Equation 4.12 and find r/L by interpolation from the table of the function $e^x K_0(x)$ in Annex 4.1;
- Knowing r/L and r, calculate L;
- Knowing Q, s_p, Δs_p, and r/L, calculate KD from Equation 4.10, using the table of the function e^{-x} in Annex 4.1, or from Equation 4.8, using the table of the function $K_0(x)$ in Annex 4.1;
- Knowing KD, t_p, r, and r/L, calculate S from Equation 4.9;
- Knowing KD and L, calculate c from the relation $c = L^2/KD$.

Remarks

- The accuracy of the calculated hydraulic characteristics depends on the accuracy

of the extrapolation of s_m. The calculations should therefore be checked by substituting the values of S, L, and KD into Equations 4.6 and 4.7.
Calculations of s should be made for different values of t. If the values of t are not too small, the values of s should fall on the observed data curve. If the calculated data deviate from the observed data, the extrapolation of s_m should be adjusted. Sometimes, the observed data curve can be drawn somewhat steeper or flatter through the plotted points, and so Δs_p can be adjusted too. With the new values of s_m and/or Δs_p, the calculation is repeated.

Example 4.4
From the pumping test 'Dalem', we use the data from the piezometer at 90 m (Table 4.2). We first plot the drawdown data of this piezometer versus t on semi-log paper (Figure 4.7) and then find the maximum (or steady-state) drawdown by extrapolation (s_m = 0.147 m). According to Equation 4.8, the drawdown at the inflection point $s_p = 0.5\ s_m = 0.0735$ m. Plotting this point on the time-drawdown curve, we obtain $t_p = 2.8 \times 10^{-2}$ d.
Through the inflection point of the curve, we draw a tangent line to the curve, which matches here with the straight portion of the curve itself. The slope of this tangent line $\Delta s_p = 0.072$ m.
Introducing these values into Equation 4.12 gives

$$2.30 \frac{s_p}{\Delta s_p} = 2.30 \times \frac{0.0735}{0.072} = 2.34 = e^{r/L} K_0(r/L)$$

Annex 4.1 gives r/L = 0.15, and because r = 90 m, it follows that L = 90/0.15 = 600 m.
Further, Q = 761 m^3/d is given, and the value of $e^{-r/L} = e^{-0.15} = 0.86$ is found from Annex 4.1. Substituting these values into Equation 4.10 yields

$$KD = \frac{2.30Q}{4\pi\Delta s_p} e^{-r/L} = \frac{2.30 \times 761}{4 \times 3.14 \times 0.072} \times 0.86 = 1665\ m^2/d$$

and consequently

$$c = \frac{L^2}{KD} = \frac{(600)^2}{1665} = 216\ d$$

Introducing the appropriate values into Equation 4.9 gives

$$S = \frac{r4KDt_p}{2Lr^2} = \frac{90}{2 \times 600} \times \frac{4 \times 1665 \times 2.8 \times 10^{-2}}{90^2} = 1.7 \times 10^{-3}$$

To verify the extrapolated steady-state drawdown, we calculate the drawdown at a chosen moment, using Equations 4.6 and 4.7. If we choose t = 0.1 d, then

$$u = \frac{r^2S}{4KDt} = \frac{90^2 \times 1.7 \times 10^{-3}}{4 \times 1665 \times 10^{-1}} = 0.02$$

According to Annex 4.2, W(u,r/L) = 3.11 (for u = 0.02 and r/L = 0.15). Thus

$$s_{(t = 0.1)} = \frac{Q}{4\pi KD} W(u,r/L) = \frac{761}{4 \times 3.14 \times 1665} \times 3.11 = 0.113\ m$$

The point t = 0.1, s = 0.113 falls on the time-drawdown curve and justifies the extrapolated value of s_m. In practice, several points should be tried.

Procedure 4.5

- On semi-log paper, plot s versus t for each piezometer (t on logarithmic scale) and draw curves through the plotted points (Figure 4.8);
- Determine the slope of the straight portion of each curve Δs;
- On semi-log paper, plot r versus Δs (Δs on logarithmic scale) and draw the best-fit straight line through the plotted points. (This line is the graphic representation of Equation 4.11);
- Determine the slope of this line Δr, i.e. the difference of r per log cycle of Δs (Figure 4.9);
- Extend the straight line until it intercepts the absciss where r = 0 and $\Delta s = (\Delta s)_o$. Read the value of $(\Delta s)_o$;
- Knowing the values of Δr and $(\Delta s)_o$, calculate L from

$$L = \frac{1}{2.30}\Delta r \qquad (4.13)$$

and KD from

$$KD = 2.30\frac{Q}{4\pi(\Delta s)_o} \qquad (4.14)$$

- Knowing KD and L, calculate c from the relation $c = L^2/KD$;

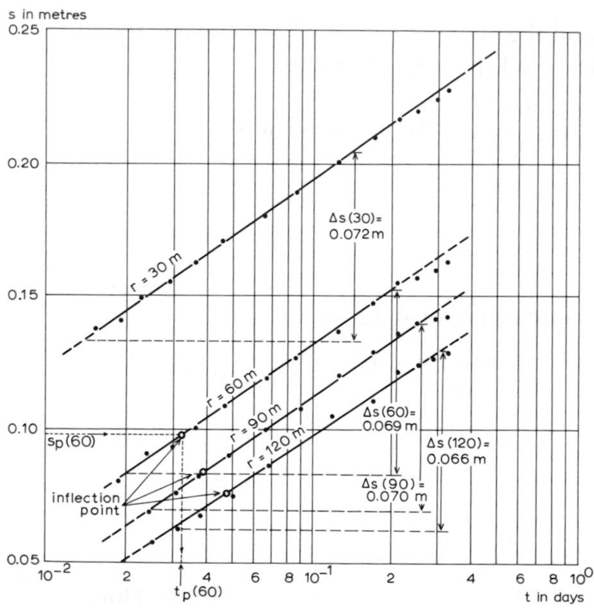

Figure 4.8 Analysis of data from pumping test 'Dalem' with Procedure 4.5 of the Hantush inflection-point method: determination of values of Δs for different values of r

- With the known values of Q, r, KD, and L, calculate s_p for each piezometer, using Equation 4.8: $s_p = (Q/4\pi KD)K_0(r/L)$ and the table for the function $K_0(x)$ in Annex 4.1;
- Plot each s_p value on its corresponding time-drawdown curve and read t_p on the absciss;
- Knowing the values of KD, r, r/L, and t_p, calculate S from Equation 4.9: $(r^2S)/(4KDt_p) = 0.5(r/L)$.

Example 4.5

From the pumping test 'Dalem', we use data from the piezometers at 30, 60, 90, and 120 m (Table 4.2). Figure 4.8 shows a time-drawdown plot for each of the piezometers on semi-log paper. Determining the slope of the straight portion of each curve, we obtain:

$\Delta s\,(30\text{ m}) = 0.072\text{ m}$ $\Delta s\,(90\text{ m}) = 0.070\text{ m}$

$\Delta s\,(60\text{ m}) = 0.069\text{ m}$ $\Delta s\,(120\text{ m}) = 0.066\text{ m}$

In Figure 4.9, the values of Δs are plotted versus r on semi-log paper and a straight line is fitted through the plotted points. Because of its steepness, the slope is measured as the difference of r over 1/20 log cycle of Δs. (If 1 log cycle measures 10 cm, 1/20 log cycle is 0.5 cm). The difference of r per 1/20 log cycle of Δs equals 120 m, or the difference of r per log cycle of Δs, i.e. Δr equals 2400 m. The straight line intersects the Δs axis where r = 0 in the point $(\Delta s)_o = 0.074$ m. Substitution of these values into Equations 4.13 and 4.14 gives

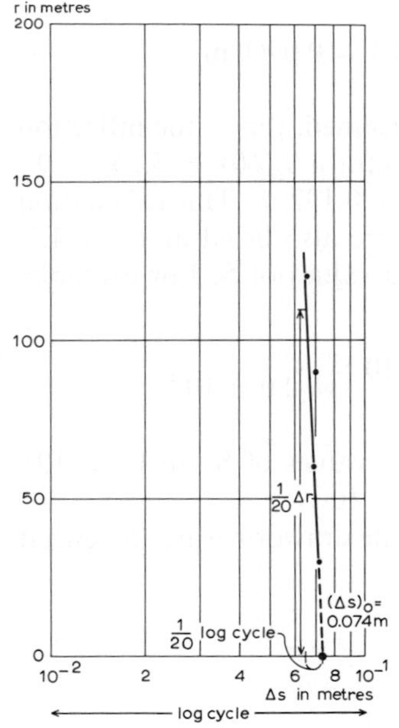

Figure 4.9 Idem: determination of the value of Δr

$$L = \frac{1}{2.30} r = \frac{1}{2.30} \times 2400 = 1043 \text{ m}$$

and because $Q = 761 \text{ m}^3/\text{d}$

$$KD = \frac{2.30Q}{4\pi(\Delta s)_o} = \frac{2.30 \times 761}{4 \times 3.14 \times 0.074} = 1883 \text{ m}^2/\text{d}$$

finally

$$c = \frac{L^2}{KD} = \frac{(1043)^2}{1883} = 578 \text{ d}$$

The value of r/L is calculated for each piezometer, and the corresponding values of $K_0(r/L)$ are found in Annex 4.1. The results are listed in Table 4.3.

Table 4.3 Data to be substituted into Equations 4.8 and 4.9

r (m)	r/L	$K_0(r/L)$	s_p (m)	t_p (d)	s_m (m)
30	0.0288	3.668	0.1180	outside figure	0.236
60	0.0575	2.984	0.0960	3.25×10^{-2}	0.192
90	0.0863	2.576	0.0829	3.85×10^{-2}	0.166
120	0.1150	2.290	0.0737	4.70×10^{-2}	0.147

The drawdown s_p at the inflection point of the curve through the observed data, as plotted in Figure 4.8 for the piezometer at 60 m, is calculated from Equation 4.8

$$s_p(60) = \frac{Q}{4\pi KD} K_0(r/L) = \frac{761}{4 \times 3.14 \times 1883} \times 2.984 = 0.0960 \text{ m}$$

The point on this curve for which $s = 0.0960$ m is determined; this is the inflection point. On the absciss, the value of t_p at the inflection point is $t_p(60) = 3.25 \times 10^{-2}$ d. From Equation 4.8, it follows that $s_m(60) = 2s_p(60) = 0.192$ m. This calculation was also made for the other piezometers. These results are also listed in Table 4.3. Substitution of the values of t_p into Equation 4.9 yields values of S. For example, for r = 60 m,

$$S = \frac{r}{2L}\frac{4KDt_p}{r^2} = \frac{60}{2 \times 1043} \times \frac{4 \times 1883 \times 3.25 \times 10^{-2}}{60^2} = 2.0 \times 10^{-3}$$

In the same way, for r = 90 m and for r = 120 m, the values of S are 1.5×10^{-3} and 1.4×10^{-3}, respectively. The average value of S is 1.6×10^{-3}.
It will be noted that the calculated values for the steady-state drawdown are somewhat higher than the extrapolated values from Table 4.1.

4.2.3 Hantush's curve-fitting method

Hantush (1960) presented a method of analysis that takes into account the storage changes in the aquitard. For small values of pumping time, he gives the following drawdown equation for unsteady flow

$$s = \frac{Q}{4\pi KD} W(u,\beta) \tag{4.15}$$

where

$$u = \frac{r^2S}{4KDt} \tag{4.16}$$

$$\beta = \frac{r}{4}\sqrt{\frac{K'/D'}{KD} \times \frac{S'}{S}} \tag{4.17}$$

S' = aquitard storativity

$$W(u,\beta) = \int_u^\infty \frac{e^{-y}}{y} \operatorname{erfc} \frac{\beta\sqrt{u}}{\sqrt{y(y-u)}} dy$$

Values of the function $W(u,\beta)$ are presented in Annex 4.3.

Hantush's curve-fitting method can be used if the following assumptions and conditions are satisfied:
- The assumptions listed at the beginning of this chapter;
- The flow to the well is in an unsteady state;
- The aquitard is compressible, i.e. the changes in aquitard storage are appreciable;
- $t < S'D'/10K'$.

Only the early-time drawdown data should be used so as to satisfy the assumption that the drawdown in the aquitard (or overlying unpumped aquifer) is negligible.

Procedure 4.6
- Using Annex 4.3, construct on log-log paper the family of type curves $W(u,\beta)$ versus $1/u$ for different values of β (Figure 4.10);
- On another sheet of log-log paper of the same scale, plot s versus t for one of the piezometers;
- Match the observed data plot with one of the type curves (Figure 4.11);
- Select an arbitrary point A on the overlapping portion of the two sheets and note the values of $W(u,\beta)$, $1/u$, s, and t for this point. Note the value of β on the selected type curve;
- Substitute the values of $W(u,\beta)$ and s and the known value of Q into Equation 4.15 and calculate KD;
- Substitute the values of KD, t, r, and the reciprocal value of $1/u$ into Equation 4.16 and solve for S;
- Substitute the values of β, KD, S, r, and D' into Equation 4.17 and solve for $K'S'$.

Remarks
- It is difficult to obtain a unique match of the two curves because the shapes of the type curves change gradually with β (β values are practically indeterminate in the range $\beta = 0 \rightarrow \beta = 0.5$, because the curves are very similar);

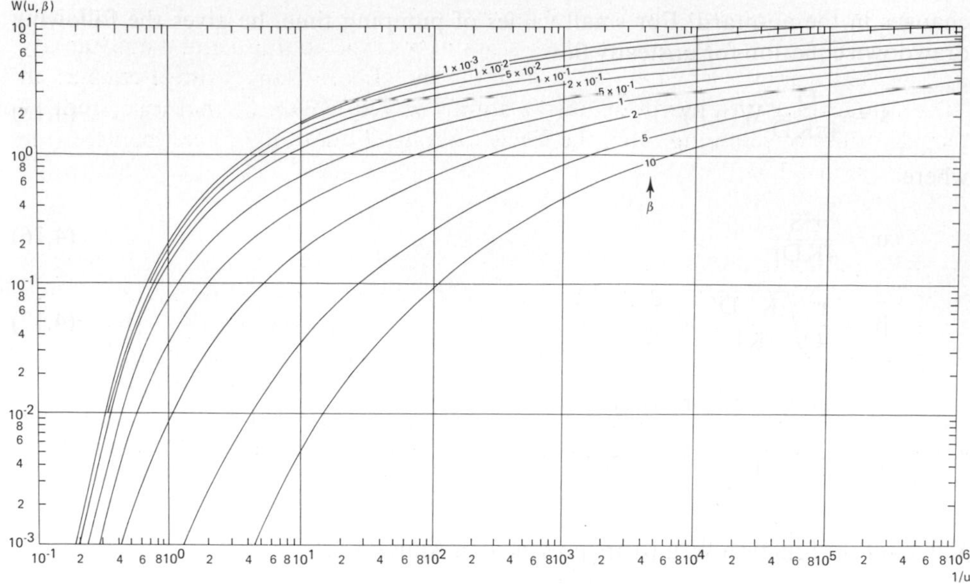

Figure 4.10 Family of Hantush's type curves W(u,β) versus 1/u for different values of β

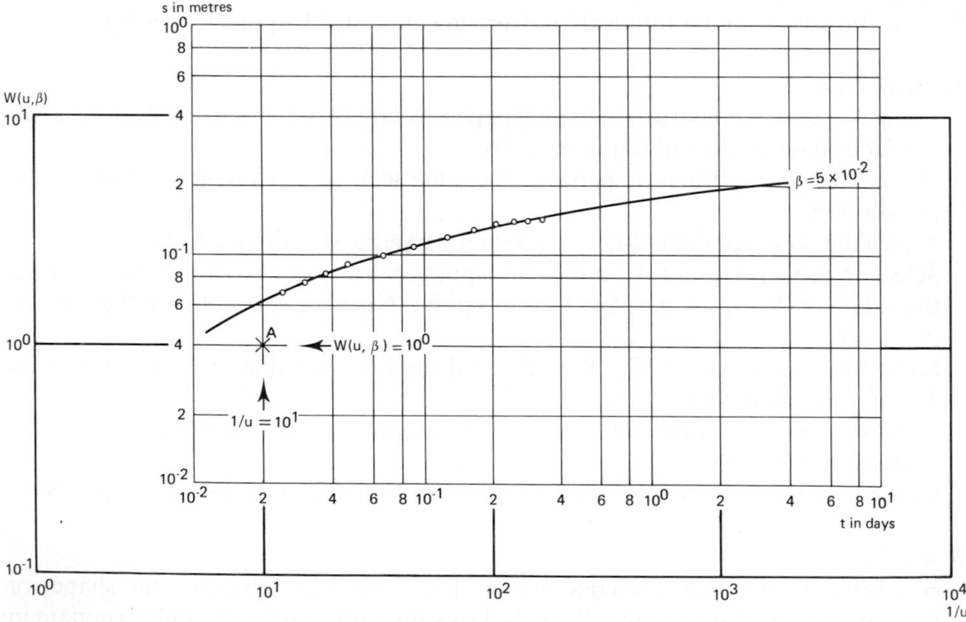

Figure 4.11 Analysis of data from pumping test 'Dalem' (r = 90 m) with the Hantush curve-fitting method

- As K′ approaches zero, the limit of Equation 4.15 is equal to the Theis equation $s = (Q/4\pi KD)W(u)$. If the ratio of the storativity of the aquitard and the storativity of the leaky aquifer is small ($S'/S < 0.01$), the effect of any storage changes in the aquitard on the drawdown in the aquifer is very small. In that case, and for small values of pumping time, the Theis formula (Equation 3.5) can be used (see also Section 4.2.4).

Example 4.6
From the pumping test 'Dalem' we use the drawdown data from the piezometer at 90 m (Table 4.2), plotting on log-log paper the drawdown data against the corresponding values of t (Figure 4.11). A comparison of the data plot with the Hantush family of type curves shows that the best fit of the plotted points is obtained with the curve $\beta = 5 \times 10^{-2}$. We choose a match point A, whose coordinates are $W(u,\beta) = 10^0$, $1/u = 10$, $s = 4 \times 10^{-2}$ m, and $t = 2 \times 10^{-2}$ d. Substituting these values, together with the values of $Q = 761$ m³/d and $r = 90$ m, into Equations 4.15, 4.16, and 4.17, we obtain

$$KD = \frac{Q}{4\pi s} W(u,\beta) = \frac{761}{4 \times 3.14 \times 4 \times 10^{-2}} 10^0 = 1515 \text{ m}^2/\text{d}$$

$$S = \frac{4KDtu}{r^2} = \frac{4 \times 1515 \times 2 \times 10^{-2} \times 10^{-1}}{90^2} = 1.5 \times 10^{-3}$$

$$\frac{K'S'}{D'} = \beta^2(4/r)^2 KDS = (5 \times 10^{-2})^2 \times (4/90)^2 \times 1515 \times 1.5 \times 10^{-3}$$
$$= 1.1 \times 10^{-5} \text{ d}^{-1}$$

The thickness of the aquitard $D' = 8$ m (Figure 4.2). Hence, $K'S' = 9 \times 10^{-5}$ m/d.

To check whether the condition $t < S'D'/10K'$ is fulfilled, we need more calculated parameters. Using the value of $c = D'/K' = 450$ d (see Section 4.3), we can calculate an approximate value of S′

$$\frac{K'S'}{D'} = 1.1 \times 10^{-5} \text{ d}^{-1}$$

$$S' = 450 \times 1.1 \times 10^{-5} = 5 \times 10^{-3}$$

Hence

$$t < 5 \times 10^{-3} \times 450 \times 0.1 \text{ or } t < 0.225 \text{ d}$$

If this time condition is to be satisfied, the drawdown data measured at $t = 2.50 \times 10^{-1}$, 2.92×10^{-1}, and 3.33×10^{-1} d should not be used in the analysis (Figure 4.11).
Note: Because the data curve matches with a type curve in the range $\beta = 0 \rightarrow \beta = 0.5$, not too much value should be attached to the exact value of β, nor to the calculated value of K′S′.

4.2.4 Neuman-Witherspoon's method

Neuman and Witherspoon (1972) developed a method for determining the hydraulic characteristics of aquitards at small values of pumping time when the drawdown in the overlying unconfined aquifer is still negligible. The method is based on a theory developed for a so-called slightly leaky aquifer (Neuman and Witherspoon 1968), where the drawdown function in the pumped aquifer is given by the Theis equation (Equation 3.5), and the drawdown in the aquitard of very low permeability is described by

$$s_c = \frac{Q}{4\pi KD} W(u,u_c) \tag{4.18}$$

where

$$W(u,u_c) = \frac{2}{\sqrt{\pi}} \int_{\sqrt{u_c}}^{\infty} -Ei\left(-\frac{uy^2}{y^2 - u_c}\right) e^{-y^2} dy$$

$$u_c = \frac{z^2 S'}{4K'D't} \tag{4.19}$$

$\frac{K'D'}{S'}$ = hydraulic diffusivity of the aquitard in m^2/d

z = vertical distance from aquifer-aquitard boundary to piezometer in the aquitard in m

At the same elapsed time and the same radial distance from the well, the ratio of the drawdown in the aquitard and the drawdown in the pumped aquifer is

$$\frac{s_c}{s} = \frac{W(u,u_c)}{W(u)}$$

Figure 4.12 shows curves of $W(u,u_c)/W(u)$ versus $1/u_c$ for different values of u. These curves have been prepared from values given by Witherspoon et al. (1967) and are presented in Annex 4.4. Knowing the ratio s_c/s from the observed drawdown data and a previously determined value of u for the aquifer, we can read a value of $1/u_c$ from Figure 4.12. By substituting the value of $1/u_c$ into Equation 4.19, we can determine the hydraulic diffusivity of the aquitard of very low permeability.

Neuman and Witherspoon (1972) showed that their ratio method, although developed for a slightly leaky aquifer, can also be used for a very leaky aquifer. The only requirement is that, in Equation 4.17, $\beta \leq 1.0$ because, as long as $\beta \leq 1.0$, the ratio s_c/s is found to be independent of β for all practical values of u_c. As β is directly proportional to the radial distance r from the well to the piezometer, r should be small ($r < 100$ m).

The Neuman-Witherspoon ratio method can be applied if the following assumptions and conditions are fulfilled:

- The assumptions listed at the beginning of this chapter;
- The flow to the well is in an unsteady state;
- The aquitard is compressible, i.e. the changes in aquitard storage are appreciable;
- $\beta < 1.0$, i.e. the radial distance from the well to the piezometers should be small ($r < 100$ m);
- $t < S'D'/10K'$.

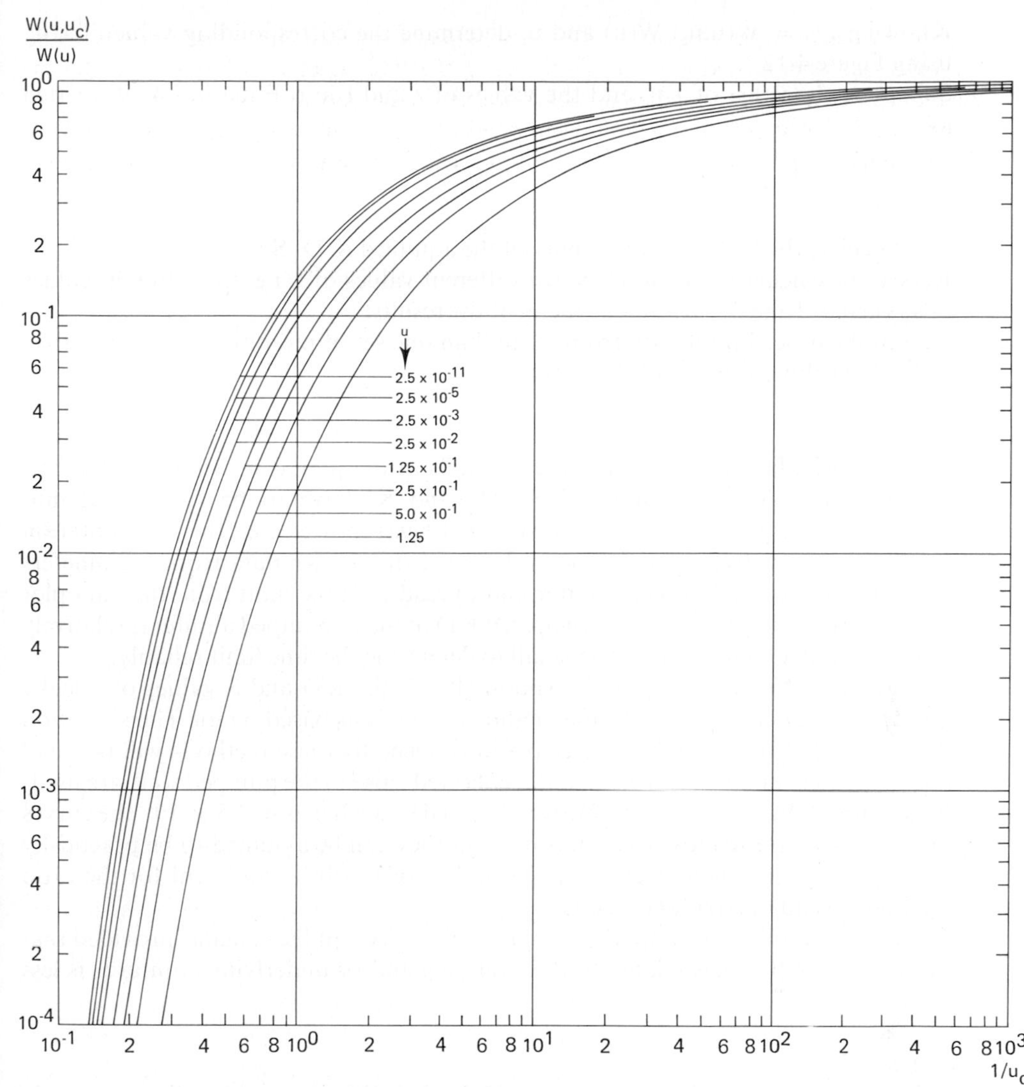

Figure 4.12 Neuman-Witherspoon's nomogram showing the relation of $W(u,u_c)/W(u)$ versus $1/u_c$ for different values of u

Procedure 4.7

- Calculate the transmissivity KD and the storativity S of the aquifer with one of the methods described in Section 4.2, using the early-time drawdown data of the aquifer;
- For a selected value of r ($r < 100$ m), prepare a table of values of the drawdown in the aquifer s, in the overlying aquitard s_c, and, if possible, in the overlying unconfined aquifer s_u for different values of t (see Remarks below);
- Select a time t and calculate for this value of t the value of the ratio s_c/s and the value of $u = r^2S/4KDt$;

- Knowing $s_c/s = W(u,u_c)/W(u)$ and u, determine the corresponding value of $1/u_c$, using Figure 4.12;
- Substitute the value of $1/u_c$ and the values of z and t into Equation 4.19, written as

$$\frac{K'D'}{S'} = \frac{1}{u_c} \times \frac{z^2}{4t}$$

and calculate the hydraulic diffusivity of the aquitard $K'D'/S'$;
- Repeat the calculation of $K'D'/S'$ for different values of t, i.e. for different values of s_c/s and u. Take the arithmetic mean of the results;
- Repeat the procedure if data from more than one set of piezometers are available. Take the arithmetic mean of the results.

Remarks
- To check whether the selected value of t falls in the period in which the method is valid, the calculated values of S′, D′, and K′ have to be substituted into $t < S'D'/10K'$. Neuman and Witherspoon (1969a) showed that this time criterion is rather conservative. It is also possible to use drawdown data from piezometers in the unpumped unconfined aquifer and to read the time limit from the data plot of s_u versus t on log-log paper. However, if KD of the unpumped aquifer is relatively large, the drawdown s_u will be too small to determine the time limit reliably;
- According to Neuman and Witherspoon (1972), the KD and S values of a leaky aquifer can be determined with the methods of analysis based on the Theis solution (Section 3.2). They state that the errors introduced by these methods will be small if the earliest available drawdown data, collected close to the pumped well, are used;
- Neuman and Witherspoon (1972) also observed that when $u < 2.5 \times 10^{-3}$ the curves in Figure 4.12 are so close to each other that they can be assumed to be practically independent of u. Then, even a crude estimate of u will be sufficient for the ratio method to yield satisfactory results;
- The ratio method is also applicable to multiple leaky aquifer systems, provided that the sum of the β values related to the overlying and/or underlying aquitards is less than 1.

Example 4.7
The data are taken from the pumping test 'Dalem'. At 30 m from the well, piezometers were placed at depths of 2 and 14 m below ground surface. The drawdowns in them at $t = 4.58 \times 10^{-2}$ d are $s_c = 0.009$ m and $s = 0.171$ m, respectively. The values of the aquifer characteristics are taken from Table 4.4: $KD = 1800$ m²/d and $S = 1.7 \times 10^{-3}$. Consequently

$$u = \frac{r^2S}{4KDt} = \frac{30^2 \times 1.7 \times 10^{-3}}{4 \times 1800 \times 4.58 \times 10^{-2}} = 4.6 \times 10^{-3}$$

and

$$\frac{s_c}{s} = \frac{0.009}{0.171} = 5.3 \times 10^{-2}$$

Plotting the value of $s_c/s = 5.3 \times 10^{-2}$ on the $W(u,u_c)/W(u)$ axis of the plot in Figure

4.12 and knowing the value of $u = 4.6 \times 10^{-3}$, we can read the value of $1/u_c$ from the horizontal axis of this plot: $1/u_c = 6.4 \times 10^{-1}$.

As the depth of the piezometer in the aquitard is 2 m below ground surface and $D' = 8$ m, it follows that $z = 6$ m. Consequently, the hydraulic diffusivity of the aquitard is

$$\frac{K'D'}{S'} = \frac{1}{u_c} \times \frac{z^2}{4t} = 6.4 \times 10^{-1} \times \frac{6^2}{4 \times 4.58 \times 10^{-2}} = 126 \text{ m}^2/\text{d}$$

The Neuman-Witherspoon method is only applicable if $t < S'D'/10K'$. From $K'D'/S' = 126$ m²/d and $D' = 8$ m, it follows that

$$t < 0.1 \left(\frac{K'D'}{S'} \times \frac{1}{(D')^2}\right)^{-1}, \text{ or } t < 0.1\,(126 \times 1/8^2)^{-1} = 0.05 \text{ d}$$

Hence, the time condition is fulfilled (the pumping time t used in the calculation was 4.58×10^{-2} d). As the radial distance of the piezometer to the well is 30 m, the condition $r < 100$ m is also satisfied.

4.3 Summary

Using data from the pumping test 'Dalem', we have illustrated the methods of analyzing steady and unsteady flow to a well in a leaky aquifer. Table 4.4 summarizes the values we obtained for the hydraulic characteristics of both the aquifer and the aquitard.

Table 4.4 Hydraulic characteristics of the leaky aquifer system at 'Dalem', calculated with the different methods

Method	Data from piezometer	KD (m²/d)	S	L (m)	c (d)	K'S' (m/d)	K'D'/S' (m²/d)
De Glee	All	2126	–	1100	569	–	–
Hantush-Jacob	All	2020	–	982	478	–	–
Walton	90	1731	1.9×10^{-3}	900	468	–	–
Hantush inflection-point 1	90	1665	1.7×10^{-3}	600	216	–	–
Hantush inflection-point 2	All	1883	1.6×10^{-3}	1043	578	–	–
Hantush curve-fitting	90	1515	1.5×10^{-3}	–	–	9×10^{-5}	–
Neuman-Witherspoon	30	–	–	–	–	–	126

We could thus conclude that the leaky aquifer system at 'Dalem' has the following (average) hydraulic characteristics:

Aquifer: KD = $1800 \, m^2/d$ Aquitard: c = 450 d
S = 1.7×10^{-3} K′D′/S′ = $126 \, m^2/d$
L = 900 m

From the aquitard characteristics, we could calculate values of K′ and S′:
$K' = D'/c = 8/450 = 1.8 \times 10^{-2}$ m/d
$S' = K'D'/126 = 1.1 \times 10^{-3}$

It will be noted that the different methods produce somewhat different results. This is due to inevitable inaccuracies in the observed and corrected or extrapolated data used in the calculations, but also, and especially, to the use of graphical methods. The steady-state drawdowns used in our examples, for instance, were extrapolated values and not measured values. These extrapolated values can be checked with Procedure 4.5 of the Hantush inflection-point method, but this requires a lot of straight lines having to be fitted through observed and calculated data that do not fall exactly on a straight line. Consequently, there are slightly different positions possible for these lines, which are still acceptable as fitted straight lines, but give different values of the hydraulic parameters.

The same difficulties are encountered when observed data plots have to be matched with a type curve or a family of type curves. In these cases too, slightly different matching positions are possible, with different match-point coordinates as a result, and thus different values for the hydraulic parameters. Because of such matching problems, the value of K′S′ in Table 4.4 is not considered to be very reliable.

Most of the methods described in this chapter only require data from the pumped aquifer. But, as already stated by Neuman and Witherspoon (1969b), such data are not sufficient to characterize a leaky system: the calculations should also be based on drawdown data from the aquitard and, if present, from the overlying unconfined unpumped aquifer, whose watertable will not remain constant, except for ideal situations, which are rare in nature.

Moreover, it should be kept in mind that, in practice, the assumptions underlying the methods are not always entirely satisfied. One of the assumptions, for instance, is that the aquifer is homogeneous, isotropic, and of uniform thickness, but it will be obvious that for an aquifer made up of alluvial sand and gravel, this assumption is not usually correct and that its hydraulic characteristics will vary from one place to another.

Summarizing, we can state that the average results of the calculations presented above are the most accurate values possible, and that, given the lithological character of the aquifer, aiming for any higher degree of accuracy would be to pursue an illusion.

5 Unconfined aquifers

Figure 5.1 shows a pumped unconfined aquifer underlain by an aquiclude. The pumping causes a dewatering of the aquifer and creates a cone of depression in the watertable. As pumping continues, the cone expands and deepens, and the flow towards the well has clear vertical components.

There are thus some basic differences between unconfined and confined aquifers when they are pumped:

- First, a confined aquifer is not dewatered during pumping; it remains fully saturated and the pumping creates a drawdown in the piezometric surface;
- Second, the water produced by a well in a confined aquifer comes from the expansion of the water in the aquifer due to a reduction of the water pressure, and from the compaction of the aquifer due to increased effective stresses;
- Third, the flow towards the well in a confined aquifer is and remains horizontal, provided, of course, that the well is a fully penetrating one; there are no vertical flow components in such an aquifer.

In unconfined aquifers, the water levels in piezometers near the well often tend to decline at a slower rate than that described by the Theis equation. Time-drawdown curves on log-log paper therefore usually show a typical S-shape, from which we can recognize three distinct segments: a steep early-time segment, a flat intermediate-time segment, and a relatively steep late-time segment (Figure 5.2). Nowadays, the widely used explanation of this S-shaped time-drawdown curve is based on the concept of 'delayed watertable response'. Boulton (1954, 1963) was the first to introduce this concept, which he called 'delayed yield'. He developed a semi-empirical solution that

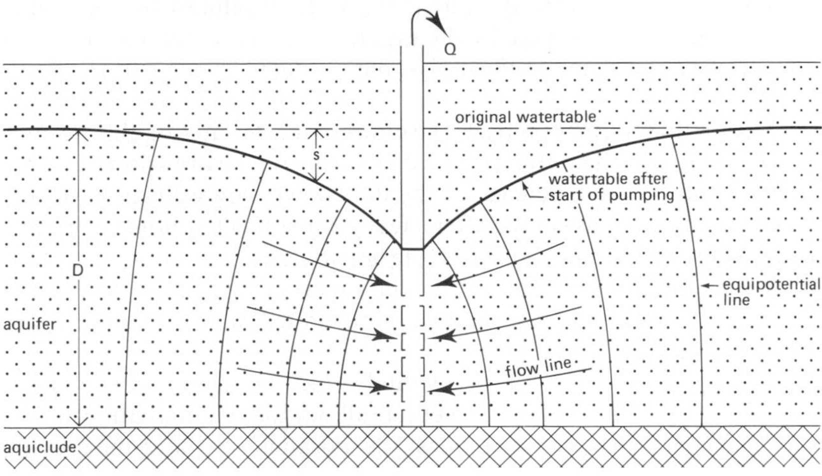

Figure 5.1 Cross-section of a pumped unconfined aquifer

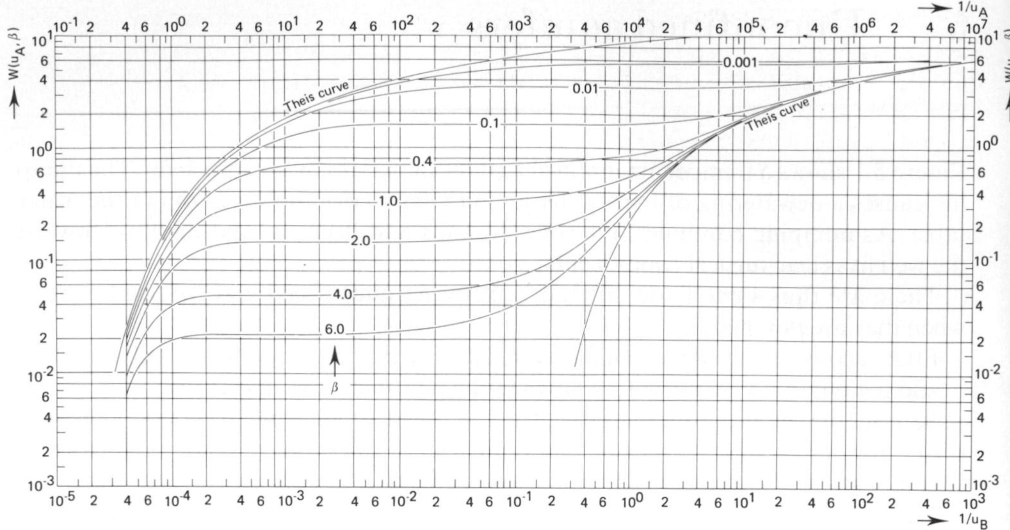

Figure 5.2 Family of Neuman type curves: $W(u_A,\beta)$ versus $1/u_A$ and $W(u_B,\beta)$ versus $1/u_B$ for different values of β

reproduced all three segments of this curve. Although useful in practice, Boulton's solution has one drawback: it requires the definition of an empirical constant, known as the Boulton's delay index, which is not clearly related to any physical phenomenon. The concept of delayed watertable response was further developed by Neuman (1972, 1973, 1979); Streltsova (1972a and b, 1973, 1976); and Gambolati (1976). According to these authors, the three time segments of the curve should be understood as follows:

- The steep early-time segment covers only a brief period after the start of pumping (often only the first few minutes). At early pumping times, an unconfined aquifer reacts in the same way as a confined aquifer: the water produced by the well is released instantaneously from storage by the expansion of the water and the compaction of the aquifer. The shape of the early-time segment is similar to the Theis type curve;
- The flat intermediate-time segment reflects the effect of the dewatering that accompanies the falling watertable. The effect of the dewatering on the drawdown is comparable to that of leakage: the increase of the drawdown slows down with time and thus deviates from the Theis curve. After a few minutes to a few hours of pumping, the time-drawdown curve may approach the horizontal;
- The relatively steep late-time segment reflects the situations where the flow in the aquifer is essentially horizontal again and the time-drawdown curve once again tends to conform to the Theis curve.

Section 5.1 presents Neuman's curve-fitting method, which is based on the concept of delayed watertable response. Neuman's method allows the determination of the horizontal and vertical hydraulic conductivities, the storativity S_A, and the specific yield S_Y.

It must be noted, however, that unreasonably low S_Y values are often obtained,

because flow in the (saturated) capillary fringe above the watertable is neglected (Van der Kamp 1985).
Under favourable conditions, the early and late-time drawdown data can also be analyzed by the methods given in Section 3.2. For example, the Theis method can be applied to the early-time segment of the time-drawdown curve, provided that data from piezometers near the well are used because the drawdown in distant piezometers during this period will often be too small to be measured. The storativity S_A computed from this segment of the curve, however, cannot be used to predict long-term drawdowns. The late-time segment of the curve may again conform closely to the Theis type curve, thus enabling the late-time drawdown data to be analyzed by the Theis equation and yielding the transmissivity and the specific yield S_Y of the aquifer. The Theis method yields a fairly realistic value of S_Y (Van der Kamp 1985).

If a pumped unconfined aquifer does not show phenomena of delayed watertable response, the time-drawdown curve only follows the late-time segment of the S-shaped curve. Because the flow pattern around the well is identical to that in a confined aquifer, the methods in Section 3.2 can be used.

True steady-state flow cannot be reached in a pumped unconfined aquifer of infinite areal extent. Nevertheless, the drawdown differences will gradually diminish with time and will eventually become negligibly small. Under these transient steady-state conditions we can use the Thiem-Dupuit method (Section 5.2).

The methods presented in this chapter are all based on the following assumptions and conditions:
- The aquifer is unconfined;
- The aquifer has a seemingly infinite areal extent;
- The aquifer is homogeneous and of uniform thickness over the area influenced by the test;
- Prior to pumping, the watertable is horizontal over the area that will be influenced by the test;
- The aquifer is pumped at a constant discharge rate;
- The well penetrates the entire aquifer and thus receives water from the entire saturated thickness of the aquifer.

In practice, the effect of flow in the unsaturated zone on the delayed watertable response can be neglected (Cooley and Case 1973; Kroszynski and Dagan 1975). According to Bouwer and Rice (1978), air entry phenomena may influence the drawdown.

Although the aquifer is assumed to be of uniform thickness, this condition is not met if the drawdown is large compared with the aquifer's original saturated thickness. A corrected value for the observed drawdown s then has to be applied. Jacob (1944) proposed the following correction

$$s' = s - (s^2/2D)$$

where
s' = corrected drawdown
s = observed drawdown
D = original saturated aquifer thickness

According to Neuman (1975), Jacob's correction is strictly applicable only to the late-time drawdown data, which fall on the Theis curve.

5.1 Unsteady-state flow

5.1.1 Neuman's curve-fitting method

Neuman (1972) developed a theory of delayed watertable response which is based on well-defined physical parameters of the unconfined aquifer. Neuman treats the aquifer as a compressible system and the watertable as a moving material boundary. He recognizes the existence of vertical flow components and his general solution of the drawdown is a function of both the distance from the well r and the elevation head. When considering an average drawdown, he is able to reduce his general solution to one that is a function of r alone. Mathematically, Neuman simulated the delayed watertable response by treating the elastic storativity S_A and the specific yield S_Y as constants.

Neuman's drawdown equation (Neuman 1975) reads

$$s = \frac{Q}{4\pi KD} W(u_A, u_B, \beta) \tag{5.1}$$

Under early-time conditions, this equation describes the first segment of the time-drawdown curve (Figure 5.2) and reduces to

$$s = \frac{Q}{4\pi KD} W(u_A, \beta) \tag{5.2}$$

where

$$u_A = \frac{r^2 S_A}{4KDt} \tag{5.3}$$

S_A = volume of water instantaneously released from storage per unit surface area per unit decline in head (= elastic early-time storativity).

Under late-time conditions, Equation 5.1 describes the third segment of the time-drawdown curve and reduces to

$$s = \frac{Q}{4\pi KD} W(u_B, \beta) \tag{5.4}$$

where

$$u_B = \frac{r^2 S_Y}{4KDt} \tag{5.5}$$

S_Y = volume of water released from storage per unit surface area per unit decline of the watertable, i.e. released by dewatering of the aquifer (= specific yield)

Neuman's parameter β is defined as

$$\beta = \frac{r^2 K_v}{D^2 K_h} \qquad (5.6)$$

where

K_v = hydraulic conductivity for vertical flow, in m/d
K_h = hydraulic conductivity for horizontal flow, in m/d

For isotropic aquifers, $K_v = K_h$, and $\beta = r^2/D^2$.

Neuman's curve-fitting method can be used if the following assumptions and conditions are satisfied:
- The assumptions listed at the beginning of this chapter;
- The aquifer is isotropic or anisotropic;
- The flow to the well is in an unsteady state;
- The influence of the unsaturated zone upon the drawdown in the aquifer is negligible;
- $S_Y/S_A > 10$;
- An observation well screened over its entire length penetrates the full thickness of the aquifer;
- The diameters of the pumped and observation wells are small, i.e. storage in them can be neglected.

As stated by Rushton and Howard (1982), fully-penetrating observation wells allow the 'short-circuiting' of vertical flow. Consequently, the water levels observed in them will not always be equivalent to the average of groundwater heads in a vertical section of the aquifer, as assumed in Neuman's theory. The theory should still be valid, however, for piezometers with short screened sections, provided that the drawdowns are averaged over the full thickness of the aquifer (Van der Kamp 1985).

Procedure 5.1
- Construct the family of Neuman type curves by plotting $W(u_A,u_B,\beta)$ versus $1/u_A$ and $1/u_B$ for a practical range of values of β on log-log paper, using Annex 5.1. The left-hand portion of Figure 5.2 shows the type A curves $[W(u_A,\beta)$ versus $1/u_A]$ and the right-hand portion the type B curves $[W(u_B,\beta)$ versus $1/u_B]$;
- Prepare the observed data curve on another sheet of log-log paper of the same scale by plotting the values of the drawdown s against the corresponding time t for a single observation well at a distance r from the pumped well;
- Match the early-time observed data plot with one of the type A curves. Note the β value of the selected type A curve;
- Select an arbitrary point A on the overlapping portion of the two sheets and note the values of s, t, $1/u_A$, and $W(u_A,\beta)$ for this point;
- Substitute these values into Equations 5.2 and 5.3 and, knowing Q and r, calculate K_hD and S_A;
- Move the observed data curve until as many as possible of the late-time observed data fall on the type B curve with the same β value as the selected type A curve;
- Select an arbitrary point B on the superimposed sheets and note the values of s, t, $1/u_B$, and $W(u_B,\beta)$ for this point;
- Substitute these values into Equations 5.4 and 5.5 and, knowing Q and r, calculate

K_hD and S_Y. The two calculations should give approximately the same value for K_hD;

- From the K_hD value and the known initial saturated thickness of the aquifer D, calculate the value of K_h;
- Substitute the numerical values of K_h, β, D, and r into Equation 5.6 and calculate K_v;
- Repeat the procedure with the observed drawdown data from any other observation well that may be available. The calculated results should be approximately the same.

Remarks

- To check whether the condition $S_Y/S_A > 10$ is fulfilled, the value of this ratio should be determined;
- Gambolati (1976) (see also Neuman 1979) pointed out that, theoretically, the effects of elastic storage and dewatering become additive at large t, the final storativity being equal to $S_A + S_Y$. However, in situations where the effect of delayed watertable response is clearly evident, $S_A \ll S_Y$ and the influence of S_A at larger times can safely be neglected.

Example 5.1

To illustrate the Neuman curve-fitting method, we shall use data from the pumping test 'Vennebulten', The Netherlands (De Ridder 1966). Figure 5.3 shows a lithostratigraphical section of the pumping test area as derived from the drilling data. The impermeable base consists of Middle Miocene marine clays. The aquifer is made up of very coarse fluvioglacial sands and coarse fluvial deposits, which grade upward into very fine sand and locally into loamy cover sand. The finer part of the aquifer is about 10 m thick. A well screen was placed between 10 and 21 m below ground surface, and piezometers were placed at distances of 10, 30, 90, and 280 m from the well at

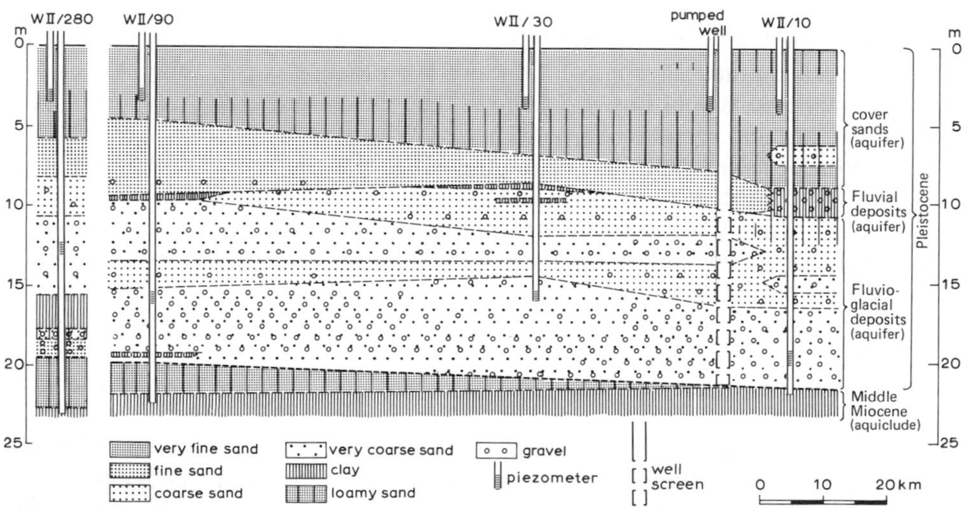

Figure 5.3 Lithostratigraphical cross-section of the pumping-test site 'Vennebulten', The Netherlands (after De Ridder 1966)

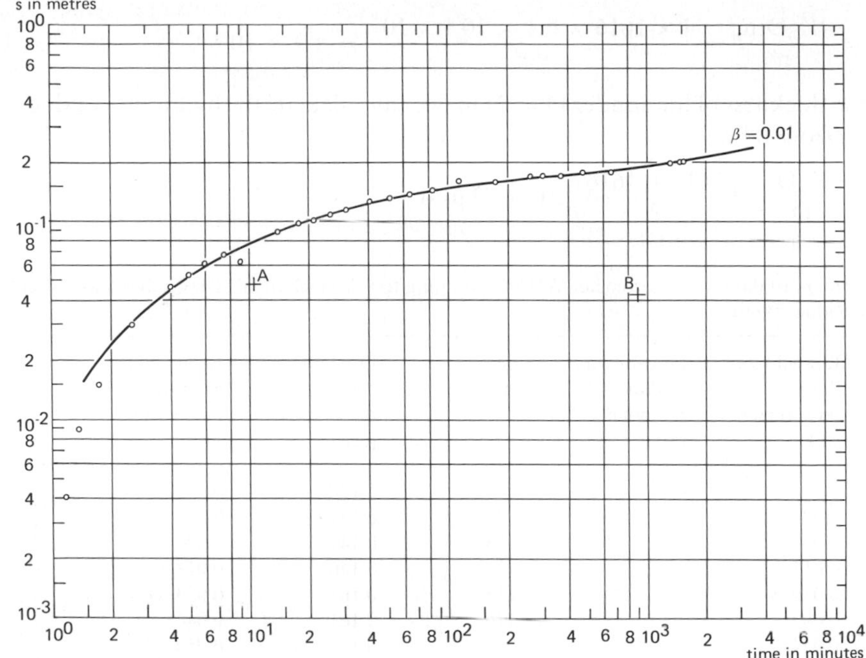

Figure 5.4 Analysis of data from pumping test 'Vennebulten', The Netherlands (r = 90 m) with the Neuman curve-fitting method

depths ranging from 12 to 19 m. Shallow piezometers (at a depth of about 3 m) were placed at the same distances. The aquifer was pumped for 25 hours at a constant discharge of 36.37 m^3/hr (or 873 m^3/d). Table 5.1 summarizes the drawdown observations in the piezometer at 90 m.

The observed time-drawdown data of Table 5.1 are plotted on log-log paper (Figure 5.4). The early-time segment of the plot gives the best match with the Neuman type A curve for $\beta = 0.01$. The match point A has the coordinates $1/u_A = 10$, $W(u_A,\beta) = 1$, $s = 4.8 \times 10^{-2}$ m, and $t = 10.5$ min $= 7.3 \times 10^{-3}$ d.

The values of K_hD and S_A are obtained from Equations 5.2 and 5.3

$$K_hD = \frac{Q}{4\pi s} W(u_A,\beta) = \frac{873}{4\pi \times 4.8 \times 10^{-2}} \times 1 = 1447 \text{ m}^2/\text{d}$$

$$S_A = \frac{4K_hDtu_A}{r^2} = \frac{4 \times 1447 \times 7.3 \times 10^{-3} \times 10^{-1}}{90^2} = 5.2 \times 10^{-4}$$

The coordinates for match point B of the observed data plot and the type B curve for $\beta = 0.01$ are $1/u_B = 10^2$, $W(u_B,\beta) = 1$, $s = 4.3 \times 10^{-2}$ m and $t = 880$ min $= 6.1 \times 10^{-1}$ d.

Calculating the values of K_hD and S_Y from Equations 5.4 and 5.5, we obtain

$$K_hD = \frac{Q}{4\pi s} W(u_B,\beta) = \frac{873}{4\pi \times 4.3 \times 10^{-2}} \times 1 = 1616 \text{ m}^2/\text{d}$$

$$S_Y = \frac{4K_hDtu_B}{r^2} = \frac{4 \times 1616 \times 6.1 \times 10^{-1} \times 10^{-2}}{90^2} = 4.9 \times 10^{-3}$$

Knowing the thickness of the aquifer D = 21 m, we can calculate the hydraulic conductivity for horizontal flow

$$K_h = \frac{K_hD}{D} = \left(\frac{1447 + 1616}{2}\right)/21 = 73 \text{ m/d}$$

Table 5.1 Summary of data from piezometer WII/90; pumping test 'Vennebulten', The Netherlands (after De Ridder 1966)

Time (min)	Drawdown deep piezometer (m)	Drawdown shallow piezometer (m)	Time (min)	Drawdown deep piezometer (m)	Drawdown shallow piezometer (m)
0	0	0	41	0.128	0.018
1.17	0.004		51	0.133	0.022
1.34	0.009		65	0.141	0.026
1.7	0.015		85	0.146	0.028
2.5	0.030		115	0.161	0.033
4.0	0.047		175	0.161	0.044
5.0	0.054		260	0.172	0.050
6.0	0.061	0.005	300	0.173	0.055
7.5	0.068		370	0.173	
9	0.064	0.006	430	0.179	
14	0.090	0.008	485	0.183	0.061
18	0.098	0.010	665	0.182	0.071
21	0.103		1.340	0.200	0.096
26	0.110	0.011	1.490	0.203	0.099
31	0.115	0.014	1.520	0.204	0.099

From Equation 5.6, the hydraulic conductivity for vertical flow can be calculated

$$K_v = \frac{\beta D^2 K_h}{r^2} = \frac{0.01 \times 21^2 \times 73}{90^2} = 4 \times 10^{-2} \text{ m/d}$$

The value of the ratio S_Y/S_A is

$$\frac{S_Y}{S_A} = \frac{4.9 \times 10^{-3}}{5.2 \times 10^{-4}} = 9.4$$

The condition of $S_Y/S_A > 10$ is therefore nearly satisfied. Note that the value of S_Y calculated by means of the 'B' curves is unreasonably low. This is in agreement with earlier observations that the determination of S_Y from 'B' curves remains a dubious procedure (Van der Kamp 1985).

5.2 Steady-state flow

When the drawdown differences have become negligibly small with time, the Thiem-

Dupuit method can be used to calculate the transmissivity of an unconfined aquifer.

5.2.1 Thiem-Dupuit's method

The Thiem-Dupuit method can be used if the following assumptions and conditions are satisfied:

- The assumptions listed in the beginning of this chapter;
- The aquifer is isotropic;
- The flow to the well is in steady state;
- The Dupuit (1863) assumptions are satisfied, i.e.:
  - The velocity of flow is proportional to the tangent of the hydraulic gradient instead of the sine as it is in reality;
  - The flow is horizontal and uniform everywhere in a vertical section through the axis of the well.

If these assumptions are met, the well discharge for steady horizontal flow to a well pumping an unconfined aquifer (Figure 5.5) can be described by

$$Q = 2\pi r K h \frac{dh}{dr}$$

After integration between r_1 and r_2 (with $r_2 > r_1$), this yields

$$Q = \pi K \frac{h_2^2 - h_1^2}{\ln(r_2/r_1)} \qquad (5.7)$$

which is known as the formula of Dupuit.

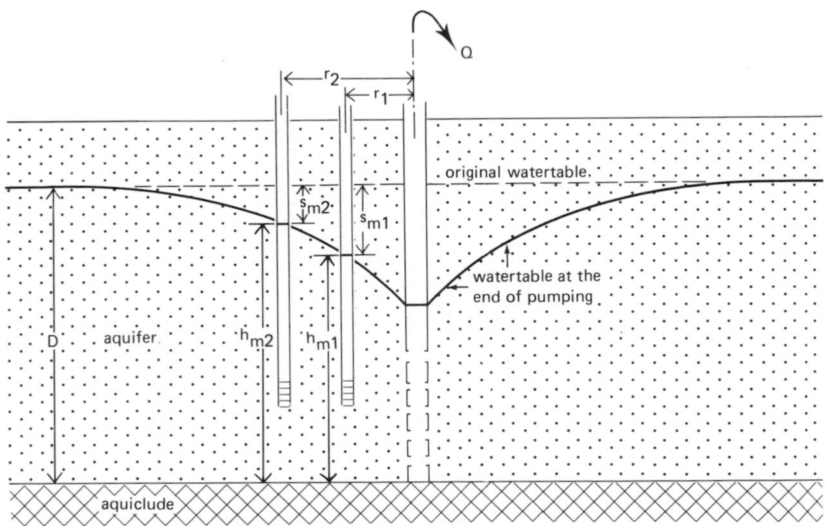

Figure 5.5 Cross-section of a pumped unconfined aquifer (steady-state flow)

Since h = D – s, Equation 5.7 can be transformed into

$$Q = \frac{\pi K[(D-s_{m2})^2 - (D-s_{m1})^2]2D/2D}{\ln(r_2/r_1)} = \frac{2\pi KD[(s_{m1} - s_{m1}^2/2D) - (s_{m2} - s_{m2}^2/2D)]}{\ln(r_2/r_1)}$$

Replacing $s - s^2/2D$ with s' = the corrected drawdown, yields

$$Q = \frac{2\pi KD(s'_{m1} - s'_{m2})}{\ln(r_2/r_1)} = \frac{2\pi KD(s'_{m1} - s'_{m2})}{2.30 \log (r_2/r_1)} \qquad (5.8)$$

This formula is identical to the Thiem formula (Equation 3.2) for a confined aquifer, so the methods in Section 3.1.1 can also be used for an unconfined aquifer.

Remarks

- The Dupuit formula (Equation 5.7) fails to give an accurate description of the drawdown curve near the well, where the strong curvature of the watertable contradicts the Dupuit assumptions. These assumptions ignore the existence of a seepage face at the well and the influence of the vertical velocity components, which reach their maximum in the vicinity of the well;
- An approximate steady-state flow condition in an unconfined aquifer will only be reached after long pumping times, i.e. when the flow in the aquifer is essentially horizontal again and the drawdown curve has followed the late-time segment of the S-shaped curve that coincides with the Theis curve for sufficiently long time.

6 Bounded aquifers

Pumping tests sometimes have to be performed near the boundary of an aquifer. A boundary may be either a recharging boundary (e.g. a river or a canal) or a barrier boundary (e.g. an impermeable valley wall). When an aquifer boundary is located within the area influenced by a pumping test, the general assumption that the aquifer is of infinite areal extent is no longer valid.

Presented in Sections 6.1 and 6.2 are methods of analysis developed for confined or unconfined aquifers with various boundaries and boundary configurations. Section 6.3 presents a method for leaky or confined aquifers bounded laterally by two parallel barrier boundaries.

To analyze the flow in bounded aquifers, we apply the principle of superposition. According to this principle, the drawdown caused by two or more wells is the sum of the drawdown caused by each separate well. So, by introducing imaginary wells, or image wells, we can transform an aquifer of finite extent into one of seemingly infinite extent, which allows us to use the methods presented in earlier chapters.

Figure 6.1A shows a fully penetrating straight canal which forms a recharging boundary with an assumed constant head. In Figure 6.1B, we replace this bounded system with an equivalent system, i.e. an imaginary system of infinite areal extent. In this system, there are two wells: the real discharging well on the left and an image recharging well on the right. The image well recharges the aquifer at a constant rate Q equal to the constant discharge of the real well. Both the real well and the image well are located on a line normal to the boundary and are equidistant from the boundary (Figure 6.1C). If we now sum the cone of depression from the real well and the cone of impression from the image well, we obtain an imaginary zero drawdown in the infinite system at the real constant-head boundary of the real bounded system.

Figure 6.1D shows a system with a straight impermeable valley wall which forms a barrier boundary. Figure 6.1E shows the real bounded system replaced by an equivalent system of infinite areal extent. The imaginary system has two wells discharging at the same constant rate: the real well on the left and an image well on the right. The image well induces a hydraulic gradient from the boundary towards the image well, which is equal to the hydraulic gradient from the boundary towards the real well. A groundwater divide thus exists at the boundary and there is no flow across the boundary. The resultant real cone of depression is the algebraic sum of the depression cones of both the real and the image well. Note that between the real well and the boundary, the real depression cone is flatter than it would be if no boundary were present, and is steeper on the opposite side away from the boundary.

If there is more than one boundary, more image wells are needed. For instance, if two boundaries are at right angles to each other, the imaginary system includes two primary image wells, both reflections of the real well, and one secondary image well, which is a reflection of the primary image wells. If the boundaries are parallel to one another, the number of image wells is theoretically infinite, but in practice it is only necessary to add pairs of image wells until the next pair would have a negligible

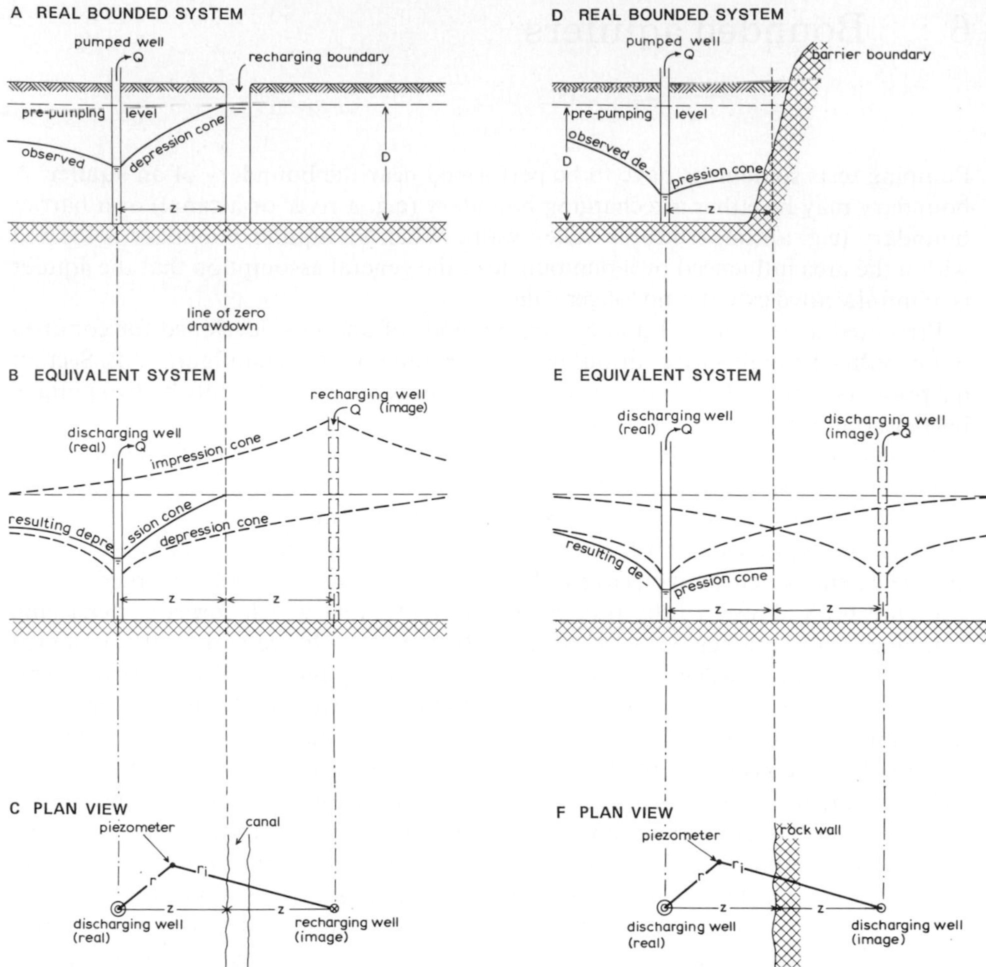

Figure 6.1 Drawdowns in the watertable of an aquifer bounded by:
A) A recharging boundary;
D) A barrier boundary.
B) and E) Equivalent systems of infinite areal extent.
C) and F) Plan views

influence on the sum of all image-well effects. Some of these boundary configurations will be discussed below.

6.1 Bounded confined or unconfined aquifers, steady-state flow

6.1.1 Dietz's method, one or more recharge boundaries

Dietz (1943) published a method of analyzing tests conducted in the vicinity of straight recharge boundaries under conditions of steady-state flow. Dietz's method, which is based on the work of Muskat (1937), uses Green's functions to describe the influence of the boundaries: in a piezometer with coordinates x_l and y_l, the steady-state drawdown caused by a well with coordinates x_w and y_w is given by

$$s_m = \frac{Q}{2\pi KD} G(x,y) \tag{6.1}$$

where $G(x,y)$ = Green's function for a certain boundary configuration.

For one straight recharge boundary (Figure 6.2A), the function reads

$$G(x,y) = \frac{1}{2} \ln \frac{(x_l + x_w)^2 + (y_l - y_w)^2}{(x_l - x_w)^2 + (y_l - y_w)^2} \tag{6.2}$$

For two straight recharge boundaries at right angles to each other (Figure 6.2B), the function reads

$$G(x,y) = \frac{1}{2} \ln \frac{[(x_l - x_w)^2 + (y_l + y_w)^2]\,[(x_l + x_w)^2 + (y_l - y_w)^2]}{[(x_l - x_w)^2 + (y_l - y_w)^2]\,[(x_l + x_w)^2 + (y_l + y_w)^2]} \tag{6.3.}$$

For two straight parallel recharge boundaries (Figure 6.2C), the function reads

$$G(x,y) = \frac{1}{2} \ln \frac{\cosh \dfrac{\pi(y_l - y_w)}{2a} + \cos \dfrac{\pi(x_l + x_w)}{2a}}{\cosh \dfrac{\pi(y_l - y_w)}{2a} - \cos \dfrac{\pi(x_l - x_w)}{2a}} \tag{6.4}$$

For a U-shaped recharge boundary (Figure 6.2D), the function reads

$$G(x,y) = \frac{1}{2} \ln \frac{\left[\cosh \dfrac{\pi(y_l - y_w)}{2a} + \cos \dfrac{\pi(x_l + x_w)}{2a}\right]}{\left[\cosh \dfrac{\pi(y_l - y_w)}{2a} - \cos \dfrac{\pi(x_l - x_w)}{2a}\right]} \times \frac{\left[\cosh \dfrac{\pi(y_l + y_w)}{2a} - \cos \dfrac{\pi(x_l - x_w)}{2a}\right]}{\left[\cosh \dfrac{\pi(y_l + y_w)}{2a} + \cos \dfrac{\pi(x_l + x_w)}{2a}\right]} \tag{6.5}$$

The assumptions and conditions underlying the Dietz method are:
- The assumptions listed at the beginning of Chapter 3, except for the first and second assumptions, which are replaced by:
  - The aquifer is confined or unconfined;
  - Within the zone influenced by the pumping test, the aquifer is crossed by one or more straight, fully penetrating recharge boundaries with a constant water level;
  - The hydraulic contact between the recharge boundaries and the aquifer is as permeable as the aquifer.

The following condition is added:
- The flow to the well is in a steady state.

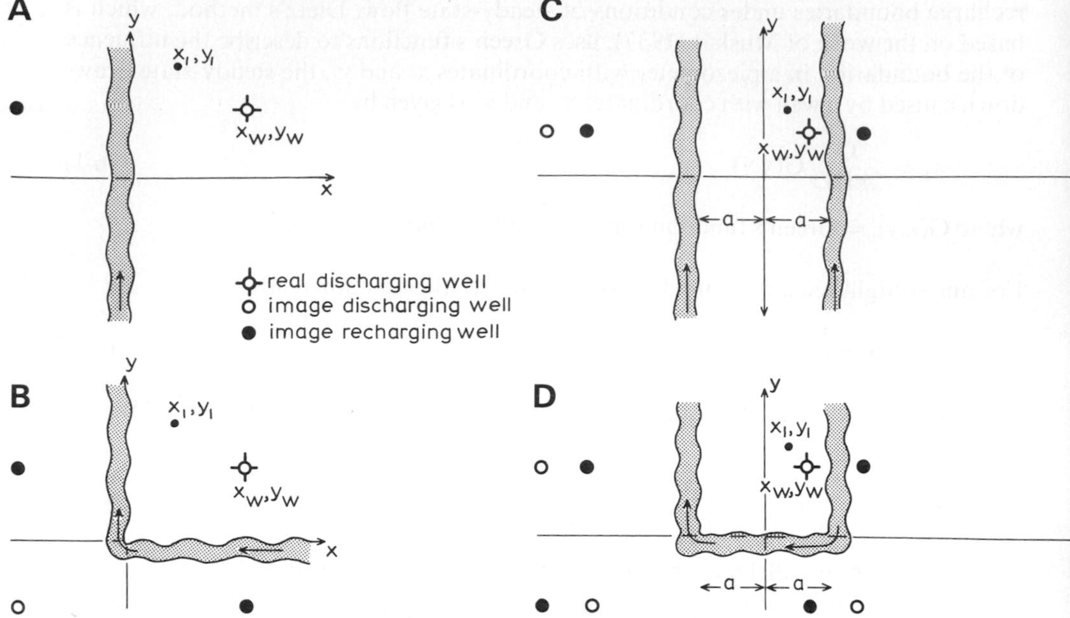

Figure 6.2 Image well systems for bounded aquifers (Dietz method)
A) One straight recharge boundary
B) Two straight recharge boundaries at right angles
C) Two straight parallel recharge boundaries
D) U-shaped recharge boundary

Procedure 6.1

- Determine the boundary configuration and substitute the appropriate Green function into Equation 6.1;
- Measure the values of x_w, y_w, x_1, and y_1 on the map of the pumping site;
- Substitute the values of Q, x_w, y_w, x_1, y_1, and s_{m1} into Equation 6.1 and calculate KD;
- Repeat this procedure for all available piezometers. The results should show a close agreement.

Remarks

- The angles in Equations 6.4 and 6.5 are expressed in radians;
- For unconfined aquifers, the maximum drawdown s_m should be replaced by $s'_m = s_m - (s^2_m/2D)$.

6.2 Bounded confined or unconfined aquifer, unsteady-state flow

6.2.1 Stallman's method, one or more boundaries

Stallman (as quoted by Ferris et al. 1962) developed a curve-fitting method for aquifers that have one or more straight recharge or barrier boundaries.

The distance between the real well and a piezometer is r; the distance between an image well and the piezometer is r_i, and their ratio is $r_i/r = r_r$.

If

$$u = \frac{r^2S}{4KDt} \tag{6.6}$$

and

$$u_i = \frac{r_i^2\,S}{4KDt} = \frac{r_r^2 r^2 S}{4KDt} = r_r^2\,u \tag{6.7}$$

the drawdown in the piezometer is described by

$$s = \frac{Q}{4\pi KD}[W(u) \pm W(r_{r1}^2u) \pm W(r_{r2}^2u) \pm ... \pm W(r_{rn}^2u)] \tag{6.8}$$

or

$$s = \frac{Q}{4\pi KD}W(u,r_{r1\rightarrow n}) \tag{6.9}$$

Numerical values of $W(r_r^2u)$ are given in Annex 6.1. In Equation 6.8, the number of terms between brackets depends on the number of image wells. If there is only one image well, there are two terms between brackets: the term $(Q/4\pi KD)\ W(u)$ describing the influence of the real well and the term $(Q/4\pi KD)\ W(r_r^2u)$ describing the influence of the image well. If there are two straight boundaries intersecting at right angles, three image wells are required, and there are consequently four terms between brackets. With parallel boundaries, the number of image wells becomes infinite, but those where $r_r > 100$ can be neglected.

A discharge well – real or image – gives terms with a positive sign; a recharge well gives terms with a negative sign. Consequently, the drawdown in a piezometer caused by a well near a boundary can be described by the following equations.

One straight boundary

One recharge boundary (Figure 6.1A-C)

$$s = \frac{Q}{4\pi KD}[W(u) - W(r_r^2u)] \tag{6.10}$$

or

$$s = \frac{Q}{4\pi KD}W_R(u,r_r) \tag{6.11}$$

One barrier boundary (Figure 6.1D-F)

$$s = \frac{Q}{4\pi KD}[W(u) + W(r_r^2u)] \tag{6.12}$$

or

$$s = \frac{Q}{4\pi KD}W_B(u,r_r) \tag{6.13}$$

Two straight boundaries at right angles to each other
One barrier boundary and one recharge boundary (Figure 6.3A)

$$s = \frac{Q}{4\pi KD}[W(u) + W(r_{r1}^2u) - W(r_{r2}^2u) - W(r_{r3}^2u)] \tag{6.14}$$

Two barrier boundaries (Figure 6.3B)

$$s = \frac{Q}{4\pi KD}[W(u) + W(r_{r1}^2u) + W(r_{r2}^2u) + W(r_{r3}^2u)] \tag{6.15}$$

Two recharge boundaries (Figure 6.3C)

$$s = \frac{Q}{4\pi KD}[W(u) - W(r_{r1}^2u) - W(r_{r2}^2u) + W(r_{r3}^2u)] \tag{6.16}$$

Two parallel boundaries
One barrier and one recharge boundary (Figure 6.4A)

$$s = \frac{Q}{4\pi KD}[W(u) + W(r_{r1}^2u) - W(r_{r2}^2u) - W(r_{r3}^2u) - ... \pm W(r_{rn}^2u)] \tag{6.17}$$

Two barrier boundaries (Figure 6.4B)

$$s = \frac{Q}{4\pi KD}[W(u) + W(r_{r1}^2u) + W(r_{r2}^2u) + W(r_{r3}^2u) + ... + W(r_{rn}^2u)] \tag{6.18}$$

Two recharge boundaries (Figure 6.4C)

$$s = \frac{Q}{4\pi KD}[W(u) - W(r_{r1}^2u) - W(r_{r2}^2u) + W(r_{r3}^2u) + ... \pm W(r_{rn}^2u)] \tag{6.19}$$

For three and four straight boundaries (Figures 6.5 and 6.6), the drawdown equations can be composed in the same way.

Stallman's method can be applied if the following assumptions and conditions are satisfied:
- The assumptions listed at the beginning of Chapter 3, with the exception of the first and second assumptions, which are replaced by:
  - The aquifer is confined or unconfined;
  - Within the zone influenced by the pumping test, the aquifer is crossed by one or more straight, fully penetrating recharge or barrier boundaries;
  - Recharge boundaries have a constant water level and the hydraulic contacts between the recharge boundaries and the aquifer are as permeable as the aquifer.

The following condition is added:
- The flow to the well is in unsteady state.

Procedure 6.2
- Determine the boundary configuration and prepare a plan of the equivalent system of image wells;
- Determine for one of the piezometers the value of r and the value or values of r_i;
- Calculate $r_r = r_i/r$ for each of the image wells and determine the sign for each of the terms between brackets in Equation 6.8;

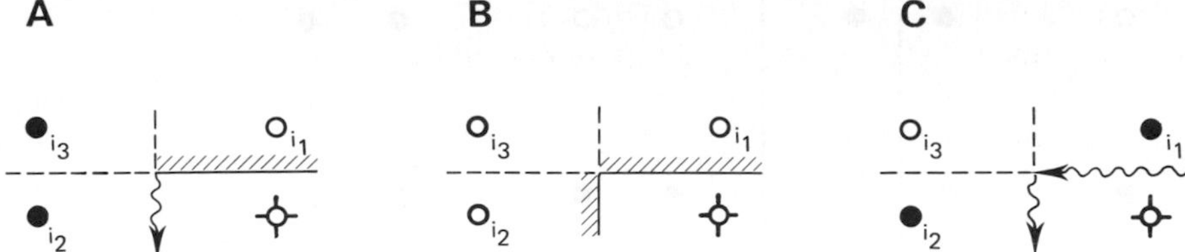

Figure 6.3 Two straight boundaries intersecting at right angles

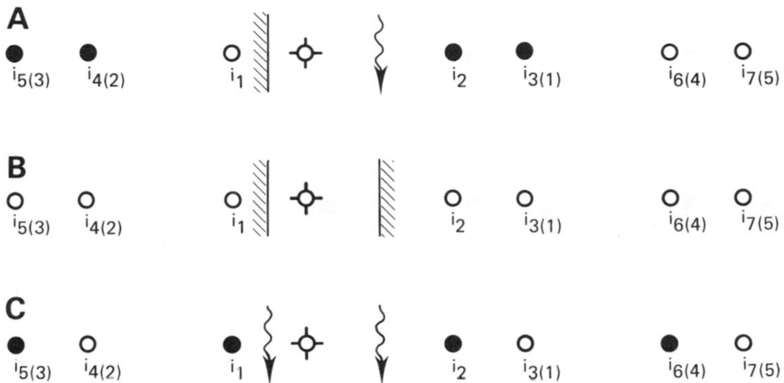

Figure 6.4 Two straight parallel boundaries

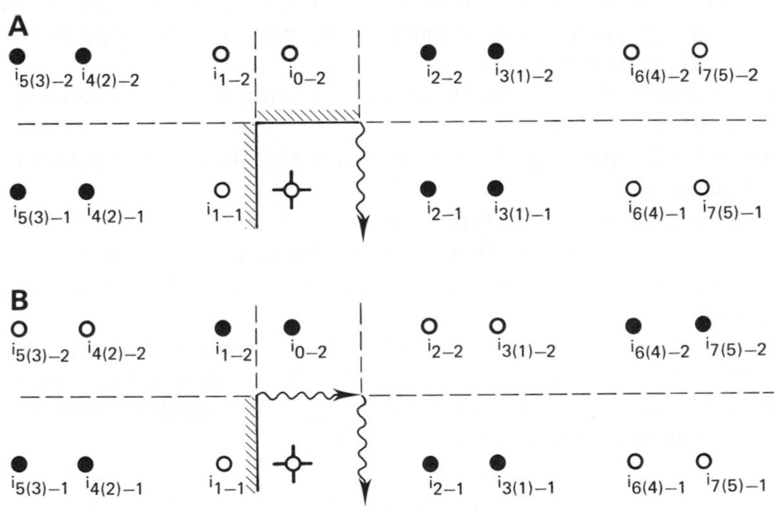

Figure 6.5 Two straight parallel boundaries intersected at right angles by a third boundary

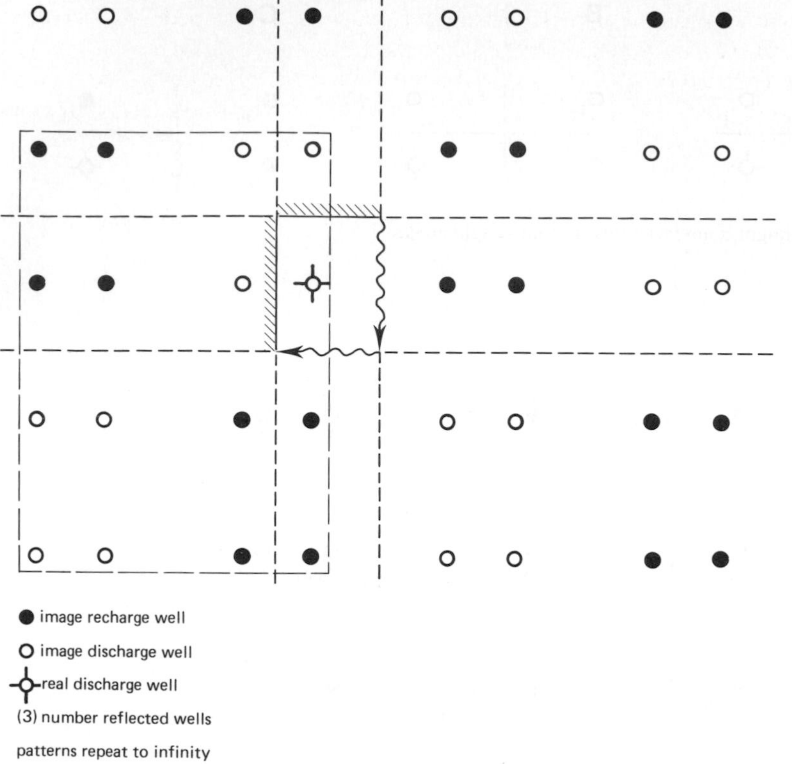

Figure 6.6 Four straight boundaries, i.e. two pairs of straight parallel boundaries intersecting at right angles

- Using Annex 6.1, calculate the numerical values of $W(u,r_{r1\to n})$ with respect to u according to the appropriate form of Equation 6.8, and plot the type curve $W(u,r_{r1\to n})$ versus u on log-log paper;
  (For one-boundary systems, the values of $W_R(u,r_r)$ and $W_B(u,r_r)$ can be read directly from Annexes 6.2 and 6.3);
- On another sheet of log-log paper, plot s as observed in the piezometer versus 1/t; this is the observed data curve;
- Match the observed data curve with the type curve;
- Select a matchpoint A and note its coordinate values u, $W(u,r_{r1\to n})$, s, and 1/t;
- Substitute these values of s and $W(u,r_{r1\to n})$ and the known value of Q into Equation 6.9 and calculate KD;
- Substitute the values of Q, r, u, KD, and 1/t into Equation 6.6 and calculate S;
- Repeat this procedure for all available piezometers. It will be noted that each piezometer has its own type curve because the value of $W(u,r_{r1\to n})$ depends on the value of the ratio $r_i/r = r_r$, which is different for each piezometer.

Remarks

- This method can also be used to analyze the drawdown data from an aquifer pumped by more than one real well, or from an aquifer that is both pumped and recharged by real wells, provided all wells operate at the same constant rate Q;

- Equation 6.8 is based on the Theis well function for confined aquifers. Stallman's method, however, is also applicable to data from unconfined aquifers as long as Assumption 7 (Chapter 3) is met, i.e. no delayed watertable response is apparent.

6.2.2 Hantush's method, one recharge boundary

The Hantush image method is useful when the effective line of recharge does not correspond with the bank or the streamline of the river or canal. This may be due to the slope of the bank, to partial penetration effects of the river or canal, or to an entrance resistance at the boundary contact. When the effects of these conditions are small but not negligible, they can be compensated for by making the distance between the pumped well and the hydraulic boundary in the equivalent system (line of zero drawdown in Figure 6.1B) greater than the distance between the pumped well and the actual boundary (Figure 6.7).

As was shown by the Stallman method, the drawdown in an aquifer limited at one side by a recharge boundary can be expressed by Equation 6.10

$$s = \frac{Q}{4\pi KD}[W(u) - W(r_r^2 u)]$$

where, according to Equation 6.6,

$$u = \frac{r^2 S}{4KDt}$$

and

$$r_r = \frac{r_i}{r}$$

$r = \sqrt{(x^2 + y^2)}$ is the distance between the piezometer and the real discharging well

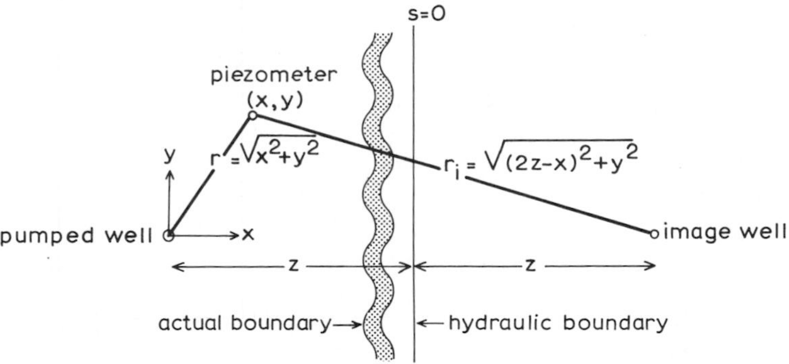

Figure 6.7 The parameters in the Hantush image method

$r_i = \sqrt{\{(2z - x)^2 + y^2\}}$ is the distance between the piezometer and the recharging well; x, y are the coordinates of the piezometer with respect to the real discharging well (see Figure 6.7)

The distance between the real discharging well and the recharging image well is 2z. The hydraulic boundary, i.e. the effective line of recharge, intersects the connecting line midway between the real well and the image well. The lines are at right angles to each other. It should be kept in mind that, especially with recharge boundaries, the hydraulic boundary does not always coincide with the bank of the river or its streamline. It is not necessary to know z beforehand, nor the location of the image well, nor the distance r_i dependent on it; neither need the relation $r_i/r = r_r$ be known beforehand.

The relation between r_r, x, r, and z is given by

$$4z^2 - 4xz - r^2(r_r^2 - 1) = 0 \tag{6.20}$$

Hantush (1959b) observed that if the drawdown s is plotted on semi-log paper versus t (with t on logarithmic scale), there is an inflection point P on the curve (Figure 6.8). At this point, the value of u is given by

$$u_p = \frac{r^2 S}{4KDt_p} = \frac{2 \ln r_r}{r_r^2 - 1} \tag{6.21}$$

The slope of the curve at this point is

$$\Delta s_p = \frac{2.30Q}{4\pi KD}\left(e^{-u_p} - r^{-r_r^2 u_p}\right) \tag{6.22}$$

and the drawdown at this point is

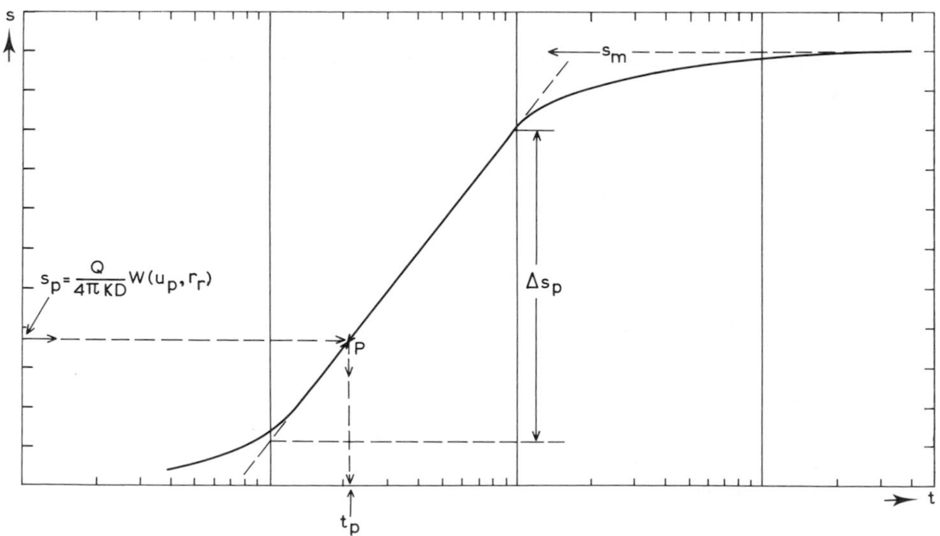

Figure 6.8 The application of the Hantush image method

$$s_p = \frac{Q}{4\pi KD} W(u_p, r_r) \qquad (6.23)$$

For values of $t > 4t_p$, the drawdown s approaches the maximum drawdown

$$s_m = \frac{Q}{2\pi KD} \ln r_r \qquad (6.24)$$

It will be noted that the ratio of s_m, as given by Equation 6.24, and Δs_p, as given by Equation 6.22, depends solely on the value of r_r. So

$$\frac{s_m}{\Delta s_p} = \frac{2 \log r_r}{e^{-u_p} - e^{-r_r^2 u_p}} = f(r_r) \qquad (6.25)$$

where u_p is given by Equation 6.21.

The Hantush image method is based on the following assumptions and conditions:
- The assumptions listed at the beginning of Chapter 3, with the exception of the first and second assumptions, which are replaced by:
  - The aquifer is confined or unconfined;
  - The aquifer is crossed by a straight recharge boundary within the zone influenced by the pumping test;
  - The recharge boundary has a constant water level, but the effective line of recharge need not necessarily be known beforehand. Entrance resistances, however, should be small, although not negligible.

The following conditions are added:
- The flow to the well is in unsteady state;
- It should be possible to extrapolate the steady-state drawdown for each of the piezometers.

Procedure 6.3
- On semi-log paper, plot s versus t for one of the piezometers (t on logarithmic scale), and draw the time-drawdown curve through the plotted points (Figure 6.8);
- Extrapolate the curve to determine the value of the maximum drawdown s_m;
- Calculate the slope Δs_p of the straight portion of the curve; this is an approximation of the slope at the inflection point P;
- Calculate the ratio $s_m/\Delta s_p$ according to Equation 6.25; this is equal to $f(r_r)$. Use Annex 6.4 to find the value of r_r from $f(r_r)$;
- Substitute the values of s_m, Q, and r_r into Equation 6.24 and calculate KD;
- Obtain the values of u_p and $W(u_p,r_r)$ from Annex 6.4;
- Substitute the values of Q, KD, and $W(u_p,r_r)$ into Equation 6.23 and calculate s_p;
- Knowing s_p, locate the inflection point on the curve and read t_p;
- Substitute the values of KD, t_p, u_p, and r into Equation 6.21 and calculate S;
- Using Equation 6.20, calculate z;
- Apply this procedure to the data from all available piezometers. The calculated values of KD and S should show a close agreement.

Remarks
- To check whether any errors have been made in the approximation of s_m and Δs_p,

the theoretical time-drawdown curve should be calculated with Equations 6.6 and 6.10, Annex 6.2, and the calculated values of r_r, KD, and S. This theoretical curve should show a close agreement with the observed time-drawdown curve. If not, the procedure should be repeated with corrected approximations of s_m and Δs_p.
- Procedure 6.3 can be applied to analyze data from unconfined aquifers when Assumption 7 (Chapter 3) is met.

6.3 Bounded leaky or confined aquifers, unsteady-state flow

6.3.1 Vandenberg's method (strip aquifer)

Leaky aquifers bounded laterally by two parallel barrier boundaries form an 'infinite strip aquifer', or a 'parallel channel aquifer'. In the analysis of such aquifers, we have to consider not only boundary effects, but also leakage effects. Vandenberg (1976; 1977) proposed a method by which the values of KD, S, and L of such aquifers can be determined.

If the distance, x, measured along the axis of the channel between the pumped well and the piezometer (Figure 6.9), is greater than the width of the channel, w, (i.e. x/w > 1), Vandenberg showed that for parallel unsteady-state flow the following drawdown function is applicable

$$s = \frac{Qx}{(2KDw)} F(u,x/L) \tag{6.26}$$

where

$$F(u,x/L) = \frac{1}{2\sqrt{\pi}} \int_u^\infty y^{-3/2} \exp(-y-x^2/4L^2y)dy \tag{6.27}$$

$$u = \frac{x^2S}{4KDt} \tag{6.28}$$

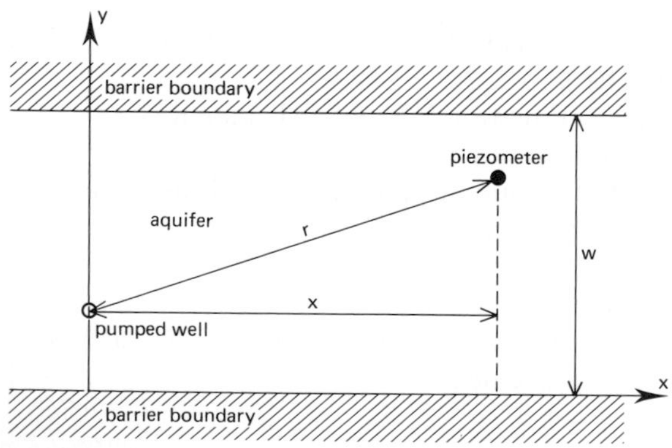

Figure 6.9 Plan view of a parallel channel aquifer

L = $\sqrt{KDc}$ = leakage factor in m (6.29)
x = projection of distance r in m between pumped well and piezometer, along the direction of the channel
w = width of the channel in m

Presented in Annex 6.5 are values of the function F(u,x/L) for different values of u and x/L, as given by Vandenberg (1976). These values can be plotted as a family of type curves (Figure 6.10).

The Vandenberg curve-fitting method can be used if the following assumptions and conditions are satisfied:

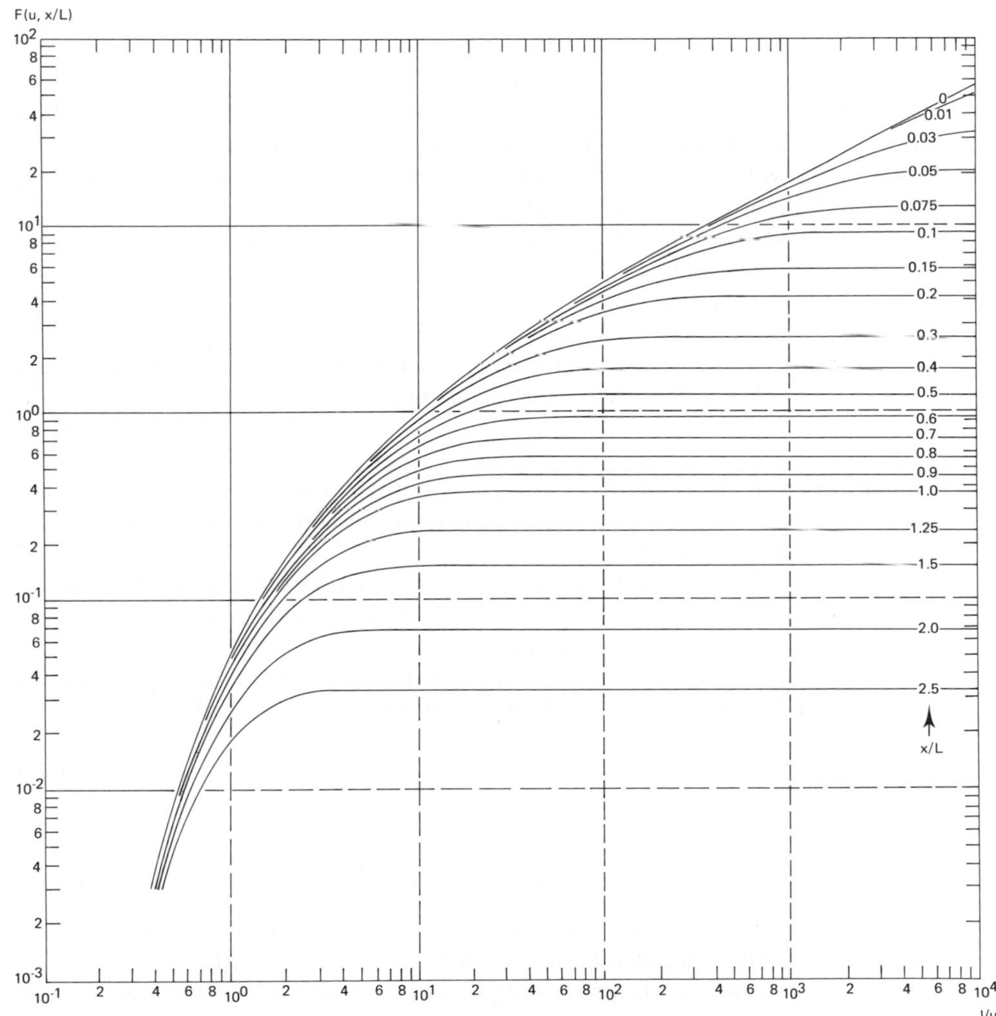

Figure 6.10 Family of Vandenberg's type curves F(u,x/L) versus 1/u for different values of x/L

- The assumptions listed at the beginning of Chapter 3, with the exception of the first and second assumptions, which are replaced by:
  - The aquifer is leaky;
  - Within the zone influenced by the pumping test, the aquifer is bounded by two straight parallel fully penetrating barrier boundaries.

The following conditions are added:
- The flow to the well is in unsteady state;
- The width and direction of the aquifer are both known with sufficient accuracy;
- $x/w > 1$.

Procedure 6.4
- Using Annex 6.5, construct on log-log paper a family of Vandenberg type curves by plotting $F(u,x/L)$ versus $1/u$ for a range of values of x/L;
- On another sheet of log-log paper of the same scale, plot s versus t for a single piezometer at a projected distance x from the pumped well;
- Match the observed data curve with one of the type curves;
- Select a match point on the superimposed sheets, and note for this point the values of $F(u,x/L)$, $1/u$, s, and t. Note also the value of x/L of the selected type curve;
- Substitute the values of $F(u,x/L)$ and s, together with the known values of Q, x, and w into Equation 6.26 and calculate KD;
- Substitute the values of u and t, together with the known values of KD and x, into Equation 6.28 and calculate S;
- Knowing x/L and x, calculate L;
- Calculate c from Equation 6.29;
- Repeat the procedure for all available piezometers ($x/w > 1$). The calculated values of KD, S, and c should show reasonable agreement.

Remarks
- If the direction of the channel is known, but not its width w, the same procedure as above can be followed, except that instead of calculating KD and S, the products KDw and Sw are calculated;
- If the direction of the channel is not known and the data from only one piezometer are available, the distance r may be used instead of x. For those cases where $r \gg w$, only a small error will be introduced;
- When $x/L = 0$, i.e. when $L \rightarrow \infty$, the drawdown function (Equation 6.26) becomes the drawdown function for parallel flow in a confined channel aquifer

$$s = \frac{Qx}{2KDw} F(u) \qquad (6.30)$$

  where

$$F(u) = \exp(-u/\sqrt{\pi u}) - \text{erfc}(\sqrt{u}) \qquad (6.31)$$

  With the type curve $F(u,x/L)$ versus $1/u$ for $x/L = 0$ (Annex 6.5), the values of KD and S of confined parallel channel aquifers can be determined;
- If $x/w < 1$, Equation 6.26 is not sufficiently accurate and the following drawdown equation for a system of real and image wells should be used (Vandenberg 1976; see also Bukhari et al. 1969)

$$s = \frac{Q}{4\pi KD}[W(u,r/L) + \sum_{i=1}^{\infty} W(u_i,r_i/L)] \qquad (6.32)$$

where W(u,r/L) is the function for radial flow towards a well in a leaky aquifer of infinite extent.

Type curves can be constructed from the exact solution of Equation 6.32. For each particular configuration of pumped well and piezometer, however, a different set of curves is required. Vandenberg (1976) provides 16 sets of type curves and gives a listing and user's guide for a Fortran program that will plot a set of type curves for any well/piezometer configuration.

7 Wedge-shaped and sloping aquifers

The standard methods of analysis are all based on the assumption that the thickness of the aquifer is constant over the area influenced by the pumping test. In wedge-shaped aquifers this assumption is not fullfilled and other methods of analysis should be used (Section 7.1). Standard methods also assume a horizontal watertable prior to a test. In some cases the watertable in unconfined aquifers is sloping and these methods cannot be used. Sections 7.2 and 7.3 present methods of analysis for unconfined aquifers with a sloping watertable.

7.1 Wedge-shaped confined aquifers, unsteady-state flow

7.1.1 Hantush's method

According to Hantush (1962), if the thickness of a confined aquifer varies exponentially in the flow direction (x-direction) while remaining constant in the y-direction (Figure 7.1), the drawdown equation for unsteady-state flow takes the form

$$s = \left[\frac{Q}{4\pi KD_w}\exp\left(\frac{r}{a}\cos\Theta\right)\right]W\left(u,\left|\frac{r}{a}\right|\right) \qquad (7.1)$$

where

D_w = thickness of the aquifer at the location of the well
Θ = the angle between the x-direction and a line through the well and a piezometer, in radians
a = constant defining the exponential variation of the aquifer thickness
$u = \frac{r^2S}{4KD_wt}$

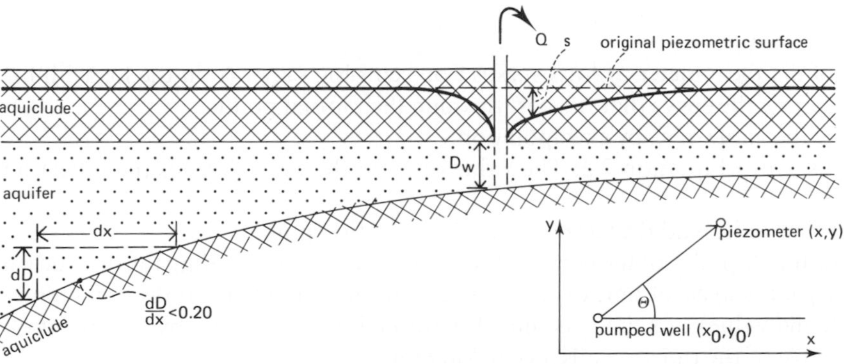

Figure 7.1 Cross-section and plan view of a pumped wedge-shaped confined aquifer

This equation has the same form as Equation 4.6, which describes the drawdown for unsteady state in a leaky aquifer of constant thickness. So, to determine the values of KD_w, S, and a of a wedge-shaped confined aquifer, we can use a method analogous to the Hantush inflection-point method for leaky aquifers of constant thickness (Procedure 4.4) (Hantush 1964).

At the inflection point P of the time-drawdown curve for a pumped confined aquifer of non-uniform thickness, Equations 4.8, 4.9, 4.10, and 4.12 become

$$s_p = \frac{1}{2} s_m = \left[\frac{Q}{4\pi KD_w} \exp\left(\frac{r}{a}\cos\Theta\right)\right] K_0\left(\left|\frac{r}{a}\right|\right) \quad (7.2)$$

$$u_p = \frac{r^2 S}{4KD_w t_p} = \frac{r}{2a} \quad (7.3)$$

The slope of the curve at the inflection point is

$$\Delta s_p = \left[\frac{2.30Q}{4\pi KD_w} \exp\left(\frac{r}{a}\cos\Theta\right)\right] e^{-r/a} \quad (7.4)$$

The relation between the drawdown and the slope of the curve is

$$2.30 \frac{s_p}{\Delta s_p} = e^{r/a} K_0\left(\left|\frac{r}{a}\right|\right) \quad (7.5)$$

Hantush's inflection-point method (Procedure 4.4) can be applied if the following assumptions and conditions are fulfilled:

- The assumptions listed at the beginning of Chapter 3, with the exception of the third assumption, which is replaced by:
  - The aquifer is homogeneous and isotropic over the area influenced by the pumping test;
  - The thickness of the aquifer varies exponentially in the direction of flow;
  - $\frac{dD}{dx} < 0.20$, i.e. $t < \frac{r_o^2 S}{20KD_w}$ with $r_o = \frac{a}{2}\ln\left(\frac{a}{10D_w}\right)$.

The following condition is added:

- The flow to the well is in an unsteady state, but the steady-state drawdown should be approximately known.

Procedure 7.1

- For one of the piezometers, plot s versus t on semi-log paper (t on the logarithmic scale) and draw the curve that fits best through the plotted points;
- Determine the value of s_m by extrapolation;
- Calculate s_p from Equation 7.2. The value of s_p on the curve locates the inflection point P;
- From the time axis, read the value of t_p at the inflection point;
- Determine the slope Δs_p of the curve at the inflection point by reading the drawdown difference per log cycle of time over the tangent to the curve at the inflection point;
- Substitute the values of s_p and Δs_p into Equation 7.5 and find r/a by interpolation from the table of the function $e^x K_o(x)$ in Annex 4.1;
- Knowing r/a and r, calculate a;

- Knowing Q, s_p, Δs_p, r/a, and cos θ, and using Annex 4.1, calculate KD_w from Equation 7.4 or Equation 7.2;
- Knowing KD_w, t_p, r, and r/a, calculate S from Equation 7.3.

Remarks
- To check whether the time condition is fulfilled, calculate the value of $(r_o^2S)/20KD_w$;
- If the well and all the piezometers are located on a single straight line, i.e. θ is the same for all piezometers, we can use a method analogous to the Hantush inflection-point method for leaky aquifers (Procedure 4.5).

7.2 Sloping unconfined aquifers, steady-state flow

7.2.1 Culmination-point method

If an unconfined aquifer with a constant saturated thickness slopes uniformly in the direction of flow (x-axis) (Figure 7.2), the slope of the watertable i is equal to the slope of the impermeable base α and the flow rate per unit width is

$$q = \frac{Q}{F} = KD\alpha \tag{7.6}$$

or

$$\alpha = \frac{q}{KD}$$

When such an aquifer is pumped at a constant discharge Q, the slope of the cone of depression along the x-axis downstream of the well is given for steady-state flow as

$$-\frac{dh}{dx} = \frac{Q}{2\pi rKD} \tag{7.7}$$

On the x-axis, there is a point where the slopes α and dh/dx are numerically the same but have opposite signs; hence the combined slope is zero. In this culmination point of the depression cone, which lies on the x-axis, the distance to the well r is designated by x_o. Consequently, a combination of Equations 7.6 and 7.7 (Huisman 1972) yields

$$\alpha = \frac{Q}{2\pi KDx_o} \tag{7.8}$$

The width of the zone from which the water is derived is $F = 2\pi x_o$.

The transmissivity can be calculated if the following assumptions and conditions are satisfied:
- The assumptions listed at the beginning of Chapter 3, with the exception of the first and fourth assumptions, which are replaced by:
  - The aquifer is unconfined;
  - Prior to pumping, the watertable slopes in the direction of flow.

The following condition is added:
- The flow to the well is in steady state.

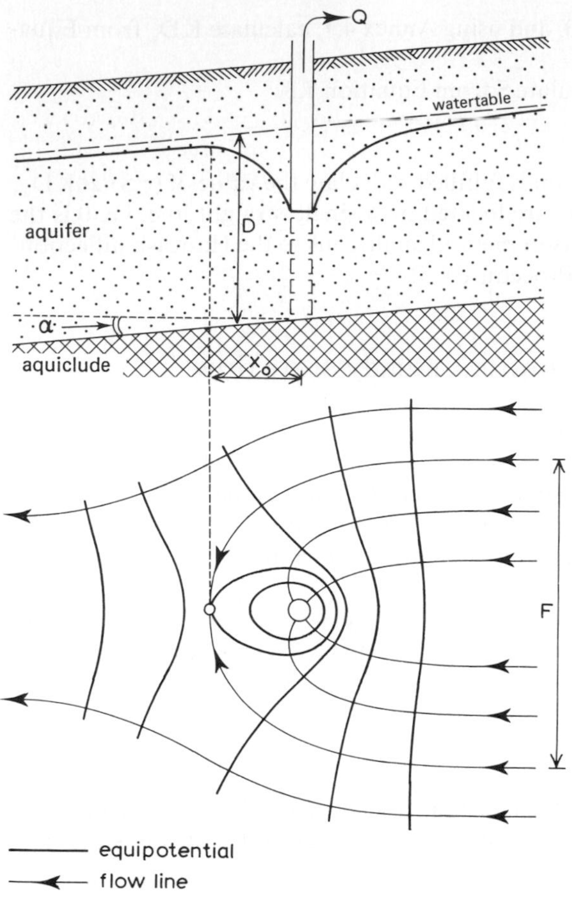

Figure 7.2 Cross-section and plan view of a pumped sloping unconfined aquifer

Procedure 7.2

- Instead of plotting the drawdown, plot the water-level elevations with reference to a horizontal datum plane versus r on arithmetic paper;
- Determine the distance x_o from the well to the point where the slope of the depression cone is zero;
- Introduce the values of Q, α, and x_o into Equation 7.8 and calculate KD.

7.3 Sloping unconfined aquifers, unsteady-state flow

7.3.1 Hantush's method

According to Hantush (1964), the unsteady-state drawdown in a sloping unconfined aquifer of constant thickness (Figure 7.2) is

$$s' = s - \frac{s^2}{2D} = \left\{\frac{Q}{4\pi KD} \exp\left(-\frac{r}{\gamma}\cos\theta\right)\right\} W\left(u,\frac{r}{\gamma}\right) \quad (7.9)$$

where

s' = corrected drawdown

s = observed drawdown

θ = the angle between the line through the well and a piezometer, and the direction of flow, in radians

$\gamma = \frac{2D}{i}$

$u = \frac{r^2S}{4KDt}$

i = slope of the watertable

This equation has the same form as Equation 4.6, which describes the drawdown for unsteady state in a leaky horizontal aquifer of constant thickness.
According to Hantush (1964), Equation 7.9 can be written alternatively as

$$s - \frac{s^2}{2D} = \left[\frac{Q}{4\pi KD} \exp\left(-\frac{r}{\gamma}\cos\theta\right)\right]\left[2K_o\left(\frac{r}{\gamma}\right) - W\left(q,\frac{r}{\gamma}\right)\right] \quad (7.10)$$

where

$$q = \frac{r^2}{4\gamma^2}\frac{1}{u} = \frac{KDt}{S\gamma^2} \quad (7.11)$$

If $q > 2\frac{r}{\gamma}$, Equation 7.10 can be approximated by

$$s'_m - s' = \frac{Q}{4\pi KD} \exp\left(-\frac{r}{\gamma}\cos\theta\right) W(q) \quad (7.12)$$

where

$$s'_m = s_m - \frac{s_m^2}{2D} = \frac{Q}{2\pi KD} \exp\left(-\frac{r}{\gamma}\cos\theta\right) K_o\left(\frac{r}{\gamma}\right) \quad (7.13)$$

s'_m = corrected maximum or steady-state drawdown

If s'_m in a piezometer at distance r from the well can be extrapolated from a plot of s′ versus t on semi-log paper (t on logarithmic scale), the drawdown at the inflection point P can be calculated ($s'_p = 0.5\, s'_m$) and t_p (the time corresponding to s'_p) can be read from the graph.

If a sufficient number of data fall within the period $t > 4t_p$, the Hantush method can be used, provided that the following assumptions and conditions are also satisfied:

- The assumptions listed at the beginning of Chapter 3, with the exception of the first and fourth assumptions, which are replaced by:
  - The aquifer is unconfined;
  - Prior to pumping, the watertable slopes in the direction of flow with a hydraulic gradient $i < 0.20$.

The following conditions are added:

- The flow to the well is in unsteady state;

$$q > 2\frac{r}{\gamma}$$

$$t > 4t_p.$$

Procedure 7.3

- For one of the piezometers, plot s' versus t on semi-log paper (t on logarithmic scale) and find the maximum drawdown s'_m by extrapolation;
- Using Annex 3.1, prepare a type curve by plotting W(q) versus q on log-log paper. This curve is identical with a plot of W(u) versus u;
- On another sheet of log-log paper of the same scale, plot the observed data curve $(s'_m - s')$ versus t. Obviously, one can only use the data of one piezometer at a time because, although q is independent of r, this is not so with $(Q/4\pi KD)$ exp $\left[-\left(\frac{r}{\gamma}\right)\cos\theta\right]$;
- Match the observed data curve with the type curve. It will be seen that the observed data in the period $t < 4t_p$ fall below the type curve because, in this period, Equation 7.12 does not apply;
- Choose a match point A on the superimposed sheets and note for A the values of $(s'_m - s')$, t, q, and W(q);
- Substitute the values of $(s'_m - s')$ and W(q) into Equation 7.12 and calculate $(Q/4\pi KD)\exp\left[-\left(\frac{r}{\gamma}\right)\cos\theta\right]$;
- Multiply this value by 2, which gives $\frac{Q}{2\pi KD}\exp\left[-\left(\frac{r}{\gamma}\right)\cos\theta\right]$. Substitute this value and that of s'_m into Equation 7.13, which gives a value of $K_o\left(\frac{r}{\gamma}\right)$. The value of $\frac{r}{\gamma}$ can be found from Annex 4.1 and, because r is known, γ can be calculated. With the values of $\frac{r}{\gamma}$ and θ known, $\left[-\left(\frac{r}{\gamma}\right)\cos\theta\right]$ can be found, and $\exp\left[-\left(\frac{r}{\gamma}\right)\cos\theta\right]$ can be obtained from Annex 4.1;
- Substitute the values of $\exp\left[-\left(\frac{r}{\gamma}\right)\cos\theta\right]$, Q, and D into $\frac{Q}{2\pi KD}\exp\left[-\left(\frac{r}{\gamma}\right)\cos\theta\right]$ and calculate K;
- Substitute the values t and q of point A and those of KD and γ into Equation 7.11 and calculate S;
- Repeat this procedure for all available piezometers.

Remarks

- When delayed watertable response phenomena are apparent (Chapter 5), the condition 'The water removed from storage is discharged instantaneously with decline of head' is not met and this Hantush method is not applicable;
- Because of the analogy between Equations 4.6 and 7.9, we can also use a method analogous to the Hantush method for horizontal leaky aquifers of constant thick-

ness (Procedure 4.4). If the well and all the piezometers are located on a single straight line, i.e. θ is the same for all piezometers, we can use a method analogous to the Hantush method for leaky aquifers (Procedure 4.5).

8 Anisotropic aquifers

The standard methods of analysis are all based on the assumption that the aquifer is isotropic, i.e. that the hydraulic conductivity is the same in all directions. Many aquifers, however, are anisotropic. In such aquifers, it is not unusual to find hydraulic conductivities that differ by a factor of between two and twenty when measured in one or another direction. Anisotropy is a common feature in water-laid sedimentary deposits (e.g. fluvial, clastic lake, deltaic and glacial outwash deposits). Aquifers that are composed of water-laid deposits may exhibit anisotropy on the horizontal plane. The hydraulic conductivity in the direction of flow tends to be greater than that perpendicular to flow. Because of the differences in hydraulic conductivity, lines of equal drawdown around a pumped well in these aquifers will form ellipses rather than concentric circles.

In addition such aquifers are often stratified, i.e. they are made up of alternating layers of coarse and fine sands, gravels, and occasional clays, with each layer possessing a unique value of K. Any layer with a low K will retard vertical flow, but horizontal flow can occur easily through any layer with relatively high K. Obviously, K_h, i.e. parallel to the bedding planes, will be much higher than K_v, and the aquifer is said to be anisotropic on the vertical plane.

Aquifers that are anisotropic on both the horizontal and vertical planes, are said to exhibit three-dimensional anisotropy, with principal axes of K in the vertical direction, the horizontal direction parallel to stream flows that prevailed in the past, and the horizontal direction at a right angle to those flows.

It will be clear that, in the analysis of pumping tests, anisotropy poses a special problem. Methods of analysis that take anisotropy on the horizontal plane into account are presented in Section 8.1 for confined aquifers and in Section 8.2 for leaky aquifers. Sections 8.3, 8.4 and 8.5 discuss anisotropy on the vertical plane in confined aquifers, leaky aquifers, and unconfined aquifers.

8.1 Confined aquifers, anisotropic on the horizontal plane

8.1.1 Hantush's method

The unsteady-state drawdown in a confined isotropic aquifer is given by the Theis equation (Equation 3.5)

$$s = \frac{Q}{4\pi KD} W(u)$$

where

$$u = \frac{r^2 S}{4KDt}$$

In a confined aquifer that is anisotropic on the horizontal plane, with the principal axes of anisotropy X and Y, the above equations, according to Hantush (1966), are replaced by

$$s = \frac{Q}{4\pi(KD)_e} W(u_{XY}) \tag{8.1}$$

where

$$u_{XY} = \frac{r^2 S}{4t(KD)_n} \tag{8.2}$$

$$(KD)_e = \sqrt{(KD)_X \times (KD)_Y} = \text{the effective transmissivity} \tag{8.3}$$

$(KD)_X$ = transmissivity in the major direction of anisotropy

$(KD)_Y$ = transmissivity in the minor direction of anisotropy

$(KD)_n$ = transmissivity in a direction that makes an angle $(\theta + \alpha)$ with the X axis (θ and α will be defined below)

If we have one or more piezometers on a ray that forms an angle $(\theta + \alpha)$ with the X axis, we can apply the methods for isotropic aquifers and obtain values for $(KD)_e$ and $S/(KD)_n$. Consequently, to calculate S and $(KD)_n$, we need data from more than one ray of piezometers.

Hantush (1966) showed that if θ is defined as the angle between the first ray of piezometers (n = 1) and the X axis and α_n as the angle between the nth ray of piezometers and the first ray of piezometers (Figures 8.1A and B), $(KD)_n$ is given by

$$(KD)_n = \frac{(KD)_X}{\cos^2(\theta + \alpha_n) + m \sin^2(\Theta + \alpha_n)} \tag{8.4}$$

where

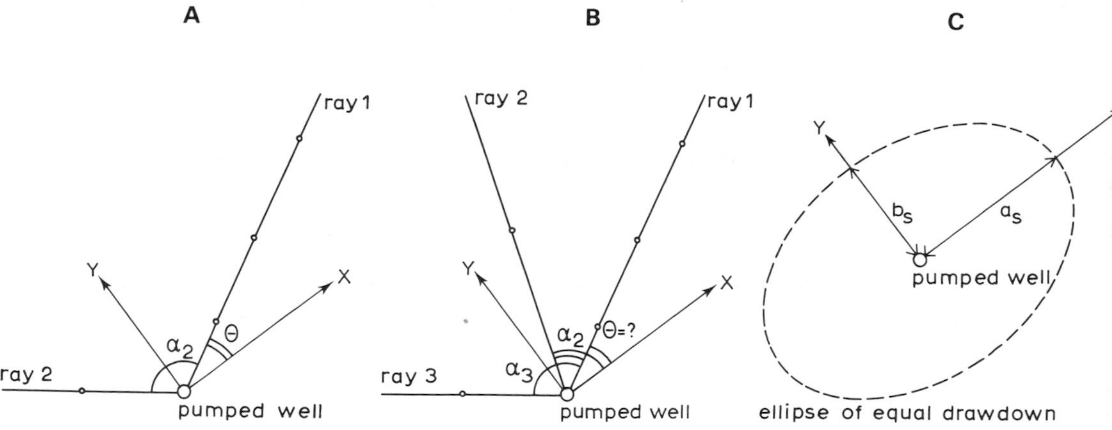

Figure 8.1 The parameters in the Hantush and the Hantush-Thomas methods for aquifers with anisotropy on the horizontal plane:
A. Principal directions of anisotropy known
B. Principal directions of anisotropy not known
C. Ellipse of equal drawdown

$$m = \frac{(KD)_X}{(KD)_Y} = \left[\frac{(KD)_e}{(KD)_Y}\right]^2 \qquad (8.5)$$

Because $\alpha_1 = 0$ for the first ray of piezometers, Equation 8.4 reduces to

$$(KD)_1 = \frac{(KD)_X}{\cos^2 \Theta + m \sin^2 \Theta} \qquad (8.6)$$

and consequently

$$a_n = \frac{(KD)_1}{(KD)_n} = \frac{\cos^2(\theta + \alpha_n) + m \sin^2(\theta + \alpha_n)}{\cos^2 \theta + m \sin^2 \theta} \qquad (8.7)$$

It goes without saying that $a_1 = 1$.
A combination of Equations 8.5 and 8.7 yields

$$m = \left[\frac{(KD)_e}{(KD)_Y}\right]^2 = \frac{a_n \cos^2 \theta - \cos^2 (\theta + \alpha_n)}{\sin^2 (\theta + \alpha_n) - a_n \sin^2 \theta} \qquad (8.8)$$

If the principal directions of anisotropy are not known, one needs at least three piezometers on different rays from the pumped well to solve Equation 8.7 for θ, using

$$\tan (2\,\theta) = -2\frac{(a_3 - 1)\sin^2\alpha_2 - (a_2 - 1)\sin^2\alpha_3}{(a_3 - 1)\sin 2\alpha_2 - (a_2 - 1)\sin 2\alpha_3} \qquad (8.9)$$

Equation 8.9 has two roots for the angle $(2\,\theta)$ in the range 0 to 2π of the XY plane. If one of the roots is δ, the other will be $\pi + \delta$. Consequently, θ has two values: $\delta/2$ and $(\pi + \delta)/2$. One of the values of θ yields $m > 1$ and the other $m < 1$. Since the X axis is assumed to be along the major axis of anisotropy, the value of θ that will make $m = (KD)_X/(KD)_Y > 1$ locates the major axis of anisotropy, X; the other value locates the minor axis of anisotropy, Y. (It should be noted that a negative value of θ indicates that the positive X axis lies to the left of the first ray of piezometers.)
The Hantush method can be applied if the following assumptions and conditions are satisfied:

- The assumptions listed at the beginning of Chapter 3, with the exception of the third assumption, which is replaced by:
  - The aquifer is homogeneous, anisotropic on the horizontal plane, and of uniform thickness over the area influenced by the pumping test.

The following conditions are added:

- The flow to the well is in unsteady state;
- If the principal directions of anisotropy are known, drawdown data from two piezometers on different rays from the pumped well will be sufficient. If the principal directions of anisotropy are not known, drawdown data must be available from at least three rays of piezometers.

Procedure 8.1 (principal directions of anisotropy known)

- Apply the methods for isotropic confined aquifers (Sections 3.2.1 and 3.2.2) to the data of each of the two rays of piezometers. This results in values for $(KD)_e$, $S/(KD)_1$, and $S/(KD)_2$;
- A combination of the last two values gives a_2 (cf. Equation 8.7). Because θ and α_2 are known, substitute the values of θ, α, a, and $(KD)_e$ into Equation 8.8 and calculate m;

- Knowing $(KD)_e$ and m, calculate $(KD)_X$ and $(KD)_Y$ from Equation 8.5;
- Substitute the values of $(KD)_X$, m, θ, and α_2 into Equations 8.6 and 8.7 and solve for $(KD)_1$ and $(KD)_2$;
- A combination of the last two values with those for $S/(KD)_1$ and $S/(KD)_2$, respectively, yields values for S, which should be essentially the same.

Procedure 8.2 (principal directions of anisotropy unknown)

- Apply the methods for isotropic confined aquifers (Sections 3.2.1 and 3.2.2) to the data from each of the three rays of piezometers. This results in values for $(KD)_e$, $S/(KD)_1$, $S/(KD)_2$, and $S/(KD)_3$;
- A combination of $S/(KD)_1$ with $S/(KD)_2$ and $S/(KD)_3$, respectively, yields values for a_2 and a_3. Because α_2 and α_3 are known, θ can be calculated from Equation 8.9;
- Substitute the values of θ, $(KD)_e$, α_2, and a_2 (or α_3 and a_3) into Equation 8.8 and calculate m;
- Knowing $(KD)_e$ and m, calculate $(KD)_X$ and $(KD)_Y$ from Equation 8.5;
- Substitute the values of $(KD)_X$, m, and θ and the values of $\alpha_1 = 0$, α_2, and α_3 into Equation 8.4 and solve for $(KD)_1$, $(KD)_2$, and $(KD)_3$;
- A combination of these values with those of $S/(KD)_1$, $S/(KD)_2$, and $S/(KD)_3$, respectively, yields values for S, which should be essentially the same.

Remarks

- The observed data should permit the use of those methods for isotropic confined aquifers that give a value for $S/(KD)_n$. Hence, the methods for steady-state flow in isotropic confined aquifers (Section 3.1) are not applicable;
- The analysis of the data from each ray of piezometers yields a value of $(KD)_e$. These values should all be essentially the same.

Example 8.1

Using Procedure 8.2, we shall analyse the drawdown data presented by Papadopulos (1965). The data are from a pumping test conducted in an anisotropic confined aquifer. During the test, the well PW was pumped at a discharge rate of 1086 m³/d. The drawdown was observed in three observation wells OW-1, OW-2, and OW-3, located as shown in Figure 8.2.

For each observation well, we plot the drawdown data on semi-log paper (Figure 8.3). The data allow the application of Jacob's straight line method (Chapter 3) to determine the values of $(KD)_e$ and $S/(KD)_1$, $S/(KD)_2$, and $S/(KD)_3$

$$(KD)_e = \frac{2.30Q}{4\pi\Delta s} = \frac{2.30 \times 1086}{4 \times 3.14 \times 1.15} = 173 \text{ m}^2/\text{d}$$

$$\frac{S}{(KD)_1} = \frac{2.25\ t_{01}}{r^2} = \frac{2.25 \times 0.37}{(28.3)^2 \times 1440} = 7.22 \times 10^{-7} \text{ d/m}^2$$

$$\frac{S}{(KD)_2} = \frac{2.25\ t_{02}}{r^2} = \frac{2.25 \times 0.72}{(9^2 + 33.5^2) \times 1440} = 9.35 \times 10^{-7} \text{ d/m}^2$$

$$\frac{S}{(KD)_3} = \frac{2.25\ t_{03}}{r^2} = \frac{2.25 \times 0.24}{(19.3^2 + 5.2^2) \times 1440} = 9.39 \times 10^{-7} \text{ d/m}^2$$

Subsequently, we calculate the values of a_2 and a_3: $a_2 = 1.295$ and $a_3 = 1.300$. The value of Θ can now be derived from Equation 8.9

$$\tan(2\Theta) = -2\left\{\frac{(1.300-1)\sin^2 75^\circ - (1.295-1)\sin^2 196^\circ}{1.300\sin(2\times 75^\circ) - (1.295-1)\sin(2\times 196^\circ)}\right\} = 82$$

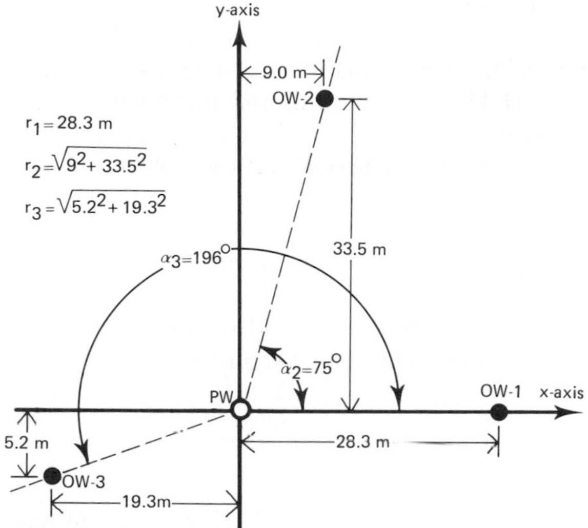

Figure 8.2 Location of the pumped well and observation wells (Papadopulos pumping test, Example 8.1)

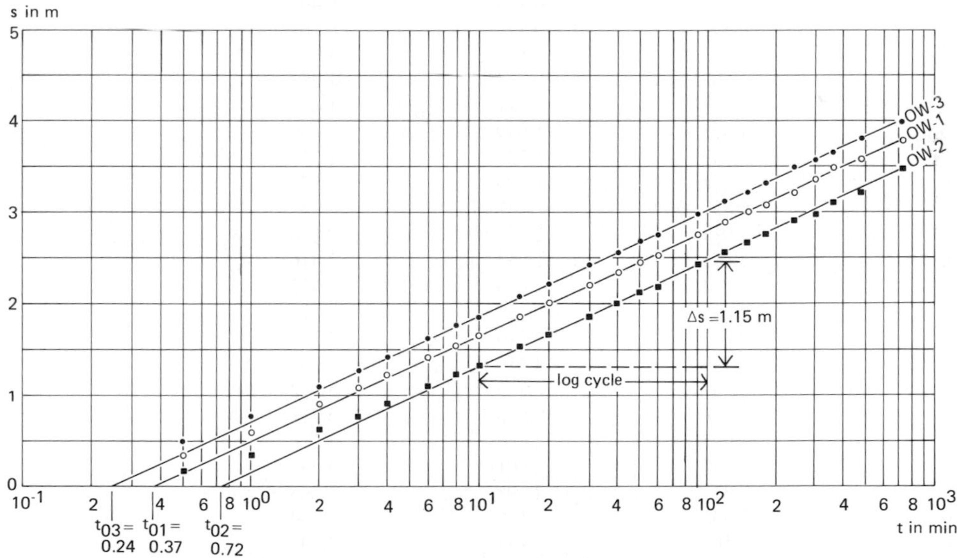

Figure 8.3 Analysis of data from the Papadopulos pumping test with the Jacob method

The two possible values of Θ are 45° and 135°.
Using $\Theta = 45°$, and subsequently $\Theta = 135°$, and the appropriate values of $(KD)_e$, α_3, and a_3 in Equation 8.8 gives the following values for m

$$\text{for } \Theta = 45°: m = \frac{1.3 \cos^2 45° - \cos^2(45° + 196°)}{\sin^2(45° + 196°) - 1.3 \sin^2 45°} = 3.6 \text{ (i.e. } m > 1)$$

$$\text{for } \Theta = 135°: m = 0.2771 \text{ (i.e. } m < 1)$$

We use $m = 3.6$ to solve $(KD)_X$ and $(KD)_Y$ from Equation 8.5. The transmissivity in the major direction of anisotropy is $(KD)_X = 328$ m²/d, and that in the minor direction of anisotropy is $(KD)_Y = 91$ m²/d.
We determine the transmissivity in the direction of each observation well from Equation 8.4

$$(KD)_1 = \frac{328}{\{\cos^2(45° + 0°) + 3.6 \sin^2(45° + 0°)\}} = 143 \text{ m}^2/\text{d}$$

and calculate in the same way $(KD)_2 = 111$ m²/d and $(KD)_3 = 110$ m²/d.
Finally, we calculate the storativity of the anisotropic confined aquifer.

$$\frac{S}{(KD)_1} = \frac{S}{143} = 7.22 \times 10^{-7}.$$

Solved for S, the equation yields $S = 1 \times 10^{-4}$.

Table 8.1 Drawdown data from the Papadopulos pumping test (from Papadopulos 1965)

Time t since pumping started (minutes)	Drawdown s (metres)		
	OW-1	OW-2	OW-3
0.5	0.335	0.153	0.492
1	0.591	0.343	0.762
2	0.911	0.611	1.089
3	1.082	0.762	1.284
4	1.215	0.911	1.419
6	1.405	1.089	1.609
8	1.549	1.225	1.757
10	1.653	1.329	1.853
15	1.853	1.531	2.071
20	2.019	1.677	2.210
30	2.203	1.853	2.416
40	2.344	2.019	2.555
50	2.450	2.123	2.670
60	2.541	2.210	2.750
90	2.750	2.416	2.963
120	2.901	2.555	3.118
150	2.998	2.670	3.218
180	3.075	2.750	3.310
240	3.235	2.901	3.455
300	3.351	2.998	3.565
360	3.438	3.118	3.649
480	3.587	3.247	3.802
720	3.784	3.455	3.996

8.1.2 Hantush-Thomas's method

In an isotropic aquifer, the lines of equal drawdown around a pumped well form concentric circles, whereas in an aquifer that is anisotropic on the horizontal plane, those lines form ellipses, which satisfy the equation

$$\frac{x^2}{a_s^2} + \frac{y^2}{b_s^2} = 1 \qquad (8.10)$$

where a_s and b_s are the lengths of the principal axes of the ellipse of equal drawdown s at the time t_s (Figure 8.1C).
It can be shown that

$$(KD)_n = (r_n^2/a_s b_s)(KD)_e \qquad (8.11)$$

$$(KD)_X = (a_s/b_s)(KD)_e \qquad (8.12)$$

$$(KD)_Y = (b_s/a_s)(KD)_e \qquad (8.13)$$

$$\frac{4\pi s(KD)_e}{Q} = W(u_{XY}) \qquad (8.14)$$

where

$$u_{XY} = \frac{r_n^2 S}{4(KD)_n t} = \frac{a_s b_s S}{4(KD)_e t_s} \qquad (8.15)$$

Hantush and Thomas (1966) stated that when $(KD)_e$, a_s, and b_s are known the other hydraulic characteristics can be calculated. Hence, it is not necessary to have values of $S/(KD)_n$, provided that one has sufficient observations to draw the ellipses of equal drawdown.

The Hantush-Thomas method can be applied if the following assumptions and conditions are satisfied:
- The assumptions listed at the beginning of Chapter 3, with the exception of the third assumption, which is replaced by:
  - The aquifer is homogeneous, anisotropic on the horizontal plane, and of uniform thickness over the area influenced by the pumping test.

The following condition is added:
- The flow to the well is in unsteady state.

Procedure 8.3
- Apply the methods for isotropic confined aquifers (Sections 3.1 and 3.2) to the data from each ray of piezometers; this yields values for $(KD)_e$ and sometimes $S/(KD)_n$. The factor $(KD)_e$ is constant for the whole flow system, and $S/(KD)_n$ is constant along each ray;
- Substitute the values of $(KD)_e$ and $S/(KD)_n$ into Equations 8.1 and 8.2 and calculate the drawdown at any desired time and at any distance along each ray of piezometers;
- Construct one or more ellipses of equal drawdown (Figure 8.1C), using observed (or calculated) data, and calculate for each ellipse a_s and b_s;
- Calculate $(KD)_n$, $(KD)_X$, and $(KD)_Y$ from Equations 8.11 to 8.13;

- Calculate the value of $W(u_{XY})$ from Equation 8.14 and find the corresponding value of u_{XY} from Annex 3.1;
  With the value of u_{XY} known, calculate S from Equation 8.15;
- Repeat this procedure for several values of s. This should produce approximately the same values for $(KD)_n$, $(KD)_X$, $(KD)_Y$, and S.

8.1.3 Neuman's extension of the Papadopulos method

In aquifers that are anisotropic on the horizontal plane, the orientation of the hydraulic-head gradients and the flow velocity seldom coincide; the flow tends to follow the direction of the highest permeability. This leads us to regard the hydraulic conductivity as a tensorial property, which is simply the mathematical translation of our observation of the non-coincidence. Regarding the hydraulic conductivity in this way, we must define the tensor K, which is a matrix of nine coefficients, symmetrical to the diagonal. This allows us to transform the components of the hydraulic gradient into components of velocity. Along the principal axes of such a tensor (X,Y), the velocity and hydraulic gradients have the same directions.

By making use of the tensor properties, Papadopulos (1965) developed an equation for the unsteady-state drawdown induced in a confined aquifer that is anisotropic on the horizontal plane

$$s = \frac{Q}{4\pi(KD)_e} W(u_{xy}) \tag{8.16}$$

where

$$(KD)_e = \sqrt{(KD)_{xx}(KD)_{yy} - (KD)^2_{xy}}$$

$$u_{xy} = \frac{S}{4t}\left(\frac{(KD)_{xx}y^2 + (KD)_{yy}x^2 - 2(KD)_{xy}xy}{(KD)_{xx}(KD)_{yy} - (KD)^2_{xy}}\right)$$

$$= \frac{S}{4t}\left(\frac{(KD)_{xx}y^2 + (KD)_{yy}x^2 - 2(KD)_{xy}xy}{(KD)^2_e}\right) \tag{8.17}$$

where x and y are local coordinates (Figure 8.4) and $(KD)_{xx}$, $(KD)_{yy}$, and $(KD)_{xy}$ are components of the transmissivity tensor.

For u < 0.01, Equation 8.16 reduces to

$$s = \frac{2.30Q}{4\pi(KD)_e} \log \frac{2.25\,t}{S}\left\{\frac{(KD)_{xx}(KD)_{yy} - (KD)^2_{xy}}{(KD)_{xx}y^2 + (KD)_{xx}x^2 - 2(KD)_{xy}xy}\right\} \tag{8.18}$$

The following relations between the principal transmissivity and the transmissivity tensors hold

$$(KD)_X = \frac{1}{2}\{(KD)_{xx} + (KD)_{yy} + \sqrt{[(KD)_{xx} - (KD)_{yy}]^2 + 4(KD)^2_{xy}}\} \tag{8.19}$$

$$(KD)_Y = \frac{1}{2}\{(KD)_{xx} + (KD)_{yy} - \sqrt{[(KD)_{xx} - (KD)_{yy}]^2 + 4(KD)^2_{xy}}\} \tag{8.20}$$

where X and Y are global coordinates of the transmissivity tensor (Figure 8.4).

The X axis is parallel to the major direction of anisotropy; the Y axis is parallel

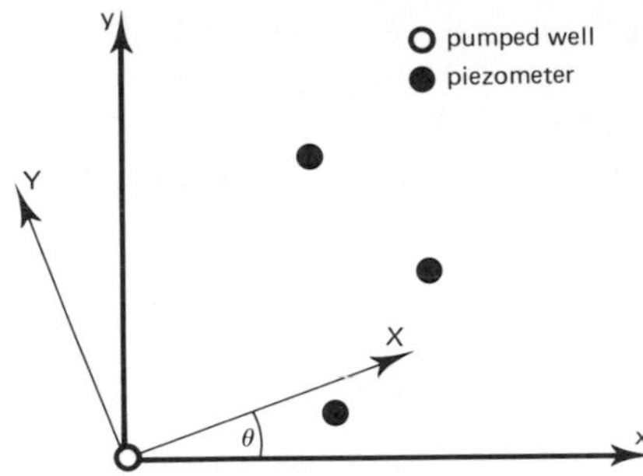

Figure 8.4 Relationship between the global coordinates (X and Y) and the local coordinates (x and y)

to the minor direction. The orientation of the X and Y axes is given by

$$\Theta = \arctan\frac{(KD)_X - (KD)_{xx}}{(KD)_{xy}} \tag{8.21}$$

where Θ is the angle between the x and the X axis ($0 \leq Z\ \Theta < \pi$). The angle of Θ is positive to the left of the axis.

If the principal directions of anisotropy are known, Equations 8.16 and 8.17 reduce to

$$s = \frac{Q}{4\pi\sqrt{(KD)_X(KD)_Y}} W(u_{XY}) \tag{8.22}$$

$$u_{XY} = \frac{S}{4t}\left(\frac{(KD)_X Y^2 + (KD)_Y X^2}{(KD)_X(KD)_Y}\right) \tag{8.23}$$

Taking the above equations as his basis, Papadopulos (1965) developed a method of determining the principal directions of anisotropy and the corresponding minimum and maximum transmissivities. This method requires drawdown data from at least three wells, other than the pumped well, all three located on different rays from the pumped well.

Neuman et al. (1984) showed that the Papadopulos method can be used with drawdown data from only three wells, provided that two pumping tests are conducted in sequence in two of those wells. When water is pumped from Well 1 at a constant rate Q_1, two sets of drawdown data, s_{12} and s_{13}, are available from Wells 2 and 3 (Figure 8.5). This is not sufficient to allow the use of the Papadopulos equations. But, if at least one other pumping test is conducted, say in Well 2, at a constant rate Q_2, and the resulting drawdown is observed at least in Well 3, these drawdown data, s_{23}, provide the third set of data needed to complete the analysis. Equation 8.17 as used in the Papadopulos method can now be replaced by

$$u_{12} = \frac{S}{4t_{12}(KD)_e^2}\left[(KD)_{xx}y_{12}^2 + (KD)_{yy}x_{12}^2 - 2(KD)_{xy}x_{12}y_{12}\right] \tag{8.24}$$

$$u_{13} = \frac{S}{4t_{13}(KD)_e^2}\left[(KD)_{xx}y_{13}^2 + (KD)_{yy}x_{13}^2 - 2(KD)_{xy}x_{13}y_{13}\right] \tag{8.25}$$

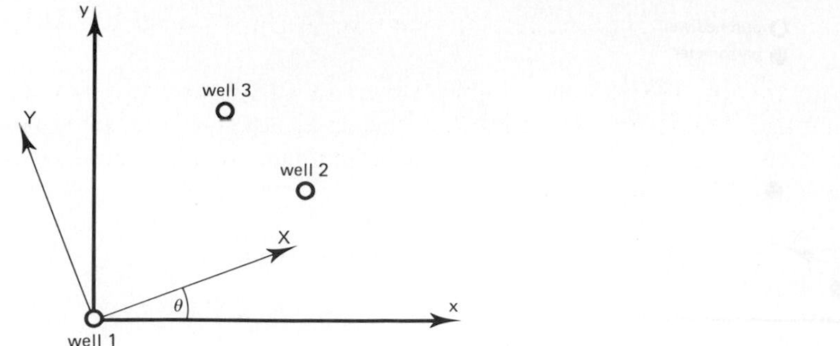

Figure 8.5 The three-well arrangement used in Neuman's extension of the Papadopulos method

$$u_{23} = \frac{S}{4t_{23}(KD)_e^2} [(KD)_{xx}y_{23}^2 + (KD)_{yy}x_{23}^2 - 2(KD)_{xy}x_{23}y_{23}] \quad (8.26)$$

Neuman's three-well method is applicable if the following assumptions and conditions are fulfilled:

- The assumptions listed at the beginning of Chapter 3, with the exception of the third assumption, which is replaced by:
  - The aquifer is homogeneous, anisotropic on the horizontal plane, and of uniform thickness over the area influenced by the pumping test.

The following conditions are added:

- The flow to the well is in an unsteady state;
- The aquifer is penetrated by three wells, which are not on one ray. Two of them are pumped in sequence.

Procedure 8.4

- Apply one of the methods for confined isotropic aquifers (Section 3.2) to the drawdown data from each well, using Equations 8.16, 8.24, 8.25, and 8.26. This results in values for $(KD)_e$, $S(KD)_{xx}$, $S(KD)_{yy}$, and $S(KD)_{xy}$;
- Knowing $(KD)_e$, $S(KD)_{xx}$, $S(KD)_{yy}$, and $S(KD)_{xy}$, calculate S from $S = \sqrt{S(KD)_{xx}S(KD)_{yy} - \{S(KD)_{xy}\}^2}/(KD)_e$
- Knowing S, $S(KD)_{xx}$, $S(KD)_{yy}$, and $S(KD)_{xy}$, calculate $(KD)_{xx}$, $(KD)_{yy}$, and $(KD)_{xy}$;
- Calculate $(KD)_X$ by substituting the known values of $(KD)_{xx}$, $(KD)_{yy}$, and $(KD)_{xy}$ into Equation 8.19;
- Calculate $(KD)_Y$ by substituting the known values of $(KD)_{xx}$, $(KD)_{yy}$, and $(KD)_{xy}$ into Equation 8.20;
- Determine the angle Θ by substituting the known values of $(KD)_X$, $(KD)_{xx}$, and $(KD)_{xy}$ into Equation 8.21.

Remarks

- The drawdown induced by the pumping test in Well 2 should be observed in Well 3 and not in the previously pumped Well 1, because s_{21} will be proportional to s_{12} under ideal conditions. Hence Equation 8.26 will not be linearly independent of

Equation 8.24 and no unique solutions can be found for the Equations 8.24, 8.25, and 8.26;

- According to Neuman et al. (1984), more reliable results can be obtained by conducting three pumping tests, pumping one well at a time and observing the drawdown in the other two wells. Equation 8.17 should then be replaced in the calculations by up to six equations of the form

$$u_{ij} = \frac{S}{4t_{ij}(KD)_e^2}\{(KD)_{xx}y_{ij}^2 + (KD)_{yy}x_{ij}^2 - 2(KD)_{xy}x_{ij}y_{ij}\}$$

  where i, j = 1, 2, 3.
  A least-squares procedure can be used to solve these equations and determine $S(KD)_{xx}$, $S(KD)_{yy}$, and $S(KD)_{xy}$. (For more information, see Neuman et al. 1984);
- If drawdown data are available from at least three piezometers or observation wells on different rays from the pumped well, the Papadopulos method can be used. The procedure is the same as Procedure 8.4, except that in the first step of Procedure 8.4, Equation 8.18 should be used instead of Equations 8.24, 8.25, and 8.26 to determine the values of $S(KD)_{xx}$, $S(KD)_{yy}$, and $S(KD)_{xy}$.

Example 8.2
We shall use the data from the Papadopulos pumping test (Example 8.1, Table 8.1, Figures 8.2 and 8.3) to illustrate the Papadopulos method, Procedure 8.4.

From Example 8.1 we know the value of the effective transmissivity: $(KD)_e = 173$ m²/d. Figure 8.3 shows the semi-log plot of the drawdown data for each observation well. The three straight lines through the plotted points intercept the t axis at $t_{01} = 0.37$ min., $t_{02} = 0.72$ min., and $t_{03} = 0.24$ min. These straight lines are described by Equation 8.18. For s = 0, Equation 8.18 reduces to

$$t_0 = \frac{S}{2.25}\left\{\frac{(KD)_{xx}y^2 + (KD)_{yy}x^2 - 2(KD)_{xy}xy}{(KD)_{xx}\,(KD)_{yy} - (KD)_{xy}^2}\right\}$$

$$= \frac{S}{2.25}\left\{\frac{(KD)_{xx}y^2 + (KD)_{yy}x^2 - 2(KD)_{xy}xy}{(KD)_e^2}\right\}$$

Hence, $2.25\,(KD)_e^2 \times t_0 = S(KD)_{xx}y^2 + S(KD)_{yy}x^2 - 2\,S(KD)_{xy}xy$.
Using this expression, we can determine $S(KD)_{xx}$, $S(KD)_{yy}$, and $S(KD)_{xy}$.
For observation well OW-1:

$$2.25 \times (KD)_e^2 \times t_{01} = 2.25 \times 173^2 \times \frac{0.37}{1440} = S(KD)_{xx} \times 0 + S(KD)_{yy} \times 28.3^2 - 2S(KD)_{xy} \times 0$$

For observation well OW-2:

$$2.25\,(KD)_e^2 \times t_{02} = 2.25 \times 173^2 \times \frac{0.72}{1440} = S(KD)_{xx} \times 33.5^2 + S(KD)_{yy} \times 9^2 - 2\,S(KD)_{xy} \times 33.5 \times 9$$

For observation well OW-3:

$$2.25\,(KD)_e^2 \times t_{03} = 2.25 \times 173^2 \times \frac{0.24}{1440}$$

$$= S(KD)_{xx} \times 5.2^2 + S(KD)_{yy} \times 19.3^2 - 2\,S(KD)_{xy} \times 19.3 \times 5.2$$

Solving these three equations gives

$$S(KD)_{xx} = 0.0215\ m^2/d$$
$$S(KD)_{yy} = 0.0216\ m^2/d$$
$$S(KD)_{xy} = -0.0219\ m^2/d$$

Substituting these values together with the value of $(KD)_e$ into

$$S = \frac{\sqrt{S(KD)_{xx}\,S(KD)_{yy} - \{S(KD)_{xy}\}^2}}{(KD)_e} \text{ yields } S = 1 \times 10^{-4}$$

The values of $(KD)_{xx}$, $(KD)_{yy}$, and $(KD)_{xy}$ can now be calculated

$$(KD)_{xx} = 215\ m^2/d$$
$$(KD)_{yy} = 216\ m^2/d$$
$$(KD)_{xy} = -129\ m^2/d$$

The transmissivity $(KD)_X$ in the principal direction of anisotropy is calculated from Equation 8.19

$$(KD)_X = \frac{1}{2}\{215 + 216 + \sqrt{(215-216)^2 + 4(-129)^2}\} = 345\ m^2/d$$

The transmissivity $(KD)_Y$ in the minor direction of anisotropy is calculated from Equation 8.20

$$(KD)_Y = \frac{1}{2}\{215 + 216 - \sqrt{(215-216)^2 + 4(-129)^2}\} = 86\ m^2/d$$

The orientation of the X and Y axes is determined from Equation 8.21

$$\Theta = \arctan\left\{\frac{(KD)_X - (KD)_{xx}}{(KD)_{xy}}\right\} = \arctan\left\{\frac{345-215}{-129}\right\} = \arctan(-1) = 135^\circ$$

The X axis is 135° to the left of the x axis (or 45° to the right of the x axis, see Example 8.1).

8.2 Leaky aquifers, anisotropic on the horizontal plane

8.2.1 Hantush's method

The flow to a well in a leaky aquifer which is anisotropic on the horizontal plane can be analyzed with a method that is essentially the same as the Hantush method for confined aquifers with anisotropy on the horizontal plane. There is, however, one more unknown parameter involved, the leakage factor L, which is given by Hantush (1966) as

$$L_n = \sqrt{(KD)_n c} \tag{8.27}$$

Because c is a constant, Equation 8.7 also gives the relationship between L_n and L_1

$$a_n = \frac{(KD)_1}{(KD)_n} = \left[\frac{L_1}{L_n}\right]^2 = \frac{\cos^2(\Theta + \alpha_n) + m \sin^2(\Theta + \alpha_n)}{\cos^2\Theta + m \sin^2\Theta} \quad (8.28)$$

The Hantush method can be applied if the following assumptions and conditions are satisfied:

- The assumptions listed at the beginning of Chapter 3, with the exception of the first and third assumptions, which are replaced by:
  - The aquifer is leaky;
  - The aquifer is homogeneous, anisotropic on the horizontal plane, and of uniform thickness over the area influenced by the pumping test.

The following condition is added:

- The flow to the well is in an unsteady state.

Procedure 8.5

This procedure is the same as Procedures 8.1 and 8.2 (the Hantush method for confined aquifers with anisotropy on the horizontal plane), except that, in the first step of Procedure 8.5, the methods for leaky isotropic aquifers (Section 4.2) are used to determine values for $(KD)_e$, $S/(KD)_n$, and L_n. Further, Equation 8.28 is used instead of Equation 8.7.

8.3 Confined aquifers, anisotropic on the vertical plane

The flow towards a well that completely penetrates a confined, horizontally stratified aquifer takes place essentially in planes parallel to the aquifer's bedding planes. Even if the hydraulic conductivities vary appreciably in horizontal and vertical directions, the effect of any anisotropy on the vertical plane may not be of any great significance.

In thick aquifers, however, wells usually penetrate only a portion of the aquifer. The flow to such partially penetrating wells is not horizontal, but three-dimensional, i.e. the flow has significant vertical components, at least in the vicinity of the well, where most observations of the drawdown are made. In aquifers with very pronounced anisotropy on the vertical plane, the yield of partially penetrating wells may be appreciably smaller than that of similar wells in isotropic aquifers.

8.3.1 Weeks's method

For large values of pumping time ($t > DS/2K_v$) in a well that partially penetrates a confined aquifer, Hantush (1961a) developed a solution for the drawdown. After modification for the influence of anisotropy on the vertical plane, this equation becomes (Hantush 1964; Weeks 1969)

$$s = \frac{Q}{4\pi KD}\left\{W(u) + f_s\left(\beta', \frac{b}{D}, \frac{d}{D}, \frac{a}{D}\right)\right\} = \frac{Q}{4\pi KD} W(u) + \delta s \quad (8.29)$$

where

W(u) = Theis well function

b, d, a = geometric parameters (Figure 8.6)

$$\beta' = \frac{r}{\Delta}\sqrt{K_v/K_h} \tag{8.30}$$

K_v = hydraulic conductivity in vertical direction

K_h = hydraulic conductivity in horizontal direction

$$f_s = \frac{4D}{\pi(b\text{-}d)} \sum_{n=1}^{\infty} \frac{1}{n} K_0(n\pi\beta') \left\{\cos \frac{n\pi a}{D}\right\}\left\{\sin \frac{n\pi b}{D} - \sin \frac{n\pi d}{D}\right\} \tag{8.31}$$

δs = difference in drawdown between the observed drawdowns and the drawdowns predicted by the Theis equation (Equation 3.5). This difference in drawdown is given by

$$\partial s = \frac{Q}{4\pi KD} f_s \tag{8.32}$$

Values of f_s for different values of β', b/D, d/D, and a/D as tabulated by Weeks (1969) are presented in Annex 8.1.

The assumptions and conditions underlying the Weeks method are:

- The assumptions listed at the beginning of Chapter 3, with the exception of the third and sixth assumptions, which are replaced by:
  - The aquifer is homogeneous, anisotropic in the vertical plane, and of uniform thickness over the area influenced by the pumping test;
  - The pumped well does not penetrate the entire thickness of the aquifer.

The following conditions are added:

- The flow to the well is in an unsteady state;
- $t > SD/2K_v$;
- Drawdown data from at least two piezometers are available; one piezometer at a distance $r > 2D\sqrt{K_h/K_v}$.

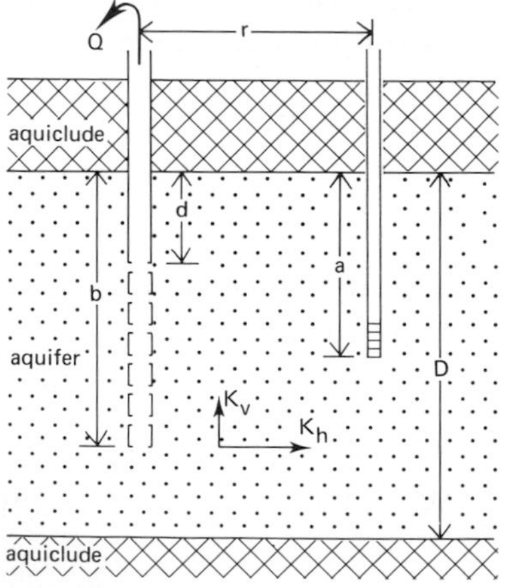

Figure 8.6 The parameters used in Weeks's method

Procedure 8.6

- Apply one of the methods for confined, fully penetrated, isotropic aquifers (Section 3.2) to the observed drawdown data of Piezometer 1 at $r > 2D\sqrt{K_h/K_v}$, and determine the values of K_hD and S;
- For Piezometer 2 at $r < 2D\sqrt{K_h/K_v}$, plot the observed drawdown s versus t on semi-log paper (t on logarithmic scale). Draw a straight line through the late-time data;
- Knowing Q, K_hD, S, and r, calculate, for different values of t, the values of s that would have occurred in Piezometer 2 if the pumped well had been fully penetrating; use Equation 3.5, $s = \frac{Q}{4\pi KD} W(u)$, and Annex 3.1;
- Plot these calculated values of s versus t on the same sheet of semi-log paper as used for the observed time-drawdown plot. Draw a straight line through the late-time data. The straight lines of the two data plots should be parallel;
- Determine the vertical distance δs between the two straight lines;
- Knowing δs, Q, and K_hD, calculate f_s from Equation 8.32;
- Knowing f_s, use Annex 8.1 to determine the value of β' for the values of b/D, d/D, and a/D nearest to the observed values for Piezometer 2;
- Knowing β' and r/D for Piezometer 2, calculate K_v/K_h from Equation 8.30;
- Knowing K_v/K_h, K_hD, and D, calculate K_h and K_v.

Remarks

- Instead of determining K_hD and S with data from a piezometer at $r > 2D\sqrt{K_h/K_v}$ from the partially penetrating well, one can, of course, also obtain these values from the data of a separate pumping test conducted in the same aquifer with a fully penetrating well;
- Whether ∂s will have a positive or a negative value depends on the location of Piezometer 2 relative to that of the screen of the partially penetrating well. When both are located at the same depth in the aquifer, the observed drawdown in Piezometer 2 will be greater than the theoretical drawdown for a fully penetrating well and consequently, ∂s will have a positive value.

8.4 Leaky aquifers, anisotropic on the vertical plane

8.4.1 Weeks's method

For large values of pumping time ($t > DS/2K_v$) in a well that partially penetrates a leaky aquifer with anisotropy on the vertical plane, the drawdown response is given by (Hantush 1964; Weeks 1969)

$$s = \frac{Q}{4\pi K_hD}\left\{W(u,r/L) + f_s\left(\beta',\frac{b}{D},\frac{d}{D},\frac{a}{D}\right)\right\} = \frac{Q}{4\pi K_hD} W(u,r/L) + \delta s \qquad (8.33)$$

where

$W(u,r/L)$ = Walton's well function

f_s, β', b, d, a, and δs are as defined in Section 8.3.1.
A procedure similar to Procedure 8.6 can be applied to leaky aquifers.

The following assumptions and conditions should be satisfied:
- The assumptions listed at the beginning of Chapter 3, with the exception of the first, third, and sixth assumptions, which are replaced by:
  - The aquifer is leaky;
  - The aquifer is homogeneous, anisotropic on the vertical plane, and of uniform thickness over the area influenced by the pumping test;
  - The pumped well does not penetrate the entire thickness of the aquifer.

The following conditions are added:
- The aquitard is incompressible;
- The flow to the well is in unsteady state;
- $t > SD/2K_v$;
- Drawdown data from at least two piezometers are available; one piezometer at a distance $r > 2D\sqrt{K_h/K_v}$.

Procedure 8.7
- Apply one of the methods for leaky, fully penetrated, isotropic aquifers (Sections 4.2.1, 4.2.2, or 4.2.3) to the observed drawdown data of Piezometer 1 at $r > 2D\sqrt{K_h/K_v}$, and determine the values of K_hD, S, and L;
- For Piezometer 2 at $r < 2D\sqrt{K_h/K_v}$, plot the observed drawdown s versus t on log-log paper;
- Knowing Q, K_hD, S, L, and r, calculate for different values of t the values of s that would have occurred in Piezometer 2 if the pumped well had been fully penetrating; use Equation 4.6

$$s = \frac{Q}{4\pi K_h D} W(u, r/L)$$

  and Annex 4.2;
- Plot these calculated values of s versus t on the same sheet of log-log paper as used for the observed time-drawdown plot. The late-time parts of the data curves should be parallel;
- Determine the vertical distance δs between the late-time parallel parts of the data curves;
- Knowing δs, Q, and K_hD, calculate f_s from Equation 8.32;
- Knowing f_s, use Annex 8.1 to determine the value of β' for the values of b/D, d/D and a/D nearest to the observed values for Piezometer 2;
- Knowing β' and r/D for Piezometer 2, calculate K_v/K_h from Equation 8.30;
- Knowing K_v/K_h, K_hD, and D, calculate K_h and K_v.

8.5 Unconfined aquifers, anisotropic on the vertical plane

The flow to a well that pumps an unconfined aquifer is considered to be three-dimensional during the time that the delayed watertable response prevails (see Chapter 5). As three-dimensional flow is affected by anisotropy on the vertical plane, one of the standard methods for unconfined aquifers already takes this anisotropy into account: Neuman's curve-fitting method (Section 5.1.1).

Apart from that standard method, there are other methods that take anisotropy on the vertical plane into account. They can be used when the well is partially penetrating. They are Streltsova's curve-fitting method (Section 10.4.1), Neuman's curve-fitting method (Section 10.4.2), and Boulton-Streltsova's curve-fitting method (Section 11.2.1).

9 Multi-layered aquifer systems

Multi-layered aquifer systems may be one of three kinds. The first consists of two or more aquifer layers, separated by aquicludes. If data on the transmissivity and storativity of thc individual aquifer layers are needed, a pumping test can be conducted in each layer, and each test can then be analyzed by the appropriate method for a single-layered aquifer.

If a well fully penetrates the aquifer system and thus pumps more than one of the aquifer layers at a time, single-layered methods are not applicable. For an aquifer system that consists of two confined aquifers, Papadopulos (1966) derived asymptotic solutions for unsteady-state flow to a well that fully penetrates the system and thus pumps both aquifers at the same time.

For an aquifer system that consists of an unconfined aquifer overlying a confined aquifer, Abdul Khader and Veerankutty (1975) derived a solution for unsteady-state flow to a fully penetrating well.

Either of these solutions allows the hydraulic characteristics of the individual aquifers to be calculated. Both, however, require the use of a computer.

The second multi-layered aquifer system consists of two or more aquifers, each with its own hydraulic characteristics, and separated by interfaces that allow unrestricted crossflow (Figure 9.1). This system's response to pumping will be analogous to that of a single-layered aquifer whose transmissivity and storativity are equal to the sum of the transmissivity and storativity of the individual layers. Hence, in an aquifer with unrestricted crossflow, the same methods as used for single-layered aquifers can be applied. One has to keep in mind, however, that only the hydraulic characteristics

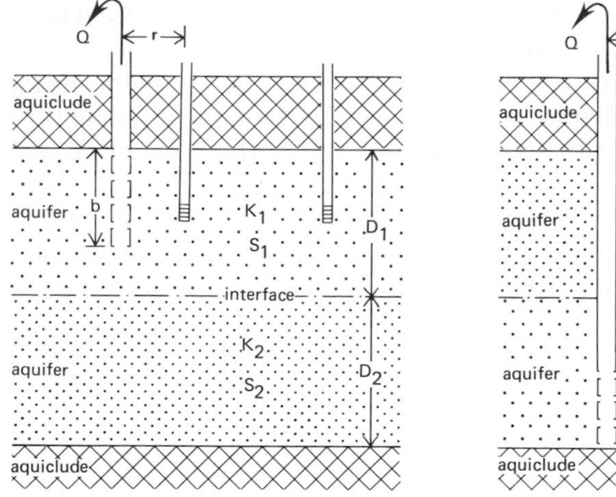

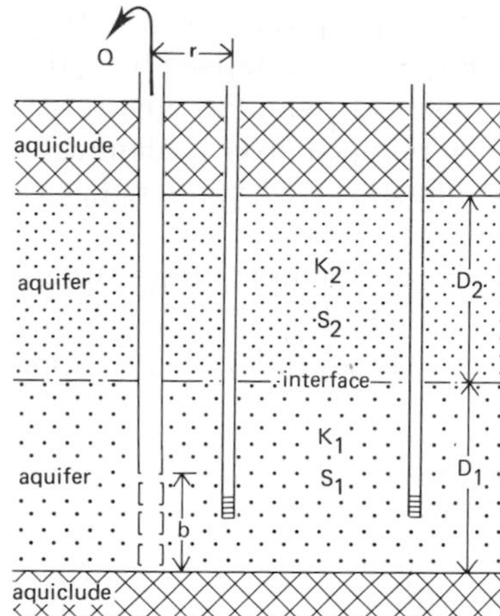

Figure 9.1 Confined two-layered aquifer system, partially penetrating well, either in the upper layer from the top downwards or in the lower layer from the bottom upwards

of the equivalent aquifer system can be determined in this way.

In a confined two-layered aquifer system with unrestricted crossflow, the hydraulic characteristics of the individual aquifers can be determined with the Javandel-Witherspoon method presented in Section 9.1.1.

The third multi-layered aquifer system consists of two or more aquifer layers, separated by aquitards. Pumping one layer of this leaky system has measurable effects in layers other than the pumped layer. The resulting drawdown in each layer is a function of several parameters, which depend on the hydraulic characteristics of the aquifer layers and those of the aquitards. Only for small values of pumping time can the drawdown in the unpumped layers be assumed to be negligible, and only then can methods for leaky single-layered aquifers (Chapter 4) be used to estimate the hydraulic characteristics of the pumped layer.

For longer pumping times, Bruggeman (1966) has developed a method for the analysis of data from leaky two-layered aquifer systems in which steady-state flow prevails. This method is presented in Section 9.2.1.

Various analytical solutions have been derived for steady and unsteady-state flow to a well pumping a leaky multi-layered aquifer system, e.g. Hantush (1967), Neuman and Whitherspoon (1969a, 1969b), and Hemker (1984, 1985). Because of the large number of unknown parameters involved, these methods require the use of a computer.

9.1 Confined two-layered aquifer systems with unrestricted crossflow, unsteady-state flow

9.1.1 Javandel-Witherspoon's method

Javandel and Witherspoon (1983) developed analytical solutions for the drawdown in both layers of a confined two-layered aquifer system pumped by a well that is partially screened, either in the upper layer from the top downwards, or in the underlying layer from the bottom upwards (Figure 9.1). Asymptotic solutions for small and large values of pumping time are derived from the general solution.

For small values of pumping time ($t \leq (D_1 - b)^2/\{(10K_1D_1)/S_1\}$), the drawdown equation for the pumped layer is identical with the equation for unsteady-state flow in a confined single-layered aquifer that is pumped by a partially penetrating well (see Section 10.2.1).

For large values of pumping time and at distances from the pumped well beyond $r \geq 1.5\,\{D_1 + (K_2D_2)/K_1\}$, the partial penetration effects of the well can be ignored and the drawdown in the pumped layer approaches the following expression

$$s_1 = \frac{Q}{4\pi(K_1D_1 + K_2D_2)} W(u) \qquad (9.1)$$

where

$$u = \frac{r^2(S_1 + S_2)}{4t(K_1D_1 + K_2D_2)} \qquad (9.2)$$

This drawdown equation has the form of the Theis equation for unsteady flow in a confined single-layered aquifer pumped by a fully penetrating well (Section 3.2.1). The response of the two-layered system reflects the hydraulic characteristics of the equivalent single-layered system:

$$KD_{eq} = K_1 D_1 + K_2 D_2$$

and

$$S_{eq} = S_1 + S_2$$

Since t is assumed to be large, u will be small. Hence, in analogy to Equation 3.7 (Jacob's method, Section 3.2.2), Equation 9.1 can be written as

$$s_1 = \frac{2.30Q}{4\pi(K_1D_1 + K_2D_2)} \log \frac{25\,(K_1D_1 + K_2D_2)t}{r^2(S_1 + S_2)} \qquad (9.3)$$

A plot on semi-log paper of s versus t will show a straight line for large values of t. The slope of this straight line is given by

$$\Delta s = \frac{2.30Q}{4\pi(K_1D_1 + K_2D_2)} \qquad (9.4)$$

The intercept t_0 of the straight line with the t axis where $s = 0$ is given by

$$t_0 = \frac{r^2\,(S_1 + S_2)}{2.25\,(K_1D_1 + K_2D_2)} \qquad (9.5)$$

The Javandel-Witherspoon method is applicable if the following assumptions and conditions are satisfied:

- The assumptions listed at the beginning of Chapter 3, with the exception of the third and sixth assumptions, which are replaced by:
  - The system consists of two aquifer layers. Each layer has its own hydraulic characteristics, is of apparent infinite areal extent, is homogeneous, isotropic, and of uniform thickness over the area influenced by the test. The interface between the two layers is an open boundary, i.e. no discontinuity of potential or its gradient is allowed across the interface;
  - The pumped well does not penetrate the entire thickness of the aquifer system, but is partially screened, either in the upper layer from the top downwards, or in the lower layer from the bottom upwards.

The following conditions are added:

- The flow to the well is in unsteady state;
- The piezometers are placed at a depth that coincides with the middle of the well screen;
- Drawdown data are available for small values of pumping time $t \leq (D_1 - b)^2/(10K_1 D_1/S_1)$ and for large values of pumping time. The late-time drawdown data are measured at $r \geq 1.5\,\{D_1 + (K_2D_2)/K_1\}$.

Procedure 9.1

- Apply the Hantush modification of the Theis method (see Section 10.2.1) to the early-time drawdown data $\{t \leq (D_1\text{-}b)^2/(10K_1D_1/S_1)\}$ and determine K_1D_1 and S_1 of the pumped layer;
- Determine K_2D_2 and S_2 of the unpumped layer with the procedure outlined for the Jacob method (Section 3.2.2):

- Plot for one of the piezometers, $r \geq 1.5\{D_1 + (K_2D_2)/K_1\}$, the observed drawdown s versus the corresponding time t on semi-log paper (t on logarithmic scale);
- Draw the best-fitting straight line through the late-time portion of the plotted points;
- Extend the straight line until it intercepts the time axis where s = 0, and read the value of t_0;
- Determine the slope of the straight line, i.e. the drawdown difference Δs per log cycle of time;
- Substitute the known values of Q, Δs, and K_1D_1 into Equation 9.4

$$K_2D_2 = \frac{2.30Q}{4\pi\Delta s} - K_1D_1$$

  and calculate K_2D_2 of the unpumped layer;
- Substitute the known values of t_0, K_1D_1, K_2D_2, r^2, and S_1 into Equation 9.5

$$S_2 = \frac{2.25t_0(K_1D_1 + K_2D_2)}{r^2} - S_1$$

  and calculate S_2.

Remarks

- To analyze the late-time drawdown data, the Theis curve-fitting method (Section 3.2.1) can be used instead of the Jacob method;
- Javandel and Witherspoon (1983) observed that the condition $r \geq 1.5\{D_1 + (K_2D_2)/K_1\}$ is on the conservative side;
- If only one piezometer at $r \geq 1.5\{D_1 + (K_2D_2)/K_1\}$ from the well is available, there may not be sufficient early-time drawdown data to determine the hydraulic characteristics of the pumped layer. Hence, only the combined hydraulic characteristics KD_{eq} $(= K_1D_1 + K_2D_2)$ and $S_{eq}(= S_1 + S_2)$ of the equivalent aquifer system can be determined;
- Javandel and Witherspoon (1980) also developed a semi-analytical solution for the drawdown distribution in both layers of a slightly different type of two-layered aquifer system with unrestricted crossflow. The upper layer of this system is bounded by an aquiclude. The lower layer is considered to be very thick compared with the upper layer. The system is pumped by a well that partially penetrates the upper layer. For more information, see the original literature.

9.2 Leaky two-layered aquifer systems with crossflow through aquitards, steady-state flow

Figure 9.2 shows a cross-section of a pumped leaky two-layered aquifer system, overlain by an aquitard, and with another aquitard separating the two aquifer layers. If the hydraulic resistance of the aquitard separating the layers is high compared with that of the overlying aquitard, and if the base layer is an aquiclude, the upper and lower parts of the system can be treated as two separate single-layered leaky aquifers.

Matters become more complicated if the hydraulic resistance of the separating aquitard is appreciably lower than that of the overlying aquitard. If the upper part of

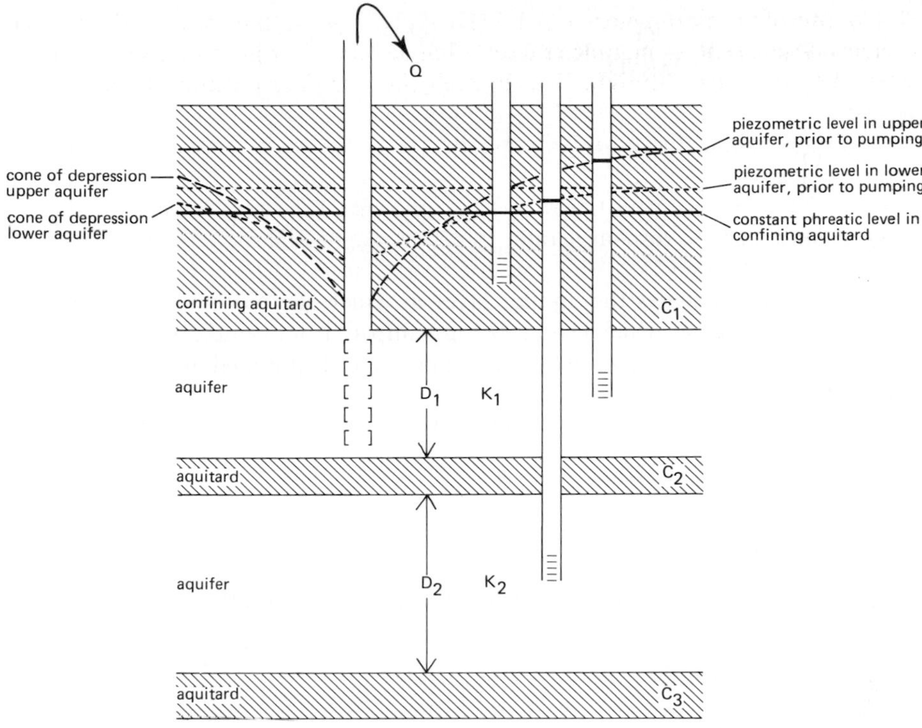

Figure 9.2 Pumped leaky two-layered aquifer system, overlain by an aquitard, and with another aquitard separating the two aquifer layers

that system is pumped, the discharged water would come from the pumped upper layer, the lower aquifer layer (through the separating aquitard), and the overlying aquitard. Bruggeman (1966) has developed a method of analysis for such a system.

9.2.1 Bruggeman's method

The Bruggeman method calls for a double pumping test in which the lower layer is pumped until a steady state is reached, and then, after complete recovery, the upper layer is pumped, again until a steady state is reached. Bruggeman (1966) does not stipulate that the aquifer system be underlain by an aquiclude; it may also be an aquitard. Bruggeman showed that the following relations are valid

$$s'_{1,1} + P_1 s'_{2,1} = \frac{Q'}{2\pi K_1 D_1} K_0(r/\lambda_1) \tag{9.6}$$

$$s'_{1,1} + P_2 s'_{2,1} = \frac{Q'}{2\pi K_1 D_1} K_0(r/\lambda_2) \tag{9.7}$$

$$s'_{1,2} + P_1 s'_{2,2} = P_1 \frac{Q'}{2\pi K_2 D_2} K_0(r/\lambda_1) \tag{9.8}$$

$$s'_{1,2} + P_2 s'_{2,2} = P_2 \frac{Q'}{2\pi K_2 D_2} K_0(r/\lambda_2) \qquad (9.9)$$

where

$$s' = \frac{Q'}{Q} s \qquad (9.10)$$

Q' = standardized discharge rate

The first index to s indicates the aquifer layer in which the piezometer is installed. The second index indicates which layer is being pumped. For example, $s'_{2,1}$ is the drawdown observed in the lower layer when the upper layer is pumped at a standardized discharge rate Q'.
Moreover

$$P_1 + P_2 = \frac{(K_2D_2/K_1D_1)(s'_{2,2} - s'_{1,1})}{s'_{1,2}} \qquad (9.11)$$

$$P_1 P_2 = -(K_2D_2/K_1D_1) \qquad (9.12)$$

where P_1, P_2, λ_1, and λ_2 are constants which are related to one another by

$$\frac{1}{\lambda_1^2} = a_1 + b_1 - a_2 P_1 \qquad (9.13)$$

$$\frac{1}{\lambda_2^2} = a_1 + b_1 - a_2 P_2 \qquad (9.14)$$

$$\frac{P_1}{\lambda_1^2} = -b_1 + b_2 P_1 + a_2 P_1 \qquad (9.15)$$

$$\frac{P_2}{\lambda_2^2} = -b_1 + b_2 P_2 + a_2 P_2 \qquad (9.16)$$

where a_1, a_2, b_1, and b_2 are also constants dependent on K_1D_1, K_2D_2, c_1, and c_2, according to the following equations

$$a_1 = \frac{1}{K_1D_1c_1} \qquad (9.17)$$

$$b_1 = \frac{1}{K_1D_1c_2} \qquad (9.18)$$

$$a_2 = \frac{1}{K_2D_2c_2} \qquad (9.19)$$

and

$$b_2 = \frac{1}{K_2D_2c_3} \qquad (9.20)$$

The Bruggeman method is based on the following assumptions and conditions:
- The assumptions listed at the beginning of Chapter 3, with the exception of the first, third and sixth assumptions, which are replaced by:

- The aquifer system consists of two aquifer layers separated by an aquitard. Each layer is homogeneous, isotropic, and of uniform thickness over the area influenced by the test. The aquifer system is overlain by an aquitard;
- The well receives water by horizontal flow from the entire thickness of the pumped layer.

The following conditions are added:

- The flow to the well is in steady state;
- r/L is small ($r/L < 0.05$);
- $c_1 > c_2$;
- $K_2D_2 > K_1D_1$;
- $c_3 \leq \infty$;
- A pumping test is first conducted in the lower layer until a steady state is reached; then after complete recovery, a pumping test is conducted in the upper layer, again until steady state is reached.

Procedure 9.2

- With Equation 9.10, transform the observed drawdown data to corrected drawdown data for an arbitrarily chosen standard discharge rate Q'. Check whether $s'_{1,2} = s'_{2,1}$ because this should be so for the application of this method;
- Plot $s'_{1,1}$ versus r on semi-log paper and calculate K_1D_1 with

$$\Delta s'_{1,1} = \frac{2.30Q'}{2\pi K_1D_1}$$

where $\Delta s'_{1,1}$ is the difference in $s'_{1,1}$ per log cycle of r;

- In the same way, calculate K_2D_2 from a plot of $s'_{2,2}$ versus r;
- Calculate P_1P_2 with Equation 9.12;
- Calculate $P_1 + P_2$ by introducing into Equation 9.11, for a given value of r, the corresponding values of $s'_{2,2}$ and $s'_{1,1}$ and the values of K_2D_2 and K_1D_1. When this is repeated for several values of r, it provides a check on the values of K_2D_2 and K_1D_1 already calculated, because $P_1 + P_2$ should be independent of r. Calculate P_1 and P_2 by combining the values of $P_1 + P_2$ and P_1P_2.
  A comparison of Equations 9.6 to 9.9 with Equation 4.1 shows the analogy between the Bruggeman equations and the De Glee equation;
- Therefore plot the curve $s'_{1,1} + P_1s'_{2,1}$ versus r on log-log paper and, using De Glee's method (Section 4.1.1, Procedure 4.1), calculate the values of λ_1. In the same way, calculate λ_2 from a plot of $s'_{1,1} + P_2s'_{2,1}$ versus r. Check the values of λ_1 and λ_2 by calculating λ_1 and λ_2 from plots on log-log paper of $(1/P_2)\, s'_{1,2} + s'_{2,2}$ versus r and $(1/P_2)\, s'_{1,2} + s'_{2,2}$ versus r with the De Glee method;
- Using Equations 9.13 to 9.16, calculate a_1, a_2, b_1, and b_2 from the known values of λ_1, λ_2, P_1, and P_2;
- Finally, calculate c_1, c_2, K_1D_1, and K_2D_2 from Equations 9.17 to 9.20. Calculating K_1D_1 and K_2D_2 in this way provides a check on the earlier calculations of K_1D_1 and K_2D_2.

10 Partially-penetrating wells

Some aquifers are so thick that it is not justified to install a fully penetrating well. Instead, the aquifer has to be pumped by a partially penetrating well. Because partial penetration induces vertical flow components in the vicinity of the well, the general assumption that the well receives water from horizontal flow (Chapter 3) is not valid. Partial penetration causes the flow velocity in the immediate vicinity of the well to be higher than it would be otherwise, leading to an extra loss of head. This effect is strongest at the well face, and decreases with increasing distance from the well. It is negligible if measured at a distance that is 1.5 to 2 times greater than the saturated thickness of the aquifer, depending on the amount of penetration. If the aquifer has obvious anisotropy on the vertical plane, the effect is negligible at distances $r > 2D\sqrt{K_h/K_v}$. Hence, the standard methods of analysis cannot be used for $r < 2D\sqrt{K_h/K_v}$ unless allowance is made for partial penetration. For long pumping times ($t > DS/2K$), the effects of partial penetration reach their maximum value for a particular well/piezometer configuration and then remain constant.

For confined and leaky aquifers under steady-state conditions, Huisman developed methods with which the observed drawdowns can be corrected for partial penetration. These are presented in Sections 10.1.1, 10.1.2, and 10.3.

For confined aquifers under unsteady-state conditions, the Hantush modification of the Theis method (Section 10.2.1) or of the Jacob method (Section 10.2.2) can be used.

For leaky aquifers under unsteady-state conditions, drawdowns can be corrected with the Weeks method (Section 10.4.1). This is based on the Walton and Hantush curve-fitting methods for horizontal flow.

Finally, for unconfined aquifers under unsteady-state conditions, the Streltsova curve-fitting method (Section 10.5.1) or the Neuman curve-fitting method (Section 10.5.2) can be used.

10.1 Confined aquifers, steady-state flow

10.1.1 Huisman's correction method I

For a confined aquifer, Huisman (in Anonymous 1964, pp. 73 and 91) presents an equation that can be used to correct the steady-state drawdown measured in a piezometer at $r < 2D$. The parameters are shown in Figure 10.1. The equation reads

$$(s_m)_{partially} - (s_m)_{fully}$$
$$= \frac{Q}{2\pi KD} \times \frac{2D}{\pi d} \sum_{n=1}^{\infty} \frac{1}{n} \left\{ \sin\left(\frac{n\pi b}{D}\right) - \sin\left(\frac{n\pi z_w}{D}\right) \right\} \cos\left(\frac{n\pi z}{D}\right) K_0\left(\frac{n\pi r}{D}\right) \qquad (10.1)$$

where

$(s_m)_{partially}$ = observed steady-state drawdown

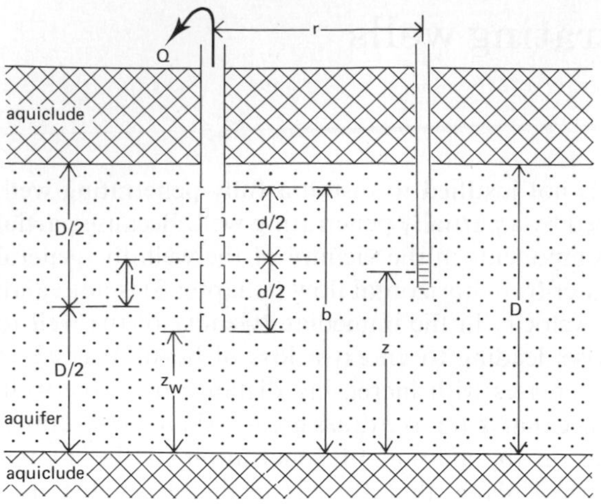

Figure 10.1 The parameters of the Huisman correction method for partial penetration

$(s_m)_{fully}$	= steady-state drawdown that would have occurred if the well had been fully penetrating
z_w	= distance from the bottom of the well screen to the underlying aquiclude
b	= distance from the top of the well screen to the underlying aquiclude
z	= distance from the middle of the piezometer screen to the underlying aquiclude
d	= length of the well screen

Note: The angles are expressed in radians

The Huisman correction method I can be used if the following assumptions and conditions are satisfied:

- The assumptions listed at the beginning of Chapter 3, with the exception of the sixth assumption, which is replaced by:
  - The well does not penetrate the entire thickness of the aquifer.

The following conditions are added:

- The flow to the well is in steady state;
- $r > r_{ew}$.

Procedure 10.1

- Calculate $(s_m)_{fully}$ from Equation 10.1, using an approximate value of KD and the observed $(s_m)_{partially}$ (see Annex 4.1 for the value of K_0);
- Calculate a corrected value of KD, using the Thiem method (Section 3.1.1);
- If there is a great difference between the corrected value of KD and its assumed value, substitute the corrected value into Equation 10.1 and repeat the procedure to get a better result.

Remarks

- This method cannot be applied in the immediate vicinity of the well; there, Huisman's correction method II (Section 10.1.2) has to be used;
- A few terms of the series behind the Σ-sign will generally suffice.

Example 10.1

For this example, we can use data from the pumping test 'Dalem' (Chapter 4) because, as will be shown in Section 10.3, the Huisman correction method can also be applied to leaky aquifers.

Numerical values for the parameters in Figure 10.1 can be read from the cross-section of the test site (Figure 4.2). For the piezometer at r = 10 m and a depth of 36 m, we derive the following data:
$D = 35$ m, $d = 8$ m, $z_w = 25$ m, $b = 33$ m, $r = 10$ m, and $z = 10$ m.
Substitution of these data, together with $Q = 761$ m^3/d and $KD \approx 2000$ m^2/d, into Equation 10.1 yields

For n = 1, the term behind the Σ-sign = − 0.1831
For n = 2, the term behind the Σ-sign = − 0.0101
For n = 3, the term behind the Σ-sign = − 0.0012
For n = 4, the term behind the Σ-sign = + 0.0044 +

− 0.1900

$$\frac{Q}{2\pi KD} \times \frac{2D}{\pi d} \times \frac{761}{2 \times 3.14 \times 2000} \times \frac{2 \times 35}{3.14 \times 8} = 0.1687 \quad \times$$

$$(s_m)_{partially} - (s_m)_{fully} = \longrightarrow -0.0320 \text{ m}$$

This means that 0.032 m has to be added to the observed drawdown to get the drawdown that would have occurred if the well had been fully penetrating.

For the piezometer at r = 10 m and a depth of 14 m, the observed data are the same as above, except that z = 30 m. This gives

For n = 1, the term behind the Σ-sign = + 0.2646
For n = 2, the term behind the Σ-sign = + 0.0284
For n = 3, the term behind the Σ-sign = + 0.0003
For n = 4, the term behind the Σ-sign = + 0.0011 +

+ 0.2944

$$\frac{Q}{2\pi KD} \times \frac{2D}{\pi d} = \longrightarrow +0.1687 \quad \times$$

$$(s_m)_{partially} - (s_m)_{fully} = \longrightarrow +0.0495 \text{ m}$$

This means that 0.05 m has to be substracted from the observed drawdown.

10.1.2 Huisman's correction method II

According to Huisman (Anonymous 1964, pp. 93), the extra drawdown at a well face induced by the eccentric position of the well screen can, for steady-state flow, be expressed by

$$(s_{wm})_{partially} - (s_{wm})_{fully} = \frac{Q}{2\pi KD}\left(\frac{1-P}{P}\right)\ln\frac{\varepsilon D}{r_{ew}} \qquad (10.2)$$

where (see Figure 10.1)

$P = \frac{d}{D}$ = the penetration ratio
d = length of the well screen
$e = \frac{l}{D}$ = amount of eccentricity
l = distance between the middle of the well screen and the middle of the aquifer
ε = function of P and e (see Annex 10.1)
r_{ew} = effective radius of the pumped well

Huisman's correction method II can be used if the following assumptions and conditions are satisfied:

- The assumptions listed at the beginning of Chapter 3, with the exception of the sixth assumption, which is replaced by:
  - The well does not penetrate the entire thickness of the aquifer.

The following conditions are added:

- The flow to the well is in a steady state;
- $r = r_{ew}$.

Procedure 10.2

- Calculate $(s_{wm})_{fully}$ from Equation 10.2, using an approximate value of KD and the observed $(s_{wm})_{partially}$;
- Calculate a corrected value of KD, applying the Thiem method (Section 3.1.1);
- If there is a great difference between the corrected value of KD and its assumed value, substitute the corrected value into Equation 10.2 and repeat the procedure to obtain a better result.

10.2 Confined aquifers, unsteady-state flow

10.2.1 Hantush's modification of the Theis method

For a relatively short period of pumping $\{t < \{(2D-b-a)^2(S_s)\}/20K$, the drawdown in a piezometer at r from a partially penetrating well is, according to Hantush (1961a; 1961b)

$$s = \frac{Q}{8\pi K(b-d)} E(u,\frac{b}{r},\frac{d}{r},\frac{a}{r}) \qquad (10.3)$$

where

$$E(u,\frac{b}{r},\frac{d}{r},\frac{a}{r}) = M(u,B_1) - M(u,B_2) + M(u,B_3) - M(u,B_4) \qquad (10.4)$$

$$u = \frac{r^2 S_s}{4Kt} \qquad (10.5)$$

$S_s = \frac{S}{D}$ = specific storage of the aquifer
$B_1 = (b + a)/r$ (for symbols b, d, and a, see Figure 10.2)
$B_2 = (d + a)/r$
$B_3 = (b - a)/r$
$B_4 = (d - a)/r$

$$M(u,B) = \int_u^\infty \frac{e^{-y}}{y} \mathrm{erf}(B\sqrt{y})dy$$

Because erf (–x) = –erf (x), it follows that M(u,–B) = –M(u,B).
Numerical values of M(u,B) are given in Annex 10.2.

The Hantush modification of the Theis method can be used if the following assumptions and conditions are satisfied:

- The assumptions listed at the beginning of Chapter 3, with the exception of the sixth assumption, which is replaced by:
  - The well does not penetrate the entire thickness of the aquifer.

The following conditions are added:

- The flow to the well is in an unsteady state;
- The time of pumping is relatively short: $t < \{(2D-b-a)^2(S_s)\}/20K$.

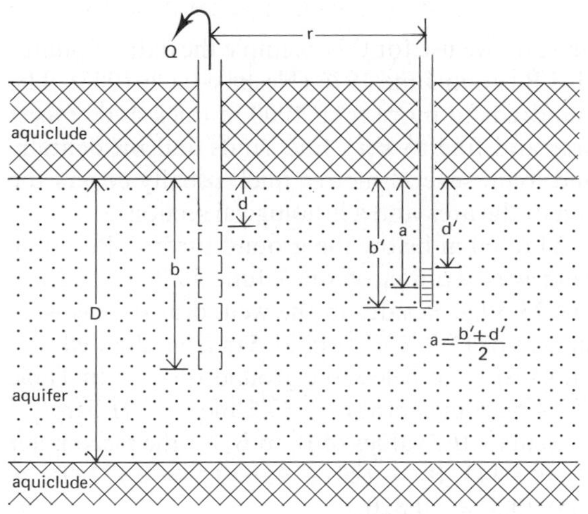

Figure 10.2 The parameters of the Hantush modification of the Theis and Jacob methods for partial penetration

Procedure 10.3

- For one of the piezometers, determine the values of B_1, B_2, B_3, and B_4 and calculate, according to Equation 10.4, its E-function for different values of u, using the tables of the function M(u,B) in Annex 10.2;
- On log-log paper, plot the values of E versus 1/u; this gives the type curve;
- On another sheet of log-log paper of the same scale, plot s versus t for the piezometer;
- Match the data curve with the type curve. It will be seen that for relatively large values of t the data curve deviates upwards from the type curve. This is to be expected because the type curve is based on the assumption that the pumping time is relatively short;
- Select a point A on the superimposed sheets in the range where the curves do not deviate, and note for A the values of s, E, 1/u, and t;
- Substitute the values of s and E into Equation 10.3 and, with Q, b, and d known, calculate K;
- Substitute the values of 1/u and t into Equation 10.5 and, with r and K known, calculate S_s;
- If the data curve departs from the type curve, note the value of 1/u at the point of departure, $1/u_{dep}$;
- Calculate D from the relation

$$D \approx 0.5\left(b + a + r\sqrt{\frac{5}{u_{dep}}}\right) \qquad (10.6)$$

- KD can now be calculated. If the data curve does not depart from the type curve within the range of observed data, record the value of 1/u at a point in the vicinity of the last observed point. If that value of 1/u is used in Equation 10.6 instead of $1/u_{dep}$, the calculated thickness of the aquifer is greater;
- Repeat this procedure for all piezometers in the vicinity of the well, i.e. all piezometers that satisfy the condition $r < 2D$.

Example 10.2

By courtesy of WAPDA, Lahore, Pakistan, we use for this example the data of pumping test BWP 9 conducted in the Indus Basin in June 1976 (Nespak-Ilaco 1985). The alluvial sediments of the basin are hundreds to more than 1000 m thick and consist of medium sand with lenses of coarse and fine to very fine sands and incidentally clay or loam. A top layer of clay and loam several metres thick usually covers the aquifer. Figure 10.3 shows the location of the area and a lithological section.

The pumped well was screened from 20 to 60 m below the ground surface. Pumping started on 1 June 1976 at 10.00 h and was terminated on 5 June 1976 at 21.20 h. The average discharge of the well was 73.5 l/s. Besides in the well, drawdowns were measured in three piezometers at distances of 15.2, 30.5, and 91.5 m from the well.

All piezometers were screened from 44 to 46 m below the ground surface. In Table 10.1 we present the drawdown data of the piezometers at $r_2 = 30.5$ and $r_3 = 91.5$ m.

Following Procedure 10.3 we first calculate the values of B_1 to B_4 for the piezometer at $r = 30.5$ m. $B_1 = (60 + 45)/30.5 = 3.443$, $B_2 = (20 + 45)/30.5 = 2.131$, $B_3 = (60–45)/30.5 = 0.492$ and $B_4 = (20–45)/30.5 = -0.820$.

With the values of B_1 to B_4 known, we now calculate the E-function of this piezometer for different values of u, using Equation 10.4 and Annex 10.2. By using the reciprocals

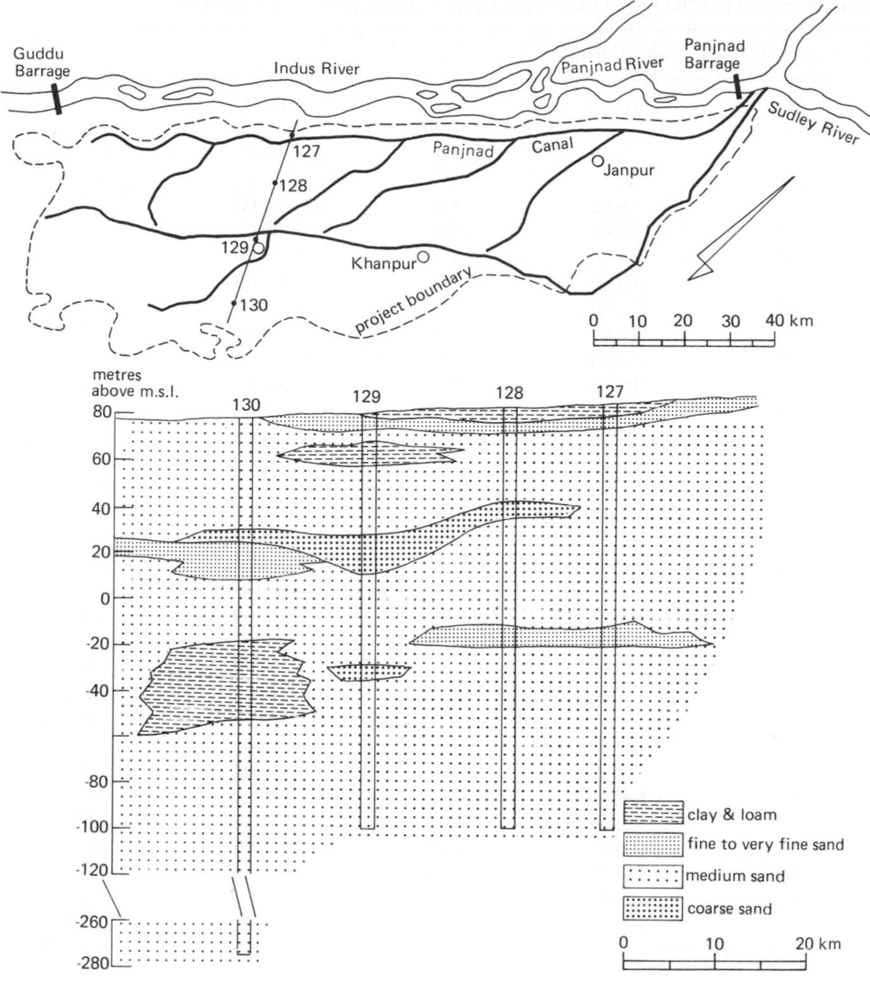

Figure 10.3 Location map of the SCARP II Project area and a representative lithological cross section (after NESPAK-ILACO 1985)

of u, we construct the type curve E versus 1/u on log-log paper. On another sheet of log-log paper, and using the data of piezometer r = 30.5 m in Table 10.1, we plot the drawdown s versus time t.

Figure 10.4 shows the result of matching the field data plot of this piezometer with the type curve. Indeed, as noted before, we observe from this diagram that for large pumping times the field data plot gradually starts to deviate from the type curve. This is not a surprise, for the method of analysis is only valid for early pumping times.

The match point A, selected on the superimposed sheets, has the following dual coordinate values: s = 0.185 m, E = 1, 1/u = 10, and t = 3.52 minutes.

Table 10.1 Data pumping test 'Janpur', Indus Plain, Pakistan (after Nespak-Ilaco 1985)

Piezometer r = 30.5 m. Screen depth 44-46 m.

t (min)	s (m)	t (min)	s (m)	t (min)	s (min)
0.00	0.000	30.00	0.518	500.00	0.613
1.00	0.177	40.00	.533	600.00	.619
2.00	.250	50.00	.543	750.00	.634
3.00	.320	60.00	.549	1000.00	.640
4.00	.344	75.00	.555	1250.00	.643
6.00	.372	100.00	.555	1500.00	.649
8.00	.427	125.00	.570	1750.00	.658
10.00	.445	150.00	.576	2000.00	.674
12.00	.457	175.00	.579	2500.00	.680
15.00	.472	200.00	.579	3000.00	.695
18.00	.488	250.00	.582	4000.00	.716
21.00	.497	300.00	.588	5000.00	.722
25.00	.509	400.00	.610	6000.00	.728

Piezometer r = 91.5 m. Screen depth 44-46 m.

t (min)	s (m)	t (min)	s (m)	t (min)	s (min)
0.00	0.000	30.00	0.168	500.00	0.253
1.00	0.010	40.00	.180	600.00	.259
2.00	.010	50.00	.186	750.00	.265
3.00	.021	60.00	.192	1000.00	.274
4.00	.034	75.00	.201	1250.00	.287
6.00	.061	100.00	.207	1500.00	.293
8.00	.088	125.00	.213	1750.00	.299
10.00	.110	150.00	.216	2000.00	.305
12.00	.122	175.00	.219	2500.00	.326
15.00	.134	200.00	.223	3000.00	.335
18.00	.143	250.00	.229	4000.00	.357
21.00	.152	300.00	.238	5000.00	.369
25.00	.158	400.00	.244	6000.00	.369

Substituting the values of s and E into Equation 10.3 and, with Q, b, and d, known, we can calculate the value of K

$$K = \frac{73.5 \times 86400 \times 10^{-3}}{8 \times 3.14 \times 0.185\,(60 - 20)} = 34.2 \text{ m/d}$$

We now substitute the known values of K, r, t, and 1/u into Equation 10.5 and find

$$S_s = \frac{4\,u\,K\,t}{r^2} = \frac{4 \times 0.1 \times 34.2 \times 3.52}{(30.5)^2 \times 1440} = 3.59 \times 10^{-5}$$

In Figure 10.4 we have indicated the time at which the data plot of piezometer r_2 gradually starts to deviate from the type curve (t = 360 minutes). From this time value and using the above values of K and S_s, we can calculate the value of 1/u (i.e. the point of departure, $1/u_{dep}$) from $1/u = 4Kt/r^2S_s$. We thus find that 1/u = 1024. This data allows us to estimate the thickness of the tested aquifer, using Equation 10.6. We thus find that

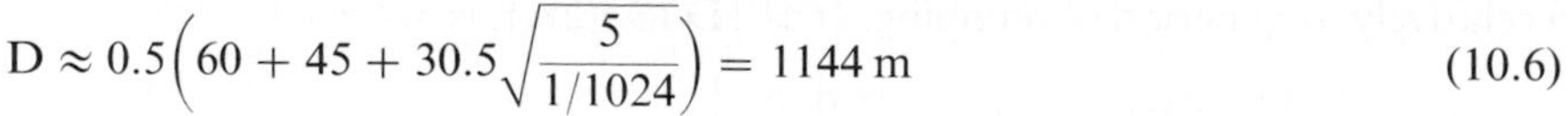

$$D \approx 0.5\left(60 + 45 + 30.5\sqrt{\frac{5}{1/1024}}\right) = 1144 \text{ m} \qquad (10.6)$$

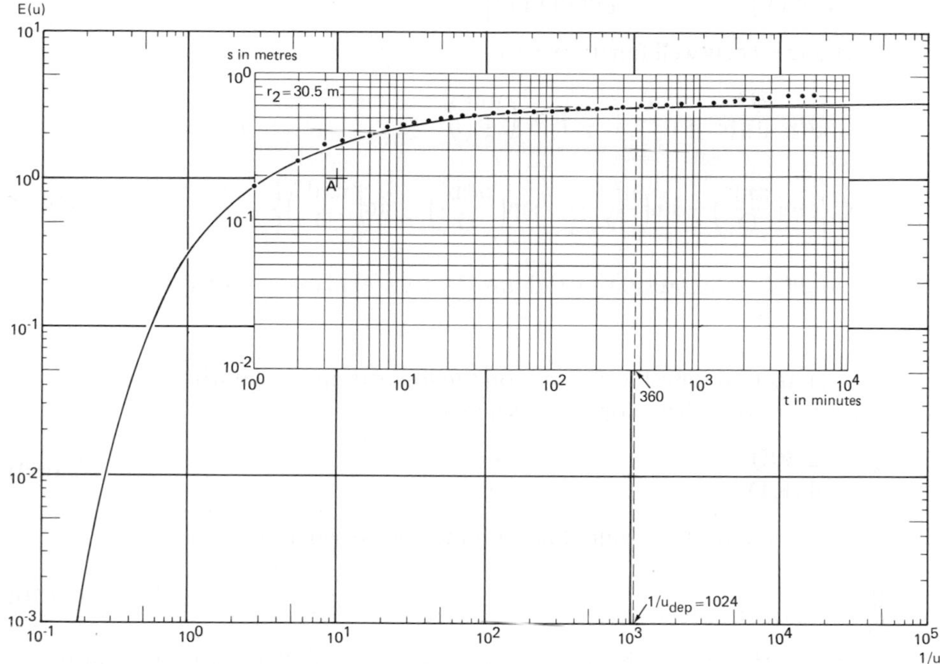

Figure 10.4 Observed-data plot of piezometer $r_2 = 30.5$ m matched with the type curve E(u) versus 1/u

We have repeated the calculations for the other piezometers and obtained the following results:

Piezometer	K (m/d)	S_s	Aquifer thickness (m)
$r_1 = 15.2$ m	31.7	3.17×10^{-5}	1145
$r_2 = 30.5$ m	34.2	3.59×10^{-5}	1144
$r_3 = 91.5$ m	34.7	4.05×10^{-5}	1178

It can be concluded that Hantush's method applied to the three piezometers yields (almost) consistent values for the hydraulic conductivity and the thickness of the aquifer, the latter being a rough estimate. The values obtained for the specific storage, however, are less consistent: they increase slightly with the distance from the well. We cannot offer a plausible explanation for this phenomenon.

10.2.2 Hantush's modification of the Jacob method

According to Hantush (1961b), the drawdown observed in an observation well for a relatively long period of pumping, $\{t > \{D^2(S_s)/2K\}$, is

$$s = \frac{Q}{4\pi KD}\left\{W(u) + f_s\left(\frac{r}{D},\frac{b}{D},\frac{d}{D},\frac{a}{D}\right)\right\} \tag{10.7}$$

where W(u) is the Theis well function, and

$$f_s = \frac{4D^2}{\pi^2(b-d)(b'-d')}\sum_{n=1}^{\infty}\left(\frac{1}{n^2}\right)K_0\left(\frac{n\pi r}{D}\right)$$

$$\times\left\{\sin\left(\frac{n\pi b}{D}\right)-\sin\left(\frac{n\pi d}{D}\right)\right\}\left\{\sin\left(\frac{n\pi b'}{D}\right)-\sin\left(\frac{n\pi d'}{D}\right)\right\} \tag{10.8}$$

Note: The angles are expressed in radians. For an explanation of the symbols, see Figure 10.2

A plot of s versus t on semi-log paper (t on the logarithmic scale) will show a straight line for large values of t. The slope of this line is

$$\Delta s = \frac{2.30Q}{4\pi KD} \tag{10.9}$$

while the intercept t_0 of the straight line with the absciss where $s = 0$ is

$$t_0 = \frac{Sr^2}{2.25KD\exp(f_s)} \tag{10.10}$$

When the difference between b′ and d′ is small $\{(b'-d') < 0.05\ D\}$, i.e. when the drawdown is observed in a piezometer, Equation 10.8 can be replaced by

$$f_s = \frac{4D}{\pi(b-d)}\sum_{n=1}^{\infty}\left(\frac{1}{n}\right)K_0\left(\frac{n\pi r}{D}\right)\left\{\cos\left(\frac{n\pi a}{D}\right)\right\}\left\{\sin\left(\frac{n\pi b}{D}\right)-\sin\left(\frac{n\pi d}{D}\right)\right\} \tag{10.11}$$

Hantush's modification of the Jacob method can be used if the following assumptions and conditions are satisfied:

- The assumptions listed at the beginning of Chapter 3, with the exception of the sixth assumption, which is replaced by:
  - The well does not penetrate the entire thickness of the aquifer.

The following conditions are added:

- The flow to the well is in an unsteady state;
- The time of pumping is relatively long: $t > D^2(S_s)/2K$.

Procedure 10.4

- On semi-log paper, plot for one of the piezometers s versus t (t on the logarithmic scale). Draw a straight line through the plotted points and extend this line until it intercepts the absciss where $s = 0$. Read the value of t_0;
- Calculate the slope of this line, Δs, i.e. the drawdown difference per log cycle of time;
- Calculate KD from Equation 10.9;
- Calculate f_s from Equation 10.8 or Equation 10.11, as is applicable (see Annex 4.1 for values of K_0, and Annex 8.1 for values of f_s defined by Equation 10.11); a few terms of the series involved are generally sufficient;

- Using Annex 4.1, calculate $\exp(f_s)$, and calculate S from Equation 10.10;
- Repeat this procedure for all piezometers at $r < 2D$.

10.3 Leaky aquifers, steady-state flow

It can be shown (Anonymous 1964) that the effect of partial penetration is, as a rule, independent of vertical replenishment, whether this be from overlying or underlying layers. This means that the Huisman correction methods I and II can also be applied to leaky aquifers if the other assumptions of Sections 10.1.1 and 10.1.2 are satisfied. The corrected steady-state drawdown data can then be used in combination with the methods in Section 4.1.

10.4 Leaky aquifers, unsteady-state flow

10.4.1 Weeks's modifications of the Walton and the Hantush curve-fitting methods

For long pumping times ($t > DS/2K$), the effects of partial penetration reach their maximum value for a particular well/piezometer configuration and then remain constant.

Analogous to the drawdown equation for confined aquifers (Equation 10.7, Section 10.2.2), the drawdown in partially penetrated leaky aquifers for $t > DS/2K$ is, according to Weeks (1969)

$$s = \frac{Q}{4\pi KD}\left\{W(u,r/L) + f_s\left(\frac{r}{D'},\frac{b}{D'},\frac{d}{D'},\frac{a}{D}\right)\right\} \tag{10.12}$$

or

$$s = \frac{Q}{4\pi KD}\left\{W(u,\beta) + f_s\left(\frac{r}{D'},\frac{b}{D'},\frac{d}{D'},\frac{a}{D}\right)\right\} \tag{10.13}$$

where

$W(u,r/L)$ = Walton's well function for unsteady-state flow in fully penetrated leaky aquifers confined by incompressible aquitard(s) (Equation 4.6, Section 4.2.1)

$W(u,\beta)$ = Hantush's well function for unsteady-state flow in fully penetrated leaky aquifers confined by compressible aquitard(s) (Equation 4.15, Section 4.2.3)

r,b,d,a = geometrical parameters given in Figure 10.2.

The value of f_s is constant for a particular well/piezometer configuration (Equations 10.8 and 10.11) and can be determined from Annex 8.1. With the value of f_s known, a family of type curves of $\{W(u,r/L) + f_s\}$ or $\{W(u,\beta) + f_s\}$ versus $1/u$ can be drawn for different values of r/L or β. These can then be matched with the data curve for $t > DS/2K$ to obtain the hydraulic characteristics of the aquifer.

The Walton curve-fitting method (Section 4.2.1) can be used if:
- $t > DS/2K$;
- The assumptions and conditions in Section 4.2.1 are satisfied;
- A corrected family of type curves $\{W(u,r/L + f_s\}$ is used instead of $W(u,r/L)$;
- Equation 10.12 is used instead of Equation 4.6.

The Hantush curve-fitting method (Section 4.2.3) can be used if:
- $t > DS/2K$;
- The assumptions and conditions in Section 4.2.3 are satisfied;
- A corrected family of type curves $\{W(u,\beta) + f_s\}$ is used instead of $W(u,\beta)$;
- Equation 10.13 is used instead of Equation 4.15.

10.5 Unconfined anisotropic aquifers, unsteady-state flow

10.5.1 Streltsova's curve-fitting method

For the early-time drawdown behaviour in a partially penetrated unconfined aquifer (Figure 10.5), Streltsova (1974) developed the following equation

$$s = \frac{Q}{4\pi K_h D(b_1/D)} W(u_A,\beta,b_1/D,b_2/D) \tag{10.14}$$

where

$$u_A = \frac{r^2 S_A}{4K_h Dt} \tag{10.15}$$

S_A = storativity of the aquifer

$$\beta = \left(\frac{r^2}{D^2}\right)\frac{K_v}{K_h} \tag{10.16}$$

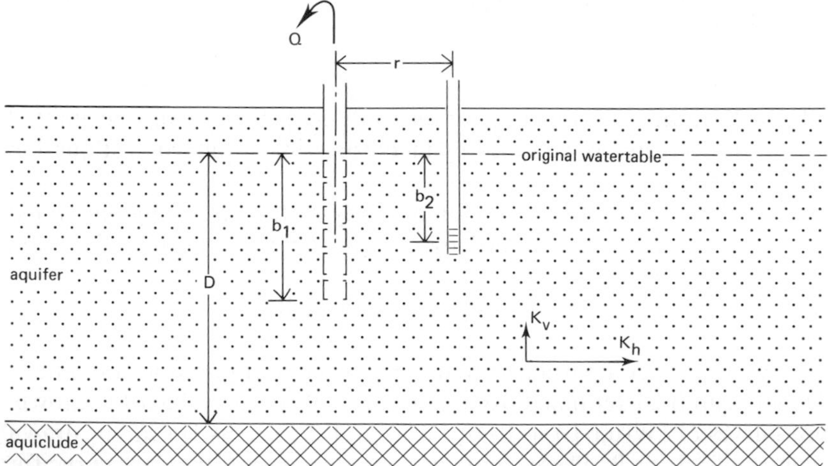

Figure 10.5 Cross-section of an unconfined anisotropic aquifer pumped by a partially penetrating well

For the late-time drawdown behaviour, Streltsova applied a modified form of the Dagan solution (Dagan 1967), written as

$$s = \frac{Q}{4\pi K_h D(b_1/D)} W(u_B,\beta,b_1/D,b_2/D) \quad (10.17)$$

$$u_B = \frac{r^2 S_Y}{4K_h Dt} \quad (10.18)$$

S_Y = specific yield of the aquifer

Values of both well functions are given in Annex 10.3 and Annex 10.4 for a selected range of parameter values. From these values, a family of type A and B curves can be drawn (Figure 10.6).

The Streltsova curve-fitting method can be used if the following assumptions and conditions are satisfied:

- The assumptions listed at the beginning of Chapter 3, with the exception of the first, third, sixth and seventh assumptions, which are replaced by:
  - The aquifer is homogeneous, anisotropic, and of uniform thickness over the area influenced by the pumping test;
  - The well does not penetrate the entire thickness of the aquifer;
  - The aquifer is unconfined and shows delayed watertable response.

The following conditions are added:

- The flow to the well is in an unsteady state;
- $S_Y/S_A > 10$.

Procedure 10.5

- On log-log paper, draw type A curves by plotting $W(u_A,\beta,b_1/D,b_2/D)$ versus $1/u_A$ for a range of values $\sqrt{\beta}$, using the table in Annex 10.3 based on values of b_1/D and b_2/D nearest to the observed values;
- On the same sheet of log-log paper, draw type B curves by plotting $W(u_B,\beta,b_1/D,b_2/$

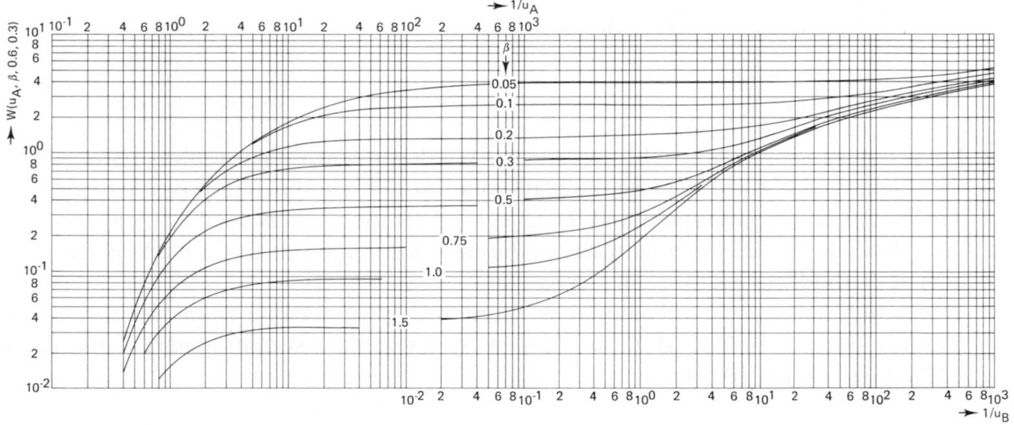

Figure 10.6 Family of Streltsova's type curves for a well partially penetrating an unconfined aquifer

D) versus $1/u_B$ for the same values of $\sqrt{\beta}$, b_1/D, and b_2/D, using Annex 10.4;
- On another sheet of log-log paper of the same scale, plot s versus t for a single piezometer at r from the well;
- Match the data curve with a type A curve and note the $\sqrt{\beta}$ value of that type curve;
- Select an arbitrary point A on the overlapping portion of the two sheets and note the values of s, t, $1/u_A$, and $W(u_A,\beta,b_1/D,b_2/D)$ for this point;
- Substitute these values into Equations 10.14 and 10.15 and, with Q, b_1/D, and r known, calculate K_hD and S_A;
- Move the data curve until as many as possible of the late-time data fall on the type B curve with the same $\sqrt{\beta}$ value as the selected type A curve;
- Select an arbitrary point B on the superimposed curves and note the values of s, t, $1/u_B$, and $W(u_B,\beta,b_1/D,b_2/D)$ for this point;
- Substitute these values into Equations 10.17 and 10.18 and, with Q, b_1/D, and r known, calculate K_hD and S_Y. The two calculations of K_hD should give approximately the same result;
- From the K_hD value and the known initial saturated thickness of the aquifer D, calculate K_h;
- Substitute the values of K_h, $\sqrt{\beta}$, D, and r into Equation 10.16 and calculate K_v;
- Repeat the procedure for each of the available piezometers. The results should be approximately the same.

10.5.2 Neuman's curve-fitting method

For the drawdown in an unconfined anisotropic aquifer pumped by a partially penetrating well (Figure 10.7), Neuman (1974, 1975; see also 1979) developed a curve-fitting method based on the following equation

$$s = \frac{Q}{4\pi K_hD} W\{u_A \text{ (or } u_B),\beta,S_A/S_Y,b/D,d/D,z/D\} \tag{10.19}$$

where

$$u_A = \frac{r^2S_A}{4K_hDt} \text{ and } u_B = \frac{r^2S_Y}{4K_hDt}$$

$$\beta = \left(\frac{r}{D}\right)^2 \frac{K_v}{K_h}$$

Equation 10.19 is expressed in terms of six independent dimensionless parameters. (See Neuman 1974 and 1975 for the exact solution.) This makes it impossible to present a sufficient number of type A and B curves to cover the range needed for field application. Neuman's method thus requires the use of a computer to develop special sets of type A and B curves for each piezometer.

Neuman's curve-fitting method is more widely applicable than the Streltsova method (Section 10.5.1). Both are limited by the same assumptions and conditions outlined in Section 10.5.1.

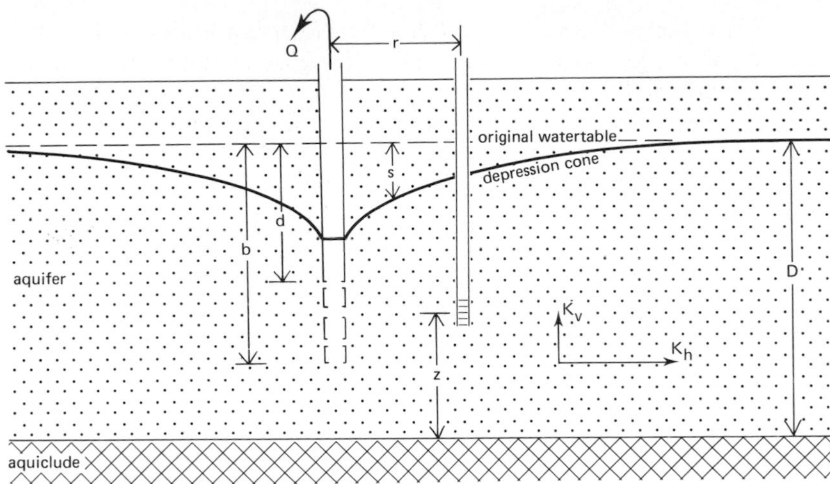

Figure 10.7 The geometric parameters of Neuman's method for a well partially penetrating an unconfined aquifer

11 Large-diameter wells

The standard methods of analysis all assume that storage in the well is negligible. This is not so in large-diameter wells, but methods have been devised that take the well storage into account.

For a large-diameter well that fully penetrates a confined aquifer, Papadopulos (1967) developed the method presented in Section 11.1.1.

For a large-diameter well that partially penetrates an unconfined anisotropic aquifer, Boulton and Streltsova (1976) developed the method presented in Section 11.2.1.

11.1 Confined aquifers, unsteady-state flow

11.1.1 Papadopulos's curve-fitting method

For unsteady-state flow to a fully penetrating, large-diameter well in a confined aquifer (Figure 11.1), Papadopulos (1967) gives the following drawdown equation

$$s = \frac{Q}{4\pi KD} F(u,\alpha,r/r_{ew}) \tag{11.1}$$

where

$$u = \frac{r^2S}{4KDt}$$

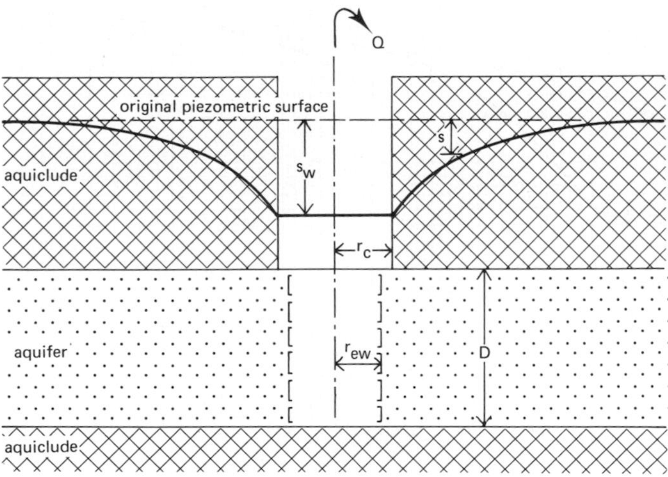

Figure 11.1 A confined aquifer pumped by a fully penetrating, large-diameter well

$$\alpha = \frac{r_{ew}{}^2 S}{r^2_c} \tag{11.2}$$

r_{ew} = effective radius of the well screen or open hole
r_c = radius of the unscreened part of the well over which the water level is changing

Numerical values of the function $F(u,\alpha,r/r_{ew})$ are given in Annex 11.1. These values can be plotted as families of type curves (Figure 11.2).

For long pumping times, i.e. when the drawdown response is no longer influenced by well storage effects, the function $F(u,\alpha,r/r_{ew})$ can be approximated by the Theis well function W(u) (Equation 3.5).

The assumptions and conditions underlying the Papadopulos curve-fitting method are:

- The assumptions listed at the beginning of Chapter 3, with the exception of the eighth assumption, which is replaced by:
  - The well diameter is not small; hence, storage in the well cannot be neglected;

The following condition is added:

- The flow to the well is in unsteady state.

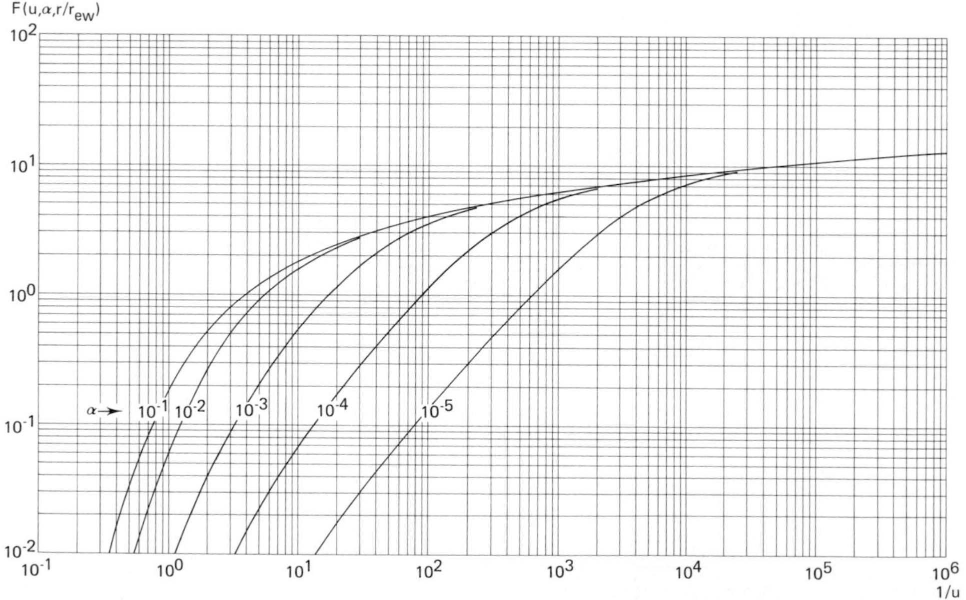

Figure 11.2 Family of Papadopulos's type curves for large-diameter wells: $F(u,\alpha,r/r_{ew})$ versus $1/u$ for different values of α ($r/r_{ew} = 20$)

Procedure 11.1

- For a single piezometer, i.e. for an estimated value of r/r_{ew}, plot a family of type curves $F(u,\alpha,r/r_{ew})$ versus $1/u$ for different values of α on log-log paper, using Annex 11.1;
- On another sheet of log-log paper of the same scale, plot the observed data curve s versus t;
- Match the observed data curve with one of the type curves and note the value of α of that type curve;
- Select an arbitrary matchpoint A on the superimposed sheets and note for this point the values of $F(u,\alpha,r/r_{ew})$, $1/u$, s, and t;
- Substitute the values of $F(u,\alpha,r/r_{ew})$ and s, together with the known value of Q, into Equation 11.1 and calculate KD;
- Calculate S by introducing the values of r, u, t, and KD into $u = r^2S/4KDt$, or by introducing the values of r_c, r_{ew}, and α into Equation 11.2.

Remarks

- If early-time drawdown data only are available, it will be difficult to obtain a unique match of the data curve and a type curve because the type curves differ only slightly in shape (Figure 11.2). The data curve can be matched equally well with more than one type curve. Moving from one type curve to another, however, results in a value of S which differs an order of magnitude. Hence, for early time, S determined by the Papadopulos curve-fitting method is of questionable reliability. The transmissivity, KD, is less sensitive to the choice of the type curve ;
- Large-diameter wells are often only partially penetrating. For long pumping times ($t > DS/2K$), the effects of partial penetration reach their maximum and then remain constant. Analogous to Equation 10.7 (Section 10.2.2), the drawdown in a confined aquifer pumped by a partially penetrating, large-diameter well can be written as

$$s = \frac{Q}{4\pi KD}\left\{F(u,\alpha,r/r_{ew}) + f_s\left(\frac{r}{D},\frac{b}{D},\frac{d}{D},\frac{a}{D}\right)\right\}$$

where b, d, and a are the geometrical parameters shown in Figure 10.2.
For a particular well/piezometer configuration, f_s is constant and can be determined from Annex 8.1. For long pumping times, a log-log set of type curves of $\{F(u,\alpha,r/r_{ew}) + f_s\}$ versus $1/u$ for different values of α can be drawn and matched with the data curve. To obtain KD, Equation 11.1 is replaced by the above equation.

11.2 Unconfined aquifers, unsteady-state flow

11.2.1 Boulton-Streltsova's curve-fitting method

In Chapter 5, we discussed the typical S-shaped time-drawdown curve representing unsteady-state flow in an unconfined aquifer. For an unconfined anisotropic aquifer pumped by a partially penetrating, large-diameter well (Figure 11.3), Boulton and Streltsova (1976) developed a well function describing the first segment of the S-curve. In an abbreviated form, this can be written as

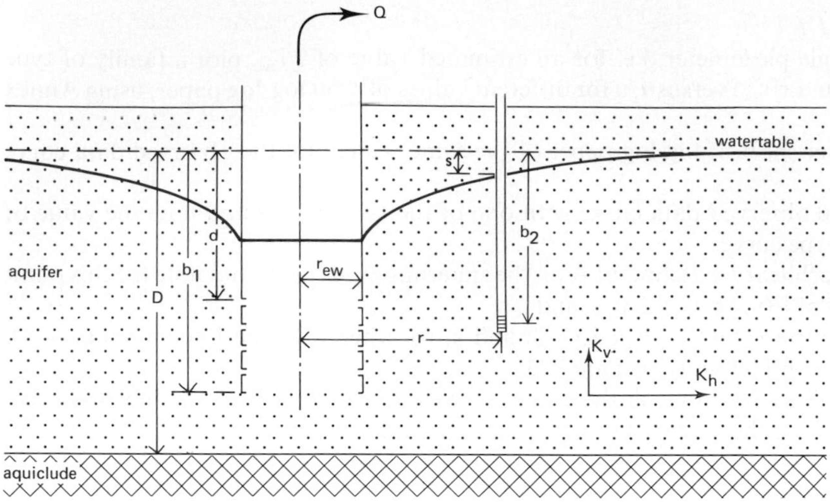

Figure 11.3 An unconfined anisotropic aquifer pumped by a partially penetrating, large-diameter well

$$s = \frac{Q}{4\pi K_h D} W\left(u_A, S_A, \beta, \frac{r}{r_{ew}}, \frac{b_1}{D}, \frac{d}{D}, \frac{b_2}{D}\right) \tag{11.3}$$

where

$$u_A = \frac{r^2 S_A}{4 K_h D t}$$

S_A = storativity of the compressible aquifer, assumed to be 10^{-3}

$$\beta = \left(\frac{r}{D}\right)^2 \frac{K_v}{K_h} \tag{11.4}$$

Because of the large number of parameters involved in this well function, only a selected range of parameter values are available with which $W(u_A,S_A,\beta,r/r_{ew},b_1/D,d/D,b_2/D)$ can be calculated for the construction of type A curves (Annex 11.2).

To analyze the late-time portion of the S-curve, the Boulton- Streltsova method applies the type B curves resulting from Streltsova's equation for a small-diameter well that partially penetrates an unconfined aquifer (Equation 10.17, Section 10.5.1). This is justified for sufficiently long pumping times when the effect of well storage is no longer pronounced.

The Boulton-Streltsova curve-fitting method can be used if the following assumptions and conditions are satisfied:

- The assumptions listed at the beginning of Chapter 3, with the exception of the first, third, sixth, seventh, and eighth assumptions, which are replaced by:
  - The aquifer is unconfined;
  - The aquifer is homogeneous, anisotropic, and of uniform thickness over the area influenced by the test;
  - The well does not penetrate the entire thickness of the aquifer;

• The well diameter is not small; hence, storage in the well cannot be neglected.
The following conditions are added:
- The flow to the well is in an unsteady state;
- $S_Y/S_A > 10$.

Procedure 11.2
- On log-log paper, draw the type A curves by plotting $W(u_A,S_A,\beta,r/r_{ew},b_1/D,d/D,b_2/D)$ versus $1/u_A$ for a range of values of $\sqrt{\beta}$, using the table in Annex 11.2 based on values of b_1/D, b_2/D, and r/r_{ew} nearest to the observed values;
- On the same sheet of log-log paper, draw the type B curves by plotting $W(u_B,\beta,b_1/D,b_2/D)$ versus $1/u_B$ for a range of values of $\sqrt{\beta}$, using the table in Annex 10.4 based on values of b_1/D and b_2/D nearest to the observed values;
- On another sheet of log-log paper of the same scale, plot s versus t for a single piezometer at r from the well;
- Match the early-time data curve with one of the type A curves and note the $\sqrt{\beta}$ value of that type curve;
- Select an arbitrary point A on the overlapping portion of the two sheets and note for this point the values of s, t, $1/u_A$, and $W(u_A,S_A,\beta,r/r_{ew},b_1/D,d/D,b_2/D)$;
- Substitute these values into Equation 11.3 and, with Q also known, calculate K_hD;
- Move the data curve until as many as possible of the late-time data fall on the type B curve with the same $\sqrt{\beta}$ value as the selected type A curve;
- Select an arbitrary point B on the superimposed curves and note for this point the values of s, t, $1/u_B$, and $W(u_B,\beta,b_1/D,b_2/D)$;
- Substitute these values into Equations 10.17 and 10.18 and, with Q, r, and b_1/D also known, calculate K_hD and S_Y. The two calculations of K_hD should give approximately the same result;
- From the K_hD value and the known initial saturated thickness of the aquifer D, calculate K_h;
- Substitute the numerical values of K_h, $\sqrt{\beta}$, D, and r into Equation 11.4 and calculate K_v;
- Repeat the procedure for each of the available piezometers. The results should be approximately the same.

12 Variable-discharge tests and tests in well fields

Aquifers are sometimes pumped at variable discharge rates. This may be done deliberately, or it may be due to the characteristics of the pump. Sometimes, aquifers are pumped step-wise (i.e. at a certain discharge from t_0 to t_1, then at another discharge from t_1 to t_2, and so on), or they may be pumped intermittently at different discharge rates. For confined aquifers that are pumped at variable discharge rates, Birsoy and Summers (1980) devised the method presented in Section 12.1.1.

It may happen that the discharge decreases with the decline of head in the well. If so, the sharpest decrease will occur soon after the start of pumping. For confined aquifers, the Aron-Scott and the Birsoy-Summers methods take this phenomenon into account. These are presented in Sections 12.1.2 and 12.1.1.

Although, strictly speaking, free-flowing wells are not pumped, the methods of analysis applied to them are very similar to those for pumped wells. Hantush's method for unsteady-state flow to a free-flowing well in a confined aquifer can be found in Section 12.2.1, and the Hantush-De Glee method for steady-state flow in a leaky aquifer in Section 12.2.2. Both methods are based on the condition that the decline of head in the well is constant and that the discharge decreases with time.

The methods presented in the previous chapters are based on analytical solutions for the drawdown response in an aquifer that is pumped by a single well. If two or more wells pump the same aquifer, the drawdown will be influenced by the combined effects of these wells. The Cooper-Jacob method (Section 12.3.1) takes such effects into account.

The principle of superposition, which was discussed in Chapter 6, is used in some of the methods in this chapter. According to this principle, two or more drawdown solutions, each for a given set of conditions for the aquifer and the well, can be summed algebraically to obtain a solution for the combined conditions.

12.1 Variable discharge

12.1.1 Confined aquifers, Birsoy-Summers's method

Birsoy and Summers (1980) present an analytical solution for the drawdown response in a confined aquifer that is pumped step-wise or intermittently at different discharge rates (Figure 12.1). Applying the principle of superposition to Jacob's approximation of the Theis equation (Equation 3.7), they obtain the following expression for the drawdown in the aquifer at time t during the nth pumping period of intermittent pumping

$$s_n = \frac{2.30Q_n}{4\pi KD} \log \left\{ \left(\frac{2.25KD}{r^2S} \right) \beta_{t(n)}(t-t_n) \right\} \tag{12.1}$$

where

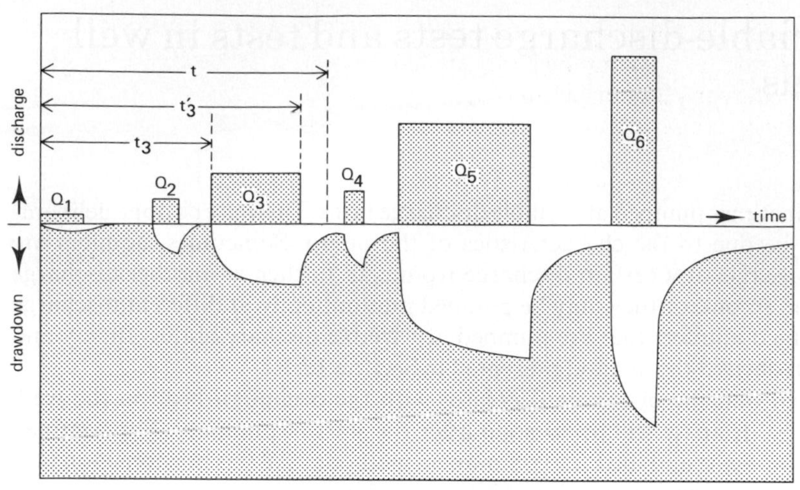

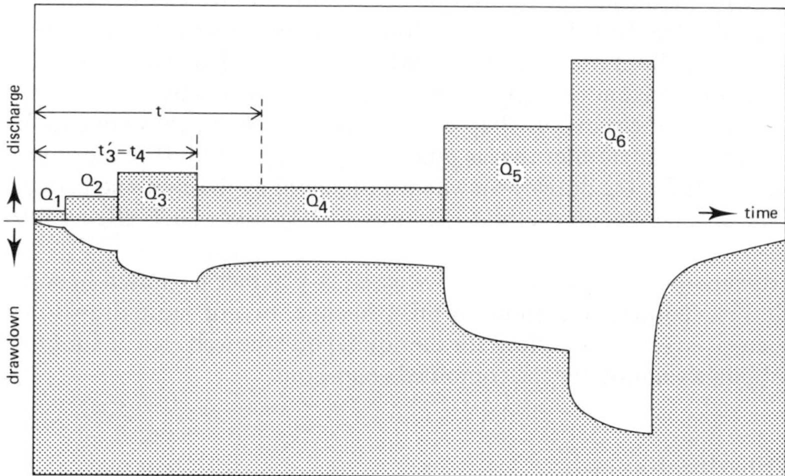

Figure 12.1 Step-wise and intermittently changing discharge rates and the resulting drawdown responses (after Birsoy and Summers 1980)

$$\beta_{t(n)} = \prod_{i=1}^{n-1} \left(\frac{t-t_i}{t-t'_i}\right)^{Q_i/Q_n}$$

$$= \left(\frac{t-t_1}{t-t'_1}\right)^{Q_1/Q_n} \times \left(\frac{t-t_2}{t-t'_2}\right)^{Q_2/Q_n} \times \cdots \times \left(\frac{t-t_{n-1}}{t-t'_{n-1}}\right)^{Q_{n-1}/Q_n} \qquad (12.2)$$

where

t_i = time at which the i-th pumping period started

$t-t_i$ = time since the i-th pumping period started
t'_i = time at which the i-th pumping period ended
$t-t'_i$ = time since the i-th pumping period ended
Q_i = constant well discharge during the i-th pumping period

For step-wise or uninterrupted pumping, $t'_{(i-1)} = t_i$, and the 'adjusted time' $\{\beta_{t(n)}(t-t_n)\}$ becomes

$$\beta_{t(n)}(t-t_n) = \prod_{i=1}^{n} (t-t_i)^{\Delta Q_i/Q_n}$$

$$= (t-t_1)^{\Delta Q_1/Q_n} \times (t-t_2)^{\Delta Q_2/Q_n} \times ... \times (t-t_n)^{\Delta Q_n/Q_n} \quad (12.3)$$

where $\Delta Q_i = Q_i - Q_{i-1}$ = discharge increment beginning at time t_i.

If the intermittent pumping rate is constant (i.e. $Q = Q_1 = Q_2 = ... = Q_n$), the adjusted time becomes

$$\beta_{t(n)}(t-t_n) = \frac{t_1}{t'_1}\frac{t_2}{t'_2}\cdots\frac{t_{n-1}}{t'_{n-1}}t_n \quad (12.4)$$

Dividing both sides of Equation 12.1 by Q_n gives an expression for the specific drawdown

$$\frac{s_n}{Q_n} = \frac{2.30}{4\pi KD}\log\left\{\frac{2.25KD}{r^2S}\beta_{t(n)}(t-t_n)\right\} \quad (12.5)$$

The Birsoy-Summers method can be used if the following assumptions and conditions are satisfied:

- The assumptions listed at the beginning of Chapter 3, with the exception of the fifth assumption, which is replaced by:
  - The aquifer is pumped step-wise or intermittently at a variable discharge rate or is intermittently pumped at a constant discharge rate.

The following conditions are added:

- The flow to the well is in an unsteady state;
- $\frac{r^2S}{4KD} \times \frac{1}{\beta_{t(n)}(t-t_n)} < 0.01$ (see also Section 3.2.2)

Procedure 12.1

- For a single piezometer, calculate the adjusted time $\beta_{t(n)}(t-t_n)$ from Equations 12.2, 12.3, or 12.4 (whichever is applicable), using all the observed discharges and the appropriate values of time;
- On semi-log paper, plot the observed specific drawdown s_n/Q_n versus the corresponding values of $\beta_{t(n)}(t-t_n)$ (the adjusted time on the logarithmic scale), and draw a straight line through the plotted points;
- Determine the slope of the straight line, $\Delta(s_n/Q_n)$, which is the difference of s_n/Q_n per log cycle of adjusted time;
- Calculate KD from $\Delta(s_n/Q_n) = 2.30/4\pi KD$;
- Extend the straight line until it intersects the $s_n/Q_n = 0$ axis and determine the value of the interception point $\{\beta_{t(n)}(t-t_n)\}_0$;

– Knowing r, KD, and $\{\beta_{t(n)}(t-t_n)\}_0$, calculate S from

$$S = \frac{2.25KD}{r^2}\{\beta_{t(n)}(t-t_n)\}_0 \tag{12.6}$$

Remarks
- Procedure 12.1 can also be applied when the well discharge changes uninterruptedly with time. In that case, however, Q versus t for a single piezometer should be plotted on arithmetic paper. The time axis is then divided into appropriate equal time intervals $t'_i - t_i$ and the average discharge Q_i for each time interval is calculated;
- Calculating the adjusted time $\beta_{t(n)}(t-t_n)$ by hand is a tedious process. Birsoy and Summers (1980) give a program for an HP-25 pocket calculator that computes $\beta_{t(n)}$ for $n < 4$ for step-wise pumping.

Example 12.1
We use drawdown data from a hypothetical pumping test conducted in a fully penetrated confined aquifer. During the test, the discharge rates changed step-wise (Table 12.1). For a piezometer at r = 5 m, the adjusted time $\beta_{t(n)}(t-t_n)$ can be calculated with Equation 12.3.
For example, for n = 3 and t = 100 min., the adjusted time is calculated as follows

$$\beta_{t(3)}(t-t_3) = (t-t_1)^{\Delta Q_1/Q_3} \times (t-t_2)^{\Delta Q_2/Q_3} \times (t-t_3)^{\Delta Q_3/Q_3}$$
$$= (100\text{-}0)^{500/600} \times (100\text{-}30)^{200/600} \times (100\text{-}80)^{-100/600} = 116 \text{ min}$$

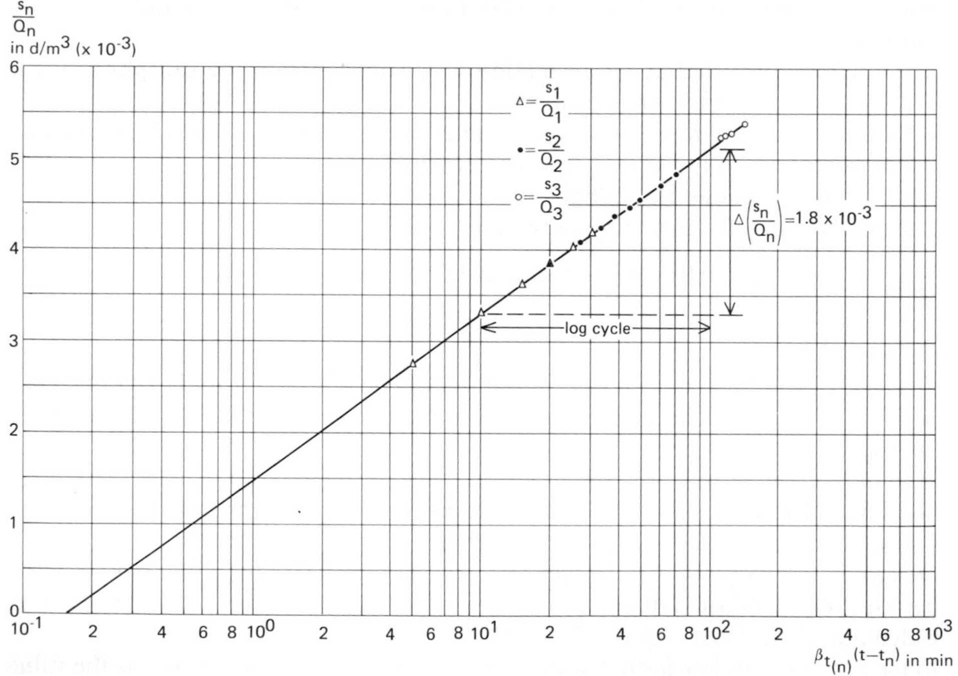

Figure 12.2 Analysis of data with the Birsoy-Summers method for variable discharge

Table 12.1 gives the results of the calculations.

The specific drawdown data (Table 12.1) are plotted against the calculated adjusted time on semi-log paper (Figure 12.2). The slope of the straight line through the plotted points $\Delta(s_n/Q_n) = 1.8 \times 10^{-3}$.
The transmissivity is

$$KD = \frac{2.30}{4\pi\Delta(s_n/Q_n)} = \frac{2.30}{4 \times 3.14 \times 1.8 \times 10^{-3}} = 102 \text{ m}^2/\text{d}$$

The straight line intersects the $s_n/Q_n = 0$ axis at $\{\beta_{t(n)}(t\text{-}t_n)\}_o = 1.5 \times 10^{-1}$ min.
Hence

$$S = \frac{2.25KD}{r^2}\{\beta_{t(n)}(t-t_n)\}_o = \frac{2.25 \times 102}{25} \times \frac{1.5 \times 10^{-1}}{1440} = 9.6 \times 10^{-4}$$

In each step, the condition $u < 0.01$ is fulfilled after $t = 8.5$ min. The less restrictive condition $u < 0.05$ (Section 3.2.2) is already fulfilled after 1.7 min., i.e. all drawdown data can be used in the analysis.

Table 12.1 Data from a pumping test with step-wise changing discharge rates

n	t min	s_n (m)	Q_n m^3/d	s_n/Q_n d/m^2	$\beta_{t(n)}(t\text{-}t_n)$ min
1	5	1.38	500	2.76×10^{-3}	5
1	10	1.65	500	3.30×10^{-3}	10
1	15	1.81	500	3.62×10^{-3}	15
1	20	1.93	500	3.86×10^{-3}	20
1	25	2.02	500	4.04×10^{-3}	25
1	30	2.09	500	4.18×10^{-3}	30
2	35	2.68	700	3.83×10^{-3}	20
2	40	2.85	700	4.07×10^{-3}	27
2	45	2.96	700	4.23×10^{-3}	33
2	50	3.05	700	4.36×10^{-3}	38
2	55	3.12	700	4.46×10^{-3}	44
2	60	3.18	700	4.54×10^{-3}	49
2	70	3.29	700	4.70×10^{-3}	60
2	80	3.38	700	4.83×10^{-3}	70
3	90	3.13	600	5.22×10^{-3}	113
3	100	3.15	600	5.25×10^{-3}	116
3	110	3.17	600	5.28×10^{-3}	123
3	130	3.23	600	5.38×10^{-3}	140

12.1.2 Confined aquifers, Aron-Scott's method

In a confined aquifer, when the head in the well declines as a result of pumping, many pumps decrease their discharge, the sharpest decrease taking place soon after the start of pumping (Figure 12.3).

An appropriate method that takes this phenomenon into account has been developed by Aron and Scott (1965). They show that when $r^2S/4KDt_n < 0.01$, the draw-

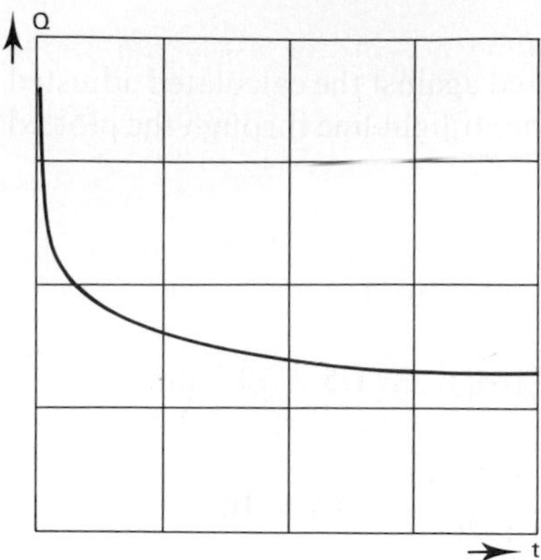

Figure 12.3 Schematic discharge-time diagram of a pump with decreasing discharge rate

down (s_n) at a certain moment t_n is approximately equal to

$$s_n \approx \left(\frac{2.30Q_n}{4\pi KD} \log \frac{2.25KDt_n}{r^2S}\right) + s_e \tag{12.7}$$

where Q_n is the discharge at time t_n, and s_e is the excess drawdown caused by the earlier higher discharge.

If $\bar{Q}_n$ is the average discharge from time 0 to t_n, the excess volume pumped is $(\bar{Q}_n - Q_n)t_n$. If the fully developed drawdown is considered to extend to the distance r_i at which $\log(2.25KDt_n/r_i^2S) = 0$, the excess drawdown s_e can be approximated by

$$s_e = \frac{(\bar{Q}_n - Q_n)t_n}{A_iS} = \frac{(\bar{Q}_n - Q_n)t_n}{S} \times \frac{S}{2.25\pi KDt_n} = \frac{\bar{Q}_n - Q_n}{2.25\pi KD} \tag{12.8}$$

where $A_i = \pi r_i^2$ = area influenced by the pumping.

If $r^2S/4KDt_n < 0.01$, a semi-log plot of s_n/Q_n versus t_n will yield a straight line. KD can then be determined by introducing the slope of the straight line, $\Delta(s_n/Q_n)$, i.e. the specific drawdown difference per log cycle of time, into

$$KD \approx \frac{2.30}{4\pi\Delta(s_n/Q_n)} \tag{12.9}$$

and S can be determined from

$$S \approx \frac{2.25KDt_0}{r^2} \tag{12.10}$$

where t_0 is the intercept of the straight line with the absciss $s_n/Q_n = \overline{s_e/Q_n}$, the latter being the average of several values of s_e/Q_n calculated from

$$\frac{s_e}{Q_n} = \frac{(\bar{Q}_n/Q_n) - 1}{2.25\pi KD} \tag{12.11}$$

The Aron-Scott method, which is analogous to the Jacob method (Section 3.2.2), can be used if the following assumptions and conditions are met:

- The assumptions listed at the beginning of Chapter 3, with the exception of the fifth assumption, which is replaced by:
  - The discharge rate decreases with time, the sharpest decrease occurring soon after the start of pumping.

The following conditions are added:
- The flow to the well is in an unsteady state;
- $r^2S/4KDt_n < 0.01$ (see also Section 3.2.2).

Procedure 12.2
- For one of the piezometers, plot s_n/Q_n versus t_n on semi-log paper (t_n on logarithmic scale). Fit a straight line through the plotted points (Figure 12.4);
- Determine the slope of the straight line, $\Delta(s_n/Q_n)$;
- Calculate KD from Equation 12.9;
- Calculate s_e/Q_n from Equation 12.11 for several values of t_n and determine the average value, $\overline{s_e/Q_n}$;
- Determine the interception point of the straight line with the absciss $s_n/Q_n = \overline{s_e/Q_n}$. The t value of this point is t_0;
- Calculate S from Equation 12.10;
- Repeat this procedure for all piezometers that satisfy the conditions. The results should show a close agreement.

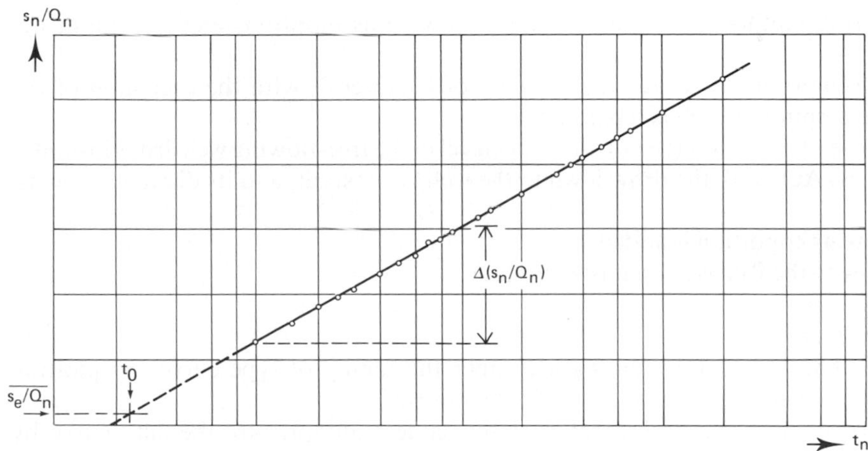

Figure 12.4 Illustration of the application of the Aron-Scott method

12.2 Free-flowing wells

The methods for free-flowing wells are based on the conditions that the drawdown in the well is constant and that the discharge decreases with time. To satisfy these conditions, the well is shut down for a period long enough for the pressure to have become static. When the well is opened up again at time t = 0, the water level in the well drops instantaneously to a constant drawdown level, which is equal to the outflow opening of the well, while the well starts discharging at a decreasing rate.

12.2.1 Confined aquifer, unsteady-state flow, Hantush's method

The unsteady-state drawdown induced by a free-flowing well in a confined aquifer is given by Hantush (1964) (see also Reed 1980) as

$$s = s_w A(u_w, r/r_{ew}) \tag{12.12}$$

where

$A(u_w,r/r_{ew})$ = Hantush's free-flowing-well function for confined aquifers

$$u_w = \frac{r_{ew}^2 S}{4KDt} \tag{12.13}$$

r_{ew} = effective radius of flowing well
s_w = constant drawdown in flowing well = difference between static head measured during shutdown of the well and the outflow opening of the well

Annex 12.1 presents values of $A(u_w,r/r_{ew})$ in tabular form for different values of $1/u_w$ and r/r_{ew}.

The Hantush method can be used if the following assumptions and conditions are satisfied:
- The assumptions listed at the beginning of Chapter 3, with the exception of the fifth assumption, which is replaced by:
  - At the start of the test ($t = 0$), the water level in the free-flowing well drops instantaneously. At $t > 0$, the drawdown in the well is constant, and its discharge is variable.

The following condition is added:
- The flow to the well is in an unsteady state.

Procedure 12.3
- Using Annex 12.1, draw on log-log paper the family of type curves by plotting $A(u_w,r/r_{ew})$ versus $1/u_w$ for a range of values of r/r_{ew};
- On another sheet of log-log paper of the same scale, prepare the data curve by plotting s/s_w against the corresponding t for a single piezometer at r from the well;
- Match the data curve with one of type curves and note the r/r_{ew} value of the type curve;
- Select an arbitrary point A on the overlapping portion of the two sheets and note for this point the values of t and $1/u_w$;
- Substitute the values of $1/u_w$, r/r_{ew}, r, and t into Equation 12.13, now written as

$$\frac{KD}{S} = \frac{1}{4}\left(\frac{1}{u_w}\right)\left(\frac{r_{ew}}{r}\right)^2\left(\frac{r^2}{t}\right)$$

and calculate the diffusivity KD/S.

Remark

- If the value of r_{ew} is known, one type curve of $A(u_w,r/r_{ew})$ versus $1/u_w$ for the known value of r/r_{ew} can be used.

12.2.2 Leaky aquifers, steady-state flow, Hantush-De Glee's method

The steady-state drawdown in a leaky aquifer tapped by a fully penetrating free-flowing well is given by Hantush (1959a) as

$$s_m = \frac{Q_m}{2\pi KD} K_0(r/L) \qquad (12.14)$$

where

s_m = steady-state drawdown in a piezometer at r from the well
Q_m = steady-state discharge (= minimum discharge) of the well

The data obtained during the steady-state phase of the free-flowing-well test can be analyzed with De Glee's method (Section 4.1.1), provided that the Hantush equation (Equation 12.14) is used instead of Equation 4.1. The following assumptions and conditions should be satisfied:

- The assumptions and conditions that underlie the standard methods for leaky aquifers (Chapter 4), with the exception of the fifth assumption, which is replaced by:
  - At the beginning of the test ($t = 0$), the water level in the well drops instantaneously. At $t > 0$, the drawdown in the well is constant, and its discharge is variable.

The following conditions are added:

- The flow to the well is in a steady state;
- $L > 3D$.

12.3 Well field

12.3.1 Cooper-Jacob's method

A modified version of the Jacob method, previously described in Section 3.2.2, can be used to resolve the effects of a well field on the drawdown (Cooper and Jacob 1946). By applying the principle of superposition and using values of specific drawdown ($s_n/\Sigma Q_i$) instead of drawdown (s), and values of the weighted logarithmic mean $\overline{(t_n/r_i^2)}$ instead of t/r^2, the same procedure as outlined for the Jacob method can be followed. The specific drawdown ($s_n/\Sigma Q_i$) is the drawdown (s_n) in a piezometer at a certain time t_n, divided by the sum of the discharges of the different pumped wells for the same time (ΣQ_i).

The assumptions and conditions underlying the Cooper-Jacob method are the same as those for the Jacob method (see Section 3.2.2) i.e.:

- The assumptions listed in Chapter 3;

- The flow to the well is in unsteady state;
- $u \dfrac{S}{4KD\overline{(t/r_i^2)n}} < 0.01.$

Procedure 12.4 (see also Section 3.2.2)
- Calculate for one of the piezometers the value of the specific drawdown $(s_n/\Sigma Q_i)$ for each corresponding time t_n;
- Determine the weighted logarithmic mean, $\overline{(t/r_i^2)}_n$, corresponding to each value of t_n in the following way:
  - Divide the elapsed time t_n by the square of the distance from each pumped well to the piezometer, r_i^2, (t_n/r_i^2);
  - Multiply the logarithm of each of those values by the individual well discharge $[Q_i \log(t_n/r_i^2)]$;
  - Sum the products algebraically $[\Sigma\, Q_i \log(t_n/r_i^2)]$;
  - Divide that sum by the sum of the discharges of the different pumping wells $[\{\Sigma\, Q_i \log(t_n/r_i^2)\}/\Sigma Q_i] = (x)$;
  - Extract the antilogarithm of the quotient $(10^{(x)})$ which is the requested value of $\overline{(t/r_i^2)}_n$;
- Plot the values of $(s_n/\Sigma Q_i)$ versus $\overline{(t/r_i^2)}_n$ on semi-log paper $\overline{(t/r_i^2}$ on the logarithmic axis). Draw a straight line through the plotted points;
- Extend the straight line till it intercepts the time-axis where $s_n/\Sigma Q_i = 0$, and read the value of $\overline{(t/r_i^2)}_0$;
- Determine the slope of the straight line, i.e. the drawdown difference $\Delta(s_n/\Sigma Q_i)$ per log cycle of $\overline{(t/r_i^2)}_n$;
- Substitute the values of $\Delta(s_n/\Sigma Q_i)$ into – a modified version of – Equation 3.13

$$KD = \frac{2.30}{4\pi\Delta(s_n/\Sigma Q_i)}$$

and solve for KD;
- With KD and $\overline{(t/r_i^2)}_0$ known, calculate S from Equation 3.12

$$S = 2.25\, KD\, \overline{(t/r_i^2)}_0$$

Remark
- The Cooper-Jacob method can also be applied if the individual wells are pumped at a variable discharge rate. Hence the discharge rate of each individual well is dependent on the elapsed time t_n, and the value of ΣQ_i will not be constant.

Example 12.2
In a hypothetical well field, the pumping started simultaneously in three wells (1, 2, 3) at constant discharge rates of $Q_1 = 150$ m³/d, $Q_2 = 200$ m³/d, and $Q_3 = 300$ m³/d. The drawdown was observed in a piezometer at a distance of $r_1 = 10$ m from Well 1, $r_2 = 20$ m from Well 2, and $r_3 = 30$ m from Well 3 (Table 12.2).

Table 12.2 gives the calculated values of $s_n/\Sigma\, Q_i$, and shows the step-by-step procedure to calculate the weighted logarithmic mean $\overline{(t/r_i^2)}_n$.

The values of $s_n/\Sigma\, Q_i$ and $\overline{(t/r_i^2)}_n$ are plotted on semi-log paper (Figure 12.5). The slope of the straight line through the plotted points $\Delta(s_n/\Sigma Q_i) = 4.75 \times 10^{-4}$. Hence

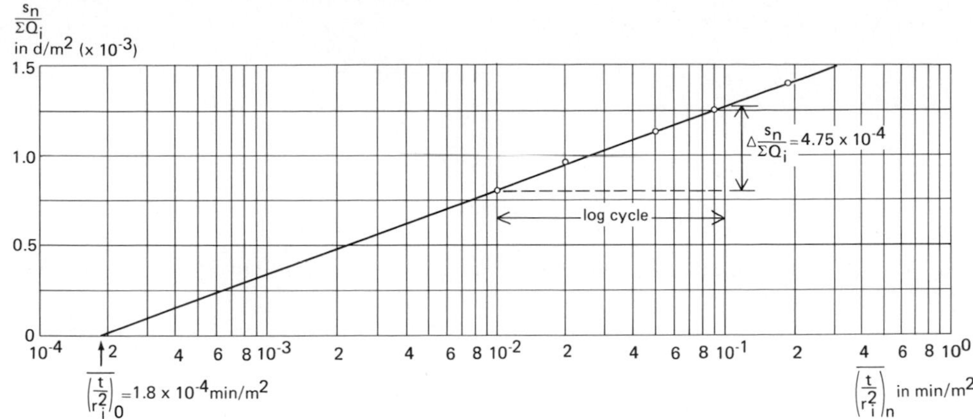

Figure 12.5 Analysis of data with the Cooper-Jacob method for well fields

$$KD = \frac{2.30}{4\pi\Delta(s_n/\Sigma Q_i)} = \frac{2.30}{4 \times 3.14 \times 4.75 \times 10^{-4}} = 386 \text{ m}^2/\text{d}$$

The interception point of the straight line with the $(s_n/\Sigma Q_i) = 0$ axis is $\overline{(t/r_i^2)}_0 = 1.8 \times 10^{-4}$ min/m^2.
S can be calculated from

$$S = 2.25\,KD\,\overline{(t/r_i^2)}_0 = 2.25 \times 386 \times 1.8 \times 10^{-4} \times \frac{1}{1440} = 10^{-4}$$

Table 12.2 Calculation of parameter $\overline{(t/r_i^2)}_n$ of the Cooper-Jacob method

	$r \rightarrow 1$	2	3	4	5
s_n (m)	0.53	0.62	0.74	0.82	0.91
ΣQ_i (m^3/d)	650	650	650	650	650
$s_n/\Sigma Q_i$(d/m^2)	8.15×10^{-4}	9.54×10^{-4}	1.13×10^{-3}	1.26×10^{-3}	1.4×10^{-3}
t_n (min)	5	10	20	40	80
$t_n/r_1^2 = t_n/100$	0.05	0.10	0.20	0.40	0.80
$t_n/r_2^2 = t_n/400$	0.0125	0.025	0.05	0.10	0.20
$t_n/r_3^2 = t_n/900$	0.0056	0.0111	0.0222	0.0444	0.0889
$Q_1 \log(t_n/r_1^2)$	− 195.2	− 150	− 104.8	− 59.7	− 14.5
$Q_2 \log(t_n/r_2^2)$	− 380.6	− 320.4	− 260.2	− 200	− 139.8
$Q_3 \log(t_n/r_3^2)$	− 676.6	− 586.3	− 496.0	− 405.7	− 315.3
	+	+	+		+
$\Sigma Q_i \log(t_n/r_i^2)$	− 1252.4	− 1056.7	− 861.0	− 665.4	− 496
$\frac{\Sigma Q_i \log(t_n/r_i^2)}{\Sigma Q_i}$	− 1.927	− 1.626	− 1.325	− 1.024	− 0.722
$\overline{(t/r_i^2)}_n$ (min/m^2)	0.01	0.02	0.05	0.09	0.19

13 Recovery tests

When the pump is shut down after a pumping test, the water levels in the well and the piezometers will start to rise. This rise in water levels is known as residual drawdown, s′. It is expressed as the difference between the original water level before the start of pumping and the water level measured at a time t′ after the cessation of pumping. Figure 13.1 shows the change in water level with time during and after a pumping test.

It is always good practice to measure the residual drawdowns during the recovery period. Recovery-test measurements allow the transmissivity of the aquifer to be calculated, thereby providing an independent check on the results of the pumping test, although costing very little in comparison with the pumping test.

Residual drawdown data are more reliable than pumping test data because recovery occurs at a constant rate, whereas a constant discharge during pumping is often difficult to achieve in the field.

The analysis of a recovery test is based on the principle of superposition, which was discussed in Chapter 6. Applying this principle, we assume that, after the pump has been shut down, the well continues to be pumped at the same discharge as before, and that an imaginary recharge, equal to the discharge, is injected into the well. The recharge and the discharge thus cancel each other, resulting in an idle well as is required for the recovery period. For any of the well-flow equations presented in the previous chapters, a corresponding 'recovery equation' can be formulated.

The Theis recovery method (Section 13.1.1) is widely used for the analysis of recovery tests. Strictly speaking, this method is only valid for confined aquifers which are fully penetrated by a well that is pumped at a constant rate. Nevertheless, if additional limiting conditions are satisfied, the Theis method can also be used for leaky aquifers (Section 13.1.2) and unconfined aquifers (Section 13.1.3), and aquifers that are only partially penetrated by a well (Section 13.1.4).

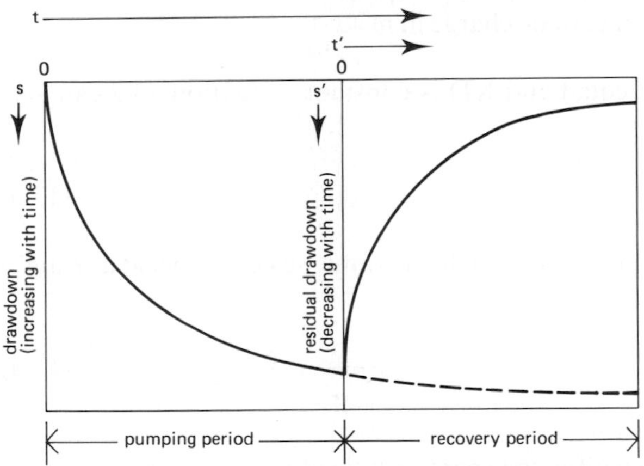

Figure 13.1 Time drawdown and residual drawdown

If the recovery test is conducted in a free-flowing well, the Theis recovery method can also be used (Section 13.2).

If the discharge rate of the pumping test was variable, the Birsoy-Summer recovery method (Section 13.3.1) can be used.

13.1 Recovery tests after constant-discharge tests

13.1.1 Confined aquifers, Theis's recovery method

According to Theis (1935), the residual drawdown after a pumping test with a constant discharge is

$$s' = \frac{Q}{4\pi KD}\{W(u) - W(u')\} \tag{13.1}$$

where

$$u = \frac{r^2S}{4KDt} \text{ and } u' = \frac{r^2S'}{4KDt'}$$

When u and u′ are sufficiently small (see Section 3.2.2 for the approximation of W(u) for $u < 0.01$), Equation 13.1 can be approximated by

$$s' = \frac{Q}{4\pi KD}\left(\ln\frac{4KDt}{r^2S} - \ln\frac{4KDt'}{r^2S'}\right) \tag{13.2}$$

where

s′ = residual drawdown in m
r = distance in m from well to piezometer
KD = transmissivity of the aquifer in m^2/d
S′ = storativity during recovery, dimensionless
S = storativity during pumping, dimensionless
t = time in days since the start of pumping
t′ = time in days since the cessation of pumping
Q = rate of recharge = rate of discharge in m^3/d

When S and S′ are constant and equal and KD is constant, Equation 13.2 can also be written as

$$s' = \frac{2.30Q}{4\pi KD}\log\frac{t}{t'} \tag{13.3}$$

A plot of s′ versus t/t′ on semi-log paper (t/t′ on logarithmic scale) will yield a straight line. The slope of the line is

$$\Delta s' = \frac{2.30Q}{4\pi KD} \tag{13.4}$$

where Δs′ is the residual drawdown difference per log cycle of t/t′.

The Theis recovery method is applicable if the following assumptions and conditions are met:

- The assumptions listed at the beginning of Chapter 3, adjusted for recovery tests.

The following conditions are added:

- The flow to the well is in an unsteady state;
- $u < 0.01$, i.e. pumping time $t_p > (25\, r^2S)/KD$
- $u' < 0.01$, i.e. $t' > (25\, r^2S)/KD$, see also Section 3.2.2.

Procedure 13.1

- For each observed value of s′, calculate the corresponding value of t/t′;
- For one of the piezometers, plot s′ versus t/t′ on semi-log paper (t/t′ on the logarithmic scale);
- Fit a straight line through the plotted points;
- Determine the slope of the straight line, i.e. the residual drawdown difference Δs′ per log cycle of t/t′;
- Substitute the known values of Q and Δs′ into Equation 13.4 and calculate KD.

Remark

- When S and S′ are constant, but unequal, the straight line through the plotted points intercepts the time axis where $s' = 0$ at a point $t/t' = (t/t')_o$. At this point, Equation 13.2 becomes

$$0 = \frac{2.30Q}{4\pi KD}\left[\log\left(\frac{t}{t'}\right)_o - \log\frac{S}{S'}\right]$$

  Because $2.30\, Q/4\pi KD \# 0$, it follows that $\log (t/t')_o - \log (S/S') = 0$. Hence $(t/t')_o = S/S'$, which determines the relative change of S.

13.1.2 Leaky aquifers, Theis's recovery method

After a constant-discharge test in a leaky aquifer, Hantush (1964), disregarding any storage effects in the confining aquitard, expresses the residual drawdown s′ at a distance r from the well as

$$s' = \frac{Q}{4\pi KD}\{W(u,r/L) - W(u',r/L)\} \qquad (13.5)$$

Taking this equation as his basis and using a digital computer, Vandenberg (1975) devised a least-squares method to determine KD, S, and L. For more information on this method, we refer the reader to the original literature.

If the pumping and recovery times are long, leakage through the confining aquitards will affect the water levels. If the times are short, i.e. if $t_p + t' \leq (L^2S)/20KD$ or $t_p + t' \leq cS/20$, the Theis recovery method (Section 13.1.1) can be used, but only the leaky aquifer's transmissivity can be determined (Uffink 1982; see also Hantush 1964).

13.1.3 Unconfined aquifers, Theis's recovery method

An unconfined aquifer's delayed watertable response to pumping (Chapter 5) is fully reversible according to Neuman's theory of delayed watertable response, because hysteresis effects do not play any part in this theory. Neuman (1975) showed that the Theis recovery method (Section 13.1.1) is applicable in unconfined aquifers, but only for late-time recovery data. At late time, the effects of elastic storage, which set in after pumping stopped, have dissipated. The residual drawdown data will then fall on a straight line in the semi-log s′ versus t/t′ plot used in the Theis recovery method.

13.1.4 Partially penetrating wells, Theis's recovery method

The Theis recovery method (Section 13.1.1) can also be used if the well is only partially penetrating. For long pumping times in such a well, i.e. $t_p > (D^2S)/2KD$, the semi-log plot of s versus t yields a straight line with a slope identical to that of a completely penetrating well (Hantush 1961b). Thus, if the straight line portion of the recovery curve is long enough, i.e. if both t_p and t′ are greater than $(10\ D^2S)/KD$, the Theis recovery method can be applied (Uffink 1982).

13.2 Recovery tests after constant-drawdown tests

If the recovery test follows a constant-drawdown test instead of a constant-discharge test, the Theis recovery method (Section 13.1.1) can be applied, provided that the discharge at the moment before the pump is shut down is used in Equation 13.4 (Rushton and Rathod 1980).

13.3 Recovery tests after variable-discharge tests

13.3.1 Confined aquifers, Birsoy-Summers's recovery method

To analyze the residual drawdown data after a pumping test with step-wise or intermittently changing discharge rates, Birsoy and Summers (1980) proposed the following expression

$$\frac{s'}{Q_n} = \frac{2.30}{4\pi KD} \log \left\{ \beta_{t(n)} \left(\frac{t-t_n}{t-t'_n} \right) \right\} \qquad (13.6)$$

where

s' = residual drawdown at $t > t'_n$
Q_n = constant discharge during the last (= n-th) pumping period
t_n = time at which the n-th pumping period started
$t-t_n$ = time since the n-th pumping period started
t'_n = time at which the n-th pumping period ended
$t-t'_n$ = time since the n-th pumping period ended
$\beta_{t(n)}$ is defined according to Equation 12.2

A semi-log plot of s'/Q_n versus the corresponding adjusted time of recovery: $\beta_{t(n)}(t-t_n/t-t'_n)$ yields a straight line. The slope of the straight line $\Delta(s'/Q_n)$ is equal to $2.30/4\pi KD$, from which the transmissivity can be determined.

The Birsoy-Summers recovery method can be used if the following assumptions and conditions are met:

- The assumptions listed at the beginning of Chapter 3, as adjusted for recovery tests, with the exception of the fifth assumption, which is replaced by:
  - Prior to the recovery test, the aquifer is pumped at a variable discharge rate.

The following conditions are added:

- The flow to the well is in an unsteady state;
- $u < 0.01$ $[u = r^2S/4KD\{\beta_{t(n)}(t_p-t_n)\}]$, see also Section 3.2.2;
- $u' < 0.01$ $[u' = r^2S/4KD\{\beta_{t(n)}(t-t_n/t-t'_n)\}]$.

Procedure 13.2

- For a single piezometer, calculate the adjusted time of recovery, $\beta_{t(n)}(t-t_n/t-t'_n)$, by applying Equation 12.2 for the calculation of $\beta_{t(n)}$, and by using all the observed values of the discharge rate and the appropriate values of time;
- On semi-log paper, plot the observed specific residual drawdown s'/Q_n versus the corresponding values of $[\beta_{t(n)}(t-t_n/t-t'_n)]$ (the adjusted time of recovery on the logarithmic scale);
- Draw a straight line through the plotted points;
- Determine the slope of the straight line, $\Delta(s'/Q_n)$, which is the difference of s'/Q_n per log cycle of adjusted time of recovery;
- Calculate KD from $\Delta(s'/Q_n) = 2.30/4\pi KD$.

Remark

- See Section 12.1 for simplified expressions of $\beta_{t(n)}(t-t_n)$ which can be introduced into the expression for the adjusted time of recovery.

14 Well-performance tests

The drawdown in a pumped well consists of two components: the aquifer losses and the well losses. A well-performance test is conducted to determine these losses.

Aquifer losses are the head losses that occur in the aquifer where the flow is laminar. They are time-dependent and vary linearly with the well discharge. In practice, the extra head loss induced, for instance, by partial penetration of a well is also included in the aquifer losses.

Well losses are divided into linear and non-linear head losses (Figure 14.1). Linear well losses are caused by damage to the aquifer during drilling and completion of the well. They comprise, for example, head losses due to compaction of the aquifer material during drilling, head losses due to plugging of the aquifer with drilling mud, which reduce the permeability near the bore hole; head losses in the gravel pack; and head losses in the screen. Amongst the non-linear well losses are the friction losses that occur inside the well screen and in the suction pipe where the flow is turbulent, and the head losses that occur in the zone adjacent to the well where the flow is usually also turbulent. All these well losses are responsible for the drawdown inside the well being much greater than one would expect on theoretical grounds.

Petroleum engineering recognizes the concept of 'skin effect' to account for the head losses in the vicinity of a well. The theory behind this concept is that the aquifer is assumed to be homogeneous up to the wall of the bore hole, while all head losses are assumed to be concentrated in a thin, resistant 'skin' against the wall of the bore hole.

In this chapter, we present two types of well-performance tests: the classical step-drawdown test (Section 14.1) and the recovery test (Section 14.2).

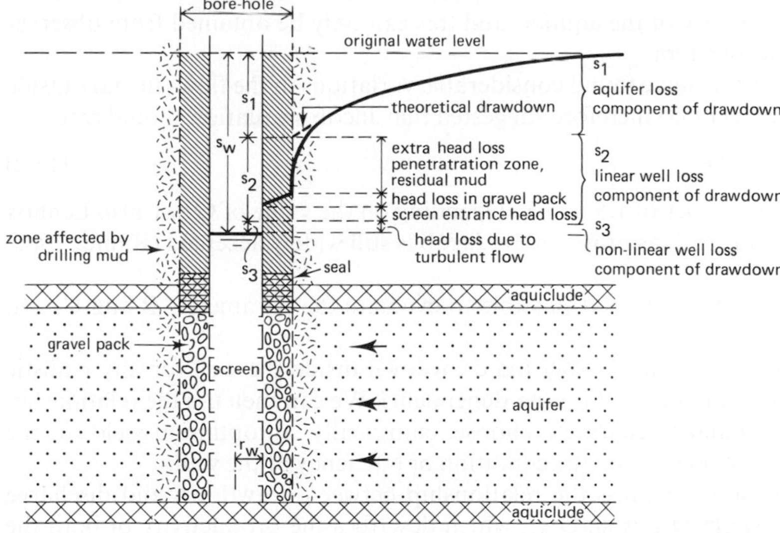

Figure 14.1 Various head losses in a pumped well

14.1 Step-drawdown test

A step-drawdown test is a single-well test in which the well is pumped at a low constant-discharge rate until the drawdown within the well stabilizes. The pumping rate is then increased to a higher constant-discharge rate and the well is pumped until the drawdown stabilizes once more. This process is repeated through at least three steps, which should all be of equal duration, say from 30 minutes to 2 hours each.

The step-drawdown test was first performed by Jacob (1947), who was primarily interested in finding out what the drawdown in a well would be if it were pumped at a rate that differs from the rate during the pumping test. For the drawdown in a pumped well, he gave the following equation

$$s_w = B(r_{ew},t)Q + CQ^2 \tag{14.1}$$

where

$B(r_{ew},t) = B_{1(r_w,t)} + B_2$

$B_{1(r_w,t)}$ = linear aquifer-loss coefficient

B_2 = linear well-loss coefficient
C = non-linear well-loss coefficient
r_{ew} = effective radius of the well
r_w = actual radius of the well
t = pumping time

Jacob combined the various linear head losses at the well into a single term, r_{ew}, the effective radius of the well. He defined this as the distance (measured radially from the axis of the well) at which the theoretical drawdown (based on the logarithmic head distribution) equals the drawdown just outside the well screen. From the data of a step-drawdown test, however, it is not possible to determine r_{ew} because one must also know the storativity of the aquifer, and this can only be obtained from observations in nearby piezometers.

Different researchers have found considerable variations in the flows in and outside of wells. Rorabaugh (1953) therefore suggested that Jacob's equation should read

$$s_w = BQ + CQ^P \tag{14.2}$$

where P can assume values of 1.5 to 3.5, depending on the value of Q (see also Lennox 1966). The value of $P = 2$, as proposed by Jacob is still widely accepted (Ramey 1982; Skinner 1988).

A step-drawdown test makes it possible to evaluate the parameters B and C, and eventually P.

Knowing B and C, we can predict the drawdown inside the well for any realistic discharge Q at a certain time t (B is time-dependent). We can then use the relationship between drawdown and discharge to choose, empirically, an optimum yield for the well, or to obtain information on the condition or efficiency of the well.

We can, for instance, express the relationship between drawdown and discharge as the specific capacity of a well, Q/s_w, which describes the productivity of both the aquifer and the well. The specific capacity is not a constant but decreases as pumping continues (Q is constant), and also decreases with increasing Q. The well efficiency, E_w, can be expressed as

$$E_w = \left\{\frac{B_1Q}{(B_1 + B_2)Q + CQ^P}\right\} \times 100\% \tag{14.3}$$

If a well exhibits no well losses, it is a perfect well. In practice, only the influence of the non-linear well losses on the efficiency can be established, because it is seldom possible to take B_1 and B_2 into account separately. As not all imperfections in well construction show up as non-linear flow resistance, the real degree of a well's imperfection cannot be determined from the well efficiency.

As used in well hydraulics, the concepts of linear and non-linear head loss components ($B_2Q + CQ^2$) relate to the concepts of skin effect and non-Darcyan flow (Ramey 1982). In well hydraulics parlance, the total drawdown inside a well due to well losses (also indicated as the apparent total skin effects) can be expressed as

$$B_2Q + CQ^2 = \frac{1}{2\pi KD}(\text{skin} + C'Q)Q \tag{14.4}$$

where

$C' = C \times 2\pi KD$ = non-linear well loss coefficient or high velocity coefficient

skin $= B_2 \times 2\pi KD$ = skin factor

Matthews and Russel (1967) relate the effective well radius, r_{ew}, to the skin factor by the equation

$$r_{ew} = r_w e^{-\text{skin}} \tag{14.5}$$

Various methods are available to analyze step-drawdown tests. The methods based on Jacob's equation (Equation 14.1) are the Hantush-Bierschenk method (Section 14.1.1) and the Eden-Hazel method (Section 14.1.2). The Hantush-Bierschenk method can determine values of B and C, and can be applied in confined, leaky, or unconfined aquifers. The Eden-Hazel method can be applied in confined aquifers and gives values of well-loss parameters as well as estimates of the transmissivity.

The methods based on Rorabaugh's equation (Equation 14.2) are the Rorabaugh trial-and-error straight line method (Section 14.1.3) and Sheahan's curve-fitting method (Section 14.1.4). They can be used in confined, leaky, or unconfined aquifers, and give values for B, C, and P. Analyzing data from a step-drawdown test does not yield separate values of B_1 and B_2. A recovery test, however, makes it possible to evaluate the skin factor (Section 14.2).

14.1.1 Hantush-Bierschenk's method

By applying the principle of superposition to Jacob's equation (Equation 14.1), Hantush (1964) expresses the drawdown $s_{w(n)}$ in a well during the n-th step of a step-drawdown test as

$$s_{w(n)} = \sum_{i=1}^{n} \Delta Q_i\, B(r_{ew},t\text{-}t_i) + CQ_n^2 \tag{14.6}$$

where

$s_{w(n)}$ = total drawdown in the well during the n-th step at time t
r_{ew} = effective radius of the well
t_i = time at which the i-th step begins ($t_1 = 0$)
Q_n = constant discharge during the n-th step
Q_i = constant discharge during the i-th step of that preceding the n-th step
$\Delta Q_i = Q_i - Q_{i-1}$ = discharge increment beginning at time t_i

The sum of increments of drawdown taken at a fixed interval of time from the beginning of each step ($t - t_i = \Delta t$) can be obtained from Equation 14.6

$$\sum_{i=1}^{n} \Delta s_{w(i)} = s_{w(n)} = B(r_{ew},\Delta t)Q_n + CQ_n^2 \qquad (14.7)$$

where

$\Delta s_{w(i)}$ = drawdown increment between the i-th step and that preceding it, taken at time $t_i + \Delta t$ from the beginning of the i-th step

Equation 14.7 can also be written as

$$\frac{s_{w(n)}}{Q_n} = B(r_{ew},\Delta t) + CQ_n \qquad (14.8)$$

A plot of $s_{w(n)}/Q_n$ versus Q_n on arithmetic paper will yield a straight line whose slope is equal to C. From Equation 14.8 and the coordinates of any point on this line, B can be calculated.

The procedure suggested by Hantush (1964) and Bierschenk (1963) is applicable if the following assumptions and conditions are satisfied:

- The assumptions listed at the beginning of Chapter 3, with the exception of the first and fifth assumptions, which are replaced by:
  - The aquifer is confined, leaky or unconfined;
  - The aquifer is pumped step-wise at increased discharge rates;

The following conditions are added:

- The flow to the well is in an unsteady state;
- The non-linear well losses are appreciable and vary according to the expression CQ^2.

Procedure 14.1

- On semi-log paper, plot the observed drawdown in the well s_w against the corresponding time t (t on the logarithmic scale) (Figure 14.2);
- Extrapolate the curve through the plotted data of each step to the end of the next step;
- Determine the increments of drawdown $\Delta s_{w(i)}$ for each step by taking the difference between the observed drawdown at a fixed time interval Δt, taken from the beginning of each step, and the corresponding drawdown on the extrapolated curve of the preceding step;
- Determine the values of $s_{w(n)}$ corresponding to the discharge Q_n from $s_{w(n)} = \Delta s_{w(1)} + \Delta s_{w(2)} + ... + \Delta s_{w(n)}$. Subsequently, calculate the ratio $s_{w(n)}/Q_n$ for each step;

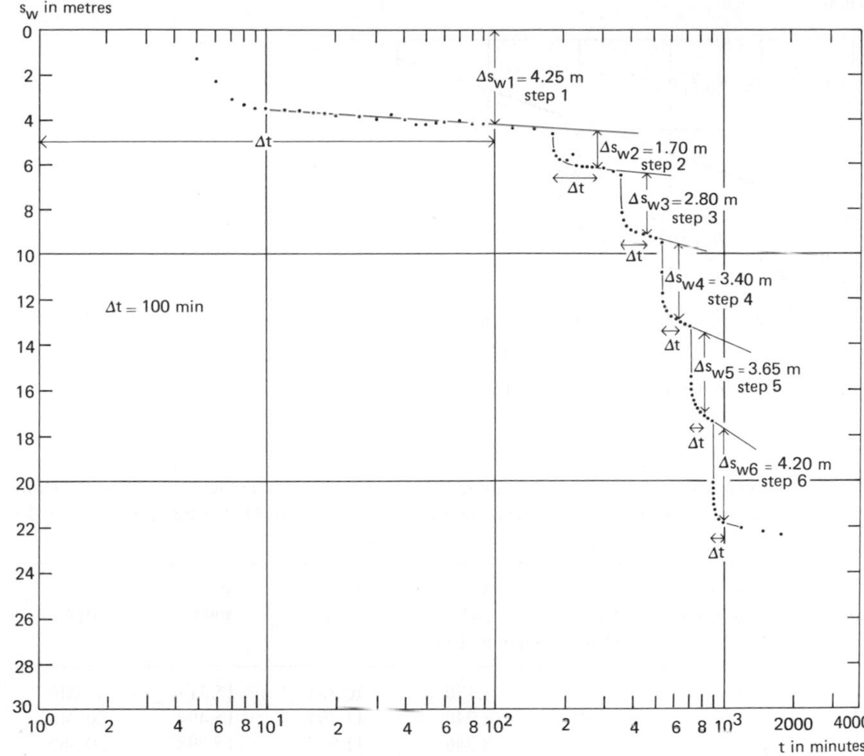

Figure 14.2 The Hantush-Bierschenk method: determination of the drawdown difference for each step

- On arithmetic paper, plot the values of $s_{w(n)}/Q_n$ versus the corresponding values of Q_n (Figure 14.3). Fit a straight line through the plotted points. (If the data do not fall on a straight line, a method based on the well loss component CQ^P should be used; see Sections 14.1.2, 14.1.3 or 14.1.4;
- Determine the slope of the straight line $\Delta(s_{w(n)}/Q_n)/\Delta Q_n$, which is the value of C;
- Extend the straight line until it intercepts the Q = 0 axis. The interception point on the $s_{w(n)}/Q_n$ axis gives the value of B.

Remarks

- The values of $\Delta s_{w(i)}$ depend on extrapolated data and are therefore subject to error;
- When a steady state is reached in each step, the drawdown in the well is no longer time-dependent. Hence, the observed steady-state drawdown and the corresponding discharge for each step can be used directly in the arithmetic plot of $s_{w(n)}/Q_n$ versus Q_n.

Example 14.1

To illustrate the Hantush-Bierschenk method, we shall use the data in Table 14.1. These data have been given by Clark (1977) for a step-drawdown test in 'Well 1', which taps a confined sandstone aquifer.

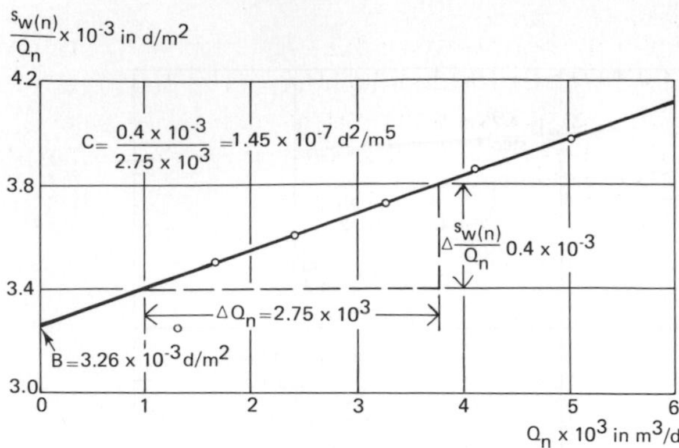

Figure 14.3 The Hantush-Bierschenk method: determination of the parameters B and C

Table 14.1 Step drawdown test data 'Well 1'. Reproduced by permission of the Geological Society from 'The analysis and planning of step-drawdown tests'. L. Clark, in Q.Jl. Engng. Geol. Vol. 10 (1977)

Time in minutes from beginning of step	Step 1 Q: 1306 (m^3/d)	2 1693 Drawdown in metres	3 2423	4 3261	5 4094	6 5019
1	–	5.458	8.170	10.881	15.318	20.036
2	–	5.529	8.240	11.797	15.494	20.248
3	–	5.564	8.346	11.902	15.598	20.389
4	–	5.599	8.451	12.008	15.740	20.529
5	1.303	5.634	8.486	12.078	15.846	20.600
6	2.289	5.669	8.557	12.149	15.881	20.660
7	3.117	5.669	8.557	12.149	15.952	20.741
8	3.345	5.705	8.592	12.184	16.022	20.811
9	3.486	5.740	8.672	12.219	16.022	20.882
10	3.521	5.740	8.672	12.325	16.093	20.917
12	3.592	5.810	8.663	12.360	16.198	20.952
14	3.627	5.810	8.698	12.395	16.268	21.022
16	3.733	5.824	8.733	12.430	16.304	21.128
18	3.768	5.845	8.839	12.430	16.374	21.163
20	3.836	5.810	8.874	12.501	16.409	21.198
25	3.873	5.824	8.874	12.508	16.586	21.304
30	4.014	5.824	8.979	12.606	16.621	21.375
35	3.803	5.881	8.979	12.712	16.691	21.480
40	4.043	5.591	8.994	12.747	16.726	21.551
45	4.261	5.591	9.050	12.783	16.776	21.619
50	4.261	6.092	9.050	12.818	16.797	21.656
55	4.190	6.092	9.120	12.853	16.902	–
60	4.120	6.176	9.120	12.853	16.938	21.663
70	4.120	6.162	9.155	12.888	16.973	21.691
80	4.226	6.176	9.191	12.923	17.079	21.762
90	4.226	6.169	9.191	12.994	17.079	21.832
100	4.226	6.169	9.226	12.994	17.114	21.903
120	4.402	6.176	9.261	13.099	17.219	22.008
150	4.402	6.374	9.367	13.205	17.325	22.184
180	4.683	6.514	9.578	13.240	17.395	22.325

Figure 14.2 shows the semi-log plot of the drawdown data versus time. From this plot, we determine the drawdown differences for each step and for a time-interval $\Delta t = 100$ min. We then calculate the specific drawdown values $s_{w(n)}/Q_n$ (Table 14.2). Plotting the $s_{w(n)}/Q_n$ values against the corresponding values of Q_n on arithmetic paper gives a straight line with a slope of 1.45×10^{-7} d^2/m^5 (= C) (Figure 14.3). The interception point of the straight line with the $Q_n = 0$ axis has a value of $s_{w(n)}/Q_n = 3.26 \times 10^{-3}$ d/m^2 (= B). Hence, we can write the drawdown equation for 'Well 1' as

$$s_w = (3.26 \times 10^{-3})\, Q + (1.45 \times 10^{-7})\, Q^2 \text{ (for t = 100 min).}$$

Table 14.2 Specific drawdown determined with the Hantush-Bierschenk method: step-drawdown test 'Well 1'

	$\Delta s_{w(n)}$	$s_{w(n)}$	Q_n	$s_{w(n)}/Q_n$
	m	m	m^3/d	d/m^2
Step 1	4.25	4.25	1306	3.25×10^{-3}
Step 2	1.70	5.95	1693	3.51×10^{-3}
Step 3	2.80	8.75	2423	3.61×10^{-3}
Step 4	3.40	12.15	3261	3.73×10^{-3}
Step 5	3.65	15.80	4094	3.86×10^{-3}
Step 6	4.20	20.00	5019	3.98×10^{-3}

($\Delta s_{w(n)}$ determined for $\Delta t = 100$ min)

14.1.2 Eden-Hazel's method (confined aquifers)

From step-drawdown tests in a fully penetrating well that taps a confined aquifer, the Eden-Hazel method (1973) can determine the well losses, and also the transmissivity of the aquifer. The method is based on Jacob's approximation of the Theis equation (Equation 3.7).

The drawdown in the well is given by the Jacob equation, now written as

$$s_w = \frac{2.30Q}{4\pi KD} \log \frac{2.25KDt}{r_{ew}^2\, S}$$

This equation can also be written as

$$s_w = (a + b \log t)Q \qquad (14.9)$$

where

$$a = \frac{2.30}{4\pi KD} \log \frac{2.25KD}{r_{ew}^2\, S} \qquad (14.10)$$

$$b = \frac{2.30}{4\pi KD} \qquad (14.11)$$

Using the principle of superposition and Equation 14.9, we derive the drawdown at time t during the n-th step from

$$s_{w(n)} = \sum_{i=1}^{n} (\Delta Q_i) \{a + b \log(t\text{-}t_i)\} \tag{14.12}$$

or

$$s_{w(n)} = aQ_n + b \sum_{i=1}^{n} \Delta Q_i \log(t\text{-}t_i) \tag{14.13}$$

where

Q_n = constant discharge during the n-th step
Q_i = constant discharge during the i-th step of that preceding the n-th step
ΔQ_i = $Q_i - Q_{i-1}$ = discharge increment beginning at time t_i
t_i = time at which the i-th step begins
t = time since the step-drawdown test started

The above equations do not account for the influence of non-linear well losses. Introducing these losses (CQ^2) into Equation 14.13 gives

$$s_{w(n)} = aQ_n + bH_n + CQ_n^2 \tag{14.14}$$

where

$$H_n = \sum_{i=1}^{n} \Delta Q_i \log(t\text{–}t_i) \tag{14.15}$$

The Eden-Hazel Procedure 14.2 can be used if the following assumptions and conditions are satisfied:
- The conditions listed at the beginning of Chapter 3, with the exception of the fifth assumption, which is replaced by:
  • The aquifer is pumped step-wise at increased discharge rates;

The following conditions are added:
- The flow to the well is in an unsteady state;
- $u < 0.01$;
- The non-linear well losses are appreciable and vary according to the expression CQ^2.

The Eden-Hazel Procedure 14.3 can be used if the last condition is replaced by:
- The non-linear well losses are appreciable and vary according to the expression CQ^P.

Procedure 14.2
- Calculate the values of H_n from Equation 14.15, using the measured discharges and times;
- On arithmetic paper, plot the observed drawdowns $s_{w(n)}$ versus the corresponding calculated values of H_n (Figure 14.4);
- Draw parallel straight lines of best fit through the plotted points, one straight line through each set of points (Figure 14.4);
- Determine the slope of the lines $\Delta s_{w(n)}/\Delta H_n$, which gives the value of b;
- Extend the lines until they intercept the $H_n = 0$ axis. The interception point (A_n) of each line is given by

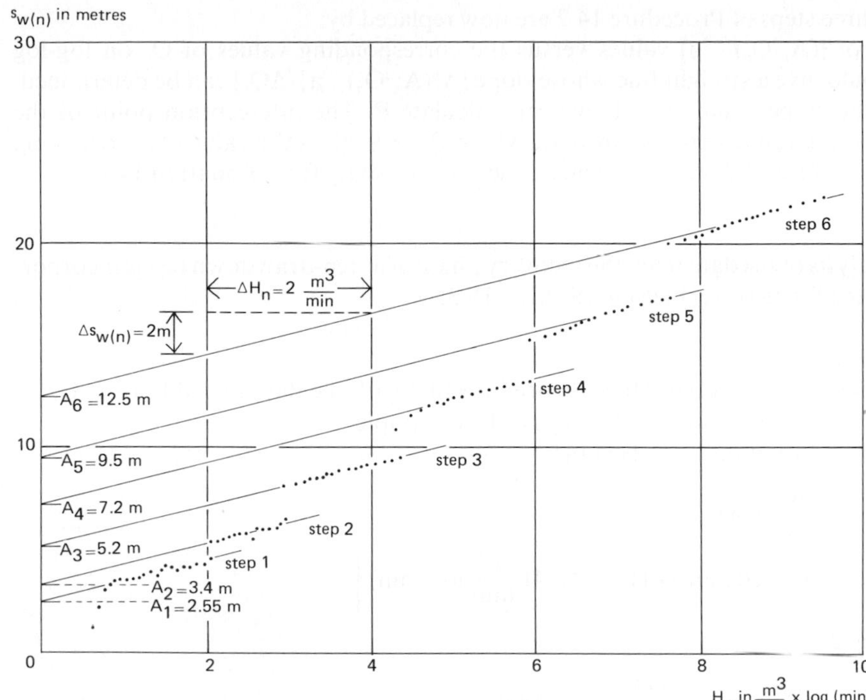

Figure 14.4 The Eden-Hazel method: arithmetic plot of $s_{w(n)}$ versus H_n

$$A_n = aQ_n + CQ_n^2, \qquad \text{or} \frac{A_n}{Q_n} = a + CQ_n \tag{14.16}$$

- Read the values of A_n;
- Calculate the ratio A_n/Q_n for each step (i.e. for each value of Q_n);
- On arithmetic paper, plot the values of A_n/Q_n versus the corresponding values of Q_n. Fit a straight line through the plotted points (Figure 14.5);
- Determine the slope of the straight line $\Delta(A_n/Q_n)/\Delta Q_n$, which is the value of C;
- Extend the straight line until it intersects the A_n/Q_n axis where $Q_n = 0$; the value of the intersection point is equal to a;
- Knowing b, calculate KD from Equation 14.11.

Procedure 14.3

- The Eden-Hazel method can also be used if the well losses vary with CQ^P, as may happen when well discharges are high (e.g. in a test to determine the maximum yield of a well). In Equations 14.14 and 14.16, CQ^2 should then be replaced by CQ^P. The adjusted Equation 14.16, after being rearranged in logarithmic form thus becomes

$$\log\left(\frac{A_n}{Q_n} - a\right) = \log C + (P-1)\log Q_n$$

The last three steps of Procedure 14.2 are now replaced by:
- A plot of $[(A_n/Q_n) - a]$ values versus the corresponding values of Q_n on log-log paper should give a straight line whose slope $[\Delta\{(A_n/Q_n) - a\}/\Delta Q_n]$ can be determined. Because the slope equals P – 1, we can calculate P. The interception point of the extended straight line with the ordinate where $Q_n = 0$, gives the value of C. Knowing b from Procedure 14.2, we can calculate the transmissivity from Equation 14.11.

Remark
- The analysis of the data from the recovery phase of a step-drawdown test is incorporated in the Eden-Hazel method (Section 15.3.3).

Example 14.2
We shall illustrate the Eden-Hazel Procedure 14.2 with the data in Table 14.1. Using Equation 14.15, we calculate values of H_n. For example:
- For Step 1, Equation 14.15 becomes

$$H_1 = \frac{1306}{1440} \log t$$

$$\left[t = 50 \text{ min} \rightarrow H_1 = 1.541 \frac{m^3}{\text{min}} \log(\text{min})\right]$$

- For Step 2

$$H_2 = \frac{1306}{1440} \log t + \frac{387}{1440} \log(t-180)$$

$$\left[t = 230 \text{ min} \rightarrow H_2 = 2.599 \frac{m^3}{\text{min}} \log(\text{min})\right]$$

- For Step 6

$$H_6 = \frac{1306}{1440} \log t + \frac{387}{1440} \log(t-180) + \frac{730}{1440} \log(t-360) + \frac{838}{1440} \log(t-540) + \frac{833}{1440} \log(t-720) + \frac{925}{1440} \log(t-900)$$

$$\left[t = 950 \text{ min} \rightarrow H_6 = 8.859 \frac{m^3}{\text{min}} \log(\text{min})\right]$$

Figure 14.4 gives the arithmetic plot of $s_{w(n)}$ versus H_n. The slope of the parallel straight lines is

$$b = \frac{\Delta s_{w(n)}}{\Delta H_n} = \frac{2}{2} \times \frac{1}{1440} = 6.9 \times 10^{-4} \text{ d/m}^2$$

Introducing b into Equation 14.11 gives $KD = 2.30/4\pi \times 6.9 \times 10^{-4} = 265$ m²/d. The values of the intersection points A_n (Figure 14.4) are: $A_1 = 2.55$ m; $A_2 = 3.4$ m; $A_3 = 5.2$ m; $A_4 = 7.2$ m; $A_5 = 9.5$ m; and $A_6 = 12.5$ m. A plot of the calculated values of A_n/Q_n versus Q_n (Figure 14.5) gives a straight line with a slope $\Delta(A_n/Q_n)/\Delta Q_n = 0.28 \times 10^{-3}/2000 = 1.4 \times 10^{-7}$. Hence, $C = 1.4 \times 10^{-7}$ d²/m⁵. At the intersection of the straight line and the ordinate where $Q_n = 0$, $a = 1.78 \times 10^{-3}$ d/m².

After being pumped at a constant discharge Q for t days, the well has a drawdown

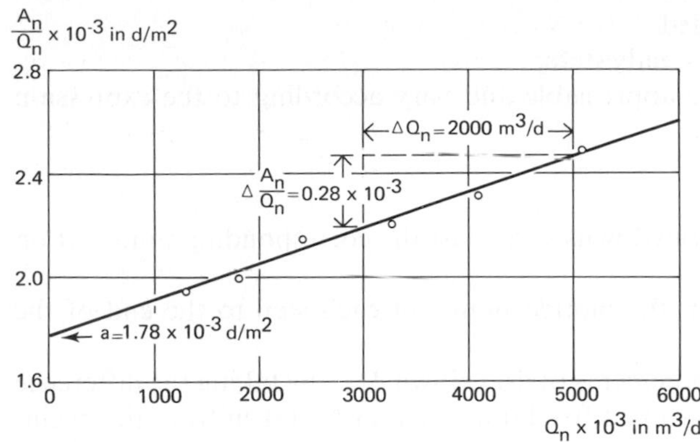

Figure 14.5 The Eden-Hazel method: arithmetic plot of A_n/Q_n versus Q_n

$s_w = \{(1.78 \times 10^{-3}) + (6.9 \times 10^{-4}) \log t\} Q + (1.4 \times 10^{-7})Q^2$. The estimated transmissivity of the aquifer $KD = 265\ m^2/d$.
Note: The separate analysis of the data from the recovery phase of the step-drawdown test on Well 1 gives $KD = 352\ m^2/d$ (Section 15.3.3). In practice, the Eden-Hazel method should be applied to both the drawdown and recovery data.

14.1.3 Rorabaugh's method

If the principle of superposition is applied to Rorabaugh's equation (Equation 14.2), the expression for the drawdown corresponding to Equation 14.7 reads

$$\sum_{i=1}^{n} \Delta s_{w(i)} = s_{w(n)} = BQ_n + CQ^P{}_n \tag{14.17}$$

which can also be written as

$$\frac{s_{w(n)}}{Q_n} = B + CQ_n^{P-1} \tag{14.18}$$

or

$$\log\left[\frac{s_{w(n)}}{Q_n} - B\right] = \log C + (P-1)\log Q_n \tag{14.19}$$

A plot of $[(s_{w(n)}/Q_n) - B]$ versus Q_n on log-log paper will yield a straight line relationship (Figure 14.6).

The assumptions and conditions underlying Rorabaugh's method are:
- The assumptions listed at the beginning of Chapter 3, with the exception of the first and fifth assumptions, which are replaced by:
  - The aquifer is confined, leaky or unconfined;
  - The aquifer is pumped step-wise at increased discharge rates.

The following conditions are added:

- The flow to the well is in an unsteady state;
- The non-linear well losses are appreciable and vary according to the expression CQ^P.

Procedure 14.4

- On semi-log paper, plot the drawdowns s_w against the corresponding times t (t on the logarithmic scale);
- Extrapolate the curve through the plotted points of each step to the end of the next step;
- For each step, determine the increments of drawdown $\Delta s_{w(i)}$ by taking the difference between the observed drawdown at a fixed time interval Δt, taken from the beginning of that step, and the corresponding drawdown on the extrapolated drawdown curve of the preceding step;
- Determine the values of $s_{w(n)}$ corresponding to the discharge Q_n from $s_{w(n)} = \Delta s_{w(1)} + \Delta s_{w(2)} + ... + \Delta s_{w(n)}$;
- Assume a value of B_i and calculate $[(s_{w(n)}/Q_n) - B_i]$ for each step;
- On log-log paper, plot the values of $[(s_{w(n)}/Q_n) - B_i]$ versus the corresponding values of Q_n. Repeat this part of the procedure for different values of B_i. The value of B_i that gives the straightest line on the plot will be the correct value of B;
- Calculate the slope of the straight line $\Delta[(s_{w(n)}/Q_n) - B]/\Delta Q_n$. This equals (P – 1), from which P can be obtained;
- Determine the value of the interception of the straight line with the $Q_n = 1$ axis. This value of $[(s_{w(n)}/Q_n) - B]$ is equal to C.

Remark

- When steady state is reached in each step, the observed steady-state drawdown and the corresponding discharge for each step can be used directly in a log-log plot of $[(s_{w(n)}/Q_n) - B_i]$ versus Q_n.

Example 14.3

To demonstrate the Rorabaugh method, we shall use the specific drawdown data and the corresponding discharge rates presented in Table 14.3 (after Sheahan 1971).

Values of $[(s_{w(n)}/Q_n) - B_i]$ have been calculated for $B_i = 0$; 0.8×10^{-3}; 1×10^{-3}; and 1.1×10^{-3} d/m² (Table 14.4). Figure 14.6 shows a log-log plot of $[(s_{w(n)}/Q)_n - B_i]$ versus Q_n. For $B_3 = 1 \times 10^{-3}$ d/m², the plotted points fall on a straight line. The slope of this line is

$$\frac{\Delta[(s_{w(n)}/Q_n) - B_3]}{\Delta Q_n} = \frac{\log 10^{-2} - \log 10^{-3}}{\log (17.500/5100)} = 1.85$$

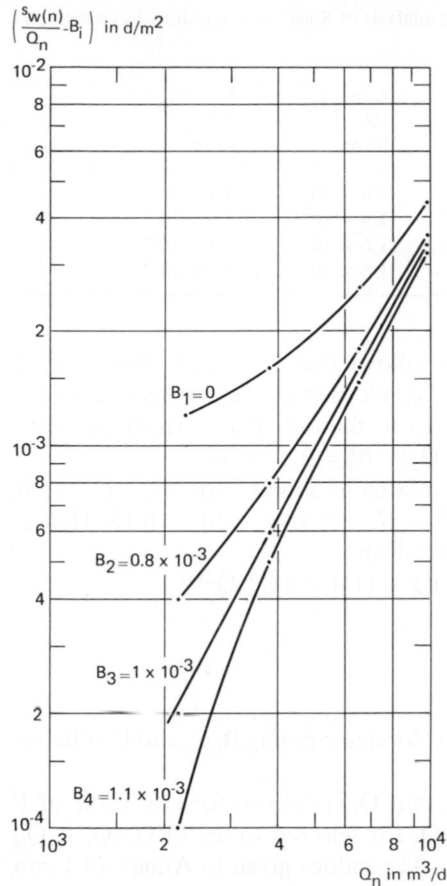

Figure 14.6 The Rorabaugh method

Table 14.3 Step-drawdown test data (from Sheahan 1971)

Total drawdown $s_{w(n)}$	Discharge Q_n	Specific drawdown $s_{w(n)}/Q_n$
(m)	(m^3/d)	(d/m^2)
2.62	2180	1.2×10^{-3}
6.10	3815	1.6×10^{-3}
17.22	6540	2.6×10^{-3}
42.98	9811	4.4×10^{-3}

Table 14.4 Values of $[(s_{w(n)}/Q_n) - B_i]$ and B_i as used in the analysis of Sheahan's step-drawdown test data with Rorabaugh's method

	$\frac{s_{w(1)}}{Q_1} - B_i$ (d/m²)	$\frac{s_{w(2)}}{Q_2} - B_i$ (d/m²)	$\frac{s_{w(3)}}{Q_3} - B_i$ (d/m²)	$\frac{s_{w(4)}}{Q_4} - B_i$ (d/m²)
$B_1 = 0$	1.2×10^{-3}	1.6×10^{-3}	2.6×10^{-3}	4.4×10^{-3}
$B_2 = 0.8 \times 10^{-3}$ d/m²	0.4×10^{-3}	0.8×10^{-3}	1.8×10^{-3}	3.6×10^{-3}
$B_3 = 1 \times 10^{-3}$ d/m²	0.2×10^{-3}	0.6×10^{-3}	1.6×10^{-3}	3.4×10^{-3}
$B_4 = 1.1 \times 10^{-3}$ d/m²	0.1×10^{-3}	0.5×10^{-3}	1.5×10^{-3}	3.3×10^{-3}

Because the slope of the line equals (P – 1), it follows that P = 2.85. The value of $[(s_{w(n)}/Q_n) - B]$ for $Q_n = 10^4$ m³/d is 3.55×10^{-3} d/m². Hence, the intersection of the line with the $Q_n = 1$ m³/d axis is four log cycles to the left. This corresponds with $4 \times 1.85 = 7.4$ log cycles below the point $[(s_{w(n)}/Q_n) - B] = 3.55 \times 10^{-3}$.

The interception point $[(s_{w(n)}/Q_n) - B]_j$ is calculated as follows: $\log [(s_{w(n)}/Q_n) - B]_j = \log 3.55 \times 10^{-3}) - \log (10^{7.4}) = -3 + 0.55 - 7 - 0.4 = -10 + 0.15$. Hence, $[(s_{w(n)}/Q_n) - B]_j = 1.4 \times 10^{-10}$, and $C = 1.4 \times 10^{-10}$ d²/m⁵.
The well drawdown equation is $s_w = (10 \times 10^{-4})Q + (1.4 \times 10^{-10})Q^{2.85}$.

14.1.4 Sheahan's method

Sheahan (1971) presented a curve-fitting method for determining B, C, and P of Rorabaugh's equation (Equation 14.18).

Assuming that B = 1, C = 1, P > 1, and that Q_i is defined for any value of P by $Q_i^{P-1} = 100$, we can calculate the ratio $s_{w(n)}/Q_n$ for selected values of Q_n ($Q_n < Q_i$) and P, using Equation 14.18 (see Annex 14.1). The values given in Annex 14.1 can be plotted on log-log paper as a family of type curves (Figure 14.7).
For those values of Q_n that equal Q_x, Equation 14.18 can be written as

$$\frac{s_{w(x)}}{Q_x} = B + CQ_x^{P-1} = 2B \tag{14.20}$$

and consequently

$$B = CQ_x^{P-1} = \frac{[s_{w(x)}/Q_x]}{2} \tag{14.21}$$

and

$$C = \frac{B}{Q_x^{P-1}} = \frac{(s_{w(x)}/Q_x)}{2Q_x^{P-1}} \tag{14.22}$$

For B = 1 and C = 1, Equation 14.21 gives $s_{w(x)}/Q_x = 2$, and from Equation 14.22 it follows that $Q_x^{P-1} = 1$, or $Q_x = 1$. Hence, for all values of P and assuming that B = 1 and C = 1, the ratio $s_{w(x)}/Q_x = 2$, and $Q_x = 1$ (see also Annex 14.1). All type curves based on the values in Annex 14.1 and plotted on log-log paper pass through the point $s_{w(n)}/Q_n = 2$; $Q_n = 1$. As this is inconvenient for the curve-matching procedure, the type curves are redrawn on plain paper in such a way that the common point expands into an 'index line', located at $s_{w(n)}/Q_n = 2$ (Figure 14.7).

Sheahan's curve-fitting method is applicable if the following assumptions and conditions are satisfied:

- The assumptions listed at the beginning of Chapter 3, with the exception of the first and fifth assumptions, which are replaced by:
  - The aquifer is confined, leaky or unconfined;
  - The aquifer is pumped step-wise at increased discharge rates.

The following conditions are added:

- The flow to the well is in an unsteady state;
- The non-linear well losses are appreciable and vary according to the expression CQ^P.

Procedure 14.5

- On a sheet of log-log paper, prepare the family of Sheahan type curves by plotting $s_{w(n)}/Q_n$ versus Q_n for different values of P, using Annex 14.1. Redraw the family

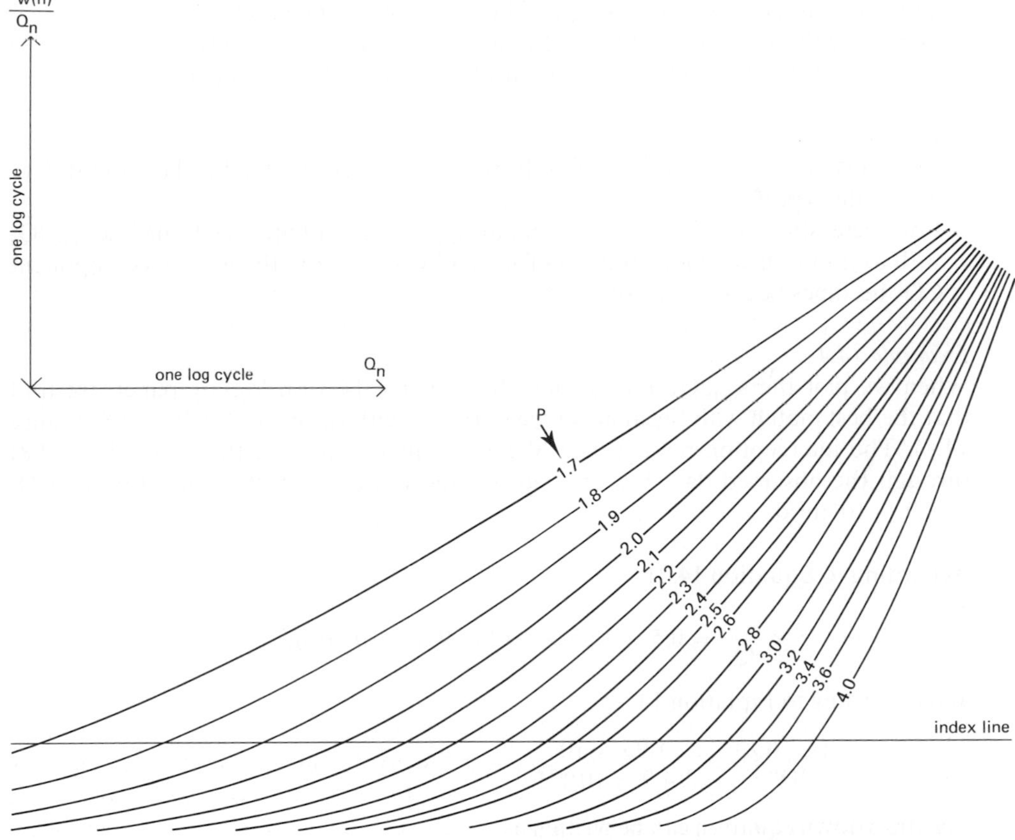

Figure 14.7 Family of Sheahan's type curves $s_{w(n)}/Q_n$ for different values of P (B = 1; C = 1; P > 1; $Q_n < Q_i$; $Q_i^{P-1} = 100$) (after Sheahan 1971)

of type curves on plain paper in such a way that the point $s_{w(n)}/Q_n = 2$; $Q_n = 1$ expands into an index line located at $s_{w(n)}/Q_n = 2$ (see Figure 14.7);

- On semi-log paper, plot the observed drawdowns in the well s_w against the corresponding times t (t on the logarithmic scale);
- Extrapolate the curve through the plotted points of each step to the end of the next step;
- Determine the increments of drawdown $\Delta s_{w(i)}$ for each step by taking the difference between the observed drawdown at a fixed time interval Δt, taken from the beginning of the step, and the corresponding drawdown on the extrapolated drawdown curve of the preceding step;
- Determine the values of $s_{w(n)}$ corresponding to the discharge Q_n from $s_{w(n)} = \Delta s_{w(1)} + \Delta s_{w(2)} + ... + \Delta s_{w(n)}$. Subsequently, calculate the ratio $s_{w(n)}/Q_n$ for each step;
- On log-log paper of the same scale as that used for the log-log plot of Sheahan's type curves, plot the calculated values of the ratio $s_{w(n)}/Q_n$ versus the corresponding values of Q_n;
- Match the data plot with one of the family of type curves and note the value of P for that type curve;
- For the intersection point of type curve and index line, read the corresponding coordinates from the data plot. This gives the values of $s_{w(x)}/Q_x$ and Q_x;
- Substitute the value of $s_{w(x)}/Q_x$ into Equation 14.21 and calculate B;
- Substitute the values of B, Q_x, and P into Equation 14.22 and calculate C.

Remarks

- The most accurate analysis of step-drawdown data is obtained if the plotted data fall on the type curve's portion of greatest curvature;
- For decreasing values of Q_n, the Sheahan type curves all approach the line $s_{w(n)}/Q_n = B$ asymptotically, indicating that for small values of Q_n, the well loss component CQ^P becomes negligibly small.

Example 14.4

When we plot the $s_{w(n)}/Q_n$ and Q_n data from Table 14.3 on log-log paper, we find that the best match with Sheahan's type curves is with the curve for P = 2.8 (Figure 14.8). The interception point (x) of Sheahan's index line and the curve (P = 2.8) through the observed data has the coordinates $s_{w(x)}/Q_x = 1.95 \times 10^{-3}$ d/m² and $Q_x = 4.9 \times 10^3$ m³/d.

According to Equation 14.21

$$B = 0.5 \times \frac{s_{w(x)}}{Q_x} = 0.5 \times 1.95 \times 10^{-3} = 9.8 \times 10^{-4} \text{ d/m}^2$$

and according to Equation 14.22

$$C = \frac{(s_{w(x)}/Q_x)}{2Q_x^{P-1}} = \frac{1.95 \times 10^{-3}}{2(4.9 \times 10^3)^{(2.8-1)}} = 2.2 \times 10^{-10} \text{ d}^2/\text{m}^5$$

The drawdown equation can be written as

$$s_w = (9.8 \times 10^{-4})Q + (2.2 \times 10^{-10})Q^{2.8}$$

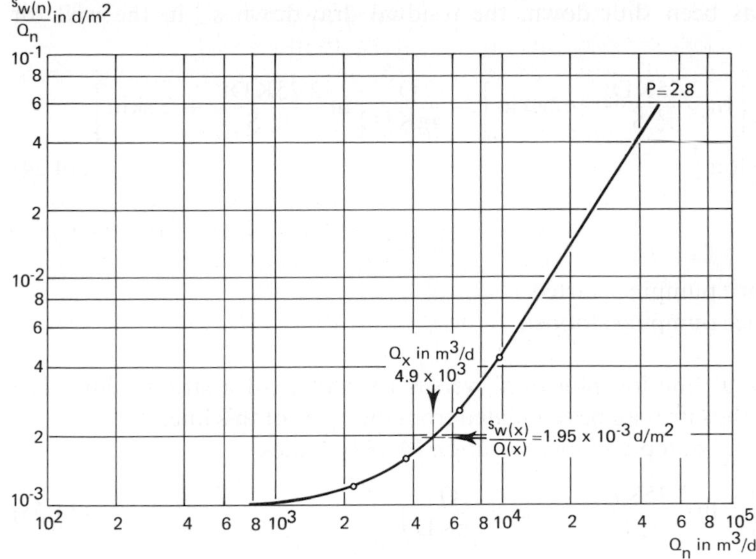

Figure 14.8 Sheahan's method

14.2 Recovery tests

14.2.1 Determination of the skin factor

If the effective radius of the well r_{ew} is larger than the real radius of the bore hole r_w, we speak of a positive skin effect. If it is smaller, the well is usually poorly developed or its screen is clogged, and we speak of a negative skin effect (De Marsily 1986).

In groundwater hydraulics, the skin effect is defined as the difference between the total drawdown observed in a well and the aquifer loss component, assuming that the non-linear well losses are negligible. Adding the skin effect to Jacob's equation (3.7) and assuming that the non-linear well losses are so small that they can be neglected, we obtain the following equation for the drawdown in a well that fully penetrates a confined aquifer and is pumped at a constant rate

$$s_w = \frac{Q}{4\pi KD} \ln \frac{2.25KDt}{r_w^2 S} + (\text{skin}) \frac{Q}{2\pi KD}$$

$$= \frac{Q}{4\pi KD} \left[\ln \frac{2.25KDt}{r_w^2 S} + 2(\text{skin}) \right] \qquad (14.23)$$

where

skin $(Q/2\pi KD)$ = skin effect in m
skin = skin factor (dimensionless)
r_w = actual radius of the well in m

After the pump has been shut down, the residual drawdown s'_w in the well for $t' > 25r_w^2S/KD$ is

$$s'_w = \frac{Q}{4\pi KD}\left[\ln\frac{2.25KDt}{r_w^2\,S} + 2\text{ skin}\right] - \frac{Q}{4\pi KD}\left[\ln\frac{2.25KDt'}{r_w^2\,S} + 2\text{ skin}\right]$$

$$= \frac{2.30Q}{4\pi KD}\log\frac{t}{t'} \tag{14.24}$$

where

t = time since pumping started
t' = time since pumping stopped

For $t' > 25r_w^2S/KD$, a semi-log plot of s'_w versus t/t' will yield a straight line. The transmissivity of the aquifer can be calculated from the slope of this line.
For time $t = t_p$ = total pumping time, Equation 14.23 becomes

$$s_w(t_p) = \frac{Q}{4\pi KD}\ln\frac{2.25KDt_p}{r_w^2\,S} + \text{skin}\left(\frac{Q}{2\pi KD}\right) \tag{14.25}$$

The difference between $s_w(t_p)$ and the residual drawdown s'_w at any time t', is

$$s_w(t_p) - s'_w = \frac{Q}{4\pi KD}\ln\frac{2.25KDt_p}{r_w^2\,S} + \text{skin}\left(\frac{Q}{2\pi KD}\right) - \frac{Q}{4\pi KD}\ln\frac{t_p + t'}{t'} \tag{14.26}$$

For $\frac{t_p + t'_i}{t'_i} = \frac{2.25KDt_p}{r_w^2\,S}$ (14.27)

Equation 14.26 reduces to

$$s_w(t_p) - s'_{wi} = \text{skin}\left(\frac{Q}{2\pi KD}\right) \tag{14.28}$$

The procedure for determining the skin factor has been described by various authors (e.g. Matthews and Russell 1967). It is applicable if the following assumptions and conditions are satisfied:
- The assumptions listed at the beginning of Chapter 3, adjusted for recovery tests.

The following conditions are added:
- The aquifer is confined, leaky or unconfined;
- The flow to the well is in an unsteady state;
- $u < 0.01$;
- $u' < 0.01$;
- The linear well losses (i.e. the skin effect) are appreciable, and the non-linear well losses are negligible.

Procedure 14.6
- Follow Procedure 13.1 or Procedure 15.8 (the Theis recovery method) to determine KD;
  • On semi-log paper, plot the residual drawdown s'_w versus corresponding values of t/t' (t/t' on logarithmic scale);
  • Fit a straight line through the plotted points;

- Determine the slope of the straight line, i.e. the residual drawdown difference $\Delta s'_w$ per log cycle of t/t';
- Substitute the known values of Q and $\Delta s'_w$ into $\Delta s'_w = 2.30Q/4\pi KD$, and calculate KD;

– Determine the ratio $(t_p + t'_i)/t'_i$ by substituting the values of the total pumping time t_p, the calculated KD, the known value of r_w, and an assumed (or known) value of S into Equation 14.27;

– Read the value of s'_{wi} corresponding to the calculated value of $(t_p + t'_i)/t'_i$ from the extrapolated straight line of the data plot s'_w versus t/t';

– Substitute the observed value of $s_w(t_p)$ corresponding to pumping time $t = t_p$, and the known values of s'_{wi}, Q, and KD into Equation 14.28 and solve for the skin factor.

15 Single-well tests with constant or variable discharges and recovery tests

A single-well test is a test in which no piezometers are used. Water-level changes during pumping or recovery are measured only in the well itself. The drawdown in a pumped well, however, is influenced by well losses (Chapter 14) and well-bore storage. In the hydraulics of well flow, the well is generally regarded as a line source or line sink, i.e. the well is assumed to have an infinitesimal radius so that the well-bore storage can be neglected. In reality, any well has a finite radius and thus a certain storage capacity. Well-bore storage is large when compared with the storage in an equal volume of aquifer material. In a single-well test, well-bore storage must be considered when analyzing the drawdown data.

Papadopulos and Cooper (1967) observed that the influence of well-bore storage on the drawdown in a well decreases with time and becomes negligible at $t > 25r_c^2/KD$, where r_c is the radius of the unscreened part of the well, where the water level is changing.

To determine whether the early-time drawdown data are dominated by well-bore storage, a log-log plot of drawdown s_w versus pumping time t should be made. If the early-time drawdowns plot as a unit-slope straight line, we can conclude that well-bore storage effects exist.

The methods presented in Sections 15.1 and 15.2 take the linear well losses (skin effects) into account by using the effective well radius r_{ew} in the equations instead of the actual well radius r_w. Most methods are based on the assumption that non-linear well losses can be neglected. If not, the drawdown data must be corrected with the methods presented in Chapter 14.

Section 15.1 presents four methods of analysis for single-well constant-discharge tests. The Papadopulos-Cooper curve-fitting method (Section 15.1.1) and Rushton-Singh's modified version of it (Section 15.1.2) are applicable for confined aquifers. Jacob's straight-line method (Section 15.1.3), does not require any corrections for non-linear well losses and can be used for confined or leaky aquifers, and so also can Hurr-Worthington's approximation method (Section 15.1.4). All four methods are applicable if the early-time data are affected by well-bore storage, provided that sufficient late-time data ($t > 25\ r_c^2/KD$) are also available.

Section 15.2 treats variable-discharge tests. Birsoy-Summers's method (Section 15.2.1) can be used for confined aquifers. A special type of variable discharge test, the free-flowing-well test, can be analyzed by Jacob-Lohman's method (Section 15.2.2) for confined aquifers and by Hantush's method (Section 15.2.3) for leaky aquifers.

A recovery test is invaluable if the pumping test is performed without the use of piezometers.

The methods for analyzing residual drawdown data (Chapter 13) are straight-line methods. The transmissivity of the aquifer is calculated from the slope of a semi-log straight-line, i.e. from differences in residual drawdown. Those influences on the residual drawdown that are or become constant with time, i.e. well losses, partial penetration, do not affect the calculation of the transmissivity. The methods presented in Chapter 13 are also applicable to single-well recovery test data (Section 15.3). In applying these methods, one must make allowance for those influences on the residual drawdown that do not become constant with time, e.g. well-bore storage.

15.1 Constant-discharge tests

15.1.1 Confined aquifers, Papadopulos-Cooper's method

For a constant-discharge test in a well that fully penetrates a confined aquifer, Papadopulos and Cooper (1967) devised a curve-fitting method that takes the storage capacity of the well into account. The method is based on the following drawdown equation

$$s_w = \frac{Q}{4\pi KD} F(u_w,\alpha) \qquad (15.1)$$

where

$$u_w = \frac{r_{ew}^2 S}{4KDt} \qquad (15.2)$$

$$\alpha = \frac{r_{ew}^2 S}{r_c^2} \qquad (15.3)$$

r_{ew} = effective radius of the screened (or otherwise open) part of the well; $r_{ew} = r_w e^{-skin}$

r_c = radius of the unscreened part of the well where the water level is changing

Values of the function $F(u_w,\alpha)$ are given in Annex 15.1.

The assumptions and conditions underlying the Papadopulos-Cooper method are:
- The assumptions listed at the beginning of Chapter 3, with the exception of the eighth assumption, which is replaced by:
  - The well diameter cannot be considered infinitesimal; hence, storage in the well cannot be neglected.

The following conditions are added:
- The flow to the well is in an unsteady state;
- The non-linear well losses are negligible.

Procedure 15.1
- On log-log paper and using Annex 15.1, plot the family of type curves $F(u_w,\alpha)$ versus $1/u_w$ for different values of α (Figure 15.1);
- On another sheet of log-log paper of the same scale, plot the data curve s_w versus t;
- Match the data curve with one of the type curves;
- Choose an arbitrary point A on the superimposed sheets and note for that point the values of $F(u_w,\alpha)$, $1/u_w$, s_w, and t; note also the value of α of the matching type curve;
- Substitute the values of $F(u_w,\alpha)$ and s_w, together with the known value of Q, into Equation 15.1 and calculate KD.

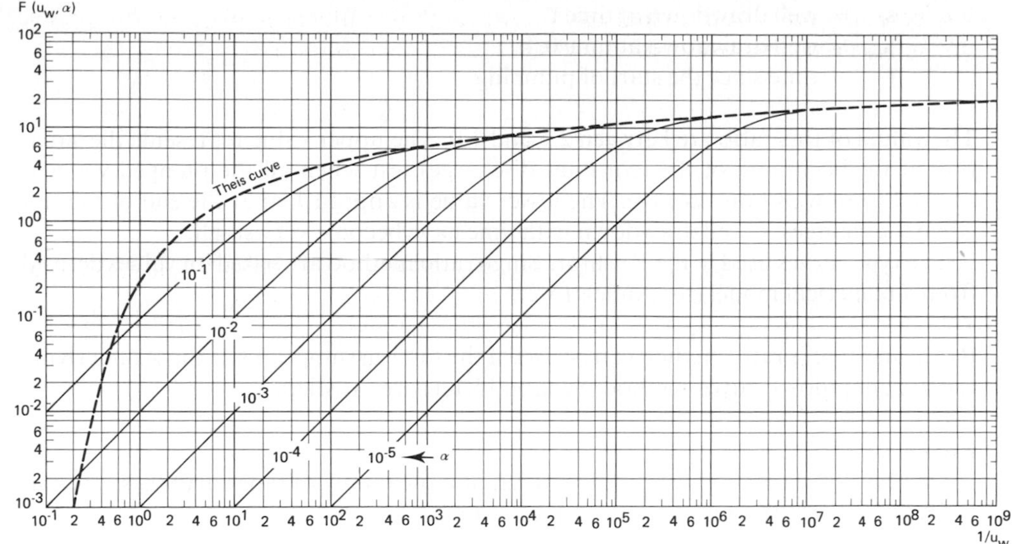

Figure 15.1 Family of Papadopulos-Cooper's type curves: $F(u_w,\alpha)$ versus $1/u_w$ for different values of α

Remarks

- The early-time, almost straight portion of the type curves corresponds to the period when most of the water is derived from storage within the well. Points on the data curve that coincide with these parts of the type curves do not adequately reflect the aquifer characteristics;
- If r_{ew} is known (i.e. if the skin factor or the linear well loss coefficient B_l is known), in theory a value of S can be calculated by introducing the values of r_{ew}, $1/u_w$, t, and KD into Equation 15.2 or by introducing the values of r_c, r_{ew}, and α into Equation 15.3. The values of S calculated in these two ways should show a close agreement. However, since the form of the type curves differs only very slightly when α differs by an order of magnitude, the value of S determined by this method has questionable reliability.

15.1.2 Confined aquifers, Rushton-Singh's ratio method

Because of the similarities of the Papadopulos-Cooper type curves (Section 15.1.1), it may be difficult to match the data curve with the appropriate type curve. To overcome this difficulty, Rushton and Singh (1983) have proposed a more sensitive curve-fitting method in which the changes in the well drawdown with time are examined. Their well-drawdown ratio is

$$\frac{s_t}{s_{0.4t}}$$

where

s_t = well drawdown at time t
$s_{0.4t}$ = well drawdown at time 0.4t
t = time since the start of pumping

The values of this ratio are between 2.5 and 1.0. The upper value represents the situation at the beginning of the (constant discharge) test when all the pumped water is derived from well-bore storage. The lower value is approached at the end of the test when the changes in well drawdown with time have become very small.

The type curves used in the Rushton-Singh ratio method are based on values derived from a numerical model (see Annex 15.2).

Rushton-Singh's ratio method can be used if the same assumptions as those underlying the Papadopulos-Cooper method (Section 15.1.1) are satisfied.

Procedure 15.2

- On semi-log paper and using Annex 15.2, plot the family of type curves $s_t/s_{0.4t}$ versus $4KDt/r^2_{ew}$ for different values of S (Figure 15.2);

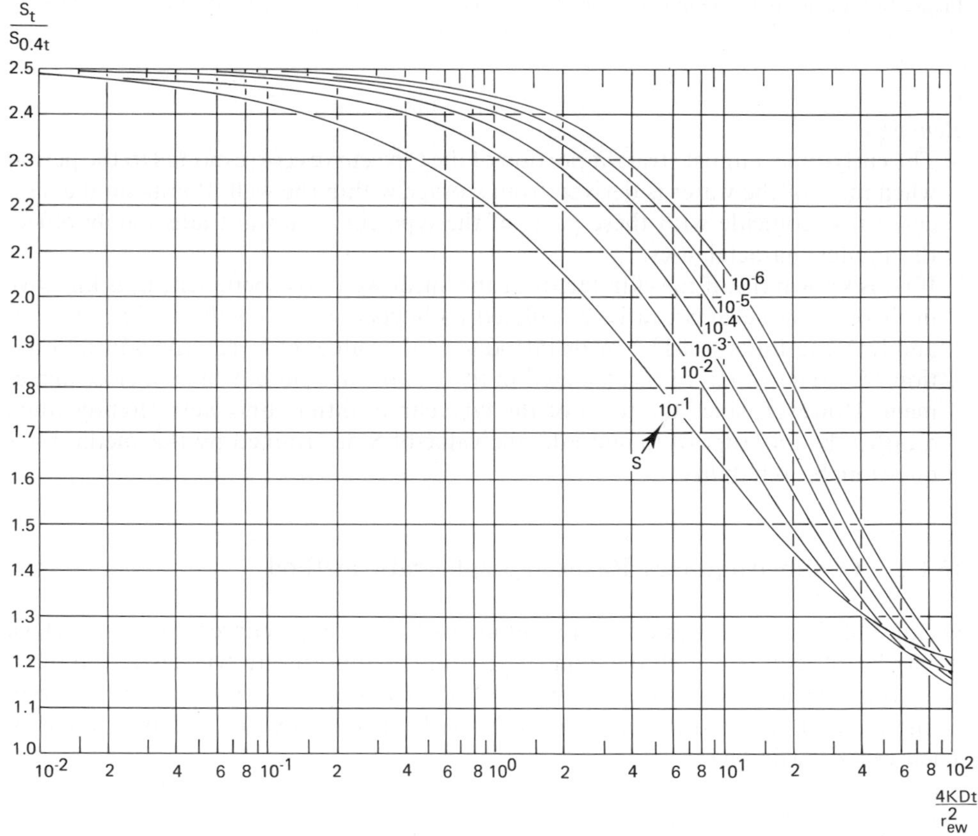

Figure 15.2 Family of Rushton-Singh's type curves for a constant discharge: $s_t/s_{0.4t}$ versus $4KDt/r^2_{ew}$ for different values of S

- Calculate the ratio $s_t/s_{0.4t}$ from the observed drawdowns for different values of t;
- On another sheet of semi-log paper of the same scale, plot the data curve ($s_t/s_{0.4t}$) versus t;
- Superimpose the data curve on the family of type curves and, with the horizontal coordinates $s_t/s_{0.4t} = 2.5$ and 1.0 of both plots coinciding, adjust until a position is found where most of the plotted points of the data curve fall on one of the type curves;
- For $4KDt/r_{ew}^2 = 1.0$, read the corresponding value of t from the time axis of the data curve;
- Substitute the value of t together with the known or estimated value of r_{ew} into $4KDt/r_{ew}^2 = 1.0$ and calculate KD;
- Read the value of S belonging to the best-matching type curve.

15.1.3 Confined and leaky aquifers, Jacob's straight-line method

Jacob's straight-line method (Section 3.2.2) can also be applied to single-well constant-discharge tests to estimate the aquifer transmissivity. However, not all the assumptions underlying the Jacob method are met if data from single-well tests are used. Therefore, the following additional conditions should also be satisfied:

- For single-well tests in confined aquifers

  $$t > 25\, r_c^2/KD$$

  If this time condition is met, the effect of well-bore storage can be neglected;
- For single-well tests in leaky aquifers

  $$\frac{25r_c^2}{KD} < t < \frac{cS}{20}\left(= \frac{L^2S}{20KD}\right)$$

  As long as $t < cS/20$, the influence of leakage is negligible.

Procedure 15.3

- On semi-log paper, plot the observed values of s_w versus the corresponding time t (t on logarithmic scale) and draw a straight line through the plotted points;
- Determine the slope of the straight line, i.e. the drawdown difference Δs_w per log cycle of time;
- Substitute the values of Q and Δs_w into $KD = 2.30Q/4\pi\Delta s_w$, and calculate KD.

Remarks

- The drawdown in the well reacts strongly to even minor variations in the discharge rate. Therefore, a constant discharge is an essential condition for the use of the Jacob method;
- There is no need to correct the observed drawdowns for well losses before applying the Jacob method; the aquifer transmissivity is determined from drawdown differences Δs_w, which are not influenced by well losses as long as the discharge is constant;
- In theory, Jacob's method can also be applied if the well is partially penetrating, provided that late-time ($t > D^2S/2KD$) data are used. According to Hantush (1964), the additional drawdown due to partial penetration will be constant for $t > D^2S/$

2KD and hence will not influence the value of Δs_w as used in Jacob's method;

- Instead of using the time condition $t > 25r_c^2/KD$ to determine when the effect of well-bore storage can be neglected, we can use the 'one and one-half log cycle rule of thumb' (Ramey 1976). On a diagnostic log-log plot, the early-time data may plot as a unit-slope straight line ($\Delta s_w/\Delta t = 1$), indicating that the drawdown data are dominated by well-bore storage. According to Ramey, the end of this unit-slope straight line is about 1.5 log cycles prior to the start of the semi-log straight line as used in the Jacob method.

Example 15.1

To illustrate the Jacob method, we shall use data from a single-well constant-discharge test conducted in a leaky aquifer in Hoogezand, The Netherlands (after Mulder 1983). Mulder's observations were made with electronic equipment that allowed very precise measurements of s_w and Q to be made every five seconds. The recorded drawdown data are given in Table 15.1.

Table 15.1 Single-well constant-discharge test 'Hoogezand', The Netherlands (from Mulder 1983)

t (s)	s_w (m)	Q m^3/hr	t	s_w	Q
1	0.108	25.893	178	1.947	29.229
5	1.064	19.991	220	1.950	29.161
10	1.484	30.431	251	1.955	29.286
15	1.721	29.551	286	1.955	28.942
20	1.791	29.248	328	1.960	29.142
25	1.820	28.891	388	1.970	28.963
30	1.843	29.003	508	1.970	28.581
45	1.895	28.547	568	1.972	29.012
60	1.909	28.446	628	1.976	28.893
75	1.916	28.186	688	1.973	28.787
90	1.919	28.135	748	1.976	28.977
148	1.939	27.765			

Figure 15.3 shows a semi-log plot of the drawdown s_w against the corresponding time, with a straight line fitted through the plotted points. The slope of this line, Δs_w, is 0.07 m per log cycle of time. The transmissivity is calculated from

$$KD = \frac{2.30Q}{4\pi\Delta s_w} = \frac{2.30 \times 28.7 \times 24}{4\pi \times 0.07} = 1800 \text{ m}^2/\text{d}$$

Jacob's straight-line method is applicable to data from single-well tests in leaky aquifers, provided that

$$\frac{25r_c^2}{KD} < t < \frac{cS}{20}$$

Substituting the value of the radius of the well ($r_c = 0.185$ m) and the calculated transmissivity into $25r_c^2/KD$ yields

$$t > \frac{25 \times 0.185^2}{1800} = 0.00048 \text{ d} \qquad \text{or } t > 41 \text{ s}$$

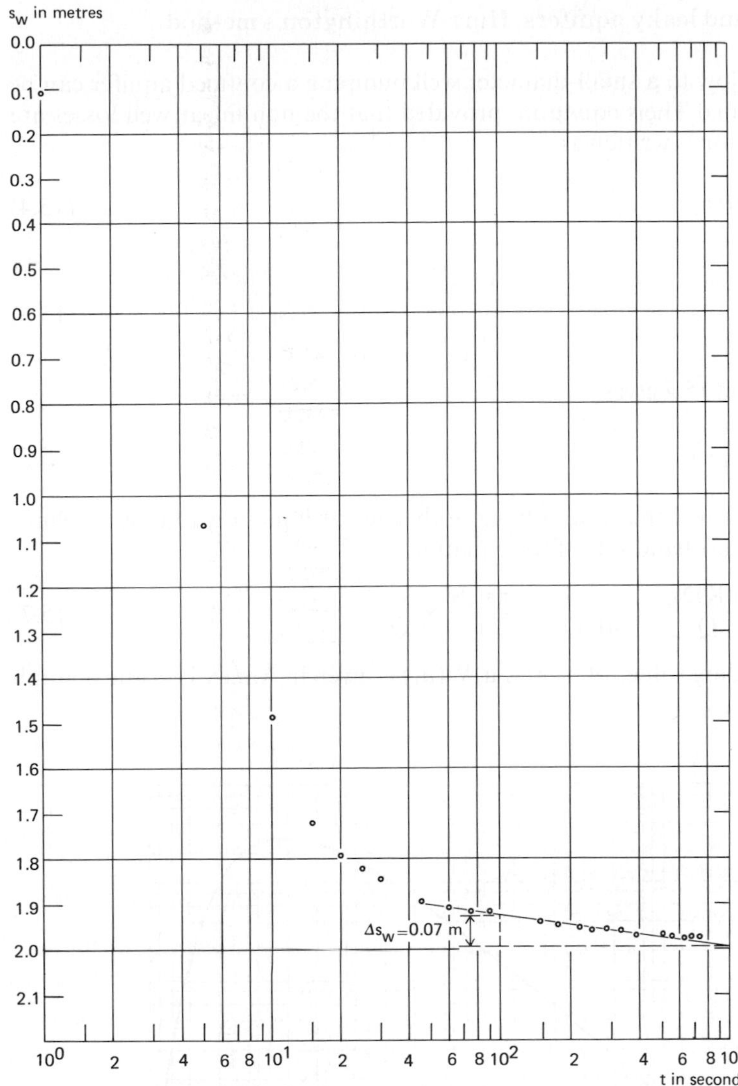

Figure 15.3 Analysis of data from the single-well constant discharge test 'Hoogezand' with the Jacob method

According to Mulder (1983), the values of c and S can be estimated at c = 1000 days and S = 4×10^{-4}. The drawdown in the well is not influenced by leakage as long as

$$t < \frac{cS}{20} = \frac{1000 \times 4 \times 10^{-4}}{20} d \qquad \text{or } t < 1728 \text{ s}$$

Hence, for t > 41 s, Jacob's method can be applied to the drawdown data from the test 'Hoogezand'.

15.1.4 Confined and leaky aquifers, Hurr-Worthington's method

The unsteady-state flow to a small-diameter well pumping a confined aquifer can be described by a modified Theis equation, provided that the non-linear well losses are negligible. The equation is written as

$$s_w = \frac{Q}{4\pi KD} W(u_w) \tag{15.4}$$

where

$$u_w = \frac{r_{ew}^2 S}{4KDt} \tag{15.5}$$

Rearranging Equation 15.4 gives

$$W(u_w) = \frac{4\pi KDs_w}{Q} \tag{15.6}$$

Hurr (1966) demonstrated that multiplying both sides of Equation 15.6 by u_w eliminates KD from the right-hand side of the equation

$$u_w W(u_w) = \frac{4\pi KDs_w}{Q} \times \frac{r_{ew}^2 S}{4KDt} = \frac{\pi r_{ew}^2 S}{t} \times \frac{s_w}{Q} \tag{15.7}$$

A table of corresponding values of u_w and $u_w W(u_w)$ is given in Annex 15.3 and a graph in Figure 15.4.

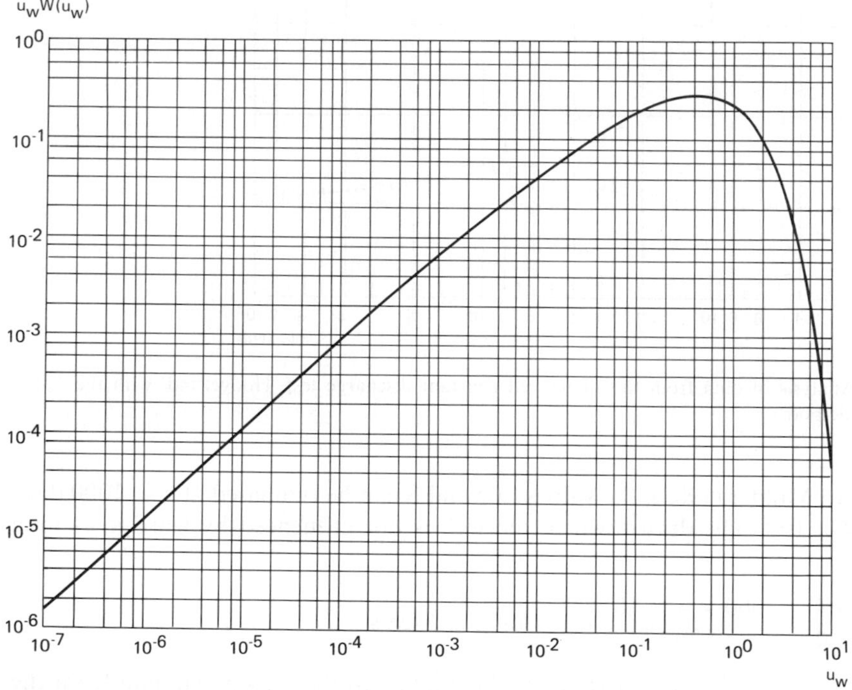

Figure 15.4 Graph of corresponding values of u_w and $u_w W(u_w)$

Hurr (1966) outlined a procedure for estimating the transmissivity of a confined aquifer from a single drawdown observation in the pumped well. In 1981, Worthington incorporated Hurr's procedure in a method for estimating the transmissivity of (thin) leaky aquifers from single-well drawdown data.

In leaky aquifers, the drawdown data can be affected by well losses, by well-bore storage phenomena during early pumping times, and by leakage during late pumping times.

According to Worthington (1981), after the drawdown data have been corrected for non-linear well losses, one can calculate 'pseudo-transmissivities' by applying Hurr's procedure to a sequence of the corrected data. Both well-bore storage effects and leakage effects reduce the drawdown in the well and will therefore lead to calculated pseudo-transmissivities that are greater than the aquifer transmissivity. A semi-log plot of pseudo-transmissivities versus time shows a minimum (Figure 15.5). A flat minimum indicates the time during which the well-bore storage effects have become negligible and leakage effects have not yet manifested themselves: the minimum value of the pseudo-transmissivity gives the value of the aquifer transmissivity. If well-bore storage and leakage effects overlap, the lowest pseudo-transmissivity is the best estimate of a leaky aquifer's transmissivity.

The unsteady-state drawdown data from confined aquifers can also be used to construct a semi-log plot of pseudo-transmissivities versus time to account for the early-time well-bore storage effects.

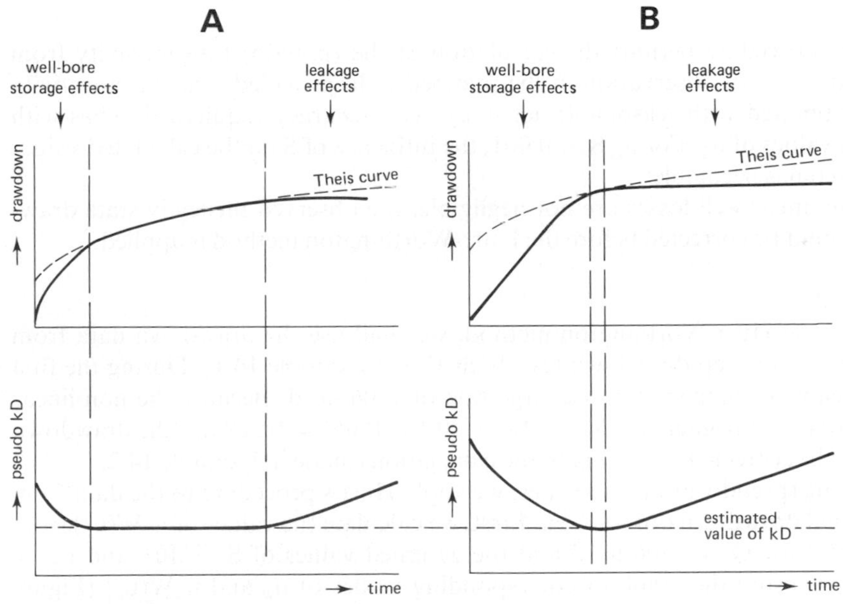

Figure 15.5 Drawdown data and calculated 'pseudo-transmissivities'
A: Moderately affected by well storage and leakage
B: Severely affected by well storage and leakage
(after Worthington 1981)

Hurr-Worthington's method is based on the following assumptions and conditions:

- The assumptions listed at the beginning of Chapter 3, with the exception of the first and eighth assumptions, which are replaced by:
  - The aquifer is confined or leaky;
  - The storage in the well cannot be neglected.

The following conditions are added:

- The flow to the well is in an unsteady-state;
- The non-linear well losses are negligible;
- The storativity is known or can be estimated with reasonable accuracy.

Procedure 15.4

- Calculate pseudo-transmissivity values by applying the following procedure proposed by Hurr to a sequence of observed drawdown data:
  - For a single drawdown observation, calculate $u_w W(u_w)$ from Equation 15.7 for known or estimated values of S and r_{ew}, and the corresponding values of t, s_w, and Q;
  - Knowing $u_w W(u_w)$, determine the corresponding value of u_w from Annex 15.3 or Figure 15.4;
  - Substitute the values of u_w, r_{ew}, t, and S into Equation 15.5 and calculate the pseudo-transmissivity;
- On semi-log paper, plot the pseudo-transmissivity values versus the corresponding t (t on the logarithmic scale). Determine the minimum value of the pseudo-transmissivity from the plot. This is the best estimate of the aquifer's transmissivity.

Remarks

- The Hurr procedure permits the calculation of the (pseudo) transmissivity from a single drawdown observation in the pumped well, provided that the storativity can be estimated with reasonable accuracy. The accuracy required declines with declining values of u_w. For $u_w/S < 0.001$, the influence of S on the calculated values of KD becomes negligible;
- If the non-linear well losses are not negligible, the observed unsteady-state drawdowns should be corrected before the Hurr-Worthington method is applied.

Example 15.2

To illustrate the Hurr-Worthington method, we shall use the drawdown data from the first step of the step-drawdown test 'Well 1' (see Example 14.1). During the first step, the well was pumped at a discharge rate of 1306 m^3/d. Because the non-linear well losses were not negligible ($CQ^2 = 1.4 \times 10^{-7} \times 1306^2 = 0.239$ m), the drawdown data have to be corrected according to the calculations made in Example 14.2.

To calculate (pseudo-)transmissivities, we apply Hurr's procedure to the data from each corrected drawdown observation. First, we calculate the values of $u_w W(u_w)$ from Equation 15.7 for Q = 1306 m^3/d and the assumed values of $S = 10^{-4}$ and r_{ew} = 0.25 m. Then, using the graph of corresponding values of u_w and $u_w W(u_w)$ (Figure 15.4) and the table in Annex 15.3, we find the corresponding values of u_w. From Equation 15.5, we calculate the pseudo-transmissivities (Table 15.2).

Table 15.2 Pseudo-transmissivity values calculated from data obtained during the first step of step-drawdown test 'Well 1'

Time (min)	s_w (m)	$(s_w)corr^*) = s_w - 0.239$ (m)	$u_wW(u_w)$	u_w	(pseudo) KD (m^2/d)
5	1.303	1.064	4.6×10^{-6}	3.2×10^{-7}	1406
6	2.289	2.050	7.4×10^{-6}	5.4×10^{-7}	694
7	3.117	2.878	8.9×10^{-6}	6.5×10^{-7}	495
8	3.345	3.106	8.4×10^{-6}	6.1×10^{-7}	461
9	3.486	3.247	7.8×10^{-6}	5.6×10^{-7}	446
10	3.521	3.282	7.1×10^{-6}	5.1×10^{-7}	441
12	3.592	3.353	6.0×10^{-6}	4.2×10^{-7}	446
14	3.627	3.388	5.2×10^{-6}	3.6×10^{-7}	446
16	3.733	3.494	4.7×10^{-6}	3.3×10^{-7}	426
18	3.768	3.529	4.2×10^{-6}	2.9×10^{-7}	431
20	3.836	3.597	3.9×10^{-6}	2.7×10^{-7}	417
25	3.873	3.634	3.1×10^{-6}	2.1×10^{-7}	429
30	4.014	3.775	2.7×10^{-6}	1.8×10^{-7}	417
35	3.803	3.564	2.2×10^{-6}	1.45×10^{-7}	443
40	4.043	3.804	2.1×10^{-6}	1.4×10^{-7}	402
45	4.261	4.022	1.9×10^{-6}	1.25×10^{-7}	400
50	4.261	4.022	1.7×10^{-6}	1.1×10^{-7}	409
55	4.190	3.951	1.6×10^{-6}	1.05×10^{-7}	390
60	4.120	3.881	1.4×10^{-6}	9×10^{-8}	417
70	4.120	3.881	1.2×10^{-6}	7.6×10^{-8}	423
80	4.226	3.987	1.1×10^{-6}	7.0×10^{-8}	402
90	4.226	3.987	9.6×10^{-7}	6.0×10^{-8}	417
100	4.226	3.987	8.6×10^{-7}	5.4×10^{-8}	417
120	4.402	4.163	7.5×10^{-7}	4.6×10^{-8}	408
150	4.402	4.163	6.0×10^{-7}	3.6×10^{-8}	417
180	4.683	4.444	5.3×10^{-7}	3.2×10^{-8}	391

* Well loss $= CQ^2 = 1.4 \times 10^{-7} \times (1306)^2 = 0.239$ m

Subsequently, we plot the calculated pseudo-transmissivities versus time on semi-log paper (Figure 15.6), from which we can see that during the first eight minutes of pumping, the drawdown in the well was clearly affected by well-bore storage effects. Our estimate of the aquifer transmissivity is 410 m^2/d.

15.2 Variable-discharge tests

15.2.1 Confined aquifers, Birsoy-Summers's method

Birsoy-Summers's method (Section 12.1.1) can also be used for analyzing single-well tests with variable discharges. The parameters s and r should be replaced by s_w and r_{ew} in all the equations.

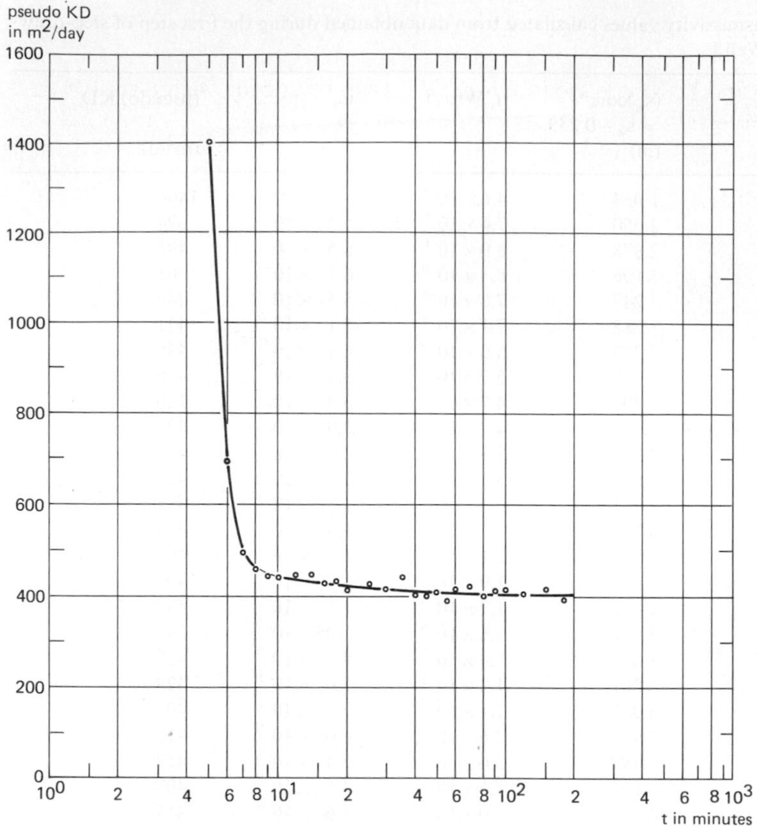

Figure 15.6 Analysis of data from the first step of the step-drawdown test 'Well 1' with the Hurr-Worthington method: determination of the aquifer's transmissivity

15.2.2 Confined aquifers, Jacob-Lohman's free-flowing-well method

Jacob and Lohman (1952) derived the following equation for the discharge of a free-flowing well

$$Q = 2\pi KDs_wG(u_w) \tag{15.8}$$

where

s_w = constant drawdown in the well (= difference between static head measured during shut-in of the well and the outflow opening of the well)

$G(u_w)$ = Jacob-Lohman's free-flowing-well discharge function for confined aquifers

$u_w = \dfrac{r_{ew}^2 S}{4KDt}$

r_{ew} = effective radius of the well

According to Jacob and Lohman, the function $G(u_w)$ can be approximated by $2/W(u_w)$ for all but extremely small values of t. If, in addition, $u_w < 0.01$, Equation 15.8 can be expressed as

$$Q = \frac{4\pi KDs_w}{2.30\log(2.25KDt/r_{ew}^2)} \text{ or } \frac{s_w}{Q} = \frac{2.30}{4\pi KD}\log\frac{2.25KDt}{r_{ew}^2} \qquad (15.9)$$

A semi-log plot of s_w/Q versus t (t on logarithmic scale) will thus yield a straight line. A method analogous to the Jacob straight-line method (Section 3.2.2) can therefore be used to analyze the data from a free-flowing well discharging from a confined aquifer.

The Jacob-Lohman method can be used if the following assumptions and conditions are satisfied:

- The assumptions listed at the beginning of Chapter 3, with the exception of the fifth assumption, which is replaced by:
  - At the beginning of the test ($t = 0$), the water level in the free-flowing well is lowered instantaneously. At $t > 0$, the drawdown in the well is constant, and its discharge is variable.

The following conditions are added:

- The flow to the well is in an unsteady state;
- $u_w < 0.01$.

Procedure 15.5

- On semi-log paper, plot the values of s_w/Q versus t (t on logarithmic scale);
- Fit a straight line through the plotted points;
- Extend the straight line until it intercepts the time-axis where $s_w/Q = 0$ at the point t_0;
- Introduce the value of the slope of the straight line $\Delta(s_w/Q)$ (i.e. the difference of s_w/Q per log cycle of time) into Equation 15.10 and solve for KD

$$KD = \frac{2.30}{4\pi\Delta(s_w/Q)} \qquad (15.10)$$

- Calculate the storativity S from

$$S = \frac{2.25KDt_0}{r_{ew}^2} \qquad (15.11)$$

Remark

- If the value of r_{ew} is not known, S cannot be determined by this method.

15.2.3 Leaky aquifers, Hantush's free-flowing-well method

The variable discharge of a free-flowing well tapping a leaky aquifer is given by Hantush (1959a) as

$$Q = 2\pi KDs_w G(u_w, r_{ew}/L) \qquad (15.12)$$

where

s_w = constant drawdown in well

$G(u_w, r_{ew}/L)$ = Hantush's free-flowing-well discharge function for leaky aquifers

$$u_w = \frac{r_{ew}^2 S}{4KDt} \qquad (15.13)$$

Annex 15.4 presents values of the function $G(u_w, r_{ew}/L)$ for different values of $1/u_w$ and r_{ew}/L, as given by Hantush (1959a, 1964; see also Reed 1980). A family of type curves can be plotted from that annex.

The Hantush method for determining a leaky aquifer's parameters KD, S, and c can be applied if the following assumptions and conditions are satisfied:

- The assumptions listed at the beginning of Chapter 4, with the exception of the fifth assumption, which is replaced by:
  - At the beginning of the test (t = 0), the water level in the free-flowing well is lowered instantaneously. At t > 0, the drawdown in the well is constant, and its discharge is variable;

The following conditions are added:

- The flow to the well is in an unsteady state;
- The aquitard is incompressible, i.e. changes in aquitard storage are negligible.

Procedure 15.6

- On log-log paper and using Annex 15.4, draw a family of type curves by plotting $G(u_w, r_{ew}/L)$ versus $1/u_w$ for a range of values of r_{ew}/L;
- On another sheet of log-log paper of the same scale, prepare the data curve by plotting the values of Q against the corresponding time t;
- Match the data plot with one of the type curves. Note the value of r_{ew}/L for that type curve;
- Select an arbitrary point A on the overlapping portion of the two sheets and note the values of $G(u_w, r_{ew}/L)$, $1/u_w$, Q, and t for that point;
- Substitute the values of Q and $G(u_w, r_{ew}/L)$ and the value of s_w into Equation 15.12 and calculate KD;
- Substitute the values of KD, t, $1/u_w$, and r_{ew} into Equation 15.13 and calculate S;
- Substitute the value of r_{ew}/L corresponding to the type curve and the values of r_{ew} and KD into $r_{ew}/L = r_{ew}/\sqrt{KDc}$, and calculate c.

Remark

- If the effective well radius r_{ew} is not known, the values of S and c cannot be obtained.

15.3 Recovery tests

15.3.1 Theis's recovery method

The Theis recovery method (Section 13.1.1) is also applicable to data from single-well

recovery tests conducted in confined, leaky or unconfined aquifers.
The method can be used if the following assumptions and conditions are met:

- The assumptions listed at the beginning of Chapter 3, adjusted for recovery tests, with the exception of the eighth assumption, which is replaced by:
  - $t_p > 25\, r_c^2/KD$;
  - $t' > 25\, r_c^2/KD$.

The following conditions are added:

- The aquifer is confined, leaky or unconfined.
  For leaky aquifers, the sum of the pumping and recovery times should be $t_p + t' \leq L^2S/20KD$ or $t_p + t' \leq cS/20$ (Section 13.1.2).
  For unconfined aquifers only late-time recovery data can be used (Section 13.1.3);
- The flow to the well is in an unsteady state;
- $u < 0.01$, i.e. $t_p > 25\, r_w^2S/KD$;
- $u' < 0.01$, i.e. $t' > 25\, r_w^2S/KD$ (see also Section 3.2.2).

Procedure 15.7

- For each observed value of s'_w, calculate the corresponding value of t/t';
- Plot s'_w versus t/t' on semi-log paper (t/t' on the logarithmic scale);
- Fit a straight line through the plotted points;
- Determine the slope of the straight line, i.e. the residual drawdown difference $\Delta s'_w$ per log cycle of t/t';
- Substitute the known values of Q and $\Delta s'_w$ into Equation 15.14 $\Delta s'_w = 2.30Q/4\pi KD$, and calculate KD.

Remarks

- Storage in the well may influence s'_w at the beginning of a recovery test. If the conditions $t_p > 25\, r_c^2/KD$ and $t' > 25\, r_c^2/KD$ are met, a semi-log plot of s'_w versus t/t' yields a straight-line and Theis's recovery method is applicable. Because the observed recovery data should plot as a straight-line for at least one log cycle of t/t', Uffink (1982) recommends that both t_p and t' should be at least $500\, r_c^2/KD$;
- If the pumped well is partially penetrating, the Theis recovery method can be used, provided that both t_p and t' are greater than $D^2S/2KD$ (Section 13.1.4);
- If the recovery test follows a constant-drawdown test instead of a constant-discharge test, the discharge at the moment before the pump is shut down should be used in Equation 15.14 (Rushton and Rathod 1980).

15.3.2 Birsoy-Summers's recovery method

Residual drawdown data from the recovery phase of single-well variable-discharge tests conducted in confined aquifers can be analyzed by the Birsoy-Summers recovery method (Section 13.3.1), provided that s' is replaced by s'_w in all equations.

15.3.3 Eden-Hazel's recovery method

The Eden-Hazel method for step-drawdown tests (Section 14.1.2) is also applicable to the data from the recovery phase of such a test.

The Eden-Hazel recovery method can be used if the following assumptions and conditions are met:

- The assumptions listed at the beginning of Chapter 3, as adjusted for recovery tests, with the exception of the fifth assumption, which is replaced by:
  - Prior to the recovery test, the aquifer is pumped step-wise.

The following conditions are added:

- The flow to the well is in unsteady state;
- $u < 0.01$ (see Section 3.2.2);
- $u' < 0.01$.

Procedure 15.8

- Calculate for the recovery phase (i.e. $t > t_n$) the values of H_n from Equation 14.15, using the measured discharges and times;
- On arithmetic paper, plot the observed residual drawdown $s'_{w(n)}$ versus the corresponding calculated values of H_n;
- Draw a straight line through the plotted points;
- Determine the slope of the straight line, $\Delta s'_{w(n)}/\Delta H_n$;
- Calculate KD from

$$\frac{\Delta s'_{w(n)}}{\Delta H_n} = \frac{2.30}{4\pi KD}$$

Example 15.3

We shall illustrate the Eden-Hazel recovery method with the data of the step-drawdown test 'Well 1' (Table 14.1 and Table 15.3).

For the recovery phase of the step-drawdown test, Equation 14.5 becomes

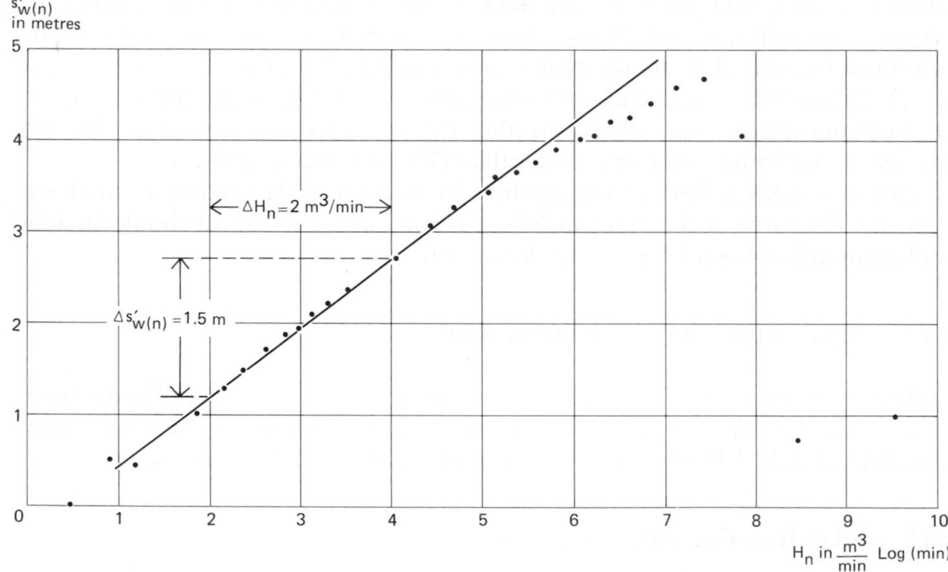

Figure 15.7 Analysis of data from the recovery phase of the step-drawdown test 'Well 1' with the Eden-Hazel recovery method

$$H_n = \left(\frac{1360}{1440}\right)\log(t) + \left(\frac{387}{1440}\right)\log(t-180) + \left(\frac{730}{1440}\right)\log(t-360)$$
$$+ \left(\frac{838}{1440}\right)\log(t-540) + \left(\frac{833}{1440}\right)\log(t-720) + \left(\frac{925}{1440}\right)\log(t-900)$$
$$- \left(\frac{5019}{1440}\right)\log(t-180)\ (m^3/min)\ \log(min)$$

Table 15.3 shows the result of the calculations for $t > t_n$.
Figure 15.7 gives the arithmetic plot of the $s'_{w(n)}$ versus H_n.
The slope of the straight line is

$$\frac{\Delta s'_{w(n)}}{\Delta H_n} = \frac{1.5}{2} \times \frac{1}{1440} = 5.2 \times 10^{-4}\ d/m^2$$

The transmissivity $KD = \dfrac{2.30}{4\pi \times 5.2 \times 10^{-4}} = 352\ m^2/d$

Table 15.3 Values of H_n calculated for the recovery phase of step-drawdown test 'Well 1'

t (min)	H_n $(m^3/min)\ \log(min)$	$s'_{w(n)}$ (m)
1081	9.515	0.599
1082	8.469	1.233
1083	7.859	4.050
1084	7.427	4.683
1085	7.092	4.578
1086	6.820	4.402
1087	6.590	4.261
1088	6.391	4.226
1089	6.216	4.050
1090	6.060	4.014
1092	5.791	3.909
1094	5.564	3.768
1096	5.369	3.662
1098	5.197	3.627
1100	5.045	3.416
1105	4.723	3.275
1110	4.463	3.064
1115	4.246	–
1120	4.059	2.711
1125	3.896	–
1130	3.752	–
1135	3.623	–
1140	3.506	2.359
1150	3.301	2.218
1160	3.127	2.078
1170	2.977	1.937
1180	2.844	1.866
1200	2.620	1.726
1230	2.356	1.479
1260	2.150	1.303
1320	1.843	1.021
1560	1.209	0.458
1800	0.914	0.528
2650	0.499	0.035

16 Slug tests

In a slug test, a small volume (or slug) of water is suddenly removed from a well, after which the rate of rise of the water level in the well is measured. Alternatively, a small slug of water is poured into the well and the rise and subsequent fall of the water level are measured. From these measurements, the aquifer's transmissivity or hydraulic conductivity can be determined.

If the water level is shallow, the slug of water can be removed with a bailer or a bucket. If not, a closed cylinder or other solid body is submerged in the well and then, after the water level has stabilized, the cylinder is pulled out. Enough water must be removed or displaced to raise or lower the water level by about 10 to 50 cm.

If the aquifer's transmissivity is higher than, say, 250 m^2/d, the water level will recover too quickly for accurate manual measurements and an automatic recording device will be needed.

No pumping is required in a slug test, no piezometers are needed, and the test can be completed within a few minutes, or at the most a few hours. No wonder that slug tests are so popular! They are invaluable in studies to evaluate regional groundwater resources; conducted on newly-constructed wells, they permit a preliminary estimate of aquifer conditions, and are also useful in areas where other wells are operating and where well interference can be expected.

But slug tests cannot be regarded as a substitute for conventional pumping tests. From a slug test, for instance, it is only possible to determine the characteristics of a small volume of aquifer material surrounding the well, and this volume may have been disturbed during well drilling and construction. Nevertheless, some authors (Ramey et al. 1975; Moench and Hsieh 1985) state that fairly accurate transmissivity values can be obtained from slug tests.

The simple slug-test technique has been further developed in recent years and has consequently become more complex and requires more equipment. In this chapter, we shall present one of these more advanced techniques: the oscillation test.

An oscillation test requires an air compressor to lower the water level in the well. After some time, when the head in the aquifer has resumed its initial value, the pressure is suddenly released. The water level in the well then resumes its initial level by a damped oscillation that can be measured, preferably with an automatic recorder.

For conventional slug tests performed in confined aquifers with fully penetrating wells, curve-fitting methods have been developed (Cooper et al. 1967; Papadopulos et al. 1973; Ramey et al. 1975). Cooper's method is presented in Section 16.1.1. For wells partially or fully penetrating unconfined aquifers, Bouwer and Rice (1976) developed the method outlined in Section 16.2.1.

All of the above methods are based on theories that neglect the forces of inertia in both the aquifer and the well: the water level in the well is assumed to return to the equilibrium level exponentially. When slug tests are performed in highly permeable aquifers or in deep wells, however, inertia effects come into play, and the water level in the well may oscillate after an instantaneous change in water level. Various methods of analyzing this response by the water level have been developed (Van der Kamp 1976; Krauss 1974; Uffink 1979, 1980; Ross 1985), but they all have the disadvantage that the aquifer transmissivity cannot be determined without a prior knowledge of the storativity. In addition, Uffink states that the skin effects also have to be taken into account and that these, too, should be known beforehand. Uffink's method is described in Section 16.1.2.

16.1 Confined aquifers, unsteady-state flow

16.1.1 Cooper's method

A volume of water (V) instantaneously withdrawn from or injected into a well of finite diameter ($2r_c$) will cause an instantaneous change of the hydraulic head in the well

$$h_o = \frac{V}{\pi r_c^2} \tag{16.1}$$

After this change, the head will gradually return to its initial head. The following solution for the rise or fall in the well's head with time was derived by Cooper et al. (1967) for a fully penetrating large-diameter well tapping a confined aquifer (Figure 16.1)

$$h_t = h_o\, F(\alpha,\beta), \quad \text{or} \quad \frac{h_t}{h_o} = F(\alpha,\beta) \tag{16.2}$$

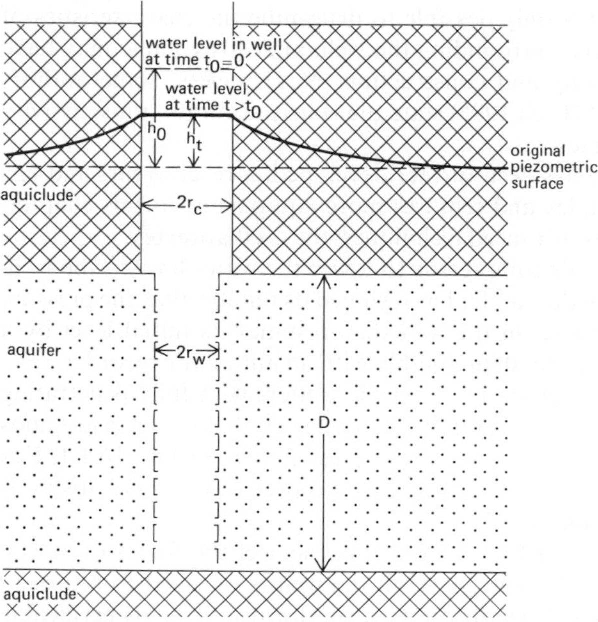

Figure 16.1 A confined aquifer, fully penetrated by a well of finite diameter into which a slug of water has been injected

where

$$\alpha = \frac{r_{ew}^2 S}{r_c^2} \tag{16.3}$$

$$\beta = \frac{KDt}{r_c^2} \tag{16.4}$$

h_o = instantaneous change of head in the well at time $t_o = 0$
h_t = head in the well at time $t > t_o$
r_c = radius of the unscreened part of the well where the head is changing
r_{ew} = effective radius of the screened (or otherwise open) part of the well

$$F(\alpha,\beta) = \frac{8\alpha}{\pi^2} \int_o^\infty \frac{\exp(-\beta u^2/\alpha)}{uf(u,\alpha)} du \tag{16.5}$$

where $f(u,\alpha) = [uJ_0(u) - 2\alpha J_1(u)]^2 + [uY_0(u) - 2\alpha Y_1(u)]^2$ and $J_0(u)$, $J_1(u)$, $Y_0(u)$, and $Y_1(u)$ are the zero and first-order Bessel functions of the first and second kind.

Annex 16.1 lists values of the function $F(\alpha,\beta)$ for different values of α and β as given by Cooper et al. (1967) and Papadopulos et al. (1973). Figure 16.2 presents these values as a family of type curves.

The Cooper curve-fitting method can be used if the following assumptions and conditions are satisfied:
- The aquifer is confined and has an apparently infinite areal extent;
- The aquifer is homogeneous, isotropic, and of uniform thickness over the area influenced by the slug test;
- Prior to the test, the piezometric surface is (nearly) horizontal over the area that will be influenced by the test;
- The head in the well is changed instantaneously at time $t_o = 0$;
- The flow to (or from) the well is in an unsteady state;
- The rate at which the water flows from the well into the aquifer (or vice versa) is equal to the rate at which the volume of water stored in the well changes as the head in the well falls (or rises);
- The inertia of the water column in the well and the non-linear well losses are negligible;
- The well penetrates the entire aquifer;
- The well diameter is finite; hence storage in the well cannot be neglected.

Procedure 16.1
- Using Tables 1 and 2 in Annex 16.1, draw a family of type curves on semi-log paper by plotting $F(\alpha,\beta)$ versus β for a range of values of α (β on the logarithmic scale) (Figure 16.2);
- Knowing the volume of water injected into or removed from the well, calculate h_o from Equation 16.1;
- Calculate the ratio h_t/h_o for different values of t;
- On another sheet of semi-log paper of the same scale, prepare the data curve by plotting the values of the ratio h_t/h_o against the corresponding time t (t on the logarithmic scale);

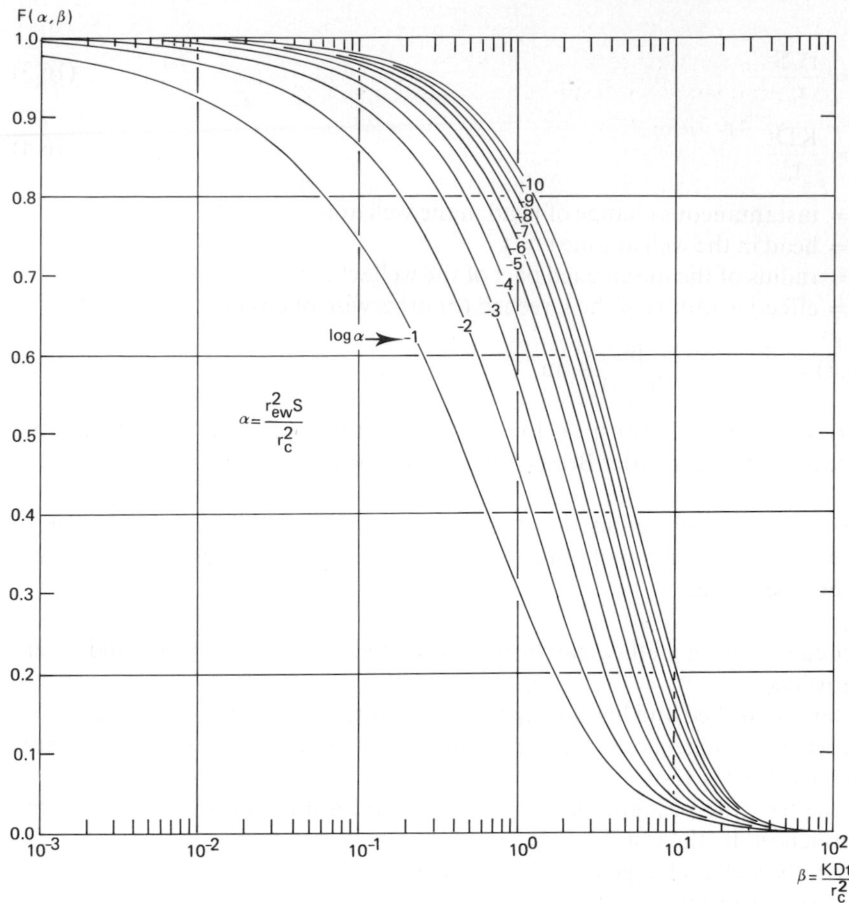

Figure 16.2 Family of Cooper's type curves F(α,β) versus β for different values of α (after Papadopulos et al. 1973)

- Superimpose the data plot on the family of type curves and, keeping the β and t axes of the two plots coinciding and moving the plots horizontally, find a position where most of the plotted points of the data curve fall on one of the type curves. Note the value of α for that type curve;
- For β = 1.0, read the corresponding value of t from the time axis of the data curve;
- Substitute this value of t together with the known value of r_c into $\beta = KDt/r_c^2 = 1$ and calculate KD;
- Knowing r_c and $\alpha = r_{ew}^2 S/r_c^2$, and provided that r_{ew} is also known or can be estimated, calculate S.

Remarks

- Because the type curves in Figure 16.2 are very similar in shape, it may be difficult to obtain a unique match of the data plot and one of the type curves. As the horizontal shift from one curve to the next is small and becomes smaller as α becomes

smaller, the error in S will be as large as the error in α, but the error in KD will still be small. Papadopulos et al. (1973) showed that, if $\alpha < 10^{-5}$, an error of two orders of magnitude in α will result in an error of less than 30 per cent in the calculated transmissivity. In addition, the effective radius of the well r_{ew} (i.e. the skin factor as $r_{ew} = r_w e^{-skin}$) will often not be known;
- The well radius r_c influences the duration of a slug test: a smaller r_c will shorten the test; this is an advantage in aquifers of low permeability;
- To analyze slug tests, Ramey et al. (1975) introduced type curves based on a function F, which has the form of an inversion integral and is expressed in terms of three independent dimensionless parameters: KDt/r_w^2S, $r_c^2/2r_w^2S$, and the skin factor. To reduce these three parameters to two, Ramey et al. showed that the concept of effective well radius ($r_{ew} = r_w e^{-skin}$) also works for slug tests. If r_{ew} is used in the function F, the two remaining independent parameters relate to Cooper's dimensionless parameters α and β. The set of type curves given by Ramey et al. (see also Earlougher 1977) are identical in appearance to Cooper's, and either set will produce approximately the same results for the aquifer transmissivity.

16.1.2 Uffink's method for oscillation tests

In an oscillation test, the well is sealed off with an inflatable packer, through which an air hose is inserted. Air is forced through the hose under high pressure, thereby forcing the water in the well through the well screen into the aquifer and lowering the head in the well. After a certain time, when the head has been lowered to, say, 50 cm and is held there by the over-pressure, the pressure is suddenly released. The response of the head in the well to this sudden change can be described as an exponentially damped harmonic oscillation (Figure 16.3), which can be measured, preferably with an automatic recorder.

This oscillation response is given by Van der Kamp (1976) and Uffink (1984) as

$$h_t = h_o\, e^{-\gamma t} \cos \omega t \qquad (16.6)$$

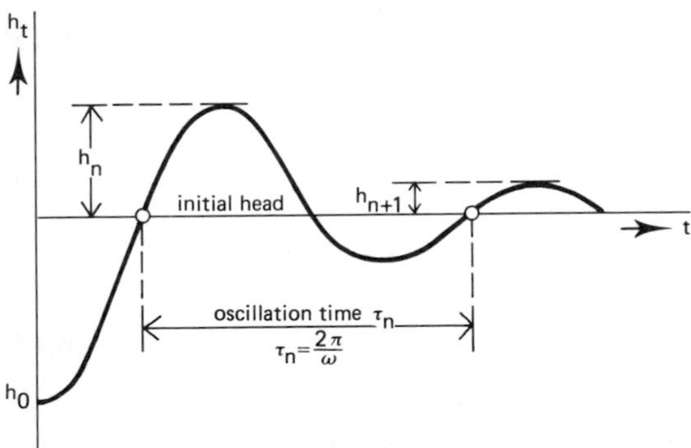

Figure 16.3 Damped harmonic oscillation

where

h_o = instantaneous change in the head at time t_o (= 0)
h_t = head in the well at time t ($t > t_o$)
γ = damping constant of head oscillation ($Time^{-1}$)
ω = angular frequency of head oscillation ($Time^{-1}$)

The damping constant, γ, and the angular frequency of oscillation, ω, can be expressed as

$$\gamma = \omega_o B \tag{16.7}$$

and

$$\omega = \omega_o \sqrt{1-B^2} \tag{16.8}$$

where

ω_o = 'damping free' frequency of head oscillation ($Time^{-1}$)
B = parameter defined by Equation 16.13 (dimensionless)

The values of γ and ω, and consequently of ω_o and B, can be derived directly from the oscillation time τ_n and the ratio between two subsequent minima or maxima, $\ln(h_n/h_{n+1}) = \delta$, of the observed oscillation

$$\gamma = \frac{\delta}{\tau_n} \tag{16.9}$$

$$\omega = \frac{2\pi}{\tau_n} \tag{16.10}$$

$$B = \frac{\delta}{\sqrt{\delta^2 + 4\pi^2}} \tag{16.11}$$

$$\omega_o = \frac{\sqrt{\delta^2 + 4\pi^2}}{\tau_n} \tag{16.12}$$

The relation between the frequency and damping of the head's oscillation and the aquifer's hydraulic characteristics can be approximated by the following equation (Uffink 1984)

$$\frac{1}{B} \ln \left\{ \frac{1.26KD}{r_c^2 \omega_o} \times \frac{1}{\alpha} \right\} + \frac{\Theta - \pi}{\sqrt{1-B^2}} = \frac{8KD}{r_c^2 \omega_o} \tag{16.13}$$

where

$$\alpha = S \frac{r_w^2}{r_c^2} e^{-2skin} \tag{16.14}$$

skin = skin factor, and

$$\Theta = \tan \left(\frac{\sqrt{1-B^2}}{B} \right) \tag{16.15}$$

The nomogram in Figure 16.4 gives the relation between the parameters B and $(r_c^2 \omega_o)/4KD$ for different values of α, as calculated by Uffink.

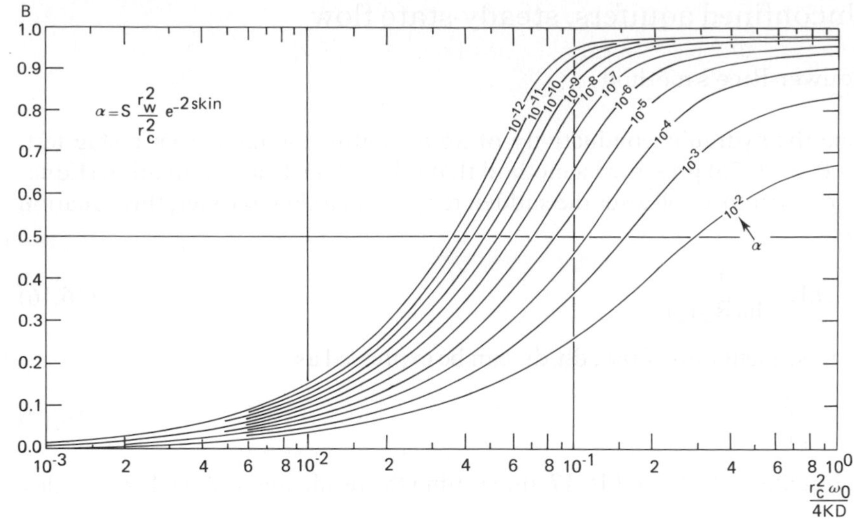

Figure 16.4 Uffink's nomogram giving the relation between B and $(r_c^2\omega_o/4KD)$ for different values of α

Oscillation tests in confined aquifers can be analyzed by Uffink's method if the following assumptions and conditions are satisfied:

- The assumptions and conditions underlying Cooper's method (Section 16.1.1), with the exception of the seventh assumption, which is replaced by:
  - The inertia of the water column in the well is not negligible; the head change in the well at time $t > t_o$ can be described as an exponentially damped cyclic fluctuation.

The following condition is added:

- The storativity S and the skin factor are already known or can be estimated with fair accuracy.

Procedure 16.2

- On arithmetic paper, plot the observed head in the well, h_t, against the corresponding time t ($t > t_o$) (see Figure 16.3);
- From the h_t versus t plot, determine the head's oscillation time τ_n;
- Read the values of two subsequent maxima (or minima) of the oscillation, h_n and h_{n+1}, and calculate δ from $\delta = \ln(h_n/h_{n+1})$;
- Knowing δ, calculate the parameter B from Equation 16.11;
- Knowing δ and B, calculate ω_o from Equation 16.12;
- Knowing B, and provided that α is also known, find the corresponding value of $r_c^2\omega_o/4KD$ from Figure 16.4;
- Knowing $r_c^2\omega_o/4KD$, r_c, and ω_o, calculate KD;
- Repeat this procedure for different sets of τ_n and $\ln(h_n/h_{n+1})$.

16.2 Unconfined aquifers, steady-state flow

16.2.1 Bouwer-Rice's method

To determine the hydraulic conductivity of an unconfined aquifer from a slug test, Bouwer and Rice (1976) presented a method that is based on Thiem's equation (Equation 3.1). For flow into a well after the sudden removal of a slug of water, this equation is written as

$$Q = 2\pi K d \frac{h_t}{\ln(R_e/r_w)} \tag{16.16}$$

The head's subsequent rate of rise, dh/dt, can be expressed as

$$\frac{dh}{dt} = -\frac{Q}{\pi r_c^2} \tag{16.17}$$

Combining Equations 16.16 and 16.17, integrating the result, and solving for K, yields

$$K = \frac{r_c^2 \ln(R_e/r_w)}{2d} \frac{1}{t} \ln \frac{h_o}{h_t} \tag{16.18}$$

where

r_c = radius of the unscreened part of the well where the head is rising
r_w = horizontal distance from well centre to undisturbed aquifer
R_e = radial distance over which the difference in head, h_o, is dissipated in the flow system of the aquifer
d = length of the well screen or open section of the well
h_o = head in the well at time $t_o = 0$
h_t = head in the well at time $t > t_o$

The geometrical parameters r_c, r_w, and d are shown in Figure 16.5.

Bouwer and Rice determined the values of R_e experimentally, using a resistance network analog for different values of r_w, d, b, and D (Figure 16.6). They derived the following empirical equations, which relate R_e to the geometry and boundary conditions of the system:

- For partially penetrating wells

$$\ln \frac{R_e}{r_w} = \left[\frac{1.1}{\ln(b/r_w)} + \frac{A + B \ln[(D-b)/r_w]^{-1}}{d/r_w}\right] \tag{16.19}$$

where A and B are dimensionless parameters, which are functions of d/r_w;

- For fully penetrating wells

$$\ln \frac{R_e}{r_w} = \left[\frac{1.1}{\ln(b/r_w)} + \frac{C}{d/r_w}\right]^{-1} \tag{16.20}$$

where C is a dimensionless parameter, which is a function of d/r_w.

Since K, r_c, r_w, R_e, and d in Equation 16.18 are constants, $(1/t)\ln(h_o/h_t)$ is also a constant. Hence, when values of h_t are plotted against t on semi-log paper (h_t on the logarithmic scale), the plotted points will fall on a straight line. With Procedure 16.3, below, this straight-line plot is used to evaluate $(1/t)\ln(h_o/h_t)$.

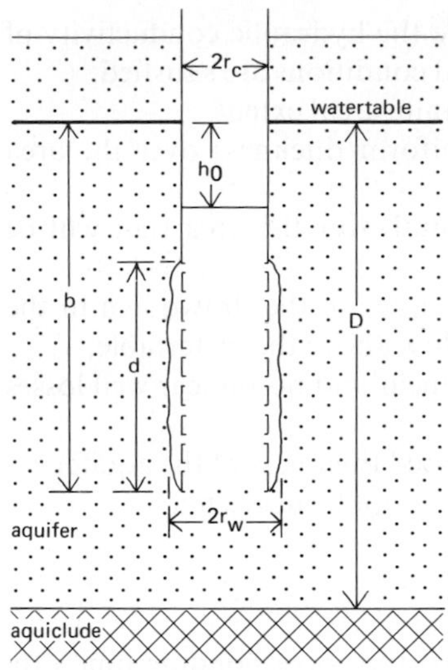

Figure 16.5 An unconfined aquifer, partially penetrated by a large-diameter well from which a slug of water has been removed

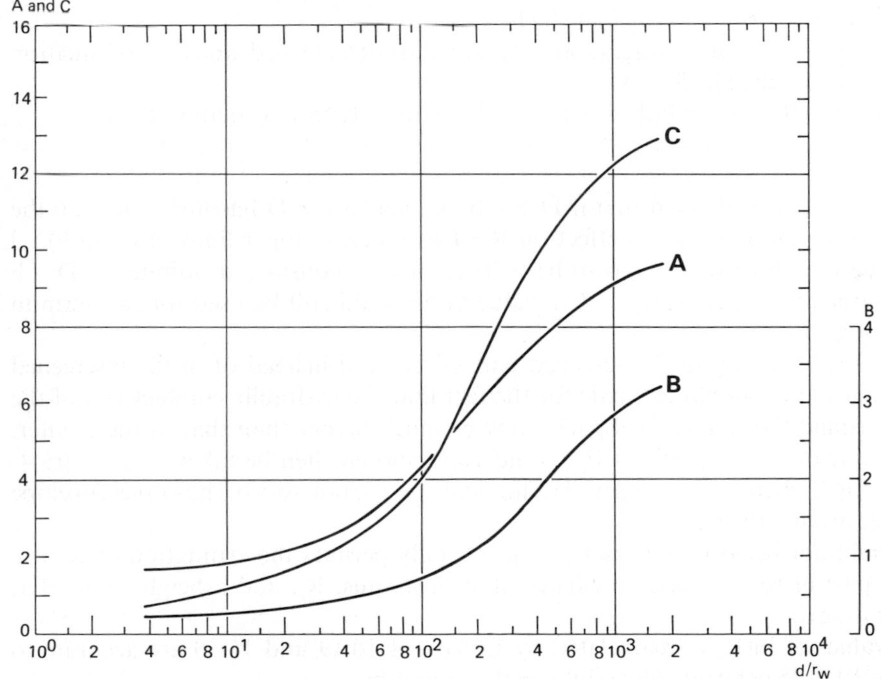

Figure 16.6 The Bouwer and Rice curves showing the relation between the parameters A, B, C, and d/r_w

The Bouwer-Rice method can be applied to determine the hydraulic conductivity of an unconfined aquifer if the following assumptions and conditions are satisfied:
- The aquifer is unconfined and has an apparently infinite areal extent;
- The aquifer is homogeneous, isotropic, and of uniform thickness over the area influenced by the slug test;
- Prior to the test, the watertable is (nearly) horizontal over the area that will be influenced by the test;
- The head in the well is lowered instantaneously at $t_o = 0$; the drawdown in the watertable around the well is negligible; there is no flow above the watertable;
- The inertia of the water column in the well and the linear and non-linear well losses are negligible;
- The well either partially or fully penetrates the saturated thickness of the aquifer;
- The well diameter is finite; hence storage in the well cannot be neglected;
- The flow to the well is in a steady state.

Procedure 16.3
- On semi-log paper, plot the observed head h_t against the corresponding time t (h_t on logarithmic scale);
- Fit a straight line through the plotted points;
- Using this straight-line plot, calculate $(1/t)\ln(h_o/h_t)$ for an arbitrarily selected value of t and its corresponding h_t;
- Knowing d/r_w, determine A and B from Figure 16.6 if the well is partially penetrating, or determine C from Figure 16.6 if the well is fully penetrating;
- If the well is partially penetrating, substitute the values of A, B, D, b, d, and r_w into Equation 16.19 and calculate $\ln(R_e/r_w)$.
  If the well is fully penetrating, substitute the values of C, D, b, d, and r_w into Equation 16.20 and calculate $\ln(R_e/r_w)$;
- Knowing $\ln(R_e/r_w)$, $(1/t)\ln(h_o/h_t)$, r_c, and d, calculate K from Equation 16.18.

Remarks
- Bouwer and Rice showed that if $D \gg b$, an increase in D has little effect on the flow system and, hence, no effect on R_e. The effective upper limit of $\ln[(D\text{-}b)/r_w]$ in Equation 16.19 was found to be 6. Thus, if D is considered infinite, or $D - b$ is so large that $\ln[(D\text{-}b)/r_w] > 6$, a value of 6 should still be used for this term in Equation 16.19;
- If the head is rising in the screened part of the well instead of in its unscreened part, allowance should be made for the fact that the hydraulic conductivity of the zone around the well (gravel pack) may be much higher than that of the aquifer. The value of r_c in Equations 16.17 and 16.18 should then be taken as $r_c = [r_a^2 + n(r_w^2 - r_a^2)]^{0.5}$, where r_a = actual well radius and n = the porosity of the gravel envelope or zone around the well;
- It should not be forgotten that a slug test only permits the estimation of K of a small part of the aquifer: a cylinder of small radius, R_e, and a height somewhat larger than d;
- The values of $\ln(R_e/r_w)$ calculated by Equations 16.19 and 16.20 are accurate to within 10 to 25 per cent, depending on the ratio d/b;
- In a highly permeable aquifer, the head in the well will rise rapidly during a slug

test. The rate of rise can be reduced by placing packers inside the well over the upper part of the screen so that groundwater can only enter through the lower part. Equations 16.19 and 16.20 can then be used to calculate $\ln(R_e/r_w)$;

- Because the watertable in the aquifer is kept constant and is taken as a plane source of water in the analog evaluations of R_e, the Bouwer and Rice method can also be used for a leaky aquifer, provided that its lower boundary is an aquiclude and its upper boundary an aquitard.

17 Uniformly-fractured aquifers, double-porosity concept

17.1 Introduction

Fractures in a rock formation strongly influence the fluid flow in that formation. Conventional well-flow equations, developed primarily for homogeneous aquifers, therefore do not adequately describe the flow in fractured rocks. An exception occurs in hard rocks of very low permeability if the fractures are numerous enough and are evenly distributed throughout the rock; then the fluid flow will only occur through the fractures and will be similar to that in an unconsolidated homogeneous aquifer.

A complicating factor in analyzing pumping tests in fractured rock is the fracture pattern, which is seldom known precisely. The analysis is therefore a matter of identifying an unknown system (Section 2.9). System identification relies on models, whose characteristics are assumed to represent the characteristics of the actual system. We must therefore search for a well-defined theoretical model to simulate the behaviour of the actual system and to produce, as closely as possible, its observed response.

In recent years, many theoretical models have been developed, all of them assuming simplified regular fracture systems that break the rock mass into blocks of equal dimensions (Figure 17.1). These models usually allow conventional type-curve matching procedures to be used. But, because the mechanism of fluid flow in fractured rocks is complex, the models are complex too, comprising, as they do, several parameters or a combination of parameters. Consequently, few of the associated well functions have been tabulated, so, for the other models, one first has to calculate a set of function values. This makes such models less attractive for our purpose.

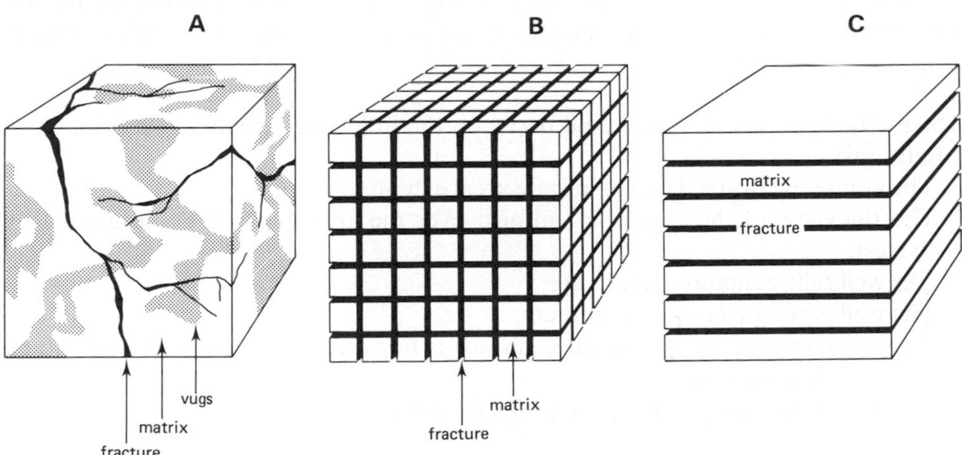

Figure 17.1 Fractured rock formations
A: A naturally fractured rock formation
B: Warren-Root's idealized three-dimensional, orthogonal fracture system
C: Idealized horizontal fracture system

Even more serious is the on-going debate about fracture flow, which indicates that the theory of fluid flow in fractured media is less well-established than that in porous media. In reviewing the literature on the subject, Streltsova-Adams (1978) states: 'Published work on well tests in fractured reservoirs clearly indicates the lack of a unified approach, which has led to contradictory results in analyzing the drawdown behaviour'. And Gringarten (1982), in his review, states: 'A careful inspection of the published analytical solutions indicates that they are essentially identical. Apparent differences come only from the definition of the various parameters used in the derivation'. Indeed, in the literature, there is an enormous overlap of equations. In this chapter, therefore, we present some practical methods that do not require lengthy tables of function values and which, when used in combination, allow a complete analysis of the data to be made.

The methods we present are all based on the double-porosity theory developed initially by Barenblatt et al. (1960). This concept regards a fractured rock formation as consisting of two media: the fractures and the matrix blocks, both of them having their own characteristic properties. Two coexisting porosities and hydraulic conductivities are thus recognized: those of primary porosity and low permeability in the matrix blocks, and those of low storage capacity and high permeability in the fractures. This concept makes it possible to explain the flow mechanism as a re-equalization of the pressure differential in the fractures and blocks by the flow of fluid from the blocks into the fractures. No variation in head within the matrix blocks is assumed. This so-called interporosity flow is in pseudo-steady state. The flow through the fractures to the well is radial and in an unsteady state.

The assumption of pseudo-steady-state interporosity flow does not have a firm theoretical justification. Transient block-to-fracture flow was therefore considered by Boulton and Streltsova (1977), Najurieta (1980), and Moench (1984). From Moench's work, it is apparent that the assumption of pseudo-steady-state interporosity flow is only justified if the faces of the matrix blocks are coated by some mineral deposit (as they often are). Only then will there be little variation in head within the blocks. The pseudo-steady-state solution is thus a special case of Moench's solution of transient interporosity flow.

The methods in this chapter are all based on the following general assumptions and conditions:

- The aquifer is confined and of infinite areal extent;
- The thickness of the aquifer is uniform over the area that will be influenced by the test;
- The well fully penetrates a fracture;
- The well is pumped at a constant rate;
- Prior to pumping, the piezometric surface is horizontal over the area that will be influenced by the test;
- The flow towards the well is in an unsteady state.

The first method in this chapter, in Section 17.2, is the Bourdet-Gringarten method and its approximation, which is more universally applicable than other methods; it uses drawdown data from observation wells. Next, in Section 17.3, we present the Kazemi et al. method; it is an extension of the method originally developed by Warren and Root (1963) for a pumped well; the Kazemi et al. method uses data from observation wells. Finally, in Section 17.4, we present the original Warren and Root method for a pumped well.

17.2 Bourdet-Gringarten's curve-fitting method (observation wells)

Bourdet and Gringarten (1980) state that, in a fractured aquifer of the double-porosity type (Figure 17.1B), the drawdown response to pumping as observed in observation wells can be expressed as

$$s = \frac{Q}{4\pi T_f} F(u^*,\lambda,\omega) \tag{17.1}$$

where

$$u^* = \frac{T_f t}{(S_f + \beta\, S_m) r^2} \tag{17.2}$$

$$\lambda = \alpha\, r^2 \frac{K_m}{K_f} \tag{17.3}$$

$$\omega = \frac{S_f}{S_f + \beta\, S_m} \tag{17.4}$$

f = of the fractures
m = of the matrix blocks
T = $\sqrt{T_{f(x)} T_{f(y)}}$ = effective transmissivity (m^2/d)
S = storativity (dimensionless)
K = hydraulic conductivity (m/d)
λ = interporosity flow coefficient (dimensionless)
α = shape factor, parameter characteristic of the geometry of the fractures and aquifer matrix of a fractured aquifer of the double-porosity type (dimension: reciprocal area)
β = factor; for early-time analysis it equals zero and for late-time analysis it equals 1/3 (orthogonal system) or 1 (strata type)
x,y = relative to the principal axes of permeability

To avoid confusion, note that our definition of the parameter λ differs from the definition of λ commonly used in the petroleum literature; $\lambda = (r/r_w)^2 \lambda_{oil}$.
Note also that for a fracture system as shown in Figure 17.1B, $\alpha = 4n(n+2)/l^2$, where n is the number of a normal set of fractures (1, 2, or 3) and l is a characteristic dimension of a matrix block. For a system of horizontal slab blocks (n = 1) as shown in Figure 17.1C, $\alpha = 12/h_m^2$, where h_m is the thickness of a matrix block. Typical values of λ and ω fall within the ranges of $10^{-3}\,(r_w/r)^2$ to $10^{-9}\,(r_w/r)^2$ for λ and 10^{-1} to 10^{-4} for ω (Serra et al. 1983).
For small values of pumping time, Equation 17.1 reduces to

$$s = \frac{Q}{4\pi T_f} W(u) \tag{17.5}$$

where

$$u = \frac{(S_f + \beta S_m)r^2}{4\ T_f t} \tag{17.6}$$

Equation 17.5 is identical to the Theis equation. It describes only the drawdown behaviour in the fracture system (β equals zero). For large values of pumping time, Equation 17.1 also reduces to the Theis equation, which now describes the drawdown behaviour in the combined fracture and block system (β equals 1/3 or 1).

According to the pseudo-steady-state interporosity flow concept, the drawdown becomes constant at intermediate pumping times when there is a transition from fracture flow to flow from fractures and matrix blocks. The drawdown at which the transition occurs is equal to

$$s = \frac{Q}{2\pi T_f} K_o(\sqrt{\lambda}) \tag{17.7}$$

where $K_o(x)$ is the modified Bessel function of the second kind and of zero order.

Bourdet and Gringarten (1980) showed that, for λ values less than 0.01, Equation 17.7 reduces to

$$s = \frac{2.30Q}{4\pi T_f} \log \frac{1.26}{\lambda} \tag{17.8}$$

The drawdown at which the transition occurs is independent of early- and late-time drawdown behaviours and is solely a function of λ.

Bourdet and Gringarten (1980) presented type curves of $F(u^*,\lambda,\omega)$ versus u^* for different values of λ and ω (Figure 17.2). These type curves are obtained as a superposition of Theis solutions labelled in ω values, with a set of curves representing the behaviour during the transitional period and depending upon λ.

As can be seen from Figure 17.2, the horizontal segment does not appear in the type curves at high values of ω. For high ω values, the type curves only have an inflection point. Numerous combinations of ω and λ values are possible, each pair yielding different type curves. But, instead of presenting extensive tables of function values required to prepare these many different type curves, we present a simplified method. It is based on matching both the early- and late-time data with the Theis type curve, which yields values of T_f and S_f, and T_f and $S_f + S_m$, respectively. From the steady-state drawdown at intermediate times, a value of λ can be estimated from Equation 17.7 or 17.8.

The Bourdet-Gringarten method can be used if, in addition to the general assumptions and conditions listed in Section 17.1, the following assumptions and conditions are satisfied:

- The aquifer is of the double-porosity type and consists of homogeneous and isotropic blocks or strata of primary porosity (the aquifer matrix), separated from each other either by an orthogonal system of continuous uniform fractures or by equally-spaced horizontal fractures;
- Any infinitesimal volume of the aquifer contains sufficient portions of both the aquifer matrix and the fracture system;

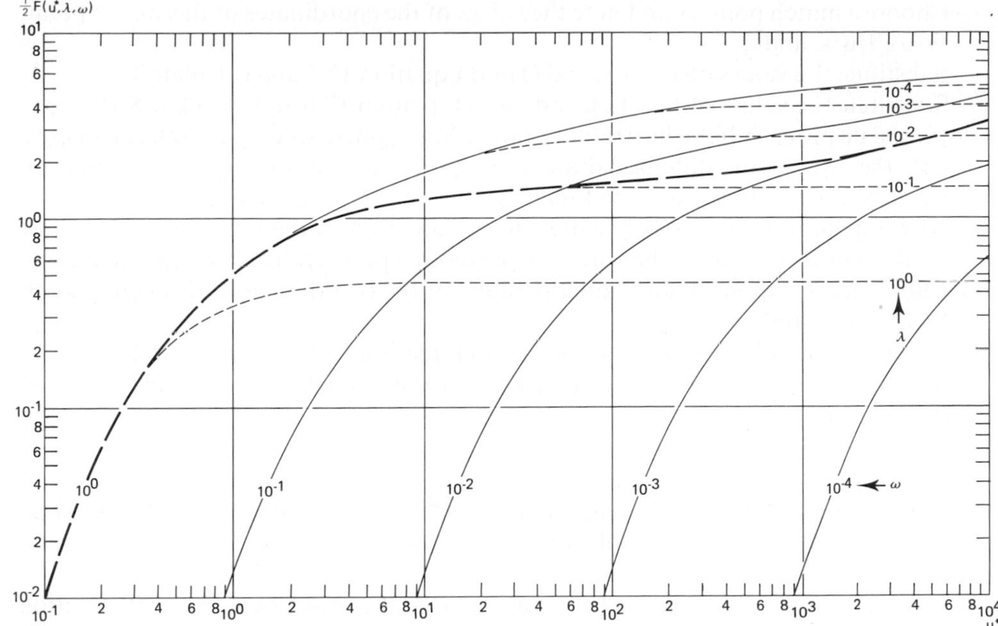

Figure 17.2 Type curves for the function F(u*,λ,ω) (after Bourdet and Gringarten 1980)

- The aquifer matrix has a lower permeability and a higher storativity than the fracture system;
- The flow from the aquifer matrix into the fractures (i.e. the interporosity flow) is in a pseudo-steady state;
- The flow to the well is entirely through the fractures, and is radial and in an unsteady state;
- The matrix blocks and the fractures are compressible;
- $\lambda < 1.78$.

Bourdet and Gringarten (1980) showed that the double-porosity behaviour of a fractured aquifer only occurs in a restricted area around the pumped well. Outside that area (i.e. for λ values greater than 1.78), the drawdown behaviour is that of an equivalent unconsolidated, homogeneous, isotropic confined aquifer, representing both the fracture and the block flow.

Procedure 17.1

- Prepare a type curve of the Theis well function on log-log paper by plotting values of W(u) versus 1/u, using Annex 3.1;
- On another sheet of log-log paper of the same scale, plot the drawdown s observed in an observation well versus the corresponding time t;
- Superimpose the data plot on the type curve and adjust until a position is found where most of the plotted points representing the early-time drawdowns fall on the type curve;

- Choose a match point A and note the values of the coordinates of this match point, W(u), 1/u, s, and t;
- Substitute the values of W(u), s, and Q into Equation 17.5 and calculate T_f;
- Substitute the values of 1/u, T_f, t, and r into Equation 17.6 and calculate S_f ($\beta = o$);
- If the data plot exhibits a horizontal straight-line segment or only an inflection point, note the value of the stabilized drawdown or that of the drawdown at the inflection point. Substitute this value into Equation 17.7 or 17.8 and calculate λ;
- Now superimpose the late-time drawdown data plot on the type curve and adjust until a position is found where most of the plotted points fall on the type curve;
- Choose a matchpoint B and note the values of the coordinates of this matchpoint, W(u), 1/u, s, and t;
- Substitute the values of W(u), s, and Q into Equation 17.5 and calculate T_f;
- Substitute the values of 1/u, T_f, t, and r into Equation 17.6 and calculate $S_f + S_m$ ($\beta = 1/3$ or 1).

Remarks

- For relatively small values of ω, matching the late-time drawdowns with the Theis type curve may not be possible and the analysis will only yield values of T_f and S_f;
- For high values of λ (i.e. for large values of r), the drawdown in an observation well no longer reflects the aquifer's double-porosity character and the analysis will only yield values of T_f and $S_f + S_m$;
- Gringarten (1982) pointed out that the Bourdet-Gringarten's type curves are identical to the time-drawdown curves for an unconsolidated unconfined aquifer with delayed yield as presented by Boulton (1963). (See also Chapter 5.) If one has no detailed knowledge of the aquifer's hydrogeology, this may lead to a misinterpretation of the pumping test data.

17.3 Kazemi et al.'s straight-line method (observation wells)

Kazemi et al. (1969) showed that the drawdown equations developed by Warren and Root (1963) for a pumped well can also be used for observation wells. Their extension of the approximation of the Warren-Root solution is, in fact, also an approximation of the general solution of Bourdet and Gringarten (1980). It can be expressed by

$$s = \frac{Q}{4\pi T_f} F(u^*,\lambda,\omega) \tag{17.1}$$

where

$$F(u^*,\lambda,\omega) = 2.3 \log(2.25\, u^*) + \mathrm{Ei}\left(-\frac{\lambda u^*}{\omega(1-\omega)}\right) - \mathrm{Ei}\left(-\frac{\lambda u^*}{(1-\omega)}\right) \tag{17.9}$$

Equation 17.9 is valid for u* values greater than 100, in analogy with Jacob's approximation of the Theis solution (Chapter 3).

A semi-log plot of the function $F(u^*,\lambda,\omega)$ versus u* (for fixed values of λ and ω) will reveal two parallel straight lines connected by a transitional curve (Figure 17.3). Consequently, the corresponding s versus t plot will theoretically show the same pattern (Figure 17.4).

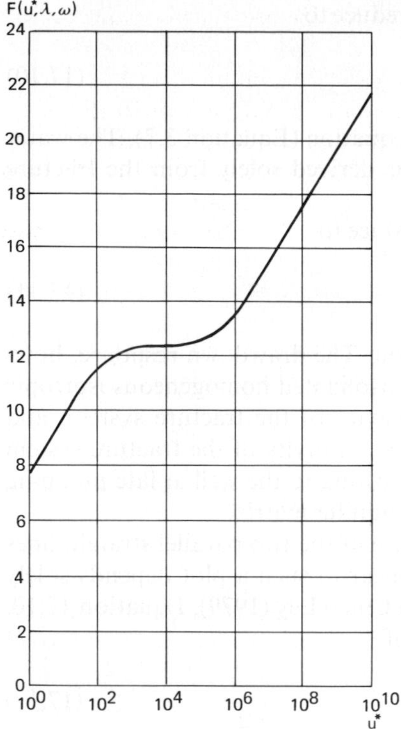

Figure 17.3 Semi-log plot of the function F(u*,λ,ω) versus u* for fixed values of λ and ω

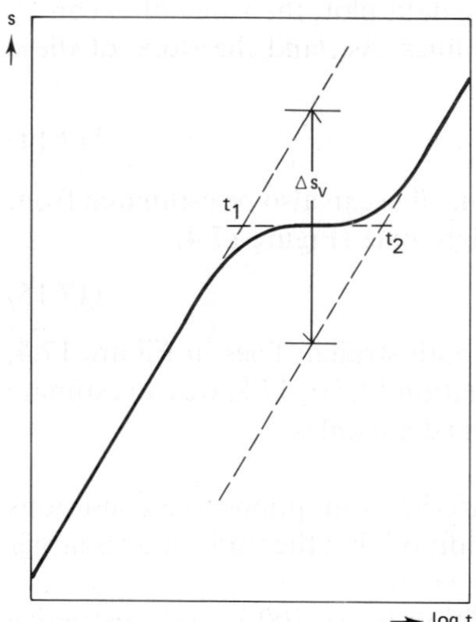

Figure 17.4 Semi-log time-drawdown plot for an observation well in a fractured rock formation of the double-porosity type

For *early pumping times*, Equations 17.1 and 17.9 reduce to

$$s = \frac{2.30Q}{4\pi T_f} \log \frac{2.25\, T_f t}{S_f r^2} \tag{17.10}$$

Equation 17.9 is identical to Jacob's straight-line equation (Equation 3.7). The water flowing to the well during early pumping times is derived solely from the fracture system ($\beta = 0$).
For *late pumping times*, Equations 17.1 and 17.9 reduce to

$$s = \frac{2.30Q}{4\pi T_f} \log \frac{2.25\, T_f t}{(S_f + \beta\, S_m) r^2} \tag{17.11}$$

Equation 17.11 is also identical to Jacob's equation. The drawdown response, however, is now equivalent to the response of an unconsolidated homogeneous isotropic aquifer whose transmissivity equals the transmissivity of the fracture system, and whose storativity equals the arithmetic sum of the storativity of the fracture system and that of the aquifer matrix. Hence, the water flowing to the well at late pumping times comes from both the fracture system and the aquifer matrix.

Kazemi et al.'s method is based on the occurrence of the two parallel straight lines in the semi-log data plot. Whether these lines appear in such a plot depends solely on the values of λ and ω. According to Mavor and Cinco Ley (1979), Equation 17.10, describing the early-time straight line, can be used if

$$u^* \leq \frac{\omega(1-\omega)}{3.6\, \lambda} \tag{17.12}$$

and Equation 17.11, describing the late-time straight line, can be used if

$$u^* \geq \frac{1-\omega}{1.3\, \lambda} \geq 100 \tag{17.13}$$

If the two parallel straight lines occur in a semi-log data plot, the value of ω can be derived from the vertical displacement of the two lines, Δs_v, and the slope of these lines, Δs (Figure 17.4).

$$\omega = 10^{-\Delta s_v/\Delta s} \tag{17.14}$$

According to Mavor and Cinco Ley (1979), the value of ω can also be estimated from the horizontal displacement of the two parallel straight lines (Figure 17.4)

$$\omega = t_1/t_2 \tag{17.15}$$

Following the procedure of the Jacob method on both straight lines in Figure 17.4, we can determine values of T_f, S_f, and S_m. Using Equation 17.7 or 17.8, we can estimate the value of λ from the constant drawdown at intermediate times.

Kazemi et al.'s method can be used if, in addition to the assumptions and conditions underlying the Bourdet-Gringarten method, the condition that the value of u^* is larger than 100 is satisfied.

According to Van Golf-Racht (1982), the condition $u^* > 100$ is very restrictive and can be replaced by $u^* > 100\,\omega$, if $\lambda \ll 1$, or by $u^* > 100 - 1/\lambda$, if $\omega \ll 1$.

Procedure 17.2

- On a sheet of semi-log paper, plot s versus t (t on logarithmic scale);
- Draw a straight line through the early-time points and another through the late-time points; the two lines should plot as parallel lines;
- Determine the slope of the lines (i.e. the drawdown difference Δs per log cycle of time);
- Substitute the values of Δs and Q into $T_f = 2.30\ Q/4\pi\ \Delta s$, and calculate T_f;
- Extend the early-time straight line until it intercepts the time axis where s = 0, and determine t_1;
- Substitute the values of T_f, t_1, and r into $S_f = 2.25\ T_f t_1/r^2$, and calculate S_f;
- Extend the late-time straight line until it intercepts the time axis where s = 0, and determine t_2;
- Substitute the values of T_f, t_2, r, and β into $S_f + \beta\ S_m = 2.25\ T_f t_2/r^2$, and calculate $S_f + S_m$;
- Calculate the separate values of S_f and S_m.

Remarks

The two parallel straight lines can only be obtained at low λ values (i.e. $\lambda < 10^{-2}$). At higher λ values, only the late-time straight line, representing the fracture and block flow, will appear, provided of course that the pumping time is long enough. The analysis then yields values of T_f and $S_f + S_m$.

To obtain separate values of S_f and S_m when only one straight line is present, Procedure 17.3 can be applied.

Procedure 17.3

- Follow Procedure 17.2 to obtain values of T_f and S_f from the first straight line, or if it is not present, values of T_f and $S_f + S_m$ from the second straight line;
- Determine the centre of the transition period of constant drawdown and determine $1/2\ \Delta s_v$;
- Calculate the value of ω using Equation 17.14;
- Substituting the values of ω and β into Equation 17.4, determine the value of S_m if S_f is known, or vice versa.

Remark

To estimate the centre of the transition period with constant drawdown, the preceding and following curved-line segments should be present in the time-drawdown plot.

17.4 Warren-Root's straight-line method (pumped well)

As Kazemi et al.'s straight-line method for observation wells is an extension of Warren-Root's straight-line method for a pumped well, we can use Equations 17.7 to 17.15 to analyze the drawdown in a pumped well if we replace the distance of the observation well to the pumped well, r, with the effective radius of the pumped well, r_w.

Following Procedure 17.2 on both straight lines in the semi-log plot of s_w versus t, we can determine T_f, S_f, and S_m, provided that there are no well losses (i.e. no skin) and that well-bore storage effects are negligible.

According to Mavor and Cinco Ley (1979), well-bore storage effects become negligible when

$$u^* > C' (60 + 3.5 \text{ skin}) \tag{17.16}$$

where, at early pumping times

$C' = C/2\pi S_f r_w^2$ (dimensionless)

C = well-bore storage constant = ratio of change in volume of water in the well and the corresponding drawdown (m^2)

For a water-level change in a perfect well (i.e. no well losses), which is pumping a homogeneous confined aquifer, the dimensionless coefficient C′ is related to the dimensionless α as defined by Papadopulos (1967) (see Section 11.1.1) by the relationship (Ramey 1982)

$$C' = 1/2\alpha$$

When well-bore storage effects are not negligible, the limiting condition for applying Equation 17.10, as expressed by Equation 17.12, should be replaced by

$$C' (60 + 3.5 \text{ skin}) < u^* < \frac{\omega(1-\omega)}{3.6\,\lambda} \tag{17.17}$$

The early-time straight line may thus be obscured by storage effects in the well and in the fractures intersecting the well. But, with Procedure 17.3, a complete analysis is then still possible.

Remarks

Well losses (skin) do not influence the calculation of T_f and ω.

If the linear well losses are not negligible, Equation 17.8 becomes (Bourdet and Gringarten 1980)

$$s_w = \frac{2.30Q}{4\pi T_f} \log \frac{1.26}{\lambda\, e^{-2\text{skin}}} \tag{17.18}$$

From the constant drawdown s_w and the calculated value of T_f, the value of $\lambda\, e^{-2\text{skin}}$ can be determined. If the well losses are known or negligible, the value of λ can be estimated.

Example 17.1

For this example, we use the time-drawdown data from Pumping Test 3 conducted on Well UE-25b#1 in the fractured Tertiary volcanic rocks of the Nevada Test Site, U.S.A., as published by Moench (1984).

The well (r_w = 0.11 m; total depth 1219 m) was drilled through thick sequences of fractured and faulted non-welded to densely welded rhyolitic ash flow and bedded tuffs to a depth below the watertable, which was struck at 470 m below the ground surface. Five major zones of water entry occurred over a depth interval of 400 m. The distance between these zones was roughly 100 m. Core samples revealed that most of the fractures dip steeply and are coated with deposits of silica, manganese, and iron oxides, and calcite. The water-producing zones, however, had mineral-filled low-angle fractures, as observed in core samples taken at 612 m below the ground surface.

The well was pumped at a constant rate of 35.8 l/s for nearly 3 days. Table 17.1 shows the time-drawdown data of the well.

Like Moench, we assume that the fractured aquifer is unconfined and of the strata type (i.e. $\beta = 1$). Figure 17.5 shows the log-log drawdown plot of the pumped well and Figure 17.6 the semi-log drawdown plot. These figures clearly reveal the double porosity of the aquifer because they show the early-time, intermediate-time, and late-time segments characteristic of double-porosity media. At early pumping times, however, well-bore storage affects the time-drawdown relationship of the well. In a log-log plot of drawdown versus time, well-bore storage is usually reflected by a straight line of slope unity. Consequently, the two parallel straight lines of the Warren and Root model do not appear in Figure 17.6. Only the late-time data plot as a straight line.

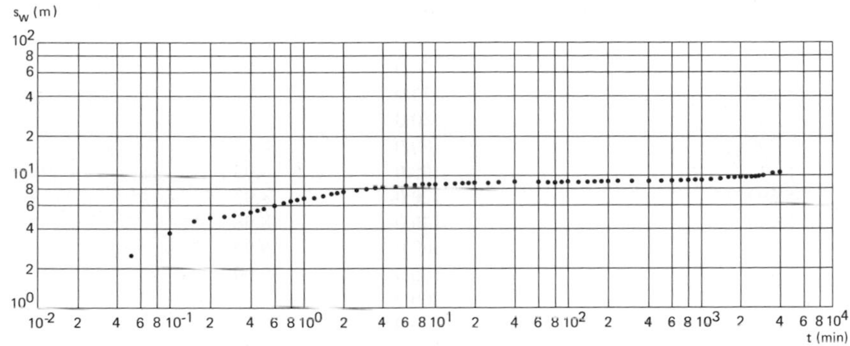

Figure 17.5 Time-drawdown log-log plot of data from the pumped well UE-25b#1 at the Nevada Test Site, U.S.A. (after Moench 1984)

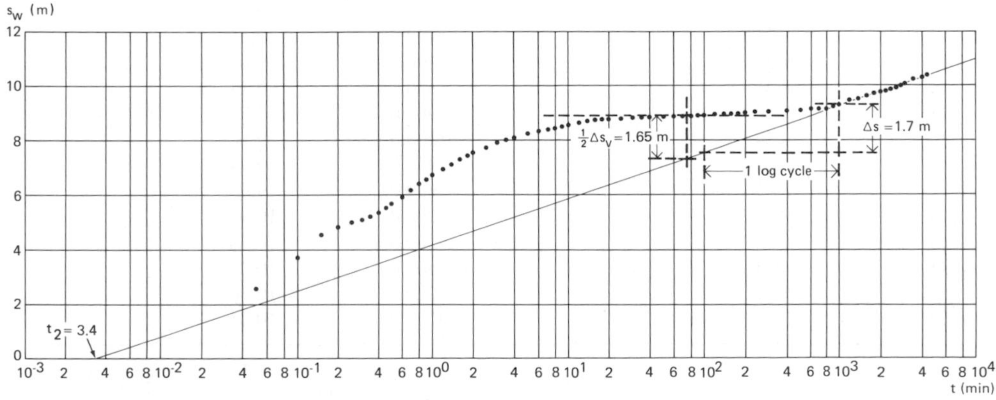

Figure 17.6 Time-drawdown semi-log plot of data from the pumped well UE-25b#1 at the Nevada Test Site, U.S.A. (after Moench 1984)

Well-bore skin effects are unlikely, because air was used when the well was being drilled, the major water-producing zones were not screened, and prior to testing the well was thoroughly developed.

To analyze the drawdown in this well, we follow Procedure 17.3. From Figure 17.6, we determine the slope of the late-time straight line, which is $\Delta s = 1.70$ m. We then calculate the fracture transmissivity from

$$T_f = \frac{2.30Q}{4\pi\Delta s} = \frac{2.3 \times 3093.12}{4 \times 3.14 \times 1.70} = 333 \text{ m}^2/\text{d}$$

Table 17.1 Drawdown data from pumped well UE-25b # 1, test 3 (after Moench 1984)

t (min)	s_w (m)	t (min)	s_w (m)
0.05	2.513	30.0	8.84
0.1	3.769	35.0	8.84
0.15	4.583	40.0	8.86
0.2	4.858	50.0	8.86
0.25	5.003	60.0	8.90
0.3	5.119	70.0	8.91
0.35	5.230	80.0	8.92
0.4	5.390	90.0	8.93
0.45	5.542	100.0	8.95
0.5	5.690	120.0	8.97
0.6	5.990	140.0	8.98
0.7	6.19	160.0	8.99
0.8	6.42	180.0	9.00
0.9	6.59	200.0	9.02
1.0	6.74	240.0	9.04
1.2	6.96	300.0	9.07
1.4	7.17	400.0	9.11
1.6	7.33	500.0	9.14
1.8	7.45	600.0	9.17
2.0	7.56	700.0	9.18
2.5	7.76	800.0	9.21
3.0	7.93	900.0	9.25
3.5	8.03	1000.0	9.30
4.0	8.12	1200.0	9.44
5.0	8.24	1400.0	9.55
6.0	8.32	1600.0	9.64
7.0	8.41	1800.0	9.74
8.0	8.46	2000.0	9.78
9.0	8.54	2200.0	9.80
10.0	8.62	2400.0	9.84
12.0	8.67	2600.0	9.93
14.0	8.70	2800.0	10.03
16.0	8.74	3000.0	10.08
18.0	8.76	3500.0	10.26
20.0	8.77	4000.0	10.30
25.0	8.81	4200.0	10.41

Extending the straight line until it intercepts the time axis where $s = 0$ yields $t_2 = 3.4 \times 10^{-3}$ min. The overall storativity is then calculated from

$$S_f + S_m = \frac{2.25\, T_f\, t_2}{r^2_w} = \frac{2.25 \times 333 \times 3.4 \times 10^{-3}}{1440\,(0.11)^2} = 0.15$$

The semi-log plot of time versus drawdown shows that the centre of the transition period is at $t \approx 75$ minutes. At $t = 75$ minutes, $1/2\ \Delta s_v = 1.65$ m. Substituting the appropriate values into Equation 17.14 yields

$$\omega = 10^{-\Delta s_v/\Delta s} = 10^{-2 \times 1.65/1.70} = 0.011$$

Substituting the appropriate values into Equation 17.4 yields

$$S_f = \omega\,(S_f + S_m) = 0.011 \times 0.146 = 0.0016$$

and

$$S_m = 0.15$$

This high value of S_m is an order of magnitude normally associated with the specific yield of unconfined aquifers. Moench (1984), however, offers an explanation for such a high value for the storativity of the fractured volcanic rock, namely that it may be due to the presence of highly compressible microfissures within the matrix blocks. We consider this a plausible explanation, because there is little reason to assume homogeneous matrix blocks, as in Figure 17.1C.

We must now check the condition that $u^* > 100$, which underlies the Warren-Root method. Substituting the appropriate values into Equation 17.2, we obtain

$$t > \frac{100\,(S_f + S_m)\, r^2_w}{T_f} = \frac{100 \times 1440 \times 0.15\,(0.11)^2}{333} = 0.8 \text{ min}$$

Hence this condition is satisfied.

Next, we must check the condition stated in Equation 17.13. For this, we need the value of λ. The constant drawdown during intermediate times is taken as 8.9 m. Using Equation 17.8, we obtain

$$\lambda = 1.26 \,/\, 10^{(4 \times 3.14 \times 333 \times 8.9)/(2.3 \times 3093.12)} = 7.3 \times 10^{-6}$$

Substituting the appropriate values into Equation 17.13 gives

$$t > \frac{(1-\omega)\, S_f + S_m)\, r_w^2}{1.3\, \lambda\, T_f} = \frac{1440\,(1-0.011)\, 0.15\,(0.11)^2}{1.3 \times 7.3 \times 10^{-6} \times 333} = 818 \text{ min}$$

The condition for the second straight-line relationship is also satisfied.

Finally, we must check our assumption that the first straight-line relationship is obscured by well-bore storage effects. Using $C' = 1/2\alpha$ and assuming $r_c = r_w$ gives us $C' = 1/2S_f$. Taking this C' value and using Equation 17.16, we get

$$t > \frac{60\, r_w^2}{2\, T_f} = \frac{1440 \times 60\,(0.11)^2}{2 \times 333} = 1.6 \text{ min}$$

So, according to Equation 17.16, after approximately 1.6 min, the drawdown data are no longer influenced by well-bore storage effects. A check of Figure 17.6 shows us that the early-time straight-line relationship would have occurred before then and is thus obscured by well-bore storage effects.

18 Single vertical fractures

18.1 Introduction

If a well intersects a single vertical fracture, the aquifer's unsteady drawdown response to pumping differs significantly from that predicted by the Theis solution (Chapter 3). This well-flow problem has long been a subject of research in the petroleum industry, especially after it had been discovered that if an oil well is artificially fractured ('hydraulic fracturing') its yield can be raised substantially. Various solutions to this problem have been proposed, but most of them produced erroneous results. A major step forward was taken when the fracture was assumed to be a plane, vertical fracture of relatively short length and infinite hydraulic conductivity. (A plane fracture is one of zero width, which means that fracture storage can be neglected.) This made it possible to analyze the system as an 'equivalent', anisotropic, homogeneous, porous medium, with a single fracture of high permeability intersected by the pumped well.

The concept underlying the analytical solutions is as follows: The aquifer is homogeneous, isotropic, and of large lateral extent, and is bounded above and below by impermeable beds. A single plane, vertical fracture of relatively short length dissects the aquifer from top to bottom (Figure 18.1A). The pumped well intersects the fracture midway. The fracture is assumed to have an infinite (or very large) hydraulic conductivity. This means that the drawdown in the fracture is uniform over its entire length at any instant of time (i.e. there is no hydraulic gradient in the fracture). This uniform drawdown induces a flow from the aquifer into the fracture. At early pumping times, this flow is one-dimensional (i.e. it is horizontal, parallel, and perpendicular to the fracture) (Figure 18.1B). All along the fracture, a uniform flux condition is assumed to exist (i.e. water from the aquifer enters the fracture at the same rate per unit area).

Groundwater hydrology recognizes a similar situation: that of a constant groundwater discharge into an open channel that fully penetrates a homogeneous unconsolidated aquifer. Solutions to this flow problem have been presented by Theis (1935), Edelman (1947; 1972), Ferris (1950), and Ferris et al. (1962). It is hardly surprising that the solutions that have been developed for early-time drawdowns in a single vertical fracture are identical to those found by the above authors (Jenkins and Prentice 1982).

As pumping continues, the flow pattern changes from parallel flow to pseudo-radial flow (Figure 18.1C), regardless of the fracture's hydraulic conductivity. During this period, most of the well discharge originates from areas farther removed from the fracture. Often, uneconomic pumping times are required to attain pseudo-radial flow, but once it has been attained, the classical methods of analysis can be applied.

The methods presented in this chapter are all based on the following general assumptions and conditions:
- The general assumptions and conditions listed in Section 17.1.

And:
- The aquifer is confined, homogeneous, and isotropic, and is fully penetrated by a single vertical fracture;

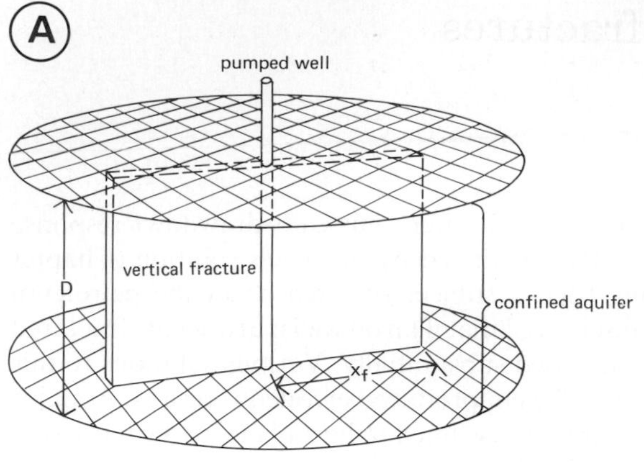

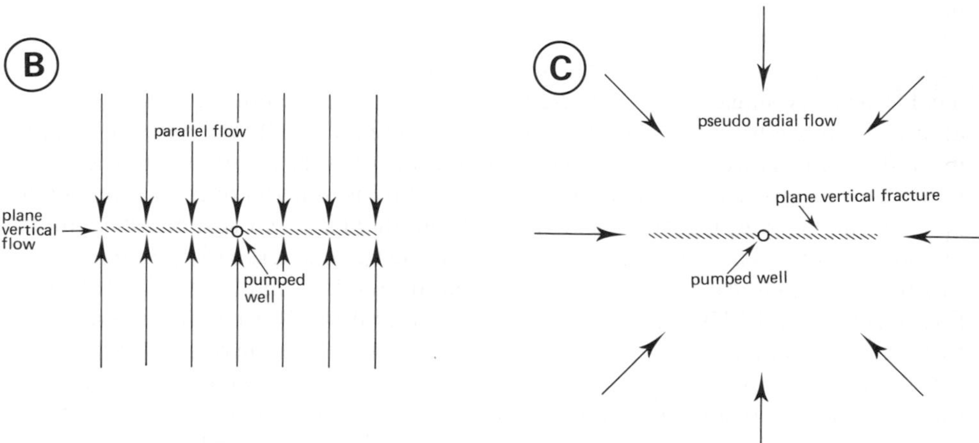

Figure 18.1 A well that intersects a single, vertical, plane fracture of finite length and infinite hydraulic conductivity
A: The well-fracture-aquifer system
B: The parallel flow system at early pumping times
C: The pseudo-radial flow system at late pumping times

- The fracture is plane (i.e. storage in the fracture can be neglected), and its horizontal extent is finite;
- The well is located on the axis of the fracture;
- With decline of head, water is instantaneously removed from storage in the aquifer;
- Water from the aquifer enters the fracture at the same rate per unit area (i.e. a uniform flux exists along the fracture, or the fracture conductivity is high although not infinite);

The first method in this chapter, in Section 18.2, is that of Gringarten and Witherspoon

(1972), which uses the drawdown data from observation wells placed at specific locations with respect to the pumped well. Next, in Section 18.3, is the method of Gringarten and Ramey (1974); it uses drawdown data from the pumped well only, neglecting well losses and well-bore storage effects. Finally, in Section 18.4, we present the Ramey and Gringarten method (1976), which allows for well-bore storage effects in the pumped well.

18.2 Gringarten-Witherspoon's curve-fitting method for observation wells

For a well pumping a single, plane, vertical fracture in an otherwise homogeneous, isotropic, confined aquifer (Figure 18.2), Gringarten and Witherspoon (1972) obtained the following general solution for the drawdown in an observation well

$$s = \frac{Q}{4\pi T} F(u_{vf}, r') \qquad (18.1)$$

where

$$u_{vf} = \frac{T\,t}{S x_f^2} \qquad (18.2)$$

$$r' = \frac{\sqrt{x^2 + y^2}}{x_f} \qquad (18.3)$$

S = storativity of the aquifer, dimensionless
T = transmissivity of the aquifer (m^2/d)
x_f = half length of the vertical fracture (m)
x,y = distance between observation well and pumped well, measured along the x and y axis, respectively (m)

From Equations 18.1 and 18.2, it can be seen that the drawdown in an observation well depends not only on the parameter u_{vf} (i.e. on the aquifer characteristics T and S, the vertical fracture half-length x_f, and the pumping time t), but also on the geometrical relationship between the location of the observation well and that of the fracture.

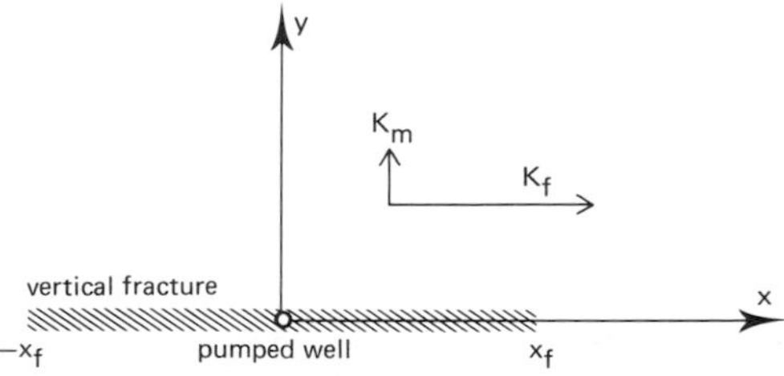

Figure 18.2 Plan view of a pumped well that intersects a plane, vertical fracture of finite length and infinite hydraulic conductivity

For observation wells in three different locations (Figure 18.3), Gringarten and Witherspoon developed simplified expressions for the drawdown derived from Equation 18.1.

For an observation well located along the x axis ($r' = x/x_f$), the drawdown function $F(u_{vf},r')$ in Equation 18.1 reads

$$F(u_{vf},r') = \frac{\sqrt{\pi}}{2} \int_0^{u_{vf}} \left[\mathrm{erf}\left(\frac{1-r'}{2\sqrt{\tau}}\right) + \mathrm{erf}\left(\frac{1+r'}{2\sqrt{\tau}}\right) \right] \frac{d\tau}{\sqrt{\tau}} \tag{18.4}$$

For an observation well located along the y axis ($r' = y/x_f$), the drawdown function $F(u_{vf},r')$ in Equation 18.1 reads

$$F(u_{vf},r') = \sqrt{\pi} \int_0^{u_{vf}} \mathrm{erf}\left(\frac{1}{2\sqrt{\tau}}\right) \exp\left[-\frac{(r')^2}{4\tau}\right] \frac{d\tau}{\sqrt{\tau}} \tag{18.5}$$

For an observation well located along a line through the pumped well and making an angle of 45° with the direction of the fracture ($r' = x\sqrt{2}/x_f = y\sqrt{2}/x_f$), the drawdown function $F(u_{vf},r')$ in Equation 18.1 reads

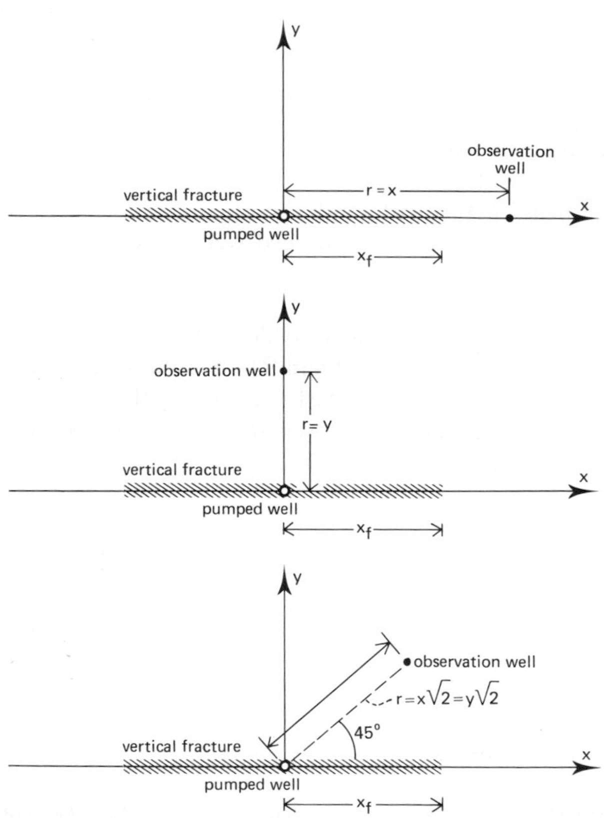

Figure 18.3 Plan view of a vertical fracture with observation wells at three different locations

$$F(u_{vf},r') = \frac{\sqrt{\pi}}{2} \int_0^{u_{vf}} \exp\left[-\frac{\left(\frac{r'}{\sqrt{2}}\right)^2}{4\tau}\right] \cdot \left[\operatorname{erf}\left(\frac{1-\left(\frac{r'}{\sqrt{2}}\right)}{2\sqrt{\tau}}\right) + \operatorname{erf}\left(\frac{1+\left(\frac{r'}{\sqrt{2}}\right)}{2\sqrt{\tau}}\right)\right] \frac{d\tau}{\sqrt{\tau}}$$

Figures 18.4, 18.5, and 18.6 show the three different families of type curves developed from Equations 18.4, 18.5, and 18.6, respectively (Gringarten and Witherspoon 1972; see also Thiery et al. 1983). For the three locations of observation well, Annex 18.1 gives values of the function $F(u_{vf},r')$ for different values of u_{vf} and r'.

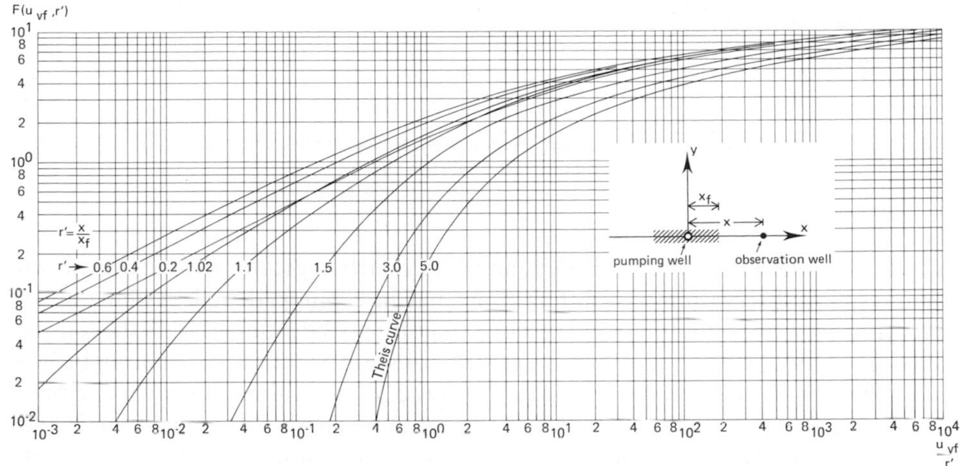

Figure 18.4 Gringarten-Witherspoon's type curves for a vertical fracture with an observation well located on the x axis (after Merton 1987)

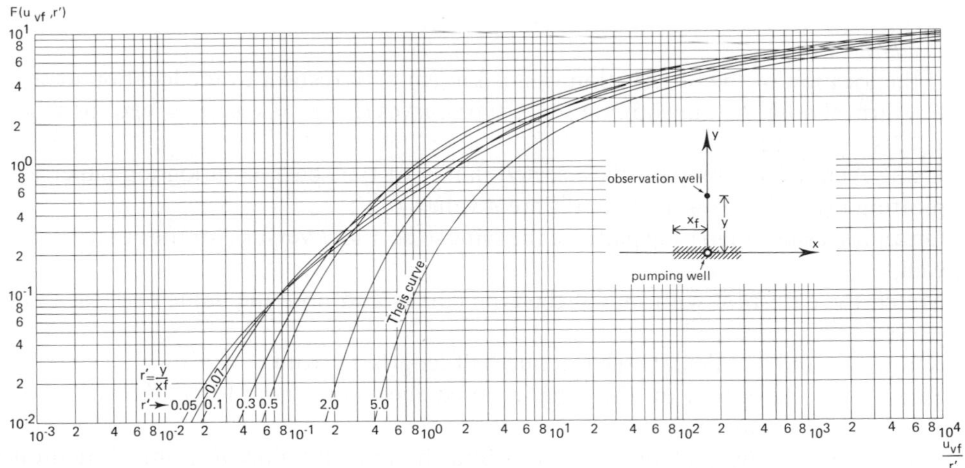

Figure 18.5 Gringarten-Witherspoon's type curves for a vertical fracture with an observation well located on the y axis (after Merton 1987)

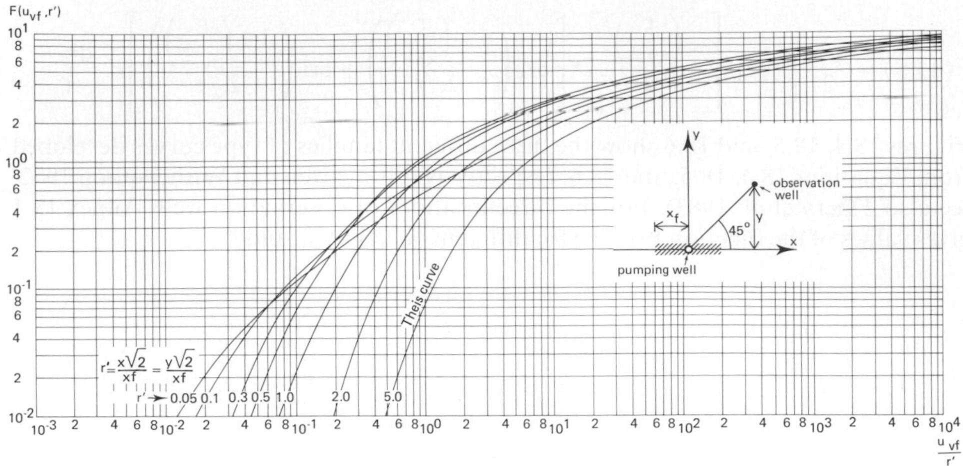

Figure 18.6 Gringarten-Witherspoon's type curves for a vertical fracture with an observation well located at 45° from the centre of the fracture (after Merton 1987)

The type curves in Figures 18.4, 18.5, and 18.6 clearly indicate that the drawdown response in an observation well differs from that in a pumped well. As long as an observation well does not intersect the same fracture as the pumped well, the log-log plot of the drawdown in the observation well does not yield an initial straight line of slope 0.5. Far enough from the pumped well (i.e. $r' > 5$), the drawdown response becomes identical to that for radial flow to a pumped well in the Theis equation (Equation 3.5). In other words, beyond a distance $r' = 5$, the influence of the fracture on the drawdown is negligible.

The Gringarten-Witherspoon curve-fitting method can be used if the assumptions and conditions listed in Section 18.1 are met.

Procedure 18.1

- If the location of the observation well is known with respect to the location of the fracture, choose the appropriate set of type curves (for $r' = x/x_f$; $r' = y/x_f$; or $r' = x\sqrt{2}/x_f = y\sqrt{2}/x_f$);
- Using Annex 18.1, prepare the selected family of type curves on log-log paper by plotting $F(u_{vf},r')$ versus u_{vf}/r' for different values of r';
- On another sheet of log-log paper of the same scale, plot s versus t for the observation well;
- Match the data plot with one of the type curves and note the value of r' for that curve;
- Knowing r and r', calculate the fracture half-length, x_f, from $r' = r/x_f$;
- Select a matchpoint A on the superimposed sheets and note for A the values of $F(u_{vf},r')$, u_{vf}/r', s, and t;
- Substitute the values of $F(u_{vf},r')$ and s and the known value of Q into Equation 18.1 and calculate T;
- Knowing u_{vf}/r' and r', calculate the value of u_{vf};

– Substitute the values of u_{vf}, t, x_f, and T into Equation 18.2 and solve for S.

If the geometrical relationship between the observation wells and the fracture is not known, a trial-and-error matching procedure will have to be applied to all three sets of type curves. Data from at least two observation wells are required for this purpose. The trial-and-error procedure should be continued until matching positions are found that yield approximations of the fracture location and its dimensions, and estimates of the aquifer parameters consistent with all available observation-well data.

Remarks

- For $r' \geq 5$, no real value of r′ (and consequently of x_f) can be found with the Gringarten-Witherspoon method alone because no separate type curves for $r' \geq 5$ can be distinguished. It will only be possible to calculate a maximum value of x_f. If data from the pumped well are also available, however, the product Sx_f^2 can be obtained (Section 18.3). Then, knowing S from the observation-well data, and also knowing Sx_f^2, one can calculate x_f. It should be noted, however, that calculated values of x_f are not precise and are often underestimated (Gringarten et al. 1975);
- For $r' \geq 5$, the observation-well data can be analyzed with the Theis method (Section 3.2.1), from which the aquifer parameters T and S can be obtained.

18.3 Gringarten et al.'s curve-fitting method for the pumped well

For a well intersecting a single, plane, vertical fracture in an otherwise homogeneous, isotropic, confined aquifer (Figure 18.1A), Gringarten and Ramey. (1974) obtained the following general solution for the drawdown in the pumped well

$$s_w = \frac{Q}{4\pi T} F(u_{vf}) \tag{18.7}$$

where

$$F(u_{vf}) = 2\sqrt{\pi u_{vf}}\, \text{erf}\left(\frac{1}{2\sqrt{u_{vf}}}\right) - \text{Ei}\left(-\frac{1}{4u_{vf}}\right) \tag{18.8}$$

and

$$-\text{Ei}(-x) = \int_0^x \frac{e^{-u}}{u}\, du = \text{the exponential integral of x}$$

Equation 18.8 is the reduced form of Equations 18.4 to 18.6 for $r' = 0$. Values of the function $F(u_{vf})$ for different values of u_{vf} are given in Annex 18.2. Figure 18.7 shows a log-log plot of $F(u_{vf})$ versus u_{vf}.

At early pumping times, when the drawdown in the well is governed by the horizontal parallel flow from the aquifer into the fracture, the drawdown can be written as

$$s_w = \frac{Q}{4\pi T} F(u_{vf}) \tag{18.7}$$

where

$$F(u_{vf}) = 2\sqrt{\pi u_{vf}} \tag{18.9}$$

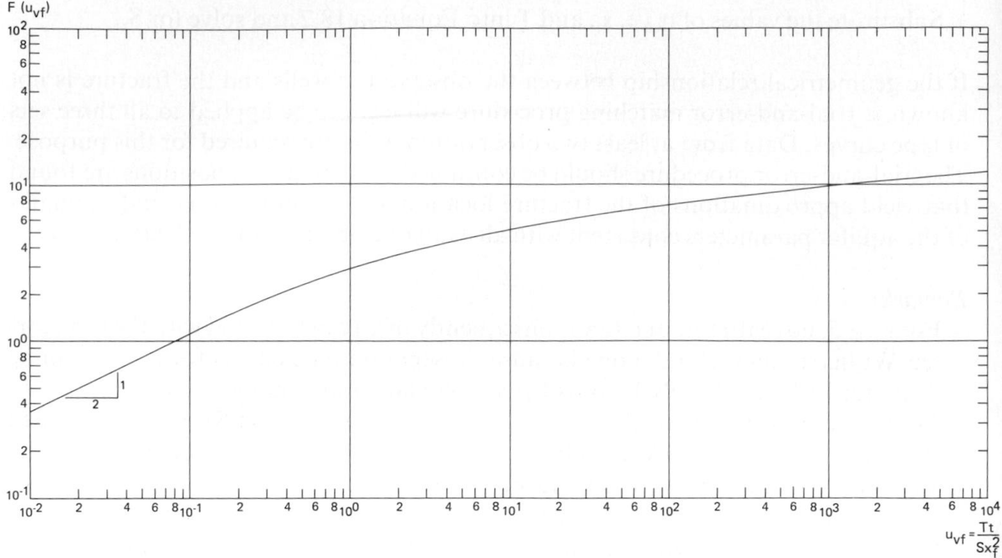

Figure 18.7 Gringarten et al.'s type curve $F(u_{vf})$ versus u_{vf} for a vertical fracture

or

$$\log F(u_{vf}) = 0.5 \log (u_{vf}) + \text{constant}$$

and consequently

$$s_w = \frac{Q}{2\sqrt{\pi T S x_f^2}} \sqrt{t} \tag{18.10}$$

or

$$\log s_w = 0.5 \log(t) + \text{constant}$$

As Equations 18.9 and 18.10 show, on a log-log plot of $F(u_{vf})$ versus u_{vf} (Figure 18.7) (and on the corresponding data plot), the early-time parallel-flow period is characterized by a straight line with a slope of 0.5. The parallel-flow period ends at approximately $u_{vf} = 1.6 \times 10^{-1}$(Gringarten and Ramey. 1975). If the aquifer has a low transmissivity and the fracture is elongated, the parallel-flow period may last relatively long.

The pseudo-radial-flow period starts at $u_{vf} = 2$ (Gringarten et al. 1975). During this period, the drawdown in the well varies according to the Theis equation for radial flow in a pumped, homogeneous, isotropic, confined aquifer (Equation 3.5), plus a constant, and can be approximated by the following expression (Gringarten and Ramey. 1974)

$$s_w = \frac{2.30Q}{4\pi T} \log \frac{16.59Tt}{Sx_f^2} \tag{18.11}$$

The log-log plot of $F(u_{vf})$ versus u_{vf} (Figure 18.7) is used as a type curve to determine T and the product Sx_f^2.

Gringarten et al.'s method is based on the following assumptions and conditions:
- The general assumptions and conditions listed in Section 18.1.

And:
- The diameter of the well is very small (i.e. well-bore storage can be neglected);
- The well losses are negligible.

Procedure 18.2
- Using Annex 18.2, prepare a type curve on log-log paper by plotting $F(u_{vf})$ versus u_{vf};
- On another sheet of log-log paper of the same scale, prepare the data curve by plotting s_w versus t;
- Match the data curve with the type curve and select a matchpoint A on the superimposed sheets; note for A the values of $F(u_{vf})$, u_{vf}, s_w, and t;
- Substitute the values of $F(u_{vf})$ and s_w and the known value of Q into Equation 18.7 and calculate T;
- Substitute the values of u_{vf} and t and the calculated value of T into Equation 18.2 and solve for the product Sx_f^2.

For large values of pumping time (i.e. for $t \geq 2Sx_f^2/T$), the data can be analyzed with Procedure 18.3, which is similar to Procedure 3.4 of the Jacob method (Section 3.2.2).

Procedure 18.3
- If the semi-log plot of s_w versus t yields a straight line, determine the slope of this line, Δs_w;
- Calculate the aquifer transmissivity from $T = 2.30Q/4\pi\Delta s_w$;
- As T is known and the value of t_o can be read from the graph, find Sx_f^2 from $Sx_f^2 = 16.59\ Tt_o$.

Remarks
- No separate values of x_f and S can be found with Gringarten et al.'s method. To obtain such values, one must have drawdown data from at least two observation wells. (See method in Section 18.2);
- Procedures 18.2 and 18.3 can only be applied to data from perfect wells (i.e. wells that have no well losses). Such wells seldom exist, but Procedure 18.3, being applied to late-time drawdown data, allows the aquifer transmissivity to be found;
- If the early-time drawdown data are influenced by well-bore storage, the initial straight line in the data plot may not have a slope of 0.5, but instead a slope of 1, which indicates a large storage volume connected with the well. This corresponds to a fracture of large dimensions rather than the assumed plane fracture. Gringarten et al.'s method will then not be applicable and the data should be analyzed by the method in Section 18.4.

18.4 Ramey-Gringarten's curve-fitting method

For a well intersecting a non-plane vertical fracture in a homogeneous, isotropic, confined aquifer, Ramey and Gringarten (1976) developed a method that takes the storage effects of the fracture into account. Their equation reads

$$s_w = \frac{Q}{4\pi T} F(u_{vf}, C'_{vf}) \tag{18.12}$$

where

$$C'_{vf} = \frac{C_{vf}}{Sx_f^2} \tag{18.13}$$

C_{vf} = a storage constant = $\Delta V/s_w$ = ratio of change in volume of water in the well plus vertical fracture, and the corresponding drawdown (m^2)

Ramey and Gringarten developed their equation by assuming a large-diameter well and a plane vertical fracture of infinite conductivity. In practice, however, the apparent storage effect, C_{vf}, is due not only to the total volume of the well, but also to the pore volume of the fracture.

The family of type curves drawn on the basis of Equation 18.12 is shown in Figure 18.8. Annex 18.3 gives a table of the values of $F(u_{vf}, C'_{vf})$ for different values of u_{vf}

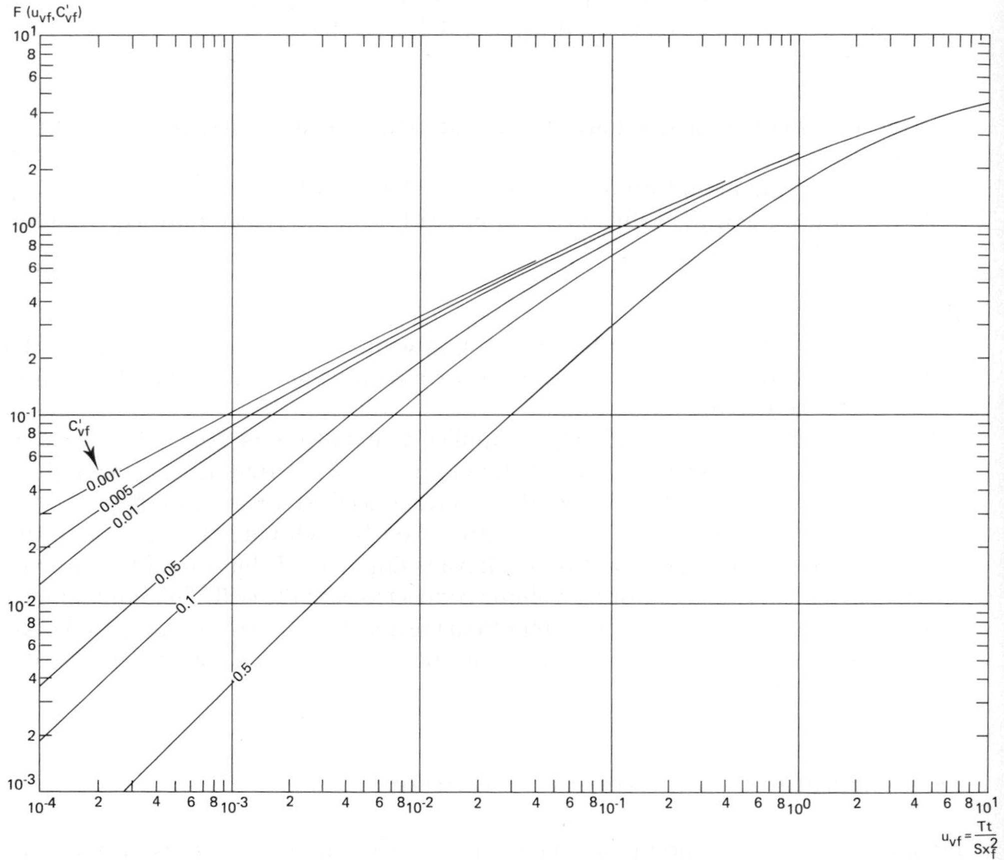

Figure 18.8 Ramey-Gringarten's family of type curves $F(u_{vf}, C'_{vf})$ versus u_{vf} for different values of C'_{vf} for a vertical fracture, taking well-bore storage effects into account

and C'_{vf}. For $C'_{vf} = 0$, the type curve is similar to the Gringarten et al. type curve (Figure 18.7) for a vertical fracture with negligible storage capacity. For values of $C'_{vf} > 0$, the type curves (and in theory also the log-log data plot) will exhibit three different segments (Figure 18.8). Initially, the curves follow a straight line of unit slope, indicating the period during which the storage effects prevail. This straight line gradually passes into another straight line with a slope of 0.5, representing the horizontal parallel-flow period. Finally, when one is using semi-log paper, a straight-line segment also appears, which corresponds to the period of pseudo-radial flow. The slope of this line is 1.15.

Ramey and Gringarten's curve-fitting method is applicable if the following assumptions and conditions are satisfied:
- The general assumptions and conditions listed in Section 18.1.

And:
- The well losses are negligible.

Procedure 18.4
- Using Annex 18.3, prepare a family of type curves on log-log paper by plotting $F(u_{vf},C'_{vf})$ versus u_{vf} for different values of C'_{vf};
- On another sheet of log-log paper of the same scale, plot s_w versus t;
- Match the data curve with one of the type curves and note the value of C'_{vf} for that type curve;
- Select a matchpoint A on the superimposed sheets and note for A the values of $F(u_{vf},C'_{vf})$, u_{vf}, s_w, and t;
- Substitute the values of $F(u_{vf},C'_{vf})$, s_w, and Q into Equation 18.12 and calculate T;
- Substitute the values of u_{vf}, t, and T into Equation 18.2, $Sx_f^2 = Tt/u_{vf}$, and calculate the product Sx_f^2;
- Knowing C'_{vf} and Sx_f^2, calculate the storage constant C_{vf} from Equation 18.13, $C_{vf} = C'_{vf} \times Sx_f^2$.

Discussion

It should not be forgotten that the above (and many other) methods have been developed primarily for a better understanding of the behaviour of hydraulically fractured geological formations in deep oil reservoirs. Although field examples are scanty in the literature, Gringarten et al. (1975) state that the type-curve approach has been successfully applied to many wells that intersect natural or hydraulic vertical fractures. Nevertheless, there are still certain problems associated with wells in fractures. Fracture storativity and fracture hydraulic conductivity cannot be determined, because, in the theoretical concept, the former is assumed to be infinitely small and the latter is assumed to be infinitely great. The assumption of an infinite hydraulic conductivity in the fracture is not very realistic, certainly not if the assumption of a plane fracture (no width) is made or if the fracture is mineral-filled, as is often so in nature. In reality, a certain hydraulic gradient will exist in the pumped fracture. The so-called uniform-flux solution must therefore be interpreted as giving the appearance of a fracture with high, but not infinite, conductivity. This solution seems, indeed, to match drawdown behaviour of wells intersecting natural fractures better than the infinite-conductivity solution does.

It has also been experienced that computed fracture lengths were far too short, which indicates that still other solutions will be necessary before fracture behaviour can be analyzed completely. Finally, naturally fractured formations that were generally broken, but not in a way as to exhibit separated planar fractures, usually do not show the characteristic early-time drawdown response that follows from the theoretical concept described above.

19 Single vertical dikes

19.1 Introduction

Dikes have long been regarded as impermeable walls in the earth's crust, but recent research has shown that dikes can be highly permeable. They become so by jointing as the magma cools, by fracturing as a result of shearing, or by weathering.

If a single, permeable, vertical dike bisects a country-rock aquifer whose transmissivity is several times less than that of the dike, a specific flow pattern will be created when the dike is pumped. Instead of a cone of depression developing around the well, as in an unconsolidated aquifer, a trough of depression develops (Figure 19.1). Conventional well-flow equations therefore cannot be used to analyze pumping tests in composite dike-aquifer systems.

The hydraulic behaviour of such systems is identical to that of single-fracture aquifer systems. Nevertheless, the concepts used for single vertical fractures in Chapter 18 (i.e. short length and zero width) are not realistic for dikes, whose length can vary from several kilometres to even hundreds of kilometres, and whose width can vary from one metre or less to tens of metres.

In this chapter, the dike is assumed to be as shown in Figure 19.1A. It is infinitely long, has a finite width and a finite hydraulic conductivity. The dike's permeability stems from a system of uniformly distributed fractures, extending downward and dying out with depth. Below the fractured zone, the dike rock is massive and impermeable. The upper part of the dike is also impermeable because of intensive weathering or a top clay layer. The water in the fractured part of the dike and in the aquifer in the country rock is thus confined.

The well in the dike is represented by a plane sink. When the well is pumped at a constant rate, three characteristic time periods can be distinguished: early time, medium time, and late time.

At early times, all the water pumped originates from storage in the dike and none is contributed from the aquifer. A log-log plot of the time-drawdown of the well yields a straight-line segment with a slope of 0.5. The governing equations are then identical with those for early times in Chapter 18, but now the parallel flow occurs in the dike instead of in the aquifer.

At medium times, all the water pumped is supplied from the aquifer and none is contributed from storage in the dike. The flow in the aquifer can be regarded as predominantly parallel, but oblique to the dike. A log-log plot of the time-drawdown data yields a straight-line segment with a slope of 0.25. In the petroleum literature, the same slope was found for fractures with a finite hydraulic conductivity (Cinco Ley et al. 1978).

At late times, the flow in the aquifer is pseudo-radial. A semi-log plot of the time-drawdown data also yields a straight-line segment.

The change in flow from one period to another is not abrupt, but gradual. During these transitional periods, a time-drawdown plot (whether a log-log plot or a semi-log plot) yields curved-line segments.

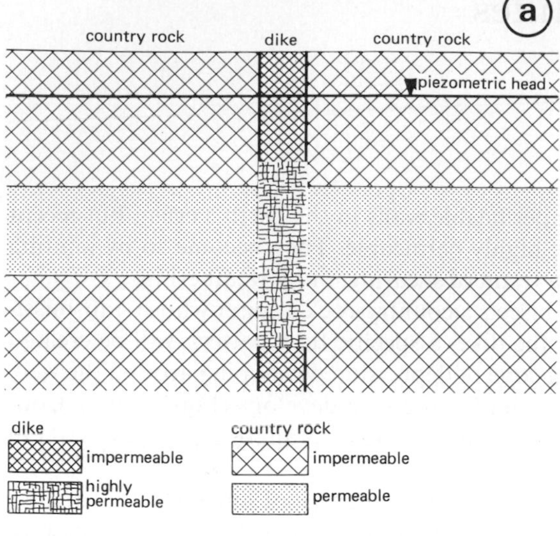

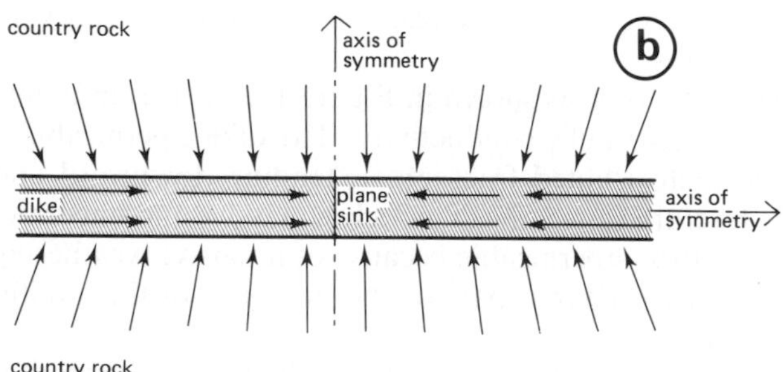

Figure 19.1 Composite dike-aquifer system:
A: Cross-section showing an aquifer of low permeability in hydraulic contact with the highly permeable, fractured part of a vertical dike;
B: Plan view: parallel flow in the pumped dike and parallel-to-near-parallel flow in the aquifer

The methods of analyzing pumping tests in composite dike-aquifer systems are based on the following general assumptions and conditions:

- The dike is vertical and of infinite extent over the length influenced by the test;
- The width of the dike is uniform and does not exceed 10 m;
- The flow through the fracture system in the dike is laminar, so Darcy's equation can be used;
- The uniformly fractured part of the dike can be replaced by a representative continuum to which spatially defined hydraulic characteristics can be assigned;
- The fractured part of the dike is bounded above by an impermeable weathered zone and below by solid rock;
- The well fully penetrates the fractured part of the dike and is represented by a plane

sink; flow through the dike towards the well is parallel;
- The country-rock aquifer, which is in hydraulic contact with the fractured part of the dike, is confined, homogeneous, isotropic, and has an apparently infinite areal extent;
- All water pumped from the well comes from storage within the composite system of dike and aquifer;
- The ratio of the hydraulic diffusivity of the dike to that of the aquifer should not be less than 25;
- Well losses and well-bore storage are negligible.

The methods we present in this chapter are based on the work of Boehmer and Boonstra (1986), Boonstra and Boehmer (1986), Boehmer and Boonstra (1987), and Boonstra and Boehmer (1989). The two methods in Section 19.2 make use of the drawdown data from observation wells placed along the dike and at specific locations in the aquifer; they are only valid for early and medium pumping times. The two methods in Section 19.3 use drawdown data from the pumped well; these methods are complementary and, when combined, cover all three characteristic time periods.

All the methods in this chapter can also be applied to single vertical fractures, provided that the fracture is relatively long.

19.2 Curve-fitting methods for observation wells

For a well in a single, vertical dike of finite width in an otherwise homogeneous, isotropic aquifer of low permeability in the country rock, partial solutions are available for the drawdown in observation wells in the dike and in the aquifer abreast of the pumped well.

19.2.1 Boonstra-Boehmer's curve-fitting method

To analyze the drawdown behaviour along the pumped dike, Boonstra and Boehmer (1986) developed the following drawdown equation for early and medium pumping times

$$s(x,t) = \frac{Q}{3.75\sqrt{T_dST/S_d}}F(\chi,\tau) \qquad (19.1)$$

where

$$F(\chi,\tau) = \frac{2}{\sqrt{\pi}}\exp(-2\sqrt{\tau})\int_0^{\sqrt{\tau}}\exp\left[2\sqrt{(\tau-\zeta^2)}-\frac{\chi^2}{4\zeta^2}\right]d\zeta \qquad (19.2)$$

$$\chi = 1.88\frac{x}{W_d\sqrt{ST/S_dT_d}} \qquad (19.3)$$

$$\tau = 3.52\frac{ST}{(W_dS_d)^2}t \qquad (19.4)$$

ζ = dummy variable of integration

$s(x,t)$ = drawdown in the dike at distance x from the pumped well (m) and pumping time t (d)
S = storativity of the aquifer, dimensionless
S_d = storativity of the dike, dimensionless
T = transmissivity of the aquifer (m^2/d)
T_d = transmissivity of the dike (m^2/d)
W_d = width of the dike (m)

Figure 19.2 shows the family of type curves developed from Equation 19.2. Values of the function $F(\chi,\tau)$ for different values of χ and τ are given in Annex 19.1.

In addition to the general assumptions and conditions listed in Section 19.1, this curve-fitting method is further based on the condition that the flow in the aquifer exhibits a near-parallel-to-parallel flow pattern, which means that the pumping time should be less than

$$t < 0.28\ S(W_dT_d)^2/4T^3$$

Procedure 19.1

- Using Annex 19.1, prepare a family of type curves on log-log paper by plotting $F(\chi,\tau)$ versus τ for different values of χ;
- Prepare the data curve by plotting the drawdown $s(x,t)$ observed in an observation well in the dike at a distance x from the pumped well versus t;
- Apply the type-curve matching procedure;

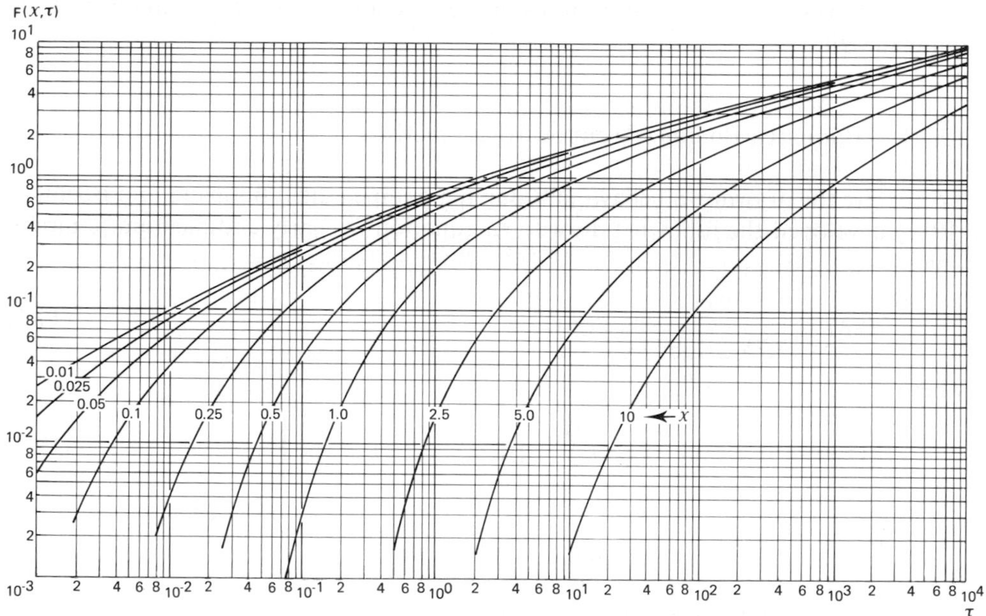

Figure 19.2 Family of type curves of the function $F(\chi,\tau)$ for different values of χ and τ (after Boonstra and Boehmer 1987)

- Substitute the values of $F(\chi,\tau)$, τ, $s(x,t)$, and t of the matchpoint A, together with the χ value of the selected type curve, the x value of the observation well, and the known value of Q into Equations 19.1, 19.3, and 19.4;
- By combining the results, calculate W_dT_d, W_dS_d, and ST.

Remark
- If data from at least two observation wells in the dike are available, W_dT_d, W_dS_d, and ST can also found from a distance-drawdown analysis.

19.2.2 Boehmer-Boonstra's curve-fitting method

To analyze the drawdown behaviour in observation wells drilled in the aquifer along a line perpendicular to the dike and abreast of the pumped well, Boehmer and Boonstra (1987) developed the following drawdown equation for early and medium pumping times

$$s(y,t) = s_w F(u_a) \tag{19.5}$$

where

$$F(u_a) = \exp(-u_a^2) - u_a\sqrt{\pi}\,[1 - \text{erf}(u_a)] \tag{19.6}$$

$$u_a = \frac{1}{2} y \sqrt{(S/T)} \frac{1}{\sqrt{T}} \tag{19.7}$$

$s(y,t)$ = drawdown in the aquifer (m)
y = distance between observation well and pumped well, measured along a line through the pumped well and perpendicular to the dike (m)

Figure 19.3 shows the type curve developed from Equation 19.6. Values of the function $F(u_a)$ for different values of $1/u_a^2$ are given in Annex 19.2.

In addition to the general assumptions and conditions listed in Section 19.1, this curve-fitting method is further based on the condition that the flow in the country-rock aquifer exhibits a near-parallel to parallel flow pattern, which means that the pumping time should be less than

$$t < 0.28\,S\,(W_dT_d)^2/4T^3$$

Procedure 19.2
- Using Annex 19.2, prepare a type curve by plotting values of $F(u_a)$ versus $1/u_a^2$ on log-log paper;
- Prepare the data curve by plotting the drawdown ratios $s(y,t)/s_w$ versus t;
- Apply the type-curve matching procedure;
- Substitute the values of $1/u_a^2$ and t of the matchpoint A, together with the y value of the observation well, into Equation 19.7 and calculate the hydraulic diffusivity T/S.

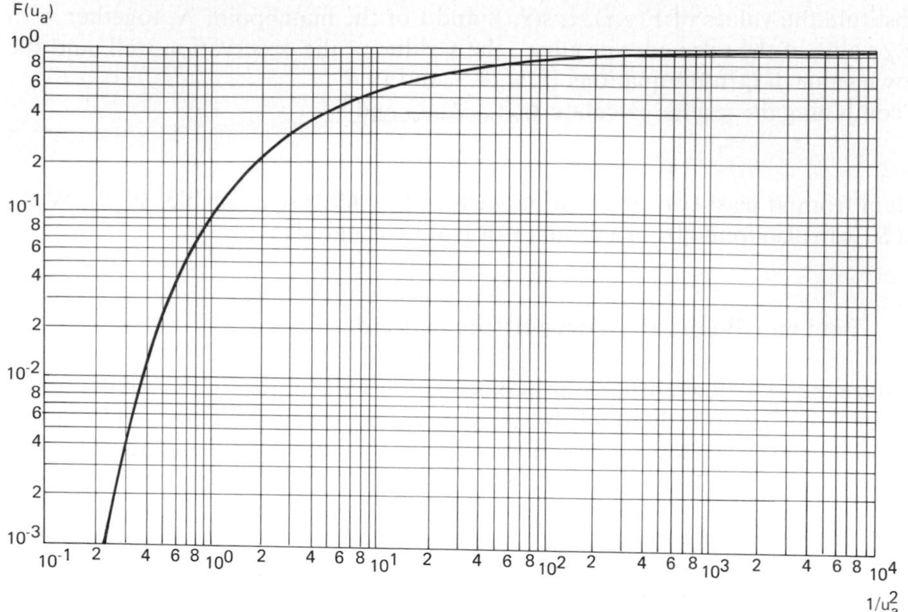

Figure 19.3 Type curve of the function $F(u_a)$ (after Boehmer and Boonstra 1987)

Remarks

- When data from at least two observation wells located in the aquifer are available, the hydraulic diffusivity T/S can also be found from a distance-drawdown analysis;
- If data are available from at least one observation well in the dike and another in the aquifer, separate values of the transmissivity and storativity of the aquifer can be found by combining the results obtained with the methods in Sections 19.2.1 and 19.2.2.

19.3 Curve-fitting methods for the pumped well

19.3.1 For early and medium pumping times

For a well in a single, vertical dike in an otherwise homogeneous, isotropic, confined, aquifer of low permeability in the country rock, Boonstra and Boehmer (1986) obtained the following solution for the drawdown in the pumped well during early and medium pumping times

$$s_w = \frac{Q}{3.75\sqrt{T_d ST/S_d}} F(\tau) \tag{19.8}$$

where

$$F(\tau) = \frac{2}{\sqrt{\pi}} \exp(-2\sqrt{\tau}) \int_0^{\sqrt{\tau}} \exp[2\sqrt{(\tau-\zeta^2)}]\, d\zeta \tag{19.9}$$

Equation 19.9 is the reduced form of Equation 19.2 for $\chi = 0$. Figure 19.4 shows the type curve developed from Equation 19.9. Values of the function $F(\tau)$ for different values of τ are given in Annex 19.3.

At early pumping times, when the drawdown behaviour in the well is predominantly governed by the water released from storage in the dike, the drawdown function in Equation 19.9 reduces to

$$F(\tau) = \frac{2}{\sqrt{\pi}}\sqrt{\tau} \qquad (19.10)$$

and consequently

$$s_w = \frac{Q}{\sqrt{\pi T_d S_d W_d^2}}\sqrt{t} \qquad (19.11)$$

As Equation 19.11 shows, a log-log plot of the early-time drawdown versus time is characterized by a straight line with a slope of 0.5. This early-time period ends at approximately $\tau = 0.003$.

At medium pumping times, when the drawdown behaviour is predominantly governed by near-parallel-to-parallel flow from the aquifer into the dike, the drawdown function in Equation 19.9 reduces to

$$F(\tau) = \sqrt[4]{\tau} \qquad (19.12)$$

and consequently

$$s_w = \frac{Q}{2.74\sqrt{W_d T_d \sqrt[4]{ST}}}\sqrt[4]{t} \qquad (19.13)$$

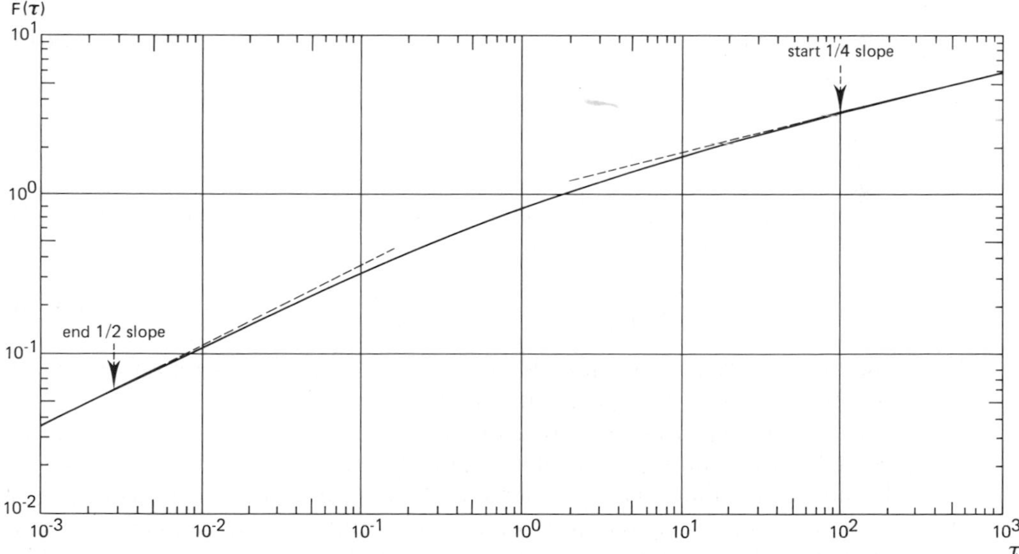

Figure 19.4 Type curve of the function $F(\tau)$ for the pumped well at early and medium pumping times (after Boonstra and Boehmer 1987)

As Equation 19.13 shows, a log-log plot of the medium-time drawdown versus time is characterized by a straight line with a slope 0.25. This period starts at approximately $\tau = 100$.

In addition to the general assumptions and conditions listed in Section 19.1, this curve-fitting method is further based on the condition that the flow in the aquifer exhibits a near-parallel-to-parallel flow pattern, which means that the pumping time should be less than

$$t < 0.28\, S(W_dT_d)^2/4T^3$$

Procedure 19.3

- Using Annex 19.3, prepare a type curve by plotting $F(\tau)$ versus τ on log-log paper;
- Prepare the data curve by plotting the drawdown s_w versus t;
- Apply the type-curve matching procedure;
- Substitute the values of $F(\tau)$, τ, s_w, and t of the chosen matchpoint A and the known value of Q into Equations 19.4 and 19.8 and calculate $(W_dS_d)(W_dT_d)$ and $(W_dT_d)\sqrt{(ST)}$.

Remark

- If the data plot only exhibits an 0.5 or an 0.25 slope straight-line segment, $(W_dT_d)(W_dS_d)$ or $(W_dT_d)\sqrt{(ST)}$ can be found from Equations 19.11 or 19.13, respectively. This yields a value for

$$(W_dT_d)(W_dS_d) = \frac{Q^2t}{\pi s_w^2} \qquad (19.14)$$

or

$$(W_dT_d)\sqrt{(ST)} = \frac{Q^2\sqrt{t}}{7.5\, s_w^2} \qquad (19.15)$$

19.3.2 For late pumping times

Boehmer and Boonstra (1986) also obtained a solution for the drawdown in the pumped well during late pumping times

$$s_w = \frac{2.30Q}{4\pi T} \log \frac{40\, T^3t}{S(W_dT_d)^2} \qquad (19.16)$$

Equation 19.16 shows that the drawdown is now a logarithmic function of time. A plot of s_w versus t on semi-log paper will thus yield a straight-line segment.

Boonstra and Boehmer (1989) showed that the solution for the drawdown in the pumped well during late times can be integrated with the corresponding solutions for early and medium times. This gives a family of type curves as a function of ST_d/S_dT (Figure 19.5). From an inspection of Figures 19.4 and 19.5, we can conclude that the log-log plot will not exhibit a straight-line with a slope of 0.25 for ST_d/S_dT values lower than 25.

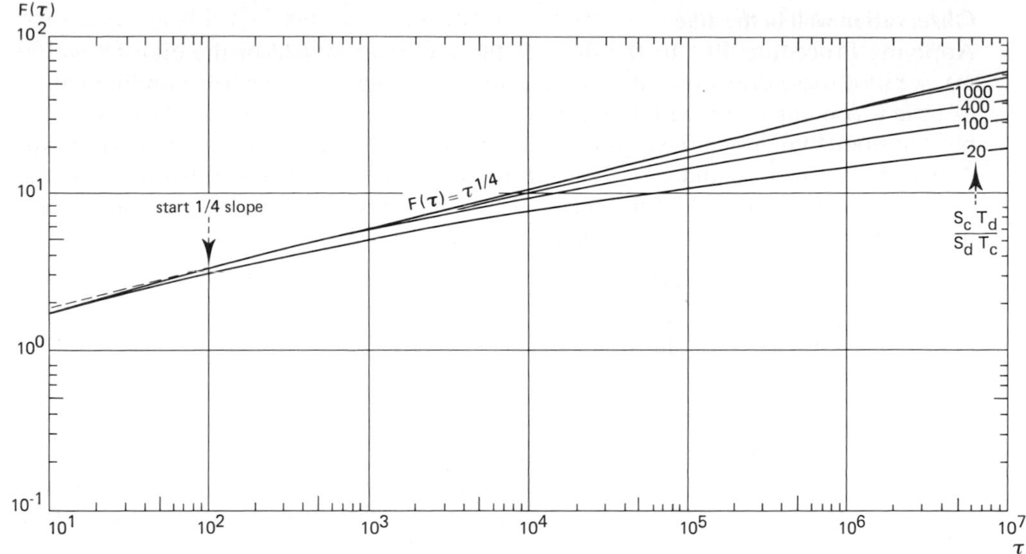

Figure 19.5 Family of type curves of the function F(τ) for the pumped well at late pumping times (after Boonstra and Boehmer 1987)

In addition to the general assumptions and conditions listed in Section 19.1, this straight-line method is further based on the condition that the flow in the aquifer exhibits a pseudo-radial flow pattern, which means that the pumping condition is

$$t > 50\, S(W_d T_d)^2/4T^3$$

Procedure 19.4
- On semi-log paper, plot the drawdown s_w versus t (t on logarithmic scale);
- Draw a straight line of best fit through the plotted points;
- Determine the slope of this line Δs and calculate $T = 2.30Q/4\pi\Delta s$).

Remark
- For a pumping test of the usual duration, the above method can only be applied to dikes not wider than a few centimetres or to fractures.

Example 19.1
Boonstra and Boehmer (1986) and Boehmer and Boonstra (1987) describe a pumping test that was conducted in a 10-m-wide fractured dolorite dike at Brandwag Tweeling, Republic of South Africa. The country rock consists of alternating layers of non-productive low-permeable sandstones, silt stones, and mudstones of the Beaufort series, which belong to the Karroo system.

The well in the dike was pumped for 2500 minutes at a constant rate of 13.9 l/s or 1200 m^3/d. Drawdowns were measured in this well and in two observation wells, one in the dike at a distance of 100 m from the pumped well and the other in the aquifer abreast of the pumped well and 20 m away from it. Table 19.1 gives the drawdown data of the three wells.

Observation well in the dike

Applying Procedure 19.1 to the data of the observation well in the dike (x = 100 m), we plot these drawdown data on log-log paper against the corresponding values of time t. A comparison with the family of type curves in Figure 19.6 shows that the plotted points fall along the type curve for $\chi = 1.0$. We choose as matchpoint, Point A, where $F(\chi,\tau) = 1$ and $\tau = 100$. On the observed data sheet, this point has the coordinates s(100,t) = 2.29 m and t = 23.5 minutes. Introducing the appropriate numerical values into Equations 19.1, 19.3, and 19.4, we obtain

$$W_dT_d = 2.6 \times 10^4 \text{ m}^3/\text{d}$$
$$W_dS_d = 4.3 \times 10^{-4} \text{ m}$$
$$S\,T = 3.2 \times 10^{-4} \text{ m}^2/\text{d}$$

Table 19.1 Drawdown data of the pumped well and two observation wells, Pumping Test Brandwag Tweeling, South Africa, after Boonstra and Boehmer (1986) and Boehmer and Boonstra (1987)

Time (min)	x = 0 (m)	x = 100 (m)	Time (min)	x = 0 (m)	x = 100 (m)	Time (min)	x = 0 (m)	x = 100 (m)
1	3.363	1.378	40	8.445	6.232	600	18.108	15.031
2	4.118	2.068	50	8.864	6.606	750	18.948	15.907
3	4.660	2.507	60	9.192	6.907	900	19.795	15.704
4	5.025	2.818	75	9.724	7.349	1050	20.253	17.813
6	5.582	3.360	100	10.366	8.031	1200	20.667	17.565
8	6.081	3.846	125	11.120	8.885	1350	21.033	17.916
10	6.470	4.224	150	11.766	9.063	1500	21.076	17.945
13	6.796	4.547	175	12.300	9.553	1700	21.389	18.285
15	7.020	4.765	200	12.874	10.045	1900	21.486	18.409
18	7.246	5.016	250	13.911	11.027	2100	–	18.483
21	7.500	5.257	300	14.643	11.672	2300	–	18.858
25	7.746	5.519	350	15.142	12.154	2500	–	19.109
30	8.102	5.700	400	16.080	12.207			
35	8.324	6.044	500	17.252	14.324			

Time (min)	y = 0 (m)	y = 20 (m)	Time (min)	y = 0 (m)	y = 20 (m)	Time (min)	y = 0 (m)	y = 20 (m)
1	3.363	0.572	30	8.102	5.630	300	14.643	11.323
2	4.118	1.249	35	8.324	3.006	350	15.142	11.766
3	4.660	1.741	40	8.445	6.110	400	16.080	12.622
4	5.025	2.540	50	8.864	6.500	500	17.252	14.847
6	5.582	2.800	60	9.192	6.815	600	18.108	14.917
8	6.081	3.422	75	9.724	7.320	750	18.948	15.421
10	6.470	3.905	100	10.366	7.858	900	19.795	16.337
13	6.796	4.286	125	11.120	8.489	1050	20.253	16.691
15	7.020	4.530	150	11.766	9.039	1200	20.667	17.125
18	7.246	4.800	175	12.300	9.457	1350	21.033	17.560
21	7.500	5.055	200	12.874	9.901	1500	21.076	17.584
25	7.746	5.375	250	13.911	10.723	1700	21.389	–

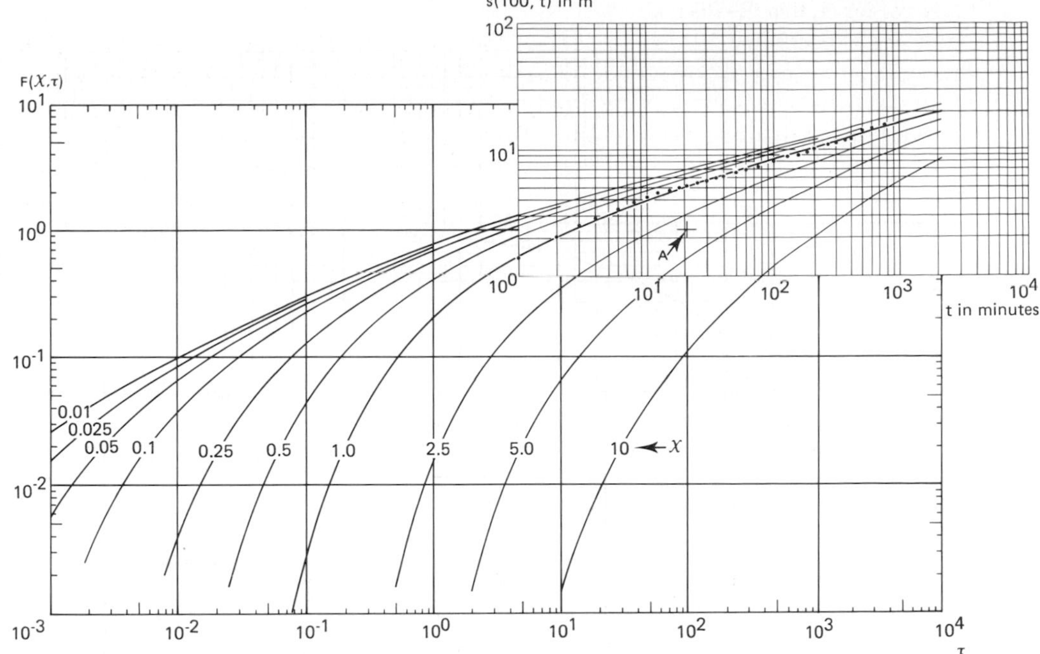

Figure 19.6 The time-drawdown data of the observation well in the dike (x = 100 m), matched with one of the curves of the family of type curves developed from Equation 19.2

Observation well in the aquifer

Applying Procedure 19.2 to the data of the observation well in the aquifer, we match the time-drawdown ratio data with the type curve $F(u_a)$, as shown in Figure 19.7. We choose as matchpoint, Point A, where $F(u_a) = 1$ and $1/u_a^2 = 10$. On the observed data sheet, this point has the coordinates $s(20,t)/s_w = 0.9$ and t = 5.3 minutes. Introducing the appropriate numerical values into Equation 19.7, we obtain

$$T/S = 2.7 \times 10^5 \text{ m}^2/\text{d}$$

Combining the results of Procedures 19.1 and 19.2, we can also obtain separate values for the transmissivity and storativity of the aquifer

$$T = 9.3 \text{ m}^2/\text{d}$$
$$S = 3.4 \times 10^{-5}$$

Pumped well

Figure 19.8 shows the time-drawdown data of the pumped well, plotted on log-log paper. This plot only exhibits a straight line with a slope of 0.25. Hence, we cannot apply Procedure 19.3. Instead, we choose an arbitrary point A on this line, with coordinates $s_w = 10.0$ m and t = 70.7 minutes. Introducing these values into Equation 19.15, we obtain

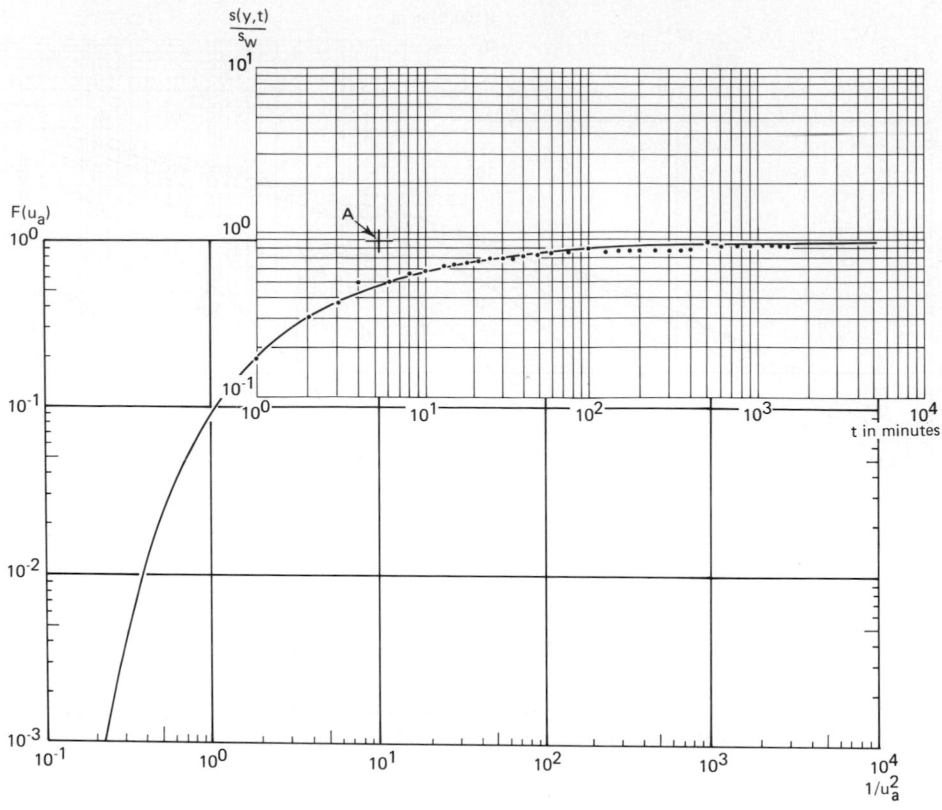

Figure 19.7 The time-drawdown ratio data of the observation well in the aquifer (y = 20 m), matched with the type curve F(u_a)

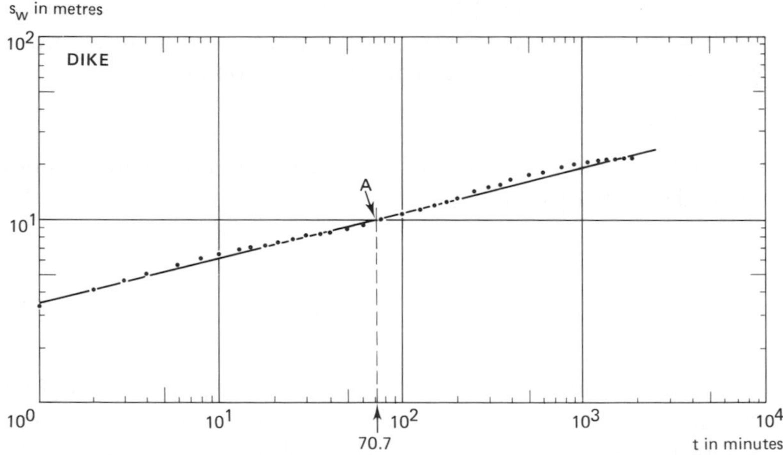

Figure 19.8 Time-drawdown relation of the pumped well, showing the characteristic straight-line slope of 0.25 for medium pumping times

$$(W_d T_d)\sqrt{(ST)} = 425 \text{ m}^4/\text{d}^{3/2}$$

Substituting the values of $W_d T_d$ and ST obtained with Procedure 19.1 into $(W_d T_d)\sqrt{(ST)}$, we get 465, which corresponds reasonably well with the value of 425 obtained with Procedure 19.3.

Annexes

Annex 2.1 Units of the International System (SI)

Basic SI units

	Name	Symbol
Length	metre	m
Time	second	s
Mass	kilogram	kg

SI-derived units

Pressure	pascal	Pa (= $kg.m^{-1}.s^{-2}$)
Viscosity	pascal-second	Pa.s
Area	square metre	m^2
Volume	cubic metre	m^3
Discharge	cubic metre per second	$m^3.s^{-1}$
Hydraulic conductivity	metre per second	$m.s^{-1}$
Transmissivity	square metre per second	$m^2.s^{-1}$
Intrinsic permeability	square metre	m^2

Annex 2.2 Conversion table

Length:

	m	cm	ft	inch
1 m	1.000	1.000×10^2	3.281	39.37
1 cm	1.000×10^{-2}	1.000	3.281×10^{-2}	0.3937
1 ft	0.3048	30.48	1.000	12.00
1 inch	2.540×10^{-2}	2.540	8.333×10^{-2}	1.000

Length reciprocals:

	m^{-1}	cm^{-1}	ft^{-1}	$inch^{-1}$
1 m^{-1}	1.000	1.000×10^{-2}	0.3048	2.540×10^{-2}
1 cm^{-1}	1.0×10^2	1.000	30.48	2.540
1 ft^{-1}	3.281	3.281×10^{-2}	1.000	8.333×10^{-2}
1 $inch^{-1}$	39.37	0.3937	12.00	1.000

Area:

	m^2	ft^2
1 m^2	1.000	10.764
1 ft^2	9.290×10^{-2}	1.000

Area reciprocals:

	m^{-2}	ft^{-2}
1 m^{-2}	1.000	9.290×10^{-2}
1 ft^{-2}	10.764	1.000

Annex 2.1 (cont.)

Volume:

	m^3	l	Imp.gal.	U.S. gal	ft^3
1 m^3	1.000	1.000×10^3	2.200×10^2	2.642×10^2	35.32
1 l	1.000×10^{-3}	1.000	0.2200	0.2642	3.532×10^{-2}
1 Imp.gal	4.546×10^{-3}	4.546	1.000	1.200	0.1605
1 U.S.gal	3.785×10^{-3}	3.785	0.8326	1.000	0.1337
1 ft^3	2.832×10^{-2}	28.32	6.229	7.481	1.000

Time:

	d	h	min	s
1 d	1.000	24.00	1.440×10^3	8.640×10^4
1 h	4.167×10^{-2}	1.000	60.00	3.600×10^3
1 min	6.944×10^{-4}	1.667×10^{-2}	1.000	60.00
1 s	1.157×10^{-5}	2.777×10^{-4}	1.667×10^{-2}	1.000

Time reciprocals:

	d^{-1}	h^{-1}	min^{-1}	s^{-1}
1 d^{-1}	1.000	4.167×10^{-2}	6.944×10^{-4}	1.157×10^{-5}
1 h^{-1}	24.00	1.000	1.667×10^{-2}	2.777×10^{-4}
1 min^{-1}	1.440×10^3	60.00	1.000	1.667×10^{-2}
1 s^{-1}	8.640×10^4	3.600×10^3	60.00	1.000

Discharge rate:

	l/s	m^3/d	m^3/s	Imp.gal/d	U.S.gal/d	ft^3/d
1 l/s	1.000	86.40	1.000×10^{-3}	1.901×10^4	2.282×10^4	3.051×10^3
1 m^3/h	0.2777	24.00	2.777×10^{-4}	5.279×10^3	6.340×10^3	8.476×10^2
1 m^3/d	1.157×10^{-2}	1.000	1.157×10^{-5}	2.200×10^2	2.642×10^2	35.32
1 m^3/s	1.000×10^3	8.640×10^4	1.000	1.901×10^7	2.282×10^7	3.051×10^6
1 Imp.gal/d	5.262×10^{-5}	4.546×10^{-3}	5.262×10^{-8}	1.000	1.201	0.1605
1 U.S.gal/d	4.381×10^{-5}	3.785×10^{-3}	4.381×10^{-8}	0.8327	1.000	0.1337
1 ft^3/d	0.3277	2.832×10^{-2}	3.277×10^{-7}	6.229	7.481	1.000

Mass:

	kg	gram	lb
kg	1.000	1.000×10^3	2.205
gram	1.000×10^{-3}	1.000	2.205×10^{-3}
lb	4.536×10^{-1}	4.536×10^2	1.000

Annex 2.1 (cont.)

Pressure:

	Pa	dyne/cm^2	kgf/cm^2	bar	atm	mm Hg (0 °C)	m H_2O (4 °C)	lbf/inch^2 (= psi)
Pa	1.000	10.000	1.020×10^{-5}	1.000×10^{-5}	9.869×10^{-6}	7.501×10^{-3}	1.020×10^{-4}	1.450×10^{-4}
dyne/cm^2	1.000×10^{-1}	1.000	1.020×10^{-6}	1.000×10^{-6}	9.869×10^{-7}	7.501×10^{-4}	1.020×10^{-5}	1.450×10^{-5}
kgf/cm^2	9.807×10^{4}	9.807×10^{5}	1.000	9.807×10^{-1}	9.68×10^{-1}	7.357×10^{2}	10.000	14.223
bar	1.000×10^{5}	1.000×10^{6}	1.020	1.000	9.869×10^{-1}	7.501×10^{2}	10.20	14.50
atm	1.013×10^{5}	1.013×10^{6}	1.033	1.013	1.000	7.60×10^{2}	10.33	14.69
mm Hg (0 °C)	1.333×10^{2}	1.333×10^{3}	1.36×10^{-3}	1.333×10^{-3}	1.316×10^{-3}	1.000	1.36×10^{-2}	1.93×10^{-2}
m H_2O (4 °C)	9.807×10^{3}	9.807×10^{4}	1.000×10^{-1}	9.807×10^{-2}	9.68×10^{-2}	73.57	1.000	1.422
lbf/inch^2 (= psi)	6.89×10^{3}	6.89×10^{4}	7.03×10^{-2}	6.89×10^{-2}	6.806×10^{-2}	51.73	7.03×10^{-1}	1.000

Viscosity:

	Pa.s	cP	lb/ft.s
Pa.s	1.000	1.000×10^{3}	6.720×10^{-1}
cP	1.000×10^{-3}	1.000	6.720×10^{-4}
lb/ft.s	1.488	1.488×10^{3}	1.000

Intrinsic permeability:

	m^2	darcy
m^2	1.000	1.013×10^{12}
darcy	9.872×10^{-11}	1.000

Hydraulic conductivity

	m/d	m/s	cm/h	Imp.gal/ d-ft^2	U.S.gal/ d-ft^2	Imp.gal/ min-ft^2	U.S.gal/ min-ft^2
1 m/d	1.000	1.157×10^{-5}	4.167	20.44	24.54	1.419×10^{-2}	1.704×10^{-2}
1 m/s	8.640×10^{4}	1.000	3.600×10^{5}	1.766×10^{6}	2.121×10^{6}	1.226×10^{3}	1.472×10^{3}
1 cm/h	0.2400	2.777×10^{-6}	1.000	4.905	5.890	3.406×10^{-3}	4.089×10^{-3}
1 Imp.gal/d-ft^2	4.893×10^{-2}	5.663×10^{-7}	0.2039	1.000	1.201	6.944×10^{-4}	8.339×10^{-4}
1 U.S.gal/d-ft^2	4.075×10^{-2}	4.716×10^{-7}	0.1698	0.8327	1.000	5.783×10^{-4}	6.944×10^{-4}
1 Imp.gal/min-ft^2	70.46	8.155×10^{-2}	2.936×10^{2}	1.440×10^{3}	1.729×10^{3}	1.000	1.201
1 U.S.gal/min-ft^2	58.67	6.791×10^{-2}	2.445×10^{2}	1.195×10^{3}	1.440×10^{3}	0.8326	1.000

Transmissivity

	m^2/d	m^2/s	Imp.gal/d-ft	U.S.gal/d-ft	Imp.gal/min-ft	U.S.gal/min-ft
1 m^2/d	1.000	1.157×10^{-5}	67.05	80.52	4.656×10^{-2}	5.592×10^{-2}
1 m^2/s	8.64×10^{4}	1.000	5.793×10^{6}	6.957×10^{6}	4.023×10^{3}	4.831×10^{3}
1 Imp.gal/d-ft	1.491×10^{-2}	1.726×10^{-7}	1.000	1.201	6.944×10^{-4}	8.339×10^{-4}
1 U.S.gal/d-ft	1.242×10^{-2}	1.437×10^{-7}	0.8326	1.000	5.783×10^{-4}	6.944×10^{-4}
1 Imp.gal/min-ft	21.48	2.486×10^{-4}	1.440×10^{3}	1.729×10^{3}	1.000	1.201
1 U.S.gal/min-ft	17.88	2.070×10^{-4}	1.199×10^{3}	1.440×10^{3}	0.8326	1.000

Annex 2.1 (cont.)

Abbreviations:

ft	= foot
l	= liter
Imp.gal	= Imperial gallon
U.S.gal	= U.S. gallon
h	= hour
lb	= pound
lbf	= pound force
kgf	= kilogram force
atm	= atmosphere
mH_2O	= metre of water
mm Hg	= millimetre of mercury
d	= day
cP	= centipoise

Care should be taken in the conversion that an approximate value does not become too exact. For example: the analysis of a pumping test may give values for the transmissivity ranging between 1833 m^2/d and 2217 m^2/d: consequently in the conclusions it is stated that the transmissivity is approximately equal to 2000 m^2/d. If this value is converted into U.S.gallons/d-ft by multiplying it by 80.52 (1 m^2/d = 80.52 U.S.gallons/d-ft) this results in

2000 m^2/d = 161 040 U.S.gal/d-ft

However

appr. 2000 m^2/d = appr. 160 000 U.S.gal/d-ft

and the variation is between

147 000 U.S.gal/d-ft and 178 000 U.S.gal/d-ft

and not between

147 593.16 and 178 512.84 U.S.gal/d-ft

Conversion coefficients that are not listed can easily be calculated. For example:

Question: 'How much is a hydraulic conductivity of 230 l/s-m^2 when expressed in U.S.gal/d-ft^2?'

Answer: 1 l/s-m^2 = 1.000×10^{-3} m/s-m^2 (= m/s)

1 m/s = $2.121 \times 10^3 \times$ U.S.gal/d-ft^2

Hence 1 l/s-m^2 = $1.000 \times 10^{-3} \times 2.121 \times 10^3$ = 2.121 U.S.gal/d-ft^2

and 230 l/s-m^2 = 230×2.121 = 487.8 U.S.gal/d-ft^2

Annex 3.1 Values of the Theis well function W(u) for confined aquifers (after Walton 1962)

		1/u=	n	n(1)	n(2)	n(3)	n(4)	n(5)	n(6)	n(7)	n(8)	n(9)	n(10)
n	N	u =	N	N(–1)	N(–2)	N(–3)	N(–4)	N(–5)	N(–6)	N(–7)	N(–8)	N(–9)	N(–10)
1.000	1.0	W(u)=	2.194(–1)	1.823	4.038	6.332	8.633	1.094(1)	1.324(1)	1.544(1)	1.784(1)	2.015(1)	2.245(1)
0.833	1.2		1.584(–1)	1.660	3.858	6.149	8.451	1.075(1)	1.306(1)	1.536(1)	1.766(1)	1.996(1)	2.227(1)
0.666	1.5		1.000(–1)	1.465	3.637	5.927	8.228	1.053(1)	1.283(1)	1.514(1)	1.744(1)	1.974(1)	2.204(1)
0.500	2.0		4.890(–2)	1.223	3.355	5.639	7.940	1.024(1)	1.255(1)	1.485(1)	1.715(1)	1.945(1)	2.176(1)
0.400	2.5		2.491(–2)	1.044	3.137	5.417	7.717	1.002(1)	1.232(1)	1.462(1)	1.693(1)	1.923(1)	2.153(1)
0.333	3.0		1.305(–2)	9.057(–1)	2.959	5.235	7.535	9.837	1.214(1)	1.444(1)	1.674(1)	1.905(1)	2.135(1)
0.286	3.5		6.970(–3)	7.942(–1)	2.810	5.081	7.381	9.683	1.199(1)	1.429(1)	1.659(1)	1.889(1)	2.120(1)
0.250	4.0		3.779(–3)	7.024(–1)	2.681	4.948	7.247	9.550	1.185(1)	1.415(1)	1.646(1)	1.876(1)	2.106(1)
0.222	4.5		2.073(–3)	6.253(–1)	2.568	4.831	7.130	9.432	1.173(1)	1.404(1)	1.634(1)	1.864(1)	2.094(1)
0.200	5.0		1.148(–3)	5.598(–1)	2.468	4.726	7.024	9.326	1.163(1)	1.393(1)	1.623(1)	1.854(1)	2.084(1)
0.166	6.0		3.601(–4)	4.544(–1)	2.295	4.545	6.842	9.144	1.145(1)	1.375(1)	1.605(1)	1.835(1)	2.066(1)
0.142	7.0		1.155(–4)	3.738(–1)	2.151	4.392	6.688	8.990	1.129(1)	1.360(1)	1.590(1)	1.820(1)	2.050(1)
0.125	8.0		3.767(–5)	3.106(–1)	2.027	4.259	6.555	8.856	1.116(1)	1.346(1)	1.576(1)	1.807(1)	2.037(1)
0.111	9.0		1.245(–5)	2.602(–1)	1.919	4.142	6.437	8.739	1.104(1)	1.334(1)	1.565(1)	1.795(1)	2.025(1)

Annex 4.1 Values of the functions e^x, e^{-x}, $K_o(x)$ and $e^xK_o(x)$ (after Hantush 1956)

x	e^x	e^{-x}	$K_o(x)$	$e^xK_o(x)$	x	e^x	e^{-x}	$K_o(x)$	$e^xK_o(x)$	x	e^x	e^{-x}	$K_o(x)$	$e^xK_o(x)$
0.010	1.010	0.990	4.721	4.769	0.040	1.041	0.961	3.336	3.473	0.070	1.072	0.932	2.780	2.981
0.011	1.011	0.989	4.626	4.677	0.041	1.042	0.960	3.312	3.450	0.071	1.074	0.931	2.766	2.969
0.012	1.012	0.988	4.539	4.594	0.042	1.043	0.959	3.288	3.429	0.072	1.075	0.930	2.752	2.957
0.013	1.013	0.987	4.459	4.517	0.043	1.044	0.958	3.264	3.408	0.073	1.076	0.930	2.738	2.945
0.014	1.014	0.986	4.385	4.447	0.044	1.045	0.957	3.241	3.387	0.074	1.077	0.929	2.725	2.934
0.015	1.015	0.985	4.316	4.381	0.045	1.046	0.956	3.219	3.367	0.075	1.078	0.928	2.711	2.923
0.016	1.016	0.984	4.251	4.320	0.046	1.047	0.955	3.197	3.348	0.076	1.079	0.927	2.698	2.911
0.017	1.017	0.983	4.191	4.263	0.047	1.048	0.954	3.176	3.329	0.077	1.080	0.926	2.685	2.900
0.018	1.018	0.982	4.134	4.209	0.048	1.049	0.953	3.155	3.310	0.078	1.081	0.925	2.673	2.889
0.019	1.019	0.981	4.080	4.158	0.049	1.050	0.952	3.134	3.292	0.079	1.082	0.924	2.660	2.879
0.020	1.020	0.980	4.028	4.110	0.050	0.051	0.951	3.114	3.274	0.080	1.083	0.923	2.647	2.868
0.021	1.021	0.979	3.980	4.064	0.051	1.052	0.950	3.094	3.256	0.081	1.084	0.922	2.635	2.857
0.022	1.022	0.978	3.933	4.021	0.052	1.053	0.949	3.075	3.239	0.082	1.085	0.921	2.623	2.847
0.023	1.023	0.977	3.889	3.979	0.053	1.054	0.948	3.056	3.223	0.083	1.086	0.920	2.611	2.837
0.024	1.024	0.976	3.846	3.940	0.054	1.055	0.947	3.038	3.206	0.084	1.088	0.919	2.599	2.827
0.025	1.025	0.975	3.806	3.902	0.055	1.056	0.946	3.019	3.190	0.085	1.089	0.918	2.587	2.817
0.026	1.026	0.974	3.766	3.866	0.056	1.058	0.945	3.001	1.174	0.086	1.090	0.918	2.576	2.807
0.027	1.027	0.973	3.729	3.831	0.057	1.059	0.945	2.984	3.159	0.087	1.091	0.917	2.564	2.798
0.028	1.028	0.972	3.692	3.797	0.058	1.060	0.944	2.967	3.144	0.088	1.092	0.916	2.553	2.788
0.029	1.029	0.971	3.657	3.765	0.059	1.061	0.943	2.950	3.129	0.089	1.093	0.915	2.542	2.779
0.030	1.030	0.970	3.623	3.734	0.060	1.062	0.942	2.933	3.114	0.090	1.094	0.914	2.531	2.769
0.031	1.031	0.969	3.591	3.704	0.061	1.063	0.941	2.916	3.100	0.091	1.095	0.913	2.520	2.760
0.032	1.032	0.968	3.559	3.675	0.062	1.064	0.940	2.900	3.086	0.092	1.096	0.912	2.509	2.751
0.033	1.034	0.967	3.528	3.647	0.063	1.065	0.939	2.884	3.072	0.093	1.097	0.911	2.499	2.742
0.034	1.035	0.967	3.499	3.620	0.064	1.066	0.938	2.869	3.058	0.094	1.099	0.910	2.488	2.733
0.035	1.036	0.966	3.470	3.593	0.065	1.067	0.937	2.853	3.045	0.095	1.100	0.909	2.478	2.725
0.036	1.037	0.965	3.442	3.568	0.066	1.068	0.936	2.838	3.032	0.096	1.101	0.908	2.467	2.716
0.037	1.038	0.964	3.414	3.543	0.067	1.069	0.935	2.823	3.019	0.097	1.102	0.908	2.457	2.707
0.038	1.039	0.963	3.388	3.519	0.068	1.070	0.934	2.809	3.006	0.098	1.103	0.907	2.447	2.699
0.039	1.040	0.962	3.362	3.495	0.069	1.071	0.933	2.794	2.994	0.099	1.104	0.906	2.437	2.691

Annex 4.1 (cont.)

x	e^x	e^{-x}	$K_o(x)$	$e^xK_o(x)$	x	e^x	e^{-x}	$K_o(x)$	$e^xK_o(x)$	x	e^x	e^{-x}	$K_o(x)$	$e^xK_o(x)$
0.10	1.105	0.905	2.427	2.682	0.40	1.492	0.670	1.114	1.663	0.70	2.014	0.497	0.660	1.330
0.11	1.116	0.896	2.333	2.605	0.41	1.507	0.664	1.093	1.647	0.71	2.034	0.492	0.650	1.322
0.12	1.127	0.887	2.248	2.534	0.42	1.522	0.657	1.072	1.632	0.72	2.054	0.487	0.640	1.315
0.13	1.139	0.878	1.169	2.471	0.43	1.537	0.650	1.052	1.617	0.73	2.075	0.482	0.630	1.307
0.14	1.150	0.869	2.097	2.412	0.44	1.553	0.644	1.032	1.602	0.74	2.096	0.477	0.620	1.300
0.15	1.162	0.861	2.030	2.358	0.45	1.568	0.638	1.013	1.589	0.75	2.117	0.472	0.611	1.293
0.16	1.173	0.852	1.967	2.309	0.46	1.584	0.631	0.994	1.575	0.76	2.138	0.468	0.601	1.285
0.17	1.185	0.844	1.909	2.262	0.47	1.600	0.625	0.976	1.562	0.77	2.160	0.463	0.592	1.278
0.18	1.197	0.835	1.854	2.219	0.48	1.616	0.619	0.958	1.549	0.78	2.181	0.458	0.583	1.272
0.19	1.209	0.827	1.802	2.179	0.49	1.632	0.613	0.941	1.536	0.79	2.203	0.454	0.574	1.265
0.20	1.221	0.819	1.753	2.141	0.50	1.649	0.606	0.924	1.524	0.80	2.225	0.449	0.565	1.258
0.21	1.234	0.811	1.706	2.105	0.51	1.665	0.600	0.908	1.512	0.81	2.248	0.445	0.557	1.252
0.22	1.246	0.802	1.662	2.071	0.52	1.682	0.594	0.892	1.501	0.82	2.270	0.440	0.548	1.245
0.23	1.259	0.794	1.620	2.039	0.53	1.699	0.589	0.877	1.489	0.83	2.293	0.436	0.540	1.239
0.24	1.271	0.787	1.580	2.008	0.54	1.716	0.583	0.861	1.478	0.84	2.316	0.432	0.532	1.233
0.25	1.284	0.779	1.541	1.979	0.55	1.733	0.577	0.847	1.467	0.85	2.340	0.427	0.524	1.226
0.26	1.297	0.771	1.505	1.952	0.56	1.751	0.571	0.832	1.457	0.86	2.363	0.423	0.516	1.220
0.27	1.310	0.763	1.470	1.925	0.57	1.768	0.565	0.818	1.446	0.87	2.387	0.419	0.509	1.214
0.28	1.323	0.756	1.436	1.900	0.58	1.786	0.560	0.804	1.436	0.88	2.411	0.415	0.501	1.209
0.29	1.336	0.748	1.404	1.876	0.59	1.804	0.554	0.791	1.426	0.89	2.435	0.411	0.494	1.203
0.30	1.350	0.741	1.372	1.853	0.60	1.822	0.549	0.777	1.417	0.90	2.460	0.407	0.487	1.197
0.31	1.363	0.733	1.342	1.830	0.61	1.840	0.543	0.765	1.407	0.91	2.484	0.402	0.480	1.192
0.32	1.377	0.726	1.314	1.809	0.62	1.859	0.538	0.752	1.398	0.92	2.509	0.398	0.473	1.186
0.33	1.391	0.719	1.286	1.788	0.63	1.878	0.533	0.740	1.389	0.93	2.534	0.395	0.466	1.181
0.34	1.405	0.712	1.259	1.768	0.64	1.896	0.527	0.728	1.380	0.94	2.560	0.391	0.459	1.175
0.35	1.419	0.705	1.233	1.749	0.65	1.915	0.522	0.716	1.371	0.95	2.586	0.387	0.452	1.170
0.36	1.433	0.698	1.207	1.731	0.66	1.935	0.517	0.704	1.363	0.96	2.612	0.383	0.446	1.165
0.37	1.448	0.691	1.183	1.713	0.67	1.954	0.512	0.693	1.354	0.97	2.638	0.379	0.440	1.159
0.38	1.462	0.684	1.160	1.696	0.68	1.974	0.507	0.682	1.346	0.98	2.664	0.375	0.433	1.154
0.39	1.477	0.677	1.137	1.679	0.69	1.994	0.502	0.671	1.338	0.99	2.691	0.372	0.427	1.149

Annex 4.1 (cont.)

x	e^x	e^{-x}	$K_o(x)$	$e^xK_o(x)$	x	e^x	e^{-x}	$K_o(x)$	$e^xK_o(x)$	x	e^x	e^{-x}	$K_o(x)$	$e^xK_o(x)$
1.0	2.718	0.368	0.421	1.144	4.0	5.460 (1)	1.83 (−2)	1.12 (−2)	0.609					
1.1	3.004	0.333	0.366	1.098	4.1	6.034 (1)	1.00 (−2)	1.00 (−2)	0.602					
1.2	3.320	0.301	0.318	1.057	4.2	6.669 (1)		8.9 (−3)	0.595					
1.3	3.669	0.272	0.278	1.021	4.3	7.370 (1)		8.0 (−3)	0.589					
1.4	4.055	0.247	0.244	0.988	4.4	8.145 (1)		7.1 (−3)	0.582					
1.5	4.482	0.223	0.214	0.958	4.5	9.002 (1)		6.4 (−3)	0.576					
1.6	4.953	0.202	0.188	0.931	4.6	9.948 (1)		5.7 (−3)	0.570					
1.7	5.474	0.183	0.165	0.906	4.7	1.099 (2)		5.1 (−3)	0.564					
1.8	6.050	0.165	0.146	0.883	4.8	1.215 (2)		4.6 (−3)	0.559					
1.9	6.686	0.150	0.129	0.861	4.9	1.343 (2)		4.1 (−3)	0.553					
2.0	7.389	0.135	0.114	0.842	5.0	1.484 (2)		3.7 (−3)	0.548					
2.1	8.166	0.122	0.101	0.823										
2.2	9.025	0.111	8.93 (−2)	0.806										
2.3	9.974	0.100	7.91 (−2)	0.789										
2.4	1.102 (1)	9.07 (−2)	7.02 (−2)	0.774										
2.5	1.218 (1)	8.21 (−2)	6.23 (−2)	0.760										
2.6	1.346 (1)	7.43 (−2)	5.54 (−2)	0.746										
2.7	1.488 (1)	6.72 (−2)	4.93 (−2)	0.733										
2.8	1.644 (1)	6.08 (−2)	4.38 (−2)	0.721										
2.9	1.817 (1)	5.50 (−2)	3.90 (−2)	0.709										
3.0	2.009 (1)	4.98 (−2)	3.47 (−2)	0.698										
3.1	2.220 (1)	4.50 (−2)	3.10 (−2)	0.687										
3.2	2.453 (1)	4.08 (−2)	2.76 (−2)	0.677										
3.3	2.711 (1)	3.69 (−2)	2.46 (−2)	0.667										
3.4	2.996 (1)	3.34 (−2)	2.20 (−2)	0.658										
3.5	3.312 (1)	3.02 (−2)	1.96 (−2)	0.649										
3.6	3.660 (1)	2.73 (−2)	1.75 (−2)	0.640										
3.7	4.045 (1)	2.47 (−2)	1.56 (−2)	0.632										
3.8	4.470 (1)	2.24 (−2)	1.40 (−2)	0.624										
3.9	4.940 (1)	2.02 (−2)	1.25 (−2)	0.617										

Annex 4.2 Values of the Walton well function W(u,r/L) for leaky aquifers (after Hantush 1956) More extensive tables can be found in HANTUSH 1956 and WALTON 1962.

u	1/u	r/L = 0	0.005	0.01	0.02	0.03	0.04	0.05	0.06	0.07	0.08	0.09
0		∞	1.08(1)	9.44	8.06	7.25	6.67	6.23	5.87	5.56	5.29	5.06
1(–6)	1.00(6)	1.32(1)										
2(–6)	5.00(5)	1.25(1)										
4(–6)	2.50(5)	1.18(1)	1.07(1)									
6(–6)	1.66(5)	1.14(1)	1.06(1)									
8(–6)	1.25(5)	1.12(1)	1.05(1)	9.43		W(u,r/L) = W(0,r/L)						
1(–5)	1.00(5)	1.09(1)	1.04(1)	9.42								
2(–5)	5.00(4)	1.02(1)	9.95	9.30								
4(–5)	2.50(4)	9.55	9.40	9.01	8.03							
6(–5)	1.66(4)	9.14	9.04	8.77	7.98	7.24						
8(–5)	1.25(4)	8.86	8.78	8.57	7.91	7.23						
1(–4)	1.00(4)	8.63	8.57	8.40	7.84	7.21						
2(–4)	5.00(3)	7.94	7.91	7.82	7.50	7.07	6.62	6.22	5.86			
4(–4)	2.50(3)	7.25	7.23	7.19	7.01	6.76	6.45	6.14	5.83	5.55		
6(–4)	1.66(3)	6.84	6.83	6.80	6.68	6.50	6.27	6.02	5.77	5.51	5.27	5.05
8(–4)	1.25(3)	6.55		6.52	6.43	6.29	6.11	5.91	5.69	5.46	5.25	5.04
1(–3)	1.00(3)	6.33		6.31	6.23	6.12	5.97	5.80	5.61	5.41	5.21	5.01
2(–3)	5.00(2)	5.64		5.63	5.59	5.53	5.45	5.35	5.24	5.12	4.89	4.85
4(–3)	2.50(2)	4.95		4.94	4.92	4.89	4.85	4.80	4.74	4.67	4.59	4.51
6(–3)	1.66(2)	4.54			4.53	4.51	4.48	4.45	4.40	4.36	4.30	4.24
8(–3)	1.25(2)	4.26			4.25	4.23	4.21	4.19	4.15	4.12	4.08	4.03
1(–2)	1.00(2)	4.04			4.03	4.02	4.00	3.98	3.95	3.92	3.89	3.85
2(–2)	5.00(1)	3.35				3.34	3.34	3.33	3.31	3.30	3.28	3.26
4(–2)	2.50(1)	2.68					2.67	2.67	2.66	2.65	2.65	2.64
6(–2)	1.66(1)	2.29							2.28	2.28	2.27	2.27
8(–2)	1.25(1)	2.03							2.02	2.01	2.01	2.01
						W(u,r/L) = W(u,0)						
1(–1)	1.00(1)	1.82								1.81	1.81	1.81
2(–1)	5.00(1)	1.22										1.22
4(–1)	2.50(1)	7.02(–1)										7.00(–1)
6(–1)	1.66(1)	4.54(–1)										
8(–1)	1.25(1)	3.11(–1)										

Annex 4.2 (cont.)

u	1/u	r/L = 0	0.1	0.2	0.3	0.4	0.6	0.8
0		∞	4.85	3.50	2.74	2.23	1.55	1.13
1(–4)	1.00(4)	8.63						
2(–4)	5.00(3)	7.94						
4(–4)	2.50(3)	7.25						
6(–4)	1.66(3)	6.84						
8(–4)	1.25(3)	6.55	4.84					
1(–3)	1.00(3)	6.33	4.83					
2(–3)	5.00(2)	5.64	4.71					
4(–3)	2.50(2)	4.95	4.42	3.48				
6(–3)	1.66(2)	4.54	4.18	3.43				
8(–3)	1.25(2)	4.26	3.98	3.36	2.73			
1(–2)	1.00(2)	4.04	3.81	3.29	2.71	2.22		
2(–2)	5.00(1)	3.35	3.24	2.95	2.57	2.18		
4(–2)	2.50(1)	2.68	2.63	2.48	2.27	2.02	1.52	
6(–2)	1.66(1)	2.29	2.26	2.17	2.02	1.84	1.46	1.11
8(–2)	1.25(1)	2.03	2.00	1.93	1.83	1.69	1.39	1.08
1(–1)	1.00(1)	1.82	1.80	1.75	1.67	1.56	1.31	1.05
2(–1)	5.00	1.22	1.21	1.19	1.16	1.11	9.96(–1)	8.58(–1)
4(–1)	2.50	7.02(–1)	7.00(–1)	6.93(–1)	6.81(–1)	6.65(–1)	6.21(–1)	5.65(–1)
6(–1)	1.66	4.54(–1)	4.53(–1)	4.50(–1)	4.44(–1)	4.36(–1)	4.15(–1)	3.87(–1)
8(–1)	1.25	3.11(–1)	3.10(–1)	3.08(–1)	3.05(–1)	3.01(–1)	2.89(–1)	2.73(–1)
1	1.00		2.19(–1)	2.18(–1)	2.16(–1)	2.14(–1)	2.07(–1)	1.97(–1)
2	5.00(–1)		4.88(–2)	4.87(–2)	4.85(–2)	4.82(–2)	4.73(–2)	4.60(–2)

W(u,r/L) = W(0,r/L)

Annex 4.2 (cont.)

u	1/u	r/L = 0	1.0	2.0	3.0	4.0	5.0	6.0
0		∞	8.42(–1)	2.28(–1)	6.95(–2)	2.23(–2)	7.4(–3)	2.5(–3)
1(–2)	1.00(2)	4.04						
2(–2)	5.00(1)	3.35						
4(–2)	2.50(1)	2.68						
6(–2)	1.66(1)	2.29	8.39(–1)					
8(–2)	1.25(1)	2.03	8.32(–1)					
1(–1)	1.00(1)	1.82	8.19(–1)					
2(–1)	5.00	1.22	7.15(–1)	2.27(–1)				
4(–1)	2.50	7.02(–1)	5.02(–1)	2.10(–1)	6.91(–2)			
6(–1)	1.66	4.54(–1)	3.54(–1)	1.77(–1)	6.64(–2)	2.22(–2)		
8(–1)	1.25	3.11(–1)	2.54(–1)	1.44(–1)	6.07(–2)	2.18(–2)		
1	1.00	2.19(–1)	1.85(–1)	1.14(–1)	5.34(–2)	2.07(–2)	7.3(–3)	
2	5.00(–1)	4.89(–2)	4.44(–2)	3.35(–2)	2.10(–2)	1.12(–2)	5.1(–3)	2.1(–3)
4	2.50(–1)	3.78(–3)	3.6 (–3)	3.1 (–3)	2.4 (–3)	1.60(–3)	1.0(–3)	6.0(–4)

W(u,r/L) = W(0,r/L)

Annex 4.3 Values of the Hantush well function W(u,β) for leaky aquifers (after Hantush 1960)

		β								
u	1/u	0.001	0.002	0.005	0.01	0.02	0.05	0.1	0.2	0.5
1(−6)	1.00 (6)	1.20 (1)	1.14 (1)	1.06 (1)	9.93 (0)	9.25 (0)	8.34 (0)	7.65 (0)	6.96 (0)	6.05 (0)
2(−6)	5.00 (5)	1.15 (1)	1.10 (1)	1.02 (1)	9.57 (0)	8.89 (0)	7.99 (0)	7.30 (0)	6.61 (0)	5.70 (0)
4(−6)	2.50 (5)	1.11 (1)	1.06 (1)	9.84 (0)	9.20 (0)	8.54 (0)	7.64 (0)	6.95 (0)	6.27 (0)	5.36 (0)
6(−6)	1.66 (5)	1.08 (1)	1.03 (1)	9.61 (0)	8.99 (0)	8.33 (0)	7.44 (0)	6.75 (0)	6.06 (0)	5.16 (0)
8(−6)	1.25 (5)	1.05 (1)	1.01 (1)	9.45 (0)	8.84 (0)	8.18 (0)	7.29 (0)	6.61 (0)	5.92 (0)	5.01 (0)
1(−5)	1.00 (5)	1.04 (1)	1.00 (1)	9.32 (0)	8.71 (0)	8.07 (0)	7.18 (0)	6.49 (0)	5.81 (0)	4.90 (0)
2(−5)	5.00 (4)	9.82 (0)	9.51 (0)	8.90 (0)	8.33 (0)	7.70 (0)	6.82 (0)	6.15 (0)	5.46 (0)	4.56 (0)
4(−5)	2.50 (4)	9.24 (0)	8.99 (0)	8.46 (0)	7.93 (0)	7.33 (0)	6.47 (0)	5.80 (0)	5.12 (0)	4.22 (0)
6(−5)	1.66 (4)	8.88 (0)	8.67 (0)	8.19 (0)	7.69 (0)	7.11 (0)	6.26 (0)	5.59 (0)	4.91 (0)	4.02 (0)
8(−5)	1.25 (4)	8.63 (0)	8.43 (0)	8.00 (0)	7.52 (0)	6.95 (0)	6.11 (0)	5.44 (0)	4.77 (0)	3.88 (0)
1(−4)	1.00 (4)	8.43 (0)	8.25 (0)	7.84 (0)	7.38 (0)	6.82 (0)	5.99 (0)	5.33 (0)	4.66 (0)	3.77 (0)
2(−4)	5.00 (3)	7.79 (0)	7.66 (0)	7.33 (0)	6.93 (0)	6.42 (0)	5.62 (0)	4.97 (0)	4.31 (0)	3.43 (0)
4(−4)	2.50 (3)	7.14 (0)	7.04 (0)	6.78 (0)	6.45 (0)	6.00 (0)	5.25 (0)	4.62 (0)	3.96 (0)	3.10 (0)
6(−4)	1.66 (3)	6.75 (0)	6.67 (0)	6.45 (0)	6.16 (0)	5.74 (0)	5.02 (0)	4.40 (0)	3.76 (0)	2.91 (0)
8(−4)	1.25 (3)	6.48 (0)	6.40 (0)	6.21 (0)	5.94 (0)	5.55 (0)	4.86 (0)	4.25 (0)	3.62 (0)	2.77 (0)
1(−3)	1.00 (3)	6.26 (0)	6.20 (0)	6.02 (0)	5.77 (0)	5.40 (0)	4.73 (0)	4.13 (0)	3.50 (0)	2.67 (0)
2(−3)	5.00 (2)	5.59 (0)	5.54 (0)	5.41 (0)	5.22 (0)	4.91 (0)	4.32 (0)	3.76 (0)	3.15 (0)	2.34 (0)
4(−3)	2.50 (2)	4.91 (0)	4.88 (0)	4.78 (0)	4.64 (0)	4.40 (0)	3.89 (0)	3.38 (0)	2.80 (0)	2.03 (0)
6(−3)	1.66 (2)	4.52 (0)	4.49 (0)	4.41 (0)	4.29 (0)	4.08 (0)	3.62 (0)	3.14 (0)	2.60 (0)	1.84 (0)
8(−3)	1.25 (2)	4.23 (0)	4.21 (0)	4.14 (0)	4.04 (0)	3.85 (0)	3.43 (0)	2.98 (0)	2.45 (0)	1.72 (0)
1(−2)	1.00 (2)	4.02 (0)	4.00 (0)	3.93 (0)	3.84 (0)	3.67 (0)	3.28 (0)	2.84 (0)	2.33 (0)	1.62 (0)
2(−2)	5.00 (1)	3.34 (0)	3.33 (0)	3.28 (0)	3.21 (0)	3.09 (0)	2.78 (0)	2.42 (0)	1.97 (0)	1.32 (0)
4(−2)	2.50 (1)	2.67 (0)	2.66 (0)	2.63 (0)	2.58 (0)	2.50 (0)	2.27 (0)	1.98 (0)	1.61 (0)	1.04 (0)
6(−2)	1.66 (1)	2.29 (0)	2.28 (0)	2.26 (0)	2.22 (0)	2.15 (0)	1.96 (0)	1.72 (0)	1.39 (0)	8.84(−1)
8(−2)	1.25 (1)	2.02 (0)	2.01 (0)	1.99 (0)	1.96 (0)	1.90 (0)	1.74 (0)	1.53 (0)	1.24 (0)	7.76(−1)
1(−1)	1.00 (1)	1.82 (0)	1.81 (0)	1.79 (0)	1.77 (0)	1.72 (0)	1.58 (0)	1.39 (0)	1.12 (0)	6.95(−1)
2(−1)	5.00 (0)	1.22 (0)	1.22 (0)	1.21 (0)	1.19 (0)	1.16 (0)	1.07 (0)	9.50(−1)	7.67(−1)	4.60(−1)
4(−1)	2.50 (0)	7.01(−1)	6.99(−1)	6.94(−1)	6.85(−1)	6.68(−1)	6.22(−1)	5.54(−1)	4.48(−1)	2.62(−1)
6(−1)	1.66 (0)	4.53(−1)	4.52(−1)	4.49(−1)	4.44(−1)	4.33(−1)	4.04(−1)	3.61(−1)	2.93(−1)	1.69(−1)
8(−1)	1.25 (0)	3.10(−1)	3.09(−1)	3.07(−1)	3.04(−1)	2.97(−1)	2.77(−1)	2.48(−1)	2.01(−1)	1.15(−1)
1 (0)	1.00 (0)	2.19(−1)	2.18(−1)	2.17(−1)	2.14(−1)	2.10(−1)	1.96(−1)	1.76(−1)	1.43(−1)	8.12(−2)
2 (0)	5.00(−1)	4.88(−2)	4.87(−2)	4.84(−2)	4.79(−2)	4.68(−2)	4.39(−2)	3.95(−2)	3.22(−2)	1.80(−2)
4 (0)	2.50(−1)	3.77(−3)	3.76(−3)	3.74(−3)	3.70(−3)	3.62(−3)	3.40(−3)	3.07(−3)	2.50(−3)	1.39(−3)

Annex 4.3 (cont.)

u	1/u	β 1	2	5	10	20	50
1(–6)	1.00 (6)	5.36 (0)	4.67 (0)	3.78 (0)	3.11 (0)	2.47 (0)	1.67 (0)
2(–6)	5.00 (5)	5.01 (0)	4.33 (0)	3.44 (0)	2.79 (0)	2.16 (0)	1.39 (0)
4(–6)	2.50 (5)	4.67 (0)	3.99 (0)	3.11 (0)	2.47 (0)	1.86 (0)	1.14 (0)
6(–6)	1.66 (5)	4.47 (0)	3.80 (0)	2.92 (0)	2.28 (0)	1.69 (0)	9.95(–1)
8(–6)	1.25 (5)	4.33 (0)	3.66 (0)	2.79 (0)	2.16 (0)	1.57 (0)	9.00(–1)
1(–5)	1.00 (5)	4.22 (0)	3.55 (0)	2.68 (0)	2.06 (0)	1.48 (0)	8.29(–1)
2(–5)	5.00 (4)	3.88 (0)	3.22 (0)	2.37 (0)	1.76 (0)	1.22 (0)	6.26(–1)
4(–5)	2.50 (4)	3.55 (0)	2.89 (0)	2.06 (0)	1.48 (0)	9.73(–1)	4.52(–1)
6(–5)	1.66 (4)	3.35 (0)	2.70 (0)	1.88 (0)	1.32 (0)	8.41(–1)	3.65(–1)
8(–5)	1.25 (4)	3.21 (0)	2.57 (0)	1.76 (0)	1.22 (0)	7.53(–1)	3.09(–1)
1(–4)	1.00 (4)	3.11 (0)	2.47 (0)	1.67 (0)	1.14 (0)	6.88(–1)	2.70(–1)
2(–4)	5.00 (3)	2.78 (0)	2.15 (0)	1.39 (0)	8.99(–1)	5.04(–1)	1.68(–1)
4(–4)	2.50 (3)	2.46 (0)	1.85 (0)	1.14 (0)	6.88(–1)	3.51(–1)	9.63(–2)
6(–4)	1.66 (3)	2.28 (0)	1.68 (0)	9.94(–1)	5.77(–1)	2.77(–1)	6.61(–2)
8(–4)	1.25 (3)	2.15 (0)	1.57 (0)	8.98(–1)	5.04(–1)	2.30(–1)	4.94(–2)
1(–3)	1.00 (3)	2.05 (0)	1.48 (0)	8.27(–1)	4.51(–1)	1.98(–1)	3.88(–2)
2(–3)	5.00 (2)	1.75 (0)	1.21 (0)	6.24(–1)	3.08(–1)	1.16(–1)	1.66(–2)
4(–3)	2.50 (2)	1.47 (0)	9.66(–1)	4.50(–1)	1.97(–1)	6.19(2)	5.88(–3)
6(–3)	1.66 (2)	1.31 (0)	8.33(–1)	3.62(–1)	1.46(–1)	4.04(–2)	2.92(–3)
8(–3)	1.25 (2)	1.20 (0)	7.44(–1)	3.06(–1)	1.16(–1)	2.90(–2)	1.69(–3)
1(–2)	1.00 (2)	1.11 (0)	6.78(–1)	2.67(–1)	9.55(–2)	2.21(–2)	1.06(–3)
2(–2)	5.00 (1)	8.68(–1)	4.91(–1)	1.65(–1)	4.87(–2)	8.31(–3)	2.03(–4)
4(–2)	2.50 (1)	6.47(–1)	3.36(–1)	9.31(–2)	2.16(–2)	2.53(–3)	2.69(–5)
6(–2)	1.66 (1)	5.30(–1)	2.59(–1)	6.30(–2)	1.24(–2)	1.12(–3)	6.55(–6)
8(–2)	1.25 (1)	4.53(–1)	2.12(–1)	4.64(–2)	7.97(–3)	5.87(–4)	2.19(–6)
1(1)	1.00 (1)	3.97(–1)	1.79(–1)	3.59(–2)	5.52(–3)	3.40(–4)	
2(–1)	5.00 (0)	2.45(–1)	9.71(–2)	1.43(–2)	1.49(–3)	4.93(–5)	
4(–1)	2.50 (0)	1.30(–1)	4.41(–2)	4.48(–3)	2.83(–4)	4.24(–6)	
6(–1)	1.66 (0)	7.99(–2)	2.47(–2)	1.95(–3)	8.73(–5)		
8(–1)	1.25 (0)	5.29(–2)	1.52(–2)	9.86(–4)	3.40(–5)		
1 (0)	1.00 (0)	3.65(–2)	9.93(–3)	5.47(–4)	1.51(–5)		
2 (0)	5.00(–1)	7.60(–3)	1.73(–3)	5.51(–5)			
4 (0)	2.50(–1)	5.58(–4)	1.08(–4)	1.89(–6)			
6 (0)	1.66(–1)	5.19(–5)	9.26(–6)				
8 (0)	1.25(–1)	5.36(–6)					

Annex 4.4 Values of the Neuman-Witherspoon function W(u,u_c) for leaky aquifers (after Witherspoon et al. 1967)

u_c	$1/u_c$	u = 1.25	6.25(−1)	5.0 (−1)	3.57(−1)	2.5 (−1)	1.25(−1)	6.25(−2)	3.57(−2)	2.5 (−2)	2.5 (−3)	2.5 (−4)	2.5 (−5)	2.5 (−6)	2.5 (−7)	2.5 (−8)	2.5 (−9)	2.5 (−10)	2.5 (−11)
1(−3)	1.00(3)	9.07(−1)	9.22(−1)	9.26(−1)	9.31(−1)	9.35(−1)	9.41(−1)	9.46(−1)	9.49(−1)	9.50(−1)	9.56(−1)	9.58(−1)	9.60(−1)	9.60(−1)	9.61(−1)	9.62(−1)	9.62(−1)	9.62(−1)	9.62(−1)
2(−3)	5.00(2)	8.70(−1)	8.91(−1)	8.96(−1)	9.03(−1)	9.09(−1)	9.18(−1)	9.24(−1)	9.28(−1)	9.30(−1)	9.37(−1)	9.41(−1)	9.43(−1)	9.44(−1)	9.45(−1)	9.46(−1)	9.46(−1)	9.47(−1)	9.47(−1)
3(−3)	3.33(2)	8.43(−1)	8.68(−1)	8.74(−1)	8.82(−1)	8.89(−1)	9.00(−1)	9.07(−1)	9.12(−1)	9.14(−1)	9.24(−1)	9.28(−1)	9.30(−1)	9.32(−1)	9.33(−1)	9.34(−1)	9.34(−1)	9.35(−1)	9.35(−1)
4(−3)	2.50(2)	8.20(−1)	8.48(−1)	8.56(−1)	8.65(−1)	8.73(−1)	8.85(−1)	8.94(−1)	8.99(−1)	9.01(−1)	9.12(−1)	9.17(−1)	9.20(−1)	9.21(−1)	9.23(−1)	9.23(−1)	9.24(−1)	9.25(−1)	9.25(−1)
5(−3)	2.00(2)	8.01(−1)	8.32(−1)	8.40(−1)	8.50(−1)	8.59(−1)	8.72(−1)	8.81(−1)	8.87(−1)	8.90(−1)	9.02(−1)	9.07(−1)	9.10(−1)	9.12(−1)	9.13(−1)	9.14(−1)	9.15(−1)	9.16(−1)	9.16(−1)
6(−3)	1.66(2)	7.84(−1)	8.17(−1)	8.26(−1)	8.36(−1)	8.46(−1)	8.60(−1)	8.71(−1)	8.77(−1)	8.80(−1)	8.93(−1)	8.99(−1)	9.02(−1)	9.04(−1)	9.05(−1)	9.06(−1)	9.07(−1)	9.08(−1)	9.08(−1)
8(−3)	1.25(2)	7.54(−1)	7.91(−1)	8.01(−1)	8.13(−1)	8.24(−1)	8.40(−1)	8.52(−1)	8.59(−1)	8.62(−1)	8.77(−1)	8.83(−1)	8.87(−1)	8.89(−1)	8.91(−1)	8.92(−1)	8.93(−1)	8.94(−1)	8.94(−1)
1(−2)	1.00(2)	7.29(−1)	7.69(−1)	7.79(−1)	7.92(−1)	8.04(−1)	8.22(−1)	8.35(−1)	8.43(−1)	8.47(−1)	8.63(−1)	8.70(−1)	8.74(−1)	8.76(−1)	8.78(−1)	8.79(−1)	8.80(−1)	8.81(−1)	8.82(−1)
2(−2)	5.00(1)	6.37(−1)	6.86(−1)	7.00(−1)	7.16(−1)	7.32(−1)	7.55(−1)	7.72(−1)	7.82(−1)	7.87(−1)	8.08(−1)	8.18(−1)	8.23(−1)	8.27(−1)	8.29(−1)	8.31(−1)	8.32(−1)	8.33(−1)	8.34(−1)
3(−2)	3.33(1)	5.73(−1)	6.28(−1)	6.43(−1)	6.62(−1)	6.79(−1)	7.06(−1)	7.25(−1)	7.37(−1)	7.43(−1)	7.68(−1)	7.79(−1)	7.85(−1)	7.89(−1)	7.92(−1)	7.94(−1)	7.95(−1)	7.97(−1)	7.98(−1)
4(−2)	2.50(1)	5.23(−1)	5.82(−1)	5.98(−1)	6.18(−1)	6.37(−1)	6.66(−1)	6.87(−1)	7.00(−1)	7.07(−1)	7.34(−1)	7.47(−1)	7.54(−1)	7.58(−1)	7.61(−1)	7.63(−1)	7.65(−1)	7.66(−1)	7.67(−1)
5(−2)	2.00(1)	4.82(−1)	5.44(−1)	5.61(−1)	5.82(−1)	6.02(−1)	6.33(−1)	6.55(−1)	6.69(−1)	6.76(−1)	7.05(−1)	7.19(−1)	7.26(−1)	7.31(−1)	7.34(−1)	7.37(−1)	7.39(−1)	7.40(−1)	7.41(−1)
6(−2)	1.66(1)	4.48(−1)	5.11(−1)	5.28(−1)	5.50(−1)	5.71(−1)	6.03(−1)	6.27(−1)	6.42(−1)	6.49(−1)	6.80(−1)	6.94(−1)	7.02(−1)	7.07(−1)	7.11(−1)	7.13(−1)	7.15(−1)	7.17(−1)	7.18(−1)
8(−2)	1.25(1)	3.92(−1)	4.56(−1)	4.75(−1)	4.98(−1)	5.20(−1)	5.54(−1)	5.79(−1)	5.95(−1)	6.03(−1)	6.36(−1)	6.51(−1)	6.60(−1)	6.65(−1)	6.69(−1)	6.72(−1)	6.74(−1)	6.76(−1)	6.77(−1)
1(−1)	1.00(1)	3.48(−1)	4.13(−1)	4.31(−1)	4.55(−1)	4.77(−1)	5.12(−1)	5.39(−1)	5.55(−1)	5.64(−1)	5.98(−1)	6.15(−1)	6.24(−1)	6.30(−1)	6.34(−1)	6.36(−1)	6.39(−1)	6.40(−1)	6.42(−1)
2(−1)	5.00(0)	2.14(−1)	2.73(−1)	2.90(−1)	3.13(−1)	3.36(−1)	3.72(−1)	3.99(−1)	4.17(−1)	4.26(−1)	4.64(−1)	4.83(−1)	4.93(−1)	4.99(−1)	5.04(−1)	5.07(−1)	5.09(−1)	5.11(−1)	5.13(−1)
3(−1)	3.33(0)	1.44(−1)	1.95(−1)	2.10(−1)	2.31(−1)	2.51(−1)	2.85(−1)	3.12(−1)	3.29(−1)	3.38(−1)	3.76(−1)	3.94(−1)	4.04(−1)	4.11(−1)	4.15(−1)	4.18(−1)	4.21(−1)	4.23(−1)	4.24(−1)
4(−1)	2.50(0)	1.02(−1)	1.45(−1)	1.58(−1)	1.76(−1)	1.95(−1)	2.25(−1)	2.50(−1)	2.66(−1)	2.75(−1)	3.11(−1)	3.28(−1)	3.38(−1)	3.44(−1)	3.48(−1)	3.51(−1)	3.54(−1)	3.56(−1)	3.57(−1)
5(−1)	2.00(0)	7.44(−2)	1.10(−1)	1.22(−1)	1.38(−1)	1.54(−1)	1.82(−1)	2.04(−1)	2.19(−1)	2.27(−1)	2.60(−1)	2.77(−1)	2.86(−1)	2.92(−1)	2.96(−1)	2.99(−1)	3.01(−1)	3.03(−1)	3.04(−1)
6(−1)	1.66(0)	5.55(−2)	8.53(−2)	9.54(−2)	1.09(−1)	1.23(−1)	1.48(−1)	1.68(−1)	1.82(−1)	1.89(−1)	2.20(−1)	2.36(−1)	2.44(−1)	2.50(−1)	2.53(−1)	2.56(−1)	2.58(−1)	2.60(−1)	2.61(−1)
8(−1)	1.25(0)	3.23(−2)	5.33(−2)	6.06(−2)	7.09(−2)	8.18(−2)	1.01(−1)	1.18(−1)	1.29(−1)	1.35(−1)	1.61(−1)	1.74(−1)	1.81(−1)	1.86(−1)	1.89(−1)	1.91(−1)	1.93(−1)	1.94(−1)	1.95(−1)
1(0)	1.00(0)	1.96(−2)	3.44(−2)	3.99(−2)	4.75(−2)	5.58(−2)	7.09(−2)	8.40(−2)	9.29(−2)	9.79(−2)	1.19(−1)	1.30(−1)	1.37(−1)	1.40(−1)	1.43(−1)	1.45(−1)	1.46(−1)	1.48(−1)	1.48(−1)
2(0)	5.00(−1)	2.29(−3)	5.14(−3)	6.34(−3)	8.19(−3)	1.03(−2)	1.46(−2)	1.87(−2)	2.16(−2)	2.33(−2)	3.11(−2)	3.52(−2)	3.76(−2)	3.90(−2)	4.00(−2)	4.08(−2)	4.13(−2)	4.18(−2)	4.21(−2)
3(0)	3.33(−1)	3.35(−4)	9.67(−4)	1.25(−3)	1.72(−3)	2.30(−3)	3.55(−3)	4.81(−3)	5.78(−3)	6.35(−3)	9.07(−3)	1.06(−2)	1.14(−2)	1.19(−2)	1.23(−2)	1.26(−2)	1.28(−2)	1.29(−2)	1.31(−2)
4(0)	2.50(−1)	6.38(−5)	2.03(−4)	2.80(−4)	4.04(−4)	5.60(−4)	9.33(−4)	1.33(−3)	1.65(−3)	1.84(−3)	2.79(−3)	3.32(−3)	3.63(−3)	3.82(−3)	3.95(−3)	4.05(−3)	4.12(−3)	4.18(−3)	4.23(−3)
5(0)	2.00(−1)	1.24(−5)	4.52(−5)	6.54(−5)	9.91(−5)	1.46(−4)	2.56(−4)	3.84(−4)	4.89(−4)	5.54(−4)	8.85(−4)	1.07(−3)	1.19(−3)	1.26(−3)	1.30(−3)	1.34(−3)	1.37(−3)	1.39(−3)	1.40(−3)
6(0)	1.66(−1)	4.10(−6)	1.08(−5)	1.59(−5)	2.60(−5)	4.06(−5)	7.80(−5)	1.17(−4)	1.50(−4)	1.73(−4)	2.87(−4)	3.55(−4)	3.95(−4)	4.21(−4)	4.38(−4)	4.50(−4)	4.60(−4)	4.68(−4)	4.74(−4)
8(0)	1.25(−1)	5.46(−9)	6.81(−7)	1.06(−6)	1.89(−6)	3.93(−6)	5.73(−6)	1.12(−5)	1.53(−5)	1.78(−5)	3.12(−5)	4.04(−5)	4.55(−5)	4.88(−5)	5.11(−5)	5.27(−5)	5.40(−5)	5.49(−5)	5.57(−5)

Annex 5.1 Values of the Neuman functions $W(u_A,\beta)$ and $W(u_B,\beta)$ for unconfined aquifers (after Neuman 1975)
Tables of values of the function $W(u_A,\beta)$

$1/u_A$	$\beta = 0.001$	$\beta = 0.004$	$\beta = 0.01$	$\beta = 0.03$	$\beta = 0.06$	$\beta = 0.1$	$\beta = 0.2$	$\beta = 0.4$	$\beta = 0.6$
4×10^{-1}	2.48×10^{-2}	2.43×10^{-2}	2.41×10^{-2}	2.35×10^{-2}	2.30×10^{-2}	2.24×10^{-2}	2.14×10^{-2}	1.99×10^{-2}	1.88×10^{-2}
8×10^{-1}	1.45×10^{-1}	1.42×10^{-1}	1.40×10^{-1}	1.36×10^{-1}	1.31×10^{-1}	1.27×10^{-1}	1.19×10^{-1}	1.08×10^{-1}	9.88×10^{-2}
1.4×10^{0}	3.58×10^{-1}	3.52×10^{-1}	3.45×10^{-1}	3.31×10^{-1}	3.18×10^{-1}	3.04×10^{-1}	2.79×10^{-1}	2.44×10^{-1}	2.17×10^{-1}
2.4×10^{0}	6.62×10^{-1}	6.48×10^{-1}	6.33×10^{-1}	6.01×10^{-1}	5.70×10^{-1}	5.40×10^{-1}	4.83×10^{-1}	4.03×10^{-1}	3.43×10^{-1}
4×10^{0}	1.02×10^{0}	9.92×10^{-1}	9.63×10^{-1}	9.05×10^{-1}	8.49×10^{-1}	7.92×10^{-1}	6.88×10^{-1}	5.42×10^{-1}	4.38×10^{-1}
8×10^{0}	1.57×10^{0}	1.52×10^{0}	1.46×10^{0}	1.35×10^{0}	1.23×10^{0}	1.12×10^{0}	9.18×10^{-1}	6.59×10^{-1}	4.97×10^{-1}
1.4×10^{1}	2.05×10^{0}	1.97×10^{0}	1.88×10^{0}	1.70×10^{0}	1.51×10^{0}	1.34×10^{0}	1.03×10^{0}	6.90×10^{-1}	5.07×10^{-1}
2.4×10^{1}	2.52×10^{0}	2.41×10^{0}	2.27×10^{0}	1.99×10^{0}	1.73×10^{0}	1.47×10^{0}	1.07×10^{0}	6.96×10^{-1}	
4×10^{1}	2.97×10^{0}	2.80×10^{0}	2.61×10^{0}	2.22×10^{0}	1.85×10^{0}	1.53×10^{0}	1.08×10^{0}		
8×10^{1}	3.56×10^{0}	3.30×10^{0}	3.00×10^{0}	2.41×10^{0}	1.92×10^{0}	1.55×10^{0}			
1.4×10^{2}	4.01×10^{0}	3.65×10^{0}	3.23×10^{0}	2.48×10^{0}	1.93×10^{0}				
2.4×10^{2}	4.42×10^{0}	3.93×10^{0}	3.37×10^{0}	2.49×10^{0}	1.94×10^{0}				
4×10^{2}	4.77×10^{0}	4.12×10^{0}	3.43×10^{0}	2.50×10^{0}					
8×10^{2}	5.16×10^{0}	4.26×10^{0}	3.45×10^{0}						
1.4×10^{3}	5.40×10^{0}	4.29×10^{0}	3.46×10^{0}						
2.4×10^{3}	5.54×10^{0}	4.30×10^{0}							
4×10^{3}	5.59×10^{0}								
8×10^{3}	5.62×10^{0}								
1.4×10^{4}	5.62×10^{0}	4.30×10^{0}	3.46×10^{0}	2.50×10^{0}	1.94×10^{0}	1.55×10^{0}	1.08×10^{0}	6.96×10^{-1}	5.07×10^{-1}

Annex 5.1 (cont.)

$1/u_A$	$\beta = 0.8$	$\beta = 1.0$	$\beta = 1.5$	$\beta = 2.0$	$\beta = 2.5$	$\beta = 3.0$	$\beta = 4.0$	$\beta = 5.0$	$\beta = 6.0$	$\beta = 7.0$
4×10^{-1}	1.79×10^{-2}	1.70×10^{-2}	1.53×10^{-2}	1.38×10^{-2}	1.25×10^{-2}	1.13×10^{-2}	9.33×10^{-3}	7.72×10^{-3}	6.39×10^{-3}	5.30×10^{-3}
8×10^{-1}	9.15×10^{-2}	8.49×10^{-2}	7.13×10^{-2}	6.03×10^{-2}	5.11×10^{-2}	4.35×10^{-2}	3.17×10^{-2}	2.34×10^{-2}	1.74×10^{-2}	1.31×10^{-2}
1.4×10^{0}	1.94×10^{-1}	1.75×10^{-1}	1.36×10^{-1}	1.07×10^{-1}	8.46×10^{-2}	6.78×10^{-2}	4.45×10^{-2}	3.02×10^{-2}	2.10×10^{-2}	1.51×10^{-2}
2.4×10^{0}	2.96×10^{-1}	2.56×10^{-1}	1.82×10^{-1}	1.33×10^{-1}	1.01×10^{-1}	7.67×10^{-2}	4.76×10^{-2}	3.13×10^{-2}	2.14×10^{-2}	1.52×10^{-2}
4×10^{0}	3.60×10^{-1}	3.00×10^{-1}	1.99×10^{-1}	1.40×10^{-1}	1.03×10^{-1}	7.79×10^{-2}	4.78×10^{-2}		2.15×10^{-2}	
8×10^{0}	3.91×10^{-1}	3.17×10^{-1}	2.03×10^{-1}	1.41×10^{-1}						
1.4×10^{1}	3.94×10^{-1}									
2.4×10^{1}										
4×10^{1}										
8×10^{1}										
1.4×10^{2}										
2.4×10^{2}										
4×10^{2}										
8×10^{2}										
1.4×10^{3}										
2.4×10^{3}										
4×10^{3}										
8×10^{3}										
1.4×10^{4}	3.94×10^{-1}	3.17×10^{-1}	2.03×10^{-1}	1.41×10^{-1}	1.03×10^{-1}	7.79×10^{-2}	4.78×10^{-2}	3.13×10^{-2}	2.15×10^{-2}	1.52×10^{-2}

Annex 5.1 (cont.)
Table of the values of the function $W(u_B,\beta)$

$1/u_B$	$\beta = 0.001$	$\beta = 0.004$	$\beta = 0.01$	$\beta = 0.03$	$\beta = 0.06$	$\beta = 0.1$	$\beta = 0.2$	$\beta = 0.4$	$\beta = 0.6$
4×10^{-4}	5.62×10^{0}	4.30×10^{0}	3.46×10^{0}	2.50×10^{0}	1.94×10^{0}	1.56×10^{0}	1.09×10^{0}	6.97×10^{-1}	5.08×10^{-1}
8×10^{-4}									
1.4×10^{-3}									
2.4×10^{-3}									
4×10^{-3}								6.97×10^{-1}	5.08×10^{-1}
8×10^{-3}								6.97×10^{-1}	5.09×10^{-1}
1.4×10^{-2}								6.98×10^{-1}	5.10×10^{-1}
2.4×10^{-2}								7.00×10^{-1}	5.12×10^{-1}
4×10^{-2}								7.03×10^{-1}	5.16×10^{-1}
8×10^{-2}						1.56×10^{0}	1.09×10^{0}	7.10×10^{-1}	5.24×10^{-1}
1.4×10^{-1}					1.94×10^{0}	1.56×10^{0}	1.10×10^{0}	7.20×10^{-1}	5.37×10^{-1}
2.4×10^{-1}				2.50×10^{0}	1.95×10^{0}	1.57×10^{0}	1.11×10^{0}	7.37×10^{-1}	5.57×10^{-1}
4×10^{-1}				2.51×10^{0}	1.96×10^{0}	1.58×10^{0}	1.13×10^{0}	7.63×10^{-1}	5.89×10^{-1}
8×10^{-1}	5.62×10^{0}	4.30×10^{0}	3.46×10^{0}	2.52×10^{0}	1.98×10^{0}	1.61×10^{0}	1.18×10^{0}	8.29×10^{-1}	6.67×10^{-1}
1.4×10^{0}	5.63×10^{0}	4.31×10^{0}	3.47×10^{0}	2.54×10^{0}	2.01×10^{0}	1.66×10^{0}	1.24×10^{0}	9.22×10^{-1}	7.80×10^{-1}
2.4×10^{0}	5.63×10^{0}	4.31×10^{0}	3.49×10^{0}	2.57×10^{0}	2.06×10^{0}	1.73×10^{0}	1.35×10^{0}	1.07×10^{0}	9.54×10^{-1}
4×10^{0}	5.63×10^{0}	4.32×10^{0}	3.51×10^{0}	2.62×10^{0}	2.13×10^{0}	1.83×10^{0}	1.50×10^{0}	1.29×10^{0}	1.20×10^{0}
8×10^{0}	5.64×10^{0}	4.35×10^{0}	3.56×10^{0}	2.73×10^{0}	2.31×10^{0}	2.07×10^{0}	1.85×10^{0}	1.72×10^{0}	1.68×10^{0}
1.4×10^{1}	5.65×10^{0}	4.38×10^{0}	3.63×10^{0}	2.88×10^{0}	2.55×10^{0}	2.37×10^{0}	2.23×10^{0}	2.17×10^{0}	2.15×10^{0}
2.4×10^{1}	5.67×10^{0}	4.44×10^{0}	3.74×10^{0}	3.11×10^{0}	2.86×10^{0}	2.75×10^{0}	2.68×10^{0}	2.66×10^{0}	2.65×10^{0}
4×10^{1}	5.70×10^{0}	4.52×10^{0}	3.90×10^{0}	3.40×10^{0}	3.24×10^{0}	3.18×10^{0}	3.15×10^{0}	3.14×10^{0}	3.14×10^{0}
8×10^{1}	5.76×10^{0}	4.71×10^{0}	4.22×10^{0}	3.92×10^{0}	3.85×10^{0}	3.83×10^{0}	3.82×10^{0}	3.82×10^{0}	3.82×10^{0}
1.4×10^{2}	5.85×10^{0}	4.94×10^{0}	4.58×10^{0}	4.40×10^{0}	4.38×10^{0}	4.38×10^{0}	4.37×10^{0}	4.37×10^{0}	4.37×10^{0}
2.4×10^{2}	5.99×10^{0}	5.23×10^{0}	5.00×10^{0}	4.92×10^{0}	4.91×10^{0}	4.91×10^{0}	4.91×10^{0}	4.91×10^{0}	4.91×10^{0}
4×10^{2}	6.16×10^{0}	5.59×10^{0}	5.46×10^{0}	5.42×10^{0}	5.42×10^{0}	5.42×10^{0}	5.42×10^{0}	5.42×10^{0}	5.42×10^{0}

Annex 5.1 (cont.)

$1/u_B$	$\beta = 0.8$	$\beta = 1.0$	$\beta = 1.5$	$\beta = 2.0$	$\beta = 2.5$	$\beta = 3.0$	$\beta = 4.0$	$\beta = 5.0$	$\beta = 6.0$	$\beta = 7.0$
4×10^{-4}	3.95×10^{-1}	3.18×10^{-1}	2.04×10^{-1}	1.42×10^{-1}	1.03×10^{-1}	7.80×10^{-2}	4.79×10^{-2}	3.14×10^{-2}	2.15×10^{-2}	1.53×10^{-2}
8×10^{-4}						7.81×10^{-2}	4.80×10^{-2}	3.15×10^{-2}	2.16×10^{-2}	1.53×10^{-2}
1.4×10^{-3}					1.03×10^{-1}	7.83×10^{-2}	4.81×10^{-2}	3.16×10^{-2}	2.17×10^{-2}	1.54×10^{-2}
2.4×10^{-3}					1.04×10^{-1}	7.85×10^{-2}	4.84×10^{-2}	3.18×10^{-2}	2.19×10^{-2}	1.56×10^{-2}
4×10^{-3}	3.95×10^{-1}	3.18×10^{-1}	2.04×10^{-1}	1.42×10^{-1}	1.04×10^{-1}	7.89×10^{-2}	4.87×10^{-2}	3.21×10^{-2}	2.21×10^{-2}	1.58×10^{-2}
8×10^{-3}	3.96×10^{-1}	3.19×10^{-1}	2.05×10^{-1}	1.43×10^{-1}	1.05×10^{-1}	7.99×10^{-2}	4.96×10^{-2}	3.29×10^{-2}	2.28×10^{-2}	1.64×10^{-2}
1.4×10^{-2}	3.97×10^{-1}	3.21×10^{-1}	2.07×10^{-1}	1.45×10^{-1}	1.07×10^{-1}	8.14×10^{-2}	5.09×10^{-2}	3.41×10^{-2}	2.39×10^{-2}	1.73×10^{-2}
2.4×10^{-2}	3.99×10^{-1}	3.23×10^{-1}	2.09×10^{-1}	1.47×10^{-1}	1.09×10^{-1}	8.38×10^{-2}	5.32×10^{-2}	3.61×10^{-2}	2.57×10^{-2}	1.89×10^{-2}
4×10^{-2}	4.03×10^{-1}	3.27×10^{-1}	2.13×10^{-1}	1.52×10^{-1}	1.13×10^{-1}	8.79×10^{-2}	5.68×10^{-2}	3.93×10^{-2}	2.86×10^{-2}	2.15×10^{-2}
8×10^{-2}	4.12×10^{-1}	3.37×10^{-1}	2.24×10^{-1}	1.62×10^{-1}	1.24×10^{-1}	9.80×10^{-2}	6.61×10^{-2}	4.78×10^{-2}	3.62×10^{-2}	2.84×10^{-2}
1.4×10^{-1}	4.25×10^{-1}	3.50×10^{-1}	2.39×10^{-1}	1.78×10^{-1}	1.39×10^{-1}	1.13×10^{-1}	8.06×10^{-2}	6.12×10^{-2}	4.86×10^{-2}	3.98×10^{-2}
2.4×10^{-1}	4.47×10^{-1}	3.74×10^{-1}	2.65×10^{-1}	2.05×10^{-1}	1.66×10^{-1}	1.40×10^{-1}	1.06×10^{-1}	8.53×10^{-2}	7.14×10^{-2}	6.14×10^{-2}
4×10^{-1}	4.83×10^{-1}	4.12×10^{-1}	3.07×10^{-1}	2.48×10^{-1}	2.10×10^{-1}	1.84×10^{-1}	1.49×10^{-1}	1.28×10^{-1}	1.13×10^{-1}	1.02×10^{-1}
8×10^{-1}	5.71×10^{-1}	5.06×10^{-1}	4.10×10^{-1}	3.57×10^{-1}	3.23×10^{-1}	2.98×10^{-1}	2.66×10^{-1}	2.45×10^{-1}	2.31×10^{-1}	2.20×10^{-1}
1.4×10^{0}	6.97×10^{-1}	6.42×10^{-1}	5.62×10^{-1}	5.17×10^{-1}	4.89×10^{-1}	4.70×10^{-1}	4.45×10^{-1}	4.30×10^{-1}	4.19×10^{-1}	4.11×10^{-1}
2.4×10^{0}	8.89×10^{-1}	8.50×10^{-1}	7.92×10^{-1}	7.63×10^{-1}	7.45×10^{-1}	7.33×10^{-1}	7.18×10^{-1}	7.09×10^{-1}	7.03×10^{-1}	6.99×10^{-1}
4×10^{0}	1.16×10^{0}	1.13×10^{0}	1.10×10^{0}	1.08×10^{0}	1.07×10^{0}	1.07×10^{0}	1.06×10^{0}	1.06×10^{0}	1.05×10^{0}	1.05×10^{0}
8×10^{0}	1.66×10^{0}	1.65×10^{0}	1.64×10^{0}	1.63×10^{0}	1.63×10^{0}	1.63×10^{0}	1.63×10^{0}	1.63×10^{0}	1.63×10^{0}	1.63×10^{0}
1.4×10^{1}	2.15×10^{0}	2.14×10^{0}	2.14×10^{0}	2.14×10^{0}	2.14×10^{0}	2.14×10^{0}	2.14×10^{0}	2.14×10^{0}	2.14×10^{0}	2.14×10^{0}
2.4×10^{1}	2.65×10^{0}	2.65×10^{0}	2.65×10^{0}	2.64×10^{0}	2.64×10^{0}	2.64×10^{0}	2.64×10^{0}	2.64×10^{0}	2.64×10^{0}	2.64×10^{0}
4×10^{1}	3.14×10^{0}	3.14×10^{0}	3.14×10^{0}	3.14×10^{0}	3.14×10^{0}	3.14×10^{0}	3.14×10^{0}	3.14×10^{0}	3.14×10^{0}	3.14×10^{0}
8×10^{1}	3.82×10^{0}	3.82×10^{0}	3.82×10^{0}	3.82×10^{0}	3.82×10^{0}	3.82×10^{0}	3.82×10^{0}	3.82×10^{0}	3.82×10^{0}	3.82×10^{0}
1.4×10^{2}	4.37×10^{0}	4.37×10^{0}	4.37×10^{0}	4.37×10^{0}	4.37×10^{0}	4.37×10^{0}	4.37×10^{0}	4.37×10^{0}	4.37×10^{0}	4.37×10^{0}
2.4×10^{2}	4.91×10^{0}	4.91×10^{0}	4.91×10^{0}	4.91×10^{0}	4.91×10^{0}	4.91×10^{0}	4.91×10^{0}	4.91×10^{0}	4.91×10^{0}	4.91×10^{0}
4×10^{2}	5.42×10^{0}	5.42×10^{0}	5.42×10^{0}	5.42×10^{0}	5.42×10^{0}	5.42×10^{0}	5.42×10^{0}	5.42×10^{0}	5.42×10^{0}	5.42×10^{0}

Annex 6.1 Values of Stallman's function $W(r_r^2,u)$ for bounded confined and unconfined aquifers

u	1/u	$r_r = 1.0$	1.1	1.2	1.3	1.4	1.5	1.6	1.7	1.8	1.9	2.0	2.2	2.4	2.6	2.8	3.0	3.3	3.6	4.0	4.5
1(−6)	1.00(6)	13.23	13.05	12.87	12.70	12.54	12.42	12.30	12.17	12.07	11.96	11.05	11.67	11.48	11.33	11.18	11.04	10.95	10.67	10.47	10.24
2(−6)	5.00(5)	12.54	12.36	12.17	12.01	11.88	11.73	11.61	11.48	11.37	11.26	11.16	10.97	10.80	10.63	10.50	10.35	10.24	9.98	9.77	9.53
4(−6)	2.50(5)	11.85	11.67	11.48	11.32	11.84	11.04	10.94	10.80	10.67	10.56	10.47	10.27	10.10	9.94	9.80	9.65	9.33	9.29	9.08	8.72
6(−6)	1.66(5)	11.45	11.26	11.08	10.93	10.75	10.64	10.53	10.40	10.27	10.15	10.06	9.87	9.68	9.53	9.39	9.25	9.15	8.88	8.67	8.45
8(−6)	1.25(5)	11.16	10.97	10.80	10.64	10.47	10.35	10.22	10.10	9.98	9.87	9.77	9.57	9.41	9.25	9.09	8.96	8.87	8.58	8.37	8.16
1(−5)	1.00(5)	10.93	10.75	10.56	10.41	10.24	10.12	9.98	9.87	9.77	9.65	9.55	9.36	9.18	9.02	8.88	8.74	8.64	8.37	8.16	7.94
2(−5)	5.00(4)	10.24	10.06	9.87	9.71	9.57	9.43	9.30	9.17	9.06	8.96	8.86	8.66	8.59	8.23	8.19	8.04	7.94	7.68	7.47	7.23
4(−5)	2.50(4)	9.549	9.367	9.178	9.019	8.882	8.739	8.633	8.494	8.371	8.262	8.163	7.965	7.800	7.640	7.502	7.353	7.024	6.985	6.777	6.426
6(−5)	1.66(4)	9.144	8.954	8.784	8.633	8.451	8.333	8.228	8.103	7.965	7.845	7.758	7.569	7.381	7.225	6.086	6.947	6.858	6.580	6.372	6.149
8(−5)	1.25(4)	8.856	8.664	8.594	8.333	8.163	8.045	7.915	7.800	7.678	7.569	7.470	7.272	7.107	6.947	6.793	6.660	6.567	6.283	6.069	5.862
1(−4)	1.00(4)	8.633	8.451	8.263	8.103	7.940	7.822	7.678	7.569	7.470	7.353	7.247	7.065	6.876	6.723	6.580	6.437	6.342	6.069	5.862	5.639
2(−4)	5.00(3)	7.940	7.758	7.569	7.410	7.272	7.129	7.004	6.876	6.762	6.660	6.554	6.362	6.193	6.032	5.944	5.745	5.639	5.378	5.171	4.935
4(−4)	2.50(3)	7.247	7.065	6.876	6.717	6.580	6.437	6.331	6.193	6.069	5.960	5.862	5.664	5.500	5.340	5.202	5.053	4.726	4.687	4.481	4.131
6(−4)	1.66(3)	6.842	6.652	6.482	6.331	6.149	6.032	5.927	5.802	5.664	5.544	5.457	5.269	5.081	4.935	4.788	4.649	4.561	4.284	4.078	3.858
8(−4)	1.25(3)	6.554	6.362	6.193	6.032	5.862	5.745	5.614	5.500	5.378	5.269	5.171	4.973	4.809	4.649	4.496	4.364	4.272	3.990	3.778	3.574
1(−3)	1.00(3)	6.331	6.149	5.961	5.802	5.639	5.522	5.378	5.269	5.171	5.053	4.948	4.666	4.578	4.427	4.284	4.142	4.048	3.778	3.574	3.355
2(−3)	5.00(2)	5.639	5.457	5.269	5.110	4.973	4.831	4.706	4.578	4.465	4.364	4.259	4.068	3.900	3.736	3.605	3.458	3.355	3.098	2.896	2.669
4(−3)	2.50(2)	4.948	4.767	4.579	4.420	4.284	4.142	4.038	3.900	3.778	3.664	3.574	3.379	3.218	3.061	2.927	2.783	2.468	2.431	2.235	1.909
6(−3)	1.66(2)	4.545	4.356	4.187	4.038	3.858	3.742	3.637	3.514	3.379	3.261	3.176	2.992	2.810	2.269	2.527	2.395	2.311	2.050	1.860	1.659
8(−3)	1.25(2)	4.259	4.068	3.900	3.742	3.574	3.458	3.330	3.218	3.098	2.992	2.896	2.706	2.547	2.395	2.249	2.125	2.039	1.784	1.589	1.409
1(−2)	1.00(2)	4.038	3.858	3.671	3.514	3.355	3.239	3.098	2.992	2.896	2.783	2.681	2.506	2.327	2.184	2.050	1.919	1.832	1.589	1.409	1.223
2(−2)	5.00(1)	3.355	3.176	2.992	2.838	2.706	2.568	2.449	2.327	2.220	2.125	2.027	1.850	1.698	1.556	1.436	1.309	1.223	1.014	0.858	0.694
4(−2)	2.50(1)	2.681	2.507	2.327	2.178	2.050	1.919	1.823	1.698	1.589	1.524	1.409	1.242	1.110	0.985	0.881	0.774	0.560	0.536	0.420	0.256
6(−2)	1.66(1)	2.295	2.117	1.960	1.823	1.659	1.556	1.464	1.358	1.243	1.145	1.076	0.931	0.794	0.694	0.598	0.514	0.464	0.322	0.235	0.158
8(−2)	1.25(1)	2.027	1.850	1.698	1.556	1.409	1.309	1.202	1.109	1.014	0.931	0.858	0.719	0.611	0.514	0.428	0.360	0.316	0.202	0.135	0.086
1(−1)	1.00(1)	1.822	1.659	1.494	1.358	1.223	1.122	1.014	0.931	0.858	0.774	0.702	0.584	0.473	0.392	0.322	0.260	0.223	0.135	0.086	0.049
2(−1)	5.00	1.222	1.076	0.930	0.815	0.719	0.625	0.548	0.473	0.411	0.360	0.311	0.231	0.172	0.126	0.091	0.065	0.049	0.022	0.010	0.003
4(−1)	2.50	0.702	0.585	0.473	0.388	0.322	0.260	0.219	0.172	0.135	0.108	0.086	0.052	0.032	0.019	0.011	0.006	0.001	0.001	0.000	
6(−1)	1.66	0.454	0.356	0.279	0.219	0.158	0.126	0.100	0.075	0.052	0.037	0.028	0.015	0.007	0.003	0.002	0.001	0.000			
8(−1)	1.25	0.311	0.231	0.172	0.125	0.086	0.065	0.045	0.032	0.022	0.014	0.010	0.004	0.002	0.001	0.000					
1	1.00	0.219	0.158	0.108	0.075	0.049	0.035	0.021	0.015	0.010	0.006	0.003	0.001	0.000							
2	5.00(−1)	4.89(−2)	2.84(−2)	1.48(−2)	0.78(−2)	0.43(−2)	0.21(−2)	0.10(−2)	0.04(−2)	0.02(−2)	0.01(−2)	0.00(−2)									
4	2.50(−1)	3.77(−3)	1.45(−3)	0.48(−3)	0.14(−3)	0.04(−3)	0.01(−3)	0.00(−3)													
6	1.66(−1)	3.60(−4)	0.87(−4)	0.19(−4)	0.04(−4)	0.00(−4)															
8	1.25(−1)	3.77(−5)	0.58(−5)	0.00(−5)			$W(r_r^2,u) = 0$														

Annex 6.1 (cont.)

u	1/u	$r_r = 5.0$	5.5	6.0	7.0	8.0	9.0	10.0	15	20	25	30	35	40	45	50	55	60	70	80	90	100
1(−6)	1.00(6)	10.02	9.84	9.65	9.34	9.08	8.86	8.63	7.82	7.25	6.80	6.44	6.15	5.86	5.64	5.42	5.23	5.05	4.75	4.48	4.26	4.04
2(−6)	5.00(5)	9.32	9.14	8.96	8.65	8.37	8.16	7.94	7.13	6.55	6.10	5.74	5.44	5.17	4.93	4.73	4.54	4.36	4.06	3.78	3.57	3.35
4(−6)	2.50(5)	8.63	8.45	8.26	7.96	7.68	7.45	7.25	6.43	5.86	5.42	5.05	4.75	4.48	4.13	4.04	3.86	3.67	3.38	3.10	2.88	2.68
6(−6)	1.66(5)	8.23	8.05	7.86	7.55	7.28	7.05	6.84	6.03	5.46	5.01	4.65	4.34	4.08	3.86	3.64	3.46	3.28	2.97	2.72	2.50	2.29
8(−6)	1.25(5)	7.94	7.75	7.75	7.27	7.00	6.76	6.55	5.74	5.17	4.73	4.36	4.05	3.78	3.57	3.35	3.18	2.99	2.70	2.45	2.22	2.03
1(−5)	1.00(5)	7.72	7.53	7.35	7.04	6.78	6.55	6.33	5.52	4.95	4.50	4.14	3.86	3.57	3.35	3.14	2.96	2.78	2.49	2.23	2.03	1.82
2(−5)	5.00(4)	7.02	6.83	6.66	6.35	6.07	5.86	5.64	4.83	4.26	3.82	3.46	3.16	2.90	2.67	2.47	2.29	2.12	1.84	1.59	1.41	1.22
4(−5)	2.50(4)	6.331	6.149	5.960	5.665	5.378	5.155	4.948	4.142	3.574	3.136	2.783	2.487	2.235	1.909	1.823	1.659	1.494	1.243	1.014	0.847	0.702
6(−5)	1.66(4)	5.927	5.745	5.567	5.251	4.986	4.756	4.545	3.742	3.176	2.743	2.395	2.105	1.860	1.659	1.464	1.309	1.168	0.917	0.728	0.578	0.454
8(−5)	1.25(4)	5.639	5.457	5.269	4.973	4.706	4.465	4.259	3.458	2.896	2.468	2.125	1.841	1.589	1.409	1.223	1.076	0.931	0.719	0.548	0.411	0.311
1(−4)	1.00(4)	5.417	5.235	5.053	4.746	4.481	4.259	4.038	3.239	2.681	2.251	1.919	1.659	1.409	1.223	1.044	0.906	0.774	0.572	0.419	0.311	0.219
2(−4)	5.00(3)	4.726	4.536	4.363	4.058	3.778	3.574	3.355	2.568	2.027	1.624	1.309	1.060	0.858	0.694	0.560	0.450	0.360	0.227	0.135	0.086	0.048
4(−4)	2.50(3)	4.038	3.858	3.671	3.379	3.098	2.881	2.681	1.919	1.409	1.044	0.774	0.572	0.420	0.256	0.219	0.158	0.108	0.052	0.022	0.009	0.004
6(−4)	1.66(3)	3.637	3.458	3.283	2.975	2.717	2.497	2.295	1.556	1.076	0.746	0.514	0.349	0.235	0.158	0.100	0.065	0.039	0.015	0.004	0.001	0.000
8(−4)	1.25(3)	3.355	3.176	2.992	2.705	2.449	2.220	2.027	1.309	0.858	0.560	0.360	0.227	0.135	0.086	0.049	0.028	0.015	0.004	0.001	0.000	
1(−3)	1.00(3)	3.136	2.959	2.783	2.487	2.235	2.027	1.823	1.122	0.702	0.432	0.260	0.158	0.086	0.049	0.025	0.013	0.006	0.001	0.000		
2(−3)	5.00(2)	2.468	2.287	2.125	1.841	1.589	1.409	1.223	0.625	0.311	0.147	0.065	0.027	0.010	0.003	0.001	0.000					
4(−3)	2.50(2)	1.823	1.659	1.494	1.243	1.014	0.847	0.702	0.260	0.086	0.025	0.006	0.001	0.000								
6(−3)	1.66(2)	1.464	1.309	1.168	0.917	0.728	0.578	0.454	0.126	0.028	0.005	0.001	0.000									
8(−3)	1.25(2)	1.223	1.076	0.931	0.719	0.548	0.411	0.311	0.065	0.010	0.001	0.000										
1(−2)	1.00(2)	1.044	0.906	0.774	0.572	0.419	0.311	0.219	0.035	0.003	0.000											
2(−2)	5.00(1)	0.560	0.450	0.360	0.227	0.135	0.086	0.048	0.002	0.000												
4(−2)	2.50(1)	0.219	0.158	0.108	0.052	0.022	0.009	0.004	0.000													
6(−2)	1.66(1)	0.100	0.065	0.039	0.015	0.004	0.001	0.000														
8(−2)	1.25(1)	0.049	0.028	0.015	0.004	0.001	0.000															
1(−1)	1.00(1)	0.025	0.013	0.006	0.001	0.000																
2(−1)	5.00	0.001	0.000				$W(r_r^2, u) = 0$															
4(−1)	2.50	0.000																				

Annex 6.2 Values of Stallman's function $W_R(u,r_r)$ for confined or unconfined aquifers with one recharge boundary

u	1/u	r_r = 1.0	1.1	1.2	1.3	1.4	1.5	1.6	1.8	2.0	2.5	3.0	4.0	6.0	8.0	10	20	30	40	60	80	100
1(−6)	1.00(6)	0.0	0.18	0.37	0.53	0.67	0.81	0.92	1.18	1.38	1.83	2.19	2.77	3.58	4.15	4.60	5.99	6.80	7.37	8.18	8.77	9.19
2(−6)	5.00(5)																			8.18	8.77	9.19
4(−6)	2.50(5)																			8.18	8.75	9.17
6(−6)	1.66(5)																			8.17	8.73	9.16
8(−6)	1.25(5)																			8.17	8.71	9.13
1(−5)	1.00(5)	0.0	0.18	0.37	0.53	0.67	0.81	0.92	1.18	1.38	1.83	2.19	2.77	3.58	4.15	4.60	5.99	6.79	7.36	8.15	8.70	9.11
2(−5)	5.00(4)																5.98	6.78	7.34	8.12	8.65	9.02
4(−5)	2.50(4)																5.97	6.77	7.31	8.04	8.54	8.85
6(−5)	1.66(4)																5.97	6.75	7.28	7.97	8.41	8.69
8(−5)	1.25(4)																5.96	6.73	7.26	7.94	8.31	8.54
1(−4)	1.00(4)	0.0	0.18	0.37	0.53	0.67	0.81	0.92	1.18	1.38	1.83	2.19	2.77	3.58	4.15	4.60	5.95	6.71	7.22	7.86	8.21	8.41
2(−4)	5.00(3)													3.58	4.15	4.59	5.91	6.63	7.08	7.58	7.81	7.89
4(−4)	2.50(3)													3.57	4.15	4.57	5.84	6.47	6.83	7.14	7.22	7.24
6(−4)	1.66(3)													3.56	4.12	4.55	5.76	6.33	6.60	6.80	6.84	
8(−4)	1.25(3)													3.56	4.11	4.52	5.69	6.19	6.42	6.54		
1(−3)	1.00(3)	0.0	0.18	0.37	0.53	0.67	0.81	0.92	1.17	1.38	1.83	2.19	2.76	3.55	4.10	4.51	5.63	6.07	6.24	6.32		
2(−3)	5.00(2)				0.53	0.67	0.81	0.92	1.17	1.38	1.83	2.18	2.74	3.52	4.05	4.42	5.33	5.57	5.63			
4(−3)	2.50(2)				0.53	0.66	0.81	0.91	1.17	1.37	1.81	2.16	2.71	3.50	3.93	4.25	4.86	4.94				
6(−3)	1.66(2)				0.52	0.66	0.80	0.91	1.17	1.36	1.80	2.15	2.68	3.48	3.81	4.19	4.51					
8(−3)	1.25(2)				0.52	0.66	0.80	0.91	1.16	1.36	1.79	2.13	2.66	3.33	3.71	3.95	4.25					
1(−2)	1.00(2)	0.0	0.18	0.36	0.52	0.66	0.80	0.90	1.15	1.36	1.77	2.12	2.63	3.27	3.62	3.82						
2(−2)	5.00(1)		0.18	0.36	0.51	0.65	0.79	0.88	1.13	1.33	1.73	1.92	2.50	2.99	3.22	3.30						
4(−2)	2.50(1)		0.17	0.36	0.50	0.63	0.77	0.86	1.09	1.27	1.64	1.80	2.26	2.57	2.66							
6(−2)	1.66(1)		0.17	0.34	0.48	0.64	0.74	0.83	1.05	1.22	1.55	1.70	2.06	2.25	2.29							
8(−2)	1.25(1)		0.17	0.34	0.47	0.62	0.72	0.82	1.01	1.17	1.47	1.60	1.87	2.01			$W_R(u,r_r) = W(u)$					
1(−1)	1.00(1)	0.0	0.16	0.33	0.46	0.57	0.70	0.79	1.01	1.12	1.38	1.54	1.74									
2(−1)	5.00		0.14	0.29	0.41	0.50	0.60	0.67	0.81	0.91	1.07	1.16	1.21									
4(−1)	2.50		0.12	0.23	0.31	0.42	0.44	0.48	0.57	0.62	0.68	0.69										
6(−1)	1.66		0.098	0.17	0.23	0.30	0.33	0.35	0.40	0.43	0.45											
8(−1)	1.25		0.079	0.14	0.19	0.22	0.25	0.27	0.29	0.30												
1	1.00	0.0	0.061	0.11	0.14	0.17	0.18	0.20	0.21	0.22												
2	5.00(−1)		2.0(−2)	3.4(−2)	4.1(−2)	4.5(−2)	4.7(−2)	4.8(−2)	4.9(−2)													
4	2.50(−1)		2.3(−3)	3.3(−3)	3.6(−3)	3.7(−3)	3.8(−3)															
6	1.66(−1)		2.8(−4)	3.4(−4)	3.6(−4)																	
8	1.25(−1)		3.2(−5)																			

Annex 6.3 Values of Stallman's function $W_B(u,r_r)$ for confined or unconfined aquifers with one barrier boundary

u	1/u	$r_r = 1.0$	1.5	2.0	3.0	4.0	6.0	8.0	10	15	20	30	40	60	80	100
1(−6)	1.00(6)	26.5	25.6	24.1	24.3	23.7	22.9	22.3	21.9	21.0	20.5	19.7	19.2	18.3	17.7	17.3
2(−6)	5.00(5)	25.1	24.3	23.7	22.9	22.3	21.5	20.9	20.5	19.7	19.1	18.3	17.7	16.9	16.3	15.9
4(−6)	2.50(5)	23.7	22.9	23.3	21.5	20.9	20.1	19.5	19.1	18.3	17.6	16.9	16.3	15.5	14.9	14.5
6(−6)	1.66(5)	22.9	22.1	21.5	20.7	20.1	19.3	18.7	18.3	17.5	16.9	16.1	15.5	14.7	14.2	13.7
8(−6)	1.25(5)	22.3	21.5	20.9	20.1	19.5	18.7	18.2	17.7	16.9	16.3	15.5	14.9	14.1	13.6	13.2
1(−5)	1.00(5)	21.9	21.0	20.5	19.7	19.1	18.3	17.7	17.3	16.4	15.9	15.1	14.5	13.7	13.2	12.7
2(−5)	5.00(4)	20.5	19.7	19.1	18.3	17.7	16.9	16.3	15.9	15.3	14.5	13.9	13.1	12.4	11.8	11.5
4(−5)	2.50(4)	19.1	18.3	17.7	16.9	16.3	15.5	14.9	14.5	13.7	13.1	12.3	11.8	11.0	10.6	10.3
6(−5)	1.66(4)	18.3	17.5	16.9	16.1	15.5	14.7	14.1	13.7	12.9	12.3	11.5	11.0	10.3	9.87	9.70
8(−5)	1.25(4)	17.7	16.9	16.3	15.5	14.9	14.1	13.6	13.1	12.3	11.8	11.0	10.4	9.70	9.40	9.17
1(−4)	1.00(4)	17.3	16.5	15.9	15.1	14.5	13.7	13.1	12.7	11.9	11.3	10.6	10.0	9.41	9.05	8.85
2(−4)	5.00(3)	15.9	15.1	14.5	13.7	13.1	12.3	11.7	11.3	10.5	9.97	9.25	8.80	8.30	8.07	7.99
4(−4)	2.50(3)	14.5	13.7	13.1	12.3	11.7	10.9	10.3	9.93	9.17	8.66	8.02	7.67	7.35	7.27	6.85
6(−4)	1.66(3)	13.7	12.9	12.3	11.5	10.9	10.1	9.56	9.14	8.40	7.92	7.36	7.08	6.88	6.84	
8(−4)	1.25(3)	13.1	12.3	11.7	10.9	10.3	9.55	9.00	8.58	7.86	7.41	6.91	6.69	6.57		
1(−3)	1.00(3)	12.7	11.8	11.3	10.5	9.90	9.11	8.57	8.15	7.45	7.03	6.59	6.42	6.34		
2(−3)	5.00(2)	11.3	10.6	9.90	9.09	8.53	7.76	7.23	6.86	6.26	5.95	5.70	5.65			
4(−3)	2.50(2)	9.90	9.09	8.52	7.73	7.18	6.44	5.96	5.65	5.21	5.03	4.55				
6(−3)	1.66(2)	9.09	8.29	7.72	6.94	6.40	5.71	5.27	5.08	4.67	4.57					
8(−3)	1.25(2)	8.52	7.72	7.15	6.38	5.85	5.19	4.81	4.57	4.32	4.27					
1(−2)	1.00(2)	8.08	7.28	6.72	5.96	5.45	4.81	4.46	4.26	4.07	4.04					
2(−2)	5.00(1)	6.71	5.92	5.38	4.66	4.21	3.71	3.49	3.40	3.36						
4(−2)	2.50(1)	5.36	4.60	4.09	3.45	3.10	2.79	2.70	2.68							
6(−2)	1.66(1)	4.59	4.24	3.76	2.81	2.53	2.33	3.30								
8(−2)	1.25(1)	4.05	3.34	2.88	2.31	2.16	2.04									
1(−1)	1.00(1)	3.64	2.94	2.52	2.08	1.91	1.83									
2(−1)	5.00	2.44	1.85	1.53	1.29	1.23										
4(−1)	2.50	1.40	0.962	0.788	0.718											
6(−1)	1.66	0.908	0.580	0.482	0.455											
8(−1)	1.25	0.622	0.376	0.321												
1	1.00	0.438	0.254	0.222												
2	5.00(−1)	9.78(−2)	5.10(−2)	$W_B(u,r_r) = W(u)$												
4	2.50(−1)	7.54(−3)	3.78(−3)													

Annex 6.4 Corresponding values of r_r, u_p, $W(u_p,r_r)$ and $f(r_r)$ for confined or unconfined aquifers with one recharge boundary (after Hantush 1959)

r_r	u_p	$W(u_p,r_r)$	$f(r_r)$	r_r	u_p	$W(u_p,r_r)$	$f(r_r)$	r_r	u_p	$W(u_p,r_r)$	$f(r_r)$	r_r	u_p	$W(u_p,r_r)$	$f(r_r)$
1.0	1.000	0.000	1.179	5.0	0.134	1.553	1.667	10	0.0466	2.534	2.115	35	0.00582	4.576	3.109
1.1	0.909	0.070	1.183	5.2	0.127	1.604	1.688	11	0.0400	2.680	2.188	36	0.00554	4.624	3.134
1.2	0.830	0.135	1.188	5.4	0.120	1.653	1.710	12	0.0348	2.815	2.251	37	0.00528	4.671	3.155
1.3	0.761	0.195	1.194	5.6	0.114	1.703	1.731	13	0.0306	2.940	2.312	38	0.00505	4.717	3.178
1.4	0.702	0.252	1.203	5.8	0.108	1.750	1.752	14	0.0271	3.057	2.367	39	0.00483	4.761	3.199
1.5	0.649	0.306	1.214	6.0	0.102	1.796	1.770	15	0.0241	3.172	2.423	40	0.00462	4.805	3.221
1.6	0.603	0.357	1.223	6.2	0.0976	1.840	1.794	16	0.0218	3.271	2.472	41	0.00443	4.847	3.242
1.7	0.562	0.407	1.235	6.4	0.0930	1.988	1.814	17	0.0203	3.342	2.520	42	0.00424	4.889	3.262
1.8	0.525	0.456	1.247	6.6	0.0888	1.927	1.833	18	0.0719	3.462	2.564	43	0.00407	4.930	3.282
1.9	0.492	0.502	1.262	6.8	0.0848	1.969	1.852	19	0.0164	3.551	2.609	44	0.00391	4.969	3.301
2.0	0.462	0.548	1.273	7.0	0.0812	2.010	1.871	20	0.0150	3.637	2.647	45	0.00376	5.008	3.321
2.2	0.411	0.635	1.301	7.2	0.0777	2.050	1.889	21	0.0138	3.716	2.687	46	0.00362	5.046	3.339
2.4	0.368	0.717	1.329	7.4	0.0745	2.089	1.908	22	0.0128	3.793	2.725	47	0.00349	5.084	3.357
2.6	0.332	0.796	1.357	7.6	0.0715	2.127	1.925	23	0.0119	3.867	2.761	48	0.00336	5.120	3.375
2.8	0.301	0.872	1.385	7.8	0.0687	2.165	1.943	24	0.0111	3.938	2.796	49	0.00325	5.156	3.393
3.0	0.275	0.945	1.413	8.0	0.0661	2.202	1.960	25	0.0103	4.007	2.837	50	0.00313	5.191	3.410
3.2	0.252	1.016	1.435	8.2	0.0636	2.238	1.977	26	0.00966	4.072	2.862	55	0.00265	5.358	3.491
3.4	0.232	1.083	1.467	8.4	0.0613	2.273	1.994	27	0.00906	4.135	2.893	60	0.00228	5.510	3.565
3.6	0.214	1.149	1.493	8.6	0.0590	2.308	2.010	28	0.00852	4.196	2.923	65	0.00198	5.650	3.634
3.8	0.199	1.212	1.500	8.8	0.0570	2.342	2.026	29	0.00803	4.256	2.952	70	0.00174	5.781	3.697
4.0	0.185	1.273	1.545	9.0	0.0550	2.376	2.041	30	0.00757	4.313	2.980	75	0.00154	5.903	3.757
4.2	0.173	1.333	1.571	9.2	0.0531	2.408	2.057	31	0.00716	4.369	3.008	80	0.00137	6.017	3.812
4.4	0.162	1.390	1.597	9.4	0.0513	2.441	2.072	32	0.00678	4.423	3.034	85	0.00123	6.124	3.864
4.6	0.152	1.447	1.619	9.6	0.0497	2.472	2.087	33	0.00643	4.475	3.059	90	0.00111	6.226	3.913
4.8	0.142	1.500	1.642	9.8	0.0481	2.503	2.102	34	0.00611	4.526	3.085	95	0.00102	6.311	3.960
5.0	0.134	1.553	1.667	10.0	0.0466	2.534	2.115	35	0.00582	4.576	3.109	100	0.00092	6.412	4.004

Annex 6.5 Values of Vandenberg's function F(u,x/L) for leaky or confined strip aquifers (after Vandenberg 1976)

u	1/u	x/L = 0.00000	.01000	.03000	.05000	.07500	.10000	.16000	.20000	.30000	.40000
1(−6)	1.00(6)	563.19015	99.00498	32.34818	19.02459	12.36991	9.04837	5.32590	4.09365	2.46939	1.67580
2(−6)	5.00(5)	397.94308	99.00493	32.34818	19.02459	12.36991	9.04837	5.32590	4.09365	2.46939	1.67580
4(−6)	2.50(5)	281.09592	98 96429	32.34818	19.02459	12.36991	9.04837	5.32590	4.09365	2.46939	1.67580
6(−6)	1.66(5)	229.33081	98.61574	32.34818	19.02459	12.36991	9.04837	5.32590	4.09365	2.46939	1.67580
8(−6)	1.25(5)	198.47274	97.76306	32.34818	19.02459	12.36991	9.04837	5.32590	4.09365	2.46939	1.67580
1(−5)	1.00(5)	177.41419	96.47027	32.34818	19.02459	12.36991	9.04837	5.32590	4.09365	2.46939	1.67580
2(−5)	5.00(4)	125.15915	87.62051	32.34811	19.02459	12.36991	9.04837	5.32590	4.09365	2.46939	1.67580
4(−5)	2.50(4)	88.20977	72.65033	32.32164	19.02459	12.36991	9.04837	5.32590	4.09365	2.46939	1.67580
6(−5)	1.66(4)	71.84093	62.87501	32.14253	19.02449	12.36991	9.04837	5.32590	4.09365	2.46939	1.67580
8(−5)	1.25(4)	62.08336	56.08700	31.75802	19.02304	12.36991	9.04837	5.32590	4.09365	2.46939	1.67580
1(−4)	1.00(4)	55.42460	51.05697	31.21844	19.01645	12.36991	9.04837	5.32590	4.09365	2.46939	1.67580
2(−4)	5.00(3)	38.90221	37.30143	27.89496	18.77624	12.36755	9.04837	5.32590	4.09365	2.46939	1.67580
4(−4)	2.50(3)	27.22076	26.64460	22.72213	17.48303	12.26315	9.04430	5.32590	4.09365	2.46939	1.67580
6(−4)	1.66(3)	22.04676	21.73137	19.46934	16.04745	11.96499	9.00947	5.32587	4.09365	2.46939	1.67580
8(−4)	1.25(3)	18.96307	18.75770	17.24494	14.80062	11.55981	8.92426	5.32550	4.09365	2.46939	1.67580
1(−3)	1.00(3)	16.85908	16.71194	15.61039	13.75650	11.12368	8.79510	5.32373	4.09361	2.46939	1.67580
2(−3)	5.00(2)	11.64088	11.58881	11.18598	10.44841	9.23116	7.91145	5.25469	4.08584	2.46939	1.67580
4(−3)	2.50(2)	7.95628	7.93793	7.79364	7.51769	7.02347	6.41876	4.86697	3.96731	2.46675	1.67578
6(−3)	1.66(2)	6.32731	6.31738	6.23877	6.08619	5.80511	5.44599	4.42861	3.75567	2.44893	1.67515
8(−3)	1.25(2)	5.35823	5.35180	5.30085	5.20123	5.01505	4.77184	4.04519	3.52749	2.41075	1.67191
1(−2)	1.00(2)	4.69822	4.69365	4.65731	4.58594	4.45141	4.27328	3.72312	3.31216	2.35727	1.66420
2(−2)	5.00(1)	3.06895	3.06737	3.05479	3.02986	2.98205	2.91698	2.70161	2.52366	2.02983	1.56373
4(−2)	2.50(1)	1.93304	1.93251	1.92825	1.91979	1.90341	1.88081	1.80325	1.73563	1.52728	1.29246
6(−2)	1.66(1)	1.44013	1.43985	1.43763	1.43322	1.42464	1.41276	1.37146	1.33479	1.21749	1.07616
8(−2)	1.25(1)	1.15219	1.15202	1.15064	1.14789	1.14254	1.13510	1.10911	1.08581	1.00982	.91502
1(−1)	1.00(1)	.95962	.95950	.95855	.95666	.95298	.94786	.92989	.91369	.86023	.79208
2(−1)	5.00(0)	.50579	.50576	.50548	.50493	.50385	.50235	.49703	.49218	.47578	.45393
4(−1)	2.50(0)	.22687	.22686	.22679	.22666	.22639	.22601	.22466	.22343	.21922	.21347
6(−1)	1.66(0)	.12641	.12641	.12638	.12633	.12622	.12607	.12553	.12504	.12335	.12103
8(−1)	1.25(0)	.07752	.07752	.07751	.07748	.07743	.07736	.07710	.07686	.07605	.07492
1(0)	1.00(0)	.05025	.05025	.05025	.05023	.05020	.05016	.05003	.04990	.04946	.04885
2(0)	5.00(−1)	.00849	.00849	.00849	.00849	.00848	.00848	.00847	.00846	.00841	.00836
4(0)	2.50(−1)	.00049	.00049	.00049	.00049	.00049	.00049	.00049	.00049	.00049	.00048

u	1/u	x/L = .50000	.60000	.70000	.80000	.90000	1.00000	1.20000	1.60000	2.0000
1(–6)	1.00(6)	1.21306	.91468	.70941	.56166	.45174	.36788	.25099	.12618	.06767
2(–6)	5.00(5)									
4(–6)	2.50(5)									
6(–6)	1.66(5)									
8(–6)	1.25(5)									
1(–5)	1.00(5)									
2(–5)	5.00(4)									
4(–5)	2.50(4)									
6(–5)	1.66(4)									
8(–5)	1.25(4)									
1(–4)	1.00(4)									
2(–4)	5.00(3)									
4(–4)	2.50(3)									
6(–4)	1.66(3)									
8(–4)	1.25(3)									
1(–3)	1.00(3)									
2(–3)	5.00(2)									
4(–3)	2.50(2)	1.21306								
6(–3)	1.66(2)	1.21305								
8(–3)	1.25(2)	1.21291	.91468	.70941						
1(–2)	1.00(2)	1.21225	.91465	.70941	.56166					
2(–2)	5.00(1)	1.18862	.91026	.70875	.56158	.45174	.36788	.25099		
4(–2)	2.50(1)	1.06317	.85989	.69097	.55601	.45017	.36749	.25098		
6(–3)	1.66(1)	.92652	.78164	.65025	.53671	.44182	.36418	.25057	.12618	
8(–2)	1.25(1)	.81015	.70338	.60123	.50808	.42622	.35623	.24889	.12615	.06767
1(–1)	1.00(1)	.71461	.63315	.55239	.47593	.40620	.34442	.24540	.12599	.06766
2(–1)	5.00(0)	.42762	.39797	.36614	.33327	.30037	.26833	.20937	.12012	.06701
4(–1)	2.50(0)	.20632	.19797	.18859	.17840	.16762	.15645	.13376	.09149	.05836
6(–1)	1.66(0)	.11812	.11466	.11072	.10636	.10165	.09666	.08613	.06469	.04549
8(–1)	1.25(0)	.07349	.07179	.06984	.06765	.06527	.06271	.05720	.04543	.03405
1(0)	1.00(0)	.04808	.04715	.04608	.04488	.04356	.04213	.03902	.03215	.02519
2(0)	5.00(–1	.00828	.00819	.00809	.00797	.00784	.00770	.00737	.00661	.00575
4(0)	2.50(–1	.00048	.00048	.00048	.00047	.00047	.00046	.00045	.00043	.00039
6(0)	1.66(–1	.00004	.00004	.00004	.00004	.00004	.00004	.00004	.00003	.00003

Annex 8.1 Values of the function $f_s(\beta',b/D,d/D,a/D$ for partially penetrated aquifers (after Weeks 1969)

Each of the tables listed below may also be used for the situation where values of the bottom and top of the pumped well screen are reversed ($b_2 = d_1$, $d_2 = D-b_1$) by reading a corrected value of a/D from the table. (a/D) corrected = 1–(a/D) observed.
For example, the first table listed could also be used to determine f_s for a well screened from the top of the aquifer down to a depth equal to 90% of the aquifer thicknes, i.e. $\frac{b_2}{D} = \frac{d_1}{D}$. If the piezometer penetrated 20% of the aquifer thickness, i.e. a/D = 0.20, the value of f_s for a given β' value would be found from $(a/D)_{corrected} = 1-0.20 = 0.80$.

Table 1 Values of f_s for b/D = 1 and d/D = 0.90

	β'												
a/D	0.05	0.10	0.15	0.20	0.25	0.30	0.40	0.50	0.60	0.80	1.00	1.20	1.50
0.0	–4.828	–3.457	–2.674	–2.134	–1.732	–1.421	–0.972	–0.673	–0.468	–0.229	–0.113	–0.056	–0.020
0.10	–4.785	–3.415	–2.633	–2.095	–1.696	–1.387	–0.944	–0.650	–0.451	–0.219	–0.108	–0.053	–0.019
0.20	–4.651	–3.284	–2.506	–1.976	–1.585	–1.284	–0.860	–0.584	–0.400	–0.191	–0.093	–0.046	–0.016
0.30	–4.408	–3.048	–2.280	–1.763	–1.388	–1.104	–0.715	–0.471	–0.315	–0.145	–0.069	–0.034	–0.012
0.40	–4.020	–2.674	–1.925	–1.434	–1.086	–0.833	–0.503	–0.312	–0.198	–0.085	–0.039	–0.018	–0.006
0.50	–3.415	–2.095	–1.387	–0.944	–0.650	–0.451	–0.219	–0.108	–0.053	–0.013	–0.003	–0.001	0.000
0.60	–2.444	–1.185	–0.566	–0.225	0.035	0.067	0.138	0.135	0.111	0.063	0.033	0.017	0.006
0.70	–0.736	0.341	0.725	0.829	0.808	0.736	0.556	0.399	0.280	0.137	0.067	0.033	0.012
0.80	2.897	3.170	2.791	2.312	1.875	1.511	0.983	0.648	0.432	0.199	0.095	0.046	0.016
0.90	13.344	8.218	5.575	3.974	2.926	2.207	1.322	0.831	0.539	2.241	0.113	0.055	0.019
1.00	21.264	11.404	7.087	4.778	3.395	2.499	1.454	0.899	0.578	0.256	0.120	0.058	0.020

Table 2 Values of f_s for b/D = 1 and d/D = 0.80

	β'												
a/D	0.05	0.10	0.15	0.20	0.25	0.30	0.40	0.50	0.60	0.80	1.00	1.20	1.50
0.00	–4.785	–3.415	–2.633	–2.095	–1.696	–1.387	–0.944	–0.650	–0.451	–0.219	–0.108	–0.053	–0.019
0.10	–4.739	–3.371	–2.590	–2.055	–1.658	–1.352	–0.916	–0.628	–0.434	–0.210	–0.103	–0.051	–0.018
0.20	–4.597	–3.232	–2.457	–1.929	–1.542	–1.246	–0.829	–0.561	–0.383	–0.182	–0.089	–0.044	–0.015
0.30	–4.336	–2.979	–2.216	–1.705	–1.335	–1.059	–0.681	–0.448	–0.299	–0.138	–0.066	–0.032	–0.011
0.40	–3.912	–2.272	–1.834	–1.354	–1.019	–0.778	–0.467	–0.290	–0.184	–0.079	–0.036	–0.017	–0.006
0.50	–3.232	–1.929	–1.246	–0.829	–0.561	–0.383	–0.182	–0.089	–0.044	–0.011	–0.003	–0.001	0.000
0.60	–2.076	–0.877	–0.331	–0.057	0.079	0.142	0.168	0.145	0.114	0.062	0.032	0.016	0.006
0.70	–0.227	0.992	1.113	1.044	0.920	0.789	0.561	0.391	0.272	0.131	0.064	0.032	0.011
0.80	6.304	4.280	3.150	2.401	1.867	1.471	0.939	0.615	0.410	0.189	0.090	0.044	0.015
0.90	12.080	7.287	4.939	3.545	2.635	2.005	1.219	0.773	0.505	0.228	0.107	0.052	0.018
1.00	13.344	8.218	5.575	3.973	2.926	2.207	1.322	0.831	0.539	0.241	0.113	0.055	0.019

Annex 8.1 (cont.)

Table 3 Values of f_s for b/D = 1 and d/D = 0.70

	β′												
a/D	0.05	0.10	0.15	0.20	0.25	0.30	0.40	0.50	0.60	0.80	1.00	1.20	1.50
0.00	–4.710	–3.342	–2.562	–2.029	–1.634	–1.330	–0.897	–0.613	–0.423	–0.204	–0.100	–0.049	–0.017
0.10	–4.659	–3.293	–2.515	–1.985	–1.593	–1.293	–0.868	–0.591	–0.406	–0.195	–0.095	–0.047	–0.017
0.20	–4.500	–3.138	–2.368	–1.848	–1.468	–1.179	–0.778	–0.523	–0.355	–0.168	–0.082	–0.040	–0.014
0.30	–4.203	–2.853	–2.100	–1.601	–1.245	–0.981	–0.626	–0.410	–0.273	–0.126	–0.060	–0.029	–0.010
0.40	–3.705	–2.381	–1.666	–1.212	–0.902	–0.683	–0.408	–0.254	–0.162	–0.071	–0.033	–0.016	–0.005
0.50	–2.853	–1.601	–0.981	–0.626	–0.410	–0.273	–0.126	–0.060	–0.029	–0.007	–0.002	–0.000	0.000
0.60	–1.189	–0.230	0.100	0.218	0.251	0.248	0.206	0.157	0.115	0.059	0.030	0.015	0.005
0.70	3.064	2.155	1.638	1.286	1.028	0.830	0.553	0.374	0.255	0.122	0.059	0.029	0.010
0.80	7.239	4.463	3.104	2.289	1.745	1.359	0.859	0.561	0.374	0.173	0.083	0.040	0.014
0.90	8.651	5.592	3.958	2.925	2.220	1.716	1.067	0.687	0.453	0.206	0.098	0.048	0.017
1.00	9.019	5.915	4.223	3.134	2.382	1.840	1.140	0.731	0.481	0.218	0.103	0.050	0.017

Table 4 Values of f_s for b/D = 1 and d/D = 0.60

	β′												
a/D	0.05	0.10	0.15	0.20	0.25	0.30	0.40	0.50	0.60	0.80	1.00	1.20	1.50
0.00	–4.597	–3.237	–2.457	–1.929	–1.542	–1.246	–0.829	–0.561	–0.383	–0.182	–0.089	–0.044	–0.015
0.10	–4.538	–3.175	–2.403	–1.880	–1.497	–1.206	–0.799	–0.538	–0.367	–0.174	–0.084	–0.041	–0.015
0.20	–4.348	–2.994	–2.233	–1.725	–1.358	–1.082	–0.705	–0.470	–0.318	–0.149	–0.072	–0.035	–0.012
0.30	–3.986	–2.650	–1.918	–1.442	–1.110	–0.868	–0.549	–0.358	–0.239	–0.110	–0.053	–0.026	–0.009
0.40	–3.336	–2.055	–1.394	–0.993	–0.731	–0.552	0.331	–0.208	–0.135	–0.060	–0.028	–0.014	–0.005
0.50	–2.055	0.993	–0.552	–0.331	–0.208	–0.135	–0.060	–0.028	–0.014	–0.003	0.001	–0.000	–0.000
0.60	1.196	0.854	0.658	0.524	0.424	0.347	0.236	0.163	0.113	0.055	0.027	0.013	0.005
0.70	4.424	2.679	1.847	1.358	1.037	0.811	0.518	0.342	0.231	0.108	0.052	0.026	0.009
0.80	5.634	3.670	2.622	1.958	1.502	1.174	0.745	0.488	0.326	0.152	0.073	0.035	0.012
0.90	6.154	4.140	3.026	2.295	1.777	1.397	0.890	0.582	0.388	0.179	0.086	0.042	0.015
1.00	6.304	4.280	3.150	2.401	1.867	1.471	0.939	0.615	0.410	0.189	0.090	0.044	0.015

Table 5 Values of f_s for b/D = 1 and d/D = 0.50

	β′												
a/D	0.05	0.10	0.15	0.20	0.25	0.30	0.40	0.50	0.60	0.80	1.00	1.20	1.50
0.00	–4.434	–3.075	–2.307	–1.791	–1.415	–1.131	–0.739	–0.493	–0.333	–0.156	–0.075	–0.037	–0.013
0.10	–4.360	–3.005	–2.243	–1.732	–1.364	–1.087	–0.707	–0.470	–0.317	–0.149	–0.072	–0.035	–0.012
0.20	–4.119	–2.777	–2.036	–1.549	–1.205	–0.951	–0.611	–0.403	–0.271	–0.127	–0.061	–0.030	–0.010
0.30	–3.626	–2.327	–1.642	–1.214	–0.924	–0.719	–0.453	–0.296	–0.198	–0.092	–0.044	–0.022	–0.008
0.40	–2.609	–1.486	–0.976	–0.691	–0.513	–0.392	–0.243	–0.157	–0.105	–0.048	–0.023	–0.011	–0.004
0.50	–0.000	–0.000	–0.000	–0.000	0.000	0.000	0.000	0.000	0.000	0.000	0.000	0.000	0.000
0.60	2.609	1.486	0.976	0.691	0.513	0.392	0.243	0.157	0.105	0.048	0.023	0.011	0.004
0.70	3.626	2.327	1.642	1.214	0.924	0.719	0.453	0.296	0.198	0.092	0.044	0.022	0.008
0.80	4.119	2.777	2.036	1.549	1.205	0.951	0.611	0.403	0.271	0.127	0.061	0.030	0.010
0.90	4.360	3.005	2.243	1.732	1.364	1.087	0.707	0.470	0.317	0.149	0.072	0.035	0.012
1.00	4.434	3.075	2.307	1.791	1.415	1.131	0.739	0.493	0.333	0.156	0.075	0.037	0.013

Annex 8.1 (cont.)

Table 6 Values of f_s for b/D = 1 and d/D = 0.40

	β′												
a/D	0.05	0.10	0.15	0.20	0.25	0.30	0.40	0.50	0.60	0.80	1.00	1.20	1.50
0.00	–4.203	–2.853	–2.100	–1.601	–1.245	–0.981	–0.626	–0.410	–0.273	–0.126	–0.060	–0.029	–0.010
0.10	–4.102	–2.760	–2.107	–1.530	–1.185	–0.931	–0.593	–0.388	–0.259	–0.120	–0.057	–0.028	–0.010
0.20	–3.756	–2.447	–1.748	–1.305	–1.002	–0.783	–0.497	–0.325	–0.218	–0.101	–0.048	–0.024	–0.008
0.30	–2.949	–1.786	–1.231	–0.905	–0.691	–0.541	–0.345	–0.228	–0.154	–0.072	–0.035	–0.017	–0.006
0.40	–0.798	–0.569	–0.439	–0.349	–0.282	–0.231	–0.157	–0.108	–0.075	–0.037	–0.018	–0.009	–0.003
0.50	1.370	0.662	0.368	0.220	0.135	0.090	0.040	0.019	0.009	0.002	0.001	0.000	0.000
0.60	2.224	1.370	0.929	0.662	0.488	0.368	0.220	0.139	0.090	0.040	0.019	0.009	0.003
0.70	2.657	1.767	1.279	0.961	0.740	0.578	0.366	0.239	0.159	0.074	0.035	0.017	0.006
0.80	2.899	1.996	1.489	1.150	0.905	0.722	0.470	0.313	0.212	0.100	0.048	0.024	0.008
0.90	3.025	2.117	1.602	1.253	0.998	0.804	0.532	0.359	0.244	0.116	0.056	0.028	0.010
1.00	3.064	2.155	1.638	1.286	1.028	0.830	0.553	0.374	0.255	0.122	0.059	0.029	0.010

Table 7 Values of f_s for b/D = 1 and d/D = 0.20

	β′												
a/D	0.05	0.10	0.15	0.20	0.25	0.30	0.40	0.50	0.60	0.80	1.00	1.20	1.50
0.00	–3.336	–2.055	–1.394	–0.993	–0.731	–0.552	–0.331	–0.208	–0.135	–0.060	–0.028	–0.014	–0.005
0.10	–3.020	–1.822	–1.235	–0.886	–0.659	–0.501	–0.305	–0.193	–0.126	–0.057	–0.027	–0.013	–0.005
0.20	–1.576	–1.070	–0.788	–0.600	–0.467	–0.368	–0.235	–0.154	–0.102	–0.047	–0.023	–0.011	–0.004
0.30	–0.057	–0.248	–0.278	–0.261	–0.230	–0.197	–0.140	–0.098	–0.068	–0.033	–0.016	–0.008	–0.003
0.40	0.519	0.219	0.083	0.014	–0.020	–0.036	–0.042	–0.036	–0.028	–0.015	–0.008	–0.004	–0.001
0.50	0.808	0.482	0.311	0.207	0.140	0.096	0.046	0.022	0.011	0.003	0.001	0.000	0.000
0.60	0.978	0.643	0.458	0.338	0.255	0.194	0.117	0.072	0.046	0.020	0.009	0.004	0.001
0.70	1.084	0.745	0.554	0.426	0.334	0.265	0.170	0.112	0.075	0.034	0.016	0.008	0.003
0.80	1.149	0.808	0.614	0.482	0.385	0.311	0.207	0.140	0.096	0.046	0.022	0.011	0.004
0.90	1.185	0.843	0.647	0.514	0.415	0.338	0.229	0.157	0.109	0.053	0.026	0.013	0.005
1.00	1.196	0.854	0.658	0.524	0.424	0.347	0.236	0.163	0.113	0.055	0.027	0.013	0.005

Table 8 Values of f_s for b/D = 0.90 and d/D = 0.80

	β′												
a/D	0.05	0.10	0.15	0.20	0.25	0.30	0.40	0.50	0.60	0.80	1.00	1.20	1.50
0.00	–4.743	–3.373	–2.592	–2.057	–1.660	–1.354	–0.916	–0.628	–0.434	–0.210	–0.103	–0.051	–0.018
0.10	–4.694	–3.326	–2.547	–2.015	–1.621	–1.318	–0.887	–0.606	–0.417	–0.201	–0.098	–0.048	–0.017
0.20	–4.542	–3.179	–2.407	–1.883	–1.499	–1.207	–0.799	–0.538	–0.366	–0.174	–0.084	–0.041	–0.015
0.30	–4.263	–2.910	–2.151	–1.646	–1.283	–1.013	–0.648	–0.425	–0.283	–0.131	–0.062	–0.030	–0.011
0.40	–3.803	–2.470	–1.742	–1.274	–0.952	–0.722	–0.431	–0.267	–0.170	–0.074	–0.034	–0.016	–0.006
0.50	–3.048	–1.763	–1.104	–0.715	–0.471	–0.315	–0.145	–0.069	–0.034	–0.008	–0.002	–0.001	0.000
0.60	–1.708	–0.569	–0.096	0.111	0.193	0.218	0.198	0.156	0.116	0.061	0.031	0.015	0.006
0.70	1.189	1.644	1.500	1.258	1.032	0.843	0.566	0.384	0.263	0.125	0.061	0.030	0.011
0.80	9.712	5.389	3.509	2.491	1.859	1.431	0.895	0.582	0.387	0.179	0.086	0.042	0.015
0.90	10.816	6.356	4.303	3.117	2.344	1.803	1.115	0.716	0.471	0.214	0.101	0.049	0.017
1.00	5.425	5.032	4.064	3.168	2.457	1.915	1.190	0.763	0.500	0.226	0.107	0.052	0.018

Annex 8.1 (cont.)

Table 9 Values of f_s for b/D = 0.90 and d/D = 0.70

	β′												
a/D	0.05	0.10	0.15	0.20	0.25	0.30	0.40	0.50	0.60	0.80	1.00	1.20	1.50
0.00	–4.651	–3.284	–2.506	–1.976	–1.585	–1.284	–0.860	–0.584	–0.400	–0.191	–0.093	–0.046	–0.016
0.10	–4.597	–3.232	–2.457	–1.929	–1.542	–1.246	–0.829	–0.561	–0.383	–0.182	–0.089	–0.044	–0.015
0.20	–4.424	–3.065	–2.299	–1.784	–1.409	–1.127	–0.737	–0.492	–0.333	–0.157	–0.076	–0.037	–0.013
0.30	–4.100	–2.755	–2.010	–1.520	–1.173	–0.919	–0.582	–0.379	–0.252	–0.116	–0.056	–0.027	–0.009
0.40	–3.547	–2.235	–1.536	–1.101	–0.810	–0.609	–0.361	–0.224	–0.144	–0.064	–0.030	–0.014	0.005
0.50	–2.572	–1.354	–0.778	–0.467	–0.290	–0.184	–0.079	–0.036	–0.017	–0.004	–0.001	–0.000	0.000
0.60	–0.562	0.248	0.433	0.439	0.395	0.339	0.240	0.168	0.117	0.057	0.028	0.014	0.005
0.70	4.965	3.061	2.094	1.515	1.138	0.878	0.551	0.362	0.243	0.114	0.055	0.027	0.009
0.80	9.410	5.109	3.260	2.277	1.680	1.283	0.796	0.517	0.344	0.160	0.076	0.037	0.013
0.90	6.304	4.280	3.150	2.401	1.867	1.471	0.939	0.615	0.410	0.189	0.090	0.044	0.015
1.00	2.897	3.170	2.791	2.312	1.875	1.511	0.983	0.648	0.432	0.199	0.095	0.046	0.016

Table 10 Values of f_s for b/D = 0.90 and d/D = 0.60

	β′												
a/D	0.05	0.10	0.15	0.20	0.25	0.30	0.40	0.50	0.60	0.80	1.00	1.20	1.50
0.00	–4.520	–3.157	–2.384	–1.861	–1.478	–1.187	–0.782	–0.524	–0.355	–0.167	–0.081	–0.039	–0.014
0.10	–4.455	–3.095	–2.326	–1.808	–1.431	–1.145	–0.750	–0.501	–0.338	–0.159	–0.077	–0.037	–0.013
0.20	–4.247	–2.897	–2.142	–1.641	–1.282	–1.015	–0.654	–0.432	–0.290	–0.136	–0.065	–0.032	–0.011
0.30	–3.845	–2.517	–1.797	–1.335	–1.017	–0.789	–0.494	–0.321	–0.213	–0.099	–0.047	–0.023	–0.008
0.40	–3.108	–1.848	1.217	–0.847	–0.613	–0.458	–0.273	–0.173	–0.114	–0.052	–0.025	–0.012	–0.004
0.50	–1.601	–0.626	–0.273	–0.126	–0.060	–0.029	–0.007	–0.002	–0.000	0.000	0.000	0.000	0.000
0.60	2.410	1.533	1.066	0.774	0.577	0.440	0.269	0.172	0.113	0.052	0.025	0.012	0.004
0.70	6.144	3.458	2.220	1.534	1.113	0.836	0.506	0.324	0.214	0.099	0.047	0.023	0.008
0.80	6.547	3.837	2.566	1.840	1.378	1.062	0.666	0.435	0.291	0.136	0.065	0.032	0.011
0.90	3.757	2.780	2.176	1.735	1.395	1.127	0.746	0.500	0.338	0.159	0.077	0.037	0.013
1.00	1.318	1.905	1.838	1.609	1.358	1.129	0.767	0.520	0.354	0.167	0.081	0.039	0.014

Table 11 Values of f_s for b/D = 0.90 and d/D = 0.50

	β′												
a/D	0.05	0.10	0.15	0.20	0.25	0.30	0.40	0.50	0.60	0.80	1.00	1.20	1.50
0.00	–4.336	–2.979	–2.216	–1.705	–1.335	–1.059	–0.681	–0.448	–0.299	–0.138	–0.066	–0.032	–0.011
0.10	–4.254	–2.902	–2.145	–1.642	–1.280	–1.012	–0.648	–0.425	–0.284	–0.131	–0.063	–0.030	–0.011
0.20	–3.986	–2.650	–1.918	–1.442	–1.110	–0.868	–0.549	–0.358	–0.239	–0.110	–0.053	–0.026	–0.009
0.30	–3.430	–0.146	–1.482	–1.076	–0.809	–0.622	–0.388	–0.253	–0.169	–0.079	–0.038	–0.019	–0.007
0.40	–2.256	–1.189	–0.739	–0.506	–0.369	–2.282	–0.177	–0.118	–0.081	–0.039	–0.019	–0.010	–0.003
0.50	0.854	0.524	0.347	0.236	0.163	0.113	0.055	0.027	0.013	0.003	0.001	0.000	0.000
0.60	3.872	2.154	1.362	0.920	0.650	0.473	0.269	0.163	0.103	0.045	0.021	0.010	0.003
0.70	4.716	2.823	1.871	1.310	0.953	0.714	0.428	0.271	0.177	0.081	0.038	0.019	0.007
0.80	4.424	2.679	1.847	1.358	1.037	0.811	0.518	0.342	0.231	0.108	0.052	0.026	0.009
0.90	2.114	1.701	1.410	1.172	0.973	0.807	0.554	0.380	0.262	0.125	0.061	0.030	0.011
1.00	0.227	0.992	1.113	1.044	0.920	0.789	0.561	0.391	0.272	0.131	0.064	0.032	0.011

Annex 8.1 (cont.)

Table 12 Values of f_s for b/D = 0.90 and d/D = 0.40

a/D	β′												
	0.05	0.10	0.15	0.20	0.25	0.30	0.40	0.50	0.60	0.80	1.00	1.20	1.50
0.00	−4.078	−2.732	−1.985	−1.494	−1.147	−0.893	−0.557	−0.357	−0.234	−0.105	−0.050	−0.024	−0.008
0.10	−3.966	−2.629	−1.894	−1.417	−1.083	−0.840	−0.523	−0.336	−0.220	−0.100	−0.047	−0.023	−0.008
0.20	−3.577	−2.279	−1.596	−1.171	−0.885	−0.683	−0.424	−0.274	−0.181	−0.083	−0.040	−0.019	−0.007
0.30	−2.658	−1.533	−1.021	−0.734	−0.552	−0.428	−0.272	−0.180	−0.122	−0.058	−0.028	−0.014	−0.005
0.40	−0.153	−0.148	−0.141	−0.132	−0.122	−0.111	−0.088	−0.068	−0.051	−0.027	−0.014	−0.007	−0.003
0.50	2.327	1.214	0.719	0.453	0.296	0.198	0.092	0.044	0.022	0.005	0.001	0.000	0.000
0.60	3.158	1.881	1.228	0.840	0.592	0.428	0.237	0.139	0.086	0.036	0.016	0.008	0.003
0.70	3.336	2.052	1.389	0.988	0.726	0.547	0.328	0.207	0.135	0.061	0.029	0.014	0.005
0.80	2.899	1.761	1.228	0.917	0.711	0.564	0.368	0.247	0.168	0.080	0.039	0.019	0.007
0.90	0.961	0.896	0.807	0.709	0.612	0.523	0.374	0.264	0.185	0.091	0.045	0.022	0.008
1.00	−0.575	0.305	0.548	0.588	0.555	0.497	0.373	0.269	0.191	0.095	0.047	0.023	0.008

Table 13 Values of f_s for b/D = 0.90 and d/D = 0.30

a/D	β′												
	0.05	0.10	0.15	0.20	0.25	0.30	0.40	0.50	0.60	0.80	1.00	1.20	1.50
0.00	−3.705	−2.381	−1.666	−1.212	−0.902	−0.683	−0.408	−0.254	−0.162	−0.071	−0.033	−0.016	−0.005
0.10	−3.528	−2.227	−1.540	−1.113	−0.827	−0.627	−0.376	−0.235	−0.151	−0.067	−0.031	−0.015	−0.005
0.20	−2.844	−1.684	−1.134	−0.815	−0.608	−0.465	−0.286	−0.183	−0.120	−0.055	−0.026	−0.013	−0.004
0.30	−0.798	−0.569	−0.439	−0.349	−0.283	−0.231	−0.157	−0.108	−0.075	−0.037	−0.018	−0.009	−0.003
0.40	1.264	0.560	0.271	0.130	0.055	0.015	−0.019	−0.026	−0.024	−0.015	−0.008	−0.004	−0.002
0.50	1.996	1.150	0.722	0.470	0.313	0.212	0.100	0.048	0.024	0.006	0.001	0.000	0.000
0.60	2.260	1.388	0.927	0.643	0.457	0.331	0.181	0.104	0.063	0.025	0.011	0.005	0.002
0.70	2.224	1.370	0.929	0.662	0.488	0.368	0.220	0.139	0.090	0.040	0.019	0.009	0.003
0.80	1.767	1.041	0.719	0.539	0.421	0.338	0.225	0.154	0.106	0.051	0.025	0.012	0.004
0.90	0.106	0.277	0.328	0.330	0.309	0.279	0.213	0.157	0.113	0.057	0.029	0.014	0.005
1.00	−1.189	−0.230	0.100	0.218	0.251	0.248	0.206	0.157	0.115	0.059	0.030	0.015	0.005

Table 14 Values of f_s for b/D = 0.90 and d/D = 0.20

a/D	β′												
	0.05	0.10	0.15	0.20	0.25	0.30	0.40	0.50	0.60	0.80	1.00	1.20	1.50
0.00	−3.123	−1.854	−1.211	−0.830	−0.588	−0.428	−0.239	−0.141	−0.087	−0.036	−0.016	−0.008	−0.003
0.10	−2.768	−1.594	−1.035	−0.714	−0.511	−0.375	−0.213	−0.128	−0.080	−0.034	−0.015	−0.007	−0.002
0.20	−1.137	−0.754	−0.542	−0.404	−0.307	−0.237	−0.145	−0.092	−0.060	−0.027	−0.013	−0.006	−0.002
0.30	0.565	0.152	0.008	−0.046	−0.065	−0.068	−0.058	−0.044	−0.033	−0.017	−0.008	−0.004	−0.002
0.40	1.167	0.633	0.370	0.221	0.133	0.078	0.024	0.003	−0.004	−0.006	−0.004	−0.002	−0.001
0.50	1.411	0.851	0.554	0.372	0.253	0.174	0.083	0.041	0.020	0.005	0.001	0.000	0.000
0.60	1.467	0.904	0.605	0.419	0.296	0.213	0.114	0.063	0.037	0.014	0.006	0.003	0.001
0.70	1.344	0.802	0.530	0.369	0.266	0.197	0.115	0.071	0.045	0.020	0.009	0.004	0.002
0.80	0.899	0.471	0.303	0.221	0.173	0.140	0.096	0.068	0.048	0.024	0.012	0.006	0.002
0.90	−0.552	−0.211	−0.056	0.020	0.056	0.071	0.073	0.061	0.047	0.026	0.013	0.007	0.002
1.00	−1.670	−0.653	−0.260	−0.084	−0.000	0.039	0.062	0.057	0.046	0.026	0.014	0.007	0.003

Annex 8.1 (cont.)

Table 15 Values of f_s for b/D = 0.90 and d/D = 0.10

	β′												
a/D	0.05	0.10	0.15	0.20	0.25	0.30	0.40	0.50	0.60	0.80	1.00	1.20	1.50
0.00	−2.055	−0.993	−0.552	−0.331	−0.208	−0.135	−0.060	−0.028	−0.014	−0.003	−0.001	−0.000	−0.000
0.10	−1.070	−0.600	−0.368	−0.235	−0.154	−0.102	−0.047	−0.023	−0.011	−0.003	−0.001	−0.000	−0.000
0.20	0.219	0.014	−0.036	−0.042	−0.036	−0.028	−0.015	−0.008	−0.004	−0.001	−0.000	−0.000	−0.000
0.30	0.643	0.338	0.194	0.117	0.072	0.046	0.020	0.009	0.004	0.001	0.000	0.000	−0.000
0.40	0.808	0.482	0.311	0.207	0.140	0.096	0.046	0.022	0.011	0.003	0.001	0.000	−0.000
0.50	0.854	0.524	0.347	0.236	0.163	0.113	0.055	0.027	0.013	0.003	0.001	0.000	0.000
0.60	0.808	0.482	0.311	0.207	0.140	0.096	0.046	0.022	0.011	0.003	0.001	0.000	0.000
0.70	0.643	0.338	0.194	0.117	0.072	0.046	0.020	0.009	0.004	0.001	0.000	0.000	0.000
0.80	0.219	0.014	−0.036	−0.042	−0.036	−0.028	−0.015	−0.008	−0.004	−0.001	−0.000	−0.000	0.000
0.90	−1.070	−0.600	−0.368	−0.235	−0.154	−0.102	−0.047	−0.023	−0.011	−0.003	−0.001	−0.000	0.000
1.00	−2.054	−0.993	−0.552	−0.331	−0.208	−0.135	−0.060	−0.028	−0.014	−0.003	−0.001	−0.000	0.000

Table 16 Values of f_s for b/D = 0.80 and d/D = 0.70

	β′												
a/D	0.05	0.10	0.15	0.20	0.25	0.30	0.40	0.50	0.60	0.80	1.00	1.20	1.50
0.00	−4.560	−3.196	−2.421	−1.895	−1.509	−1.215	−0.803	−0.539	−0.366	−0.172	−0.083	−0.041	−0.014
0.10	−4.500	−3.137	−2.366	−1.844	−1.463	−1.174	−0.771	−0.516	−0.349	−0.164	−0.079	−0.039	−0.014
0.20	−4.306	−2.952	−2.192	−1.685	−1.320	−1.047	−0.676	−0.447	−0.300	−0.140	−0.067	−0.033	−0.012
0.30	−3.937	−2.601	−1.868	−1.393	−1.063	−0.825	−0.515	−0.334	−0.221	−0.102	−0.049	−0.024	−0.008
0.40	−3.292	−1.999	−1.330	−0.927	−0.668	−0.495	−0.290	−0.182	−0.119	−0.054	−0.026	−0.013	−0.004
0.50	−2.095	−0.944	−0.451	−0.219	−0.108	−0.053	−0.013	−0.003	−0.001	0.000	0.000	0.000	0.000
0.60	0.584	1.065	0.962	0.768	0.596	0.460	0.282	0.180	0.118	0.054	0.026	0.013	0.004
0.70	8.740	4.479	2.688	1.772	1.244	0.913	0.537	0.339	0.223	0.102	0.049	0.024	0.008
0.80	9.109	4.830	3.012	2.063	1.500	1.135	0.698	0.452	0.302	0.140	0.067	0.033	0.012
0.90	1.792	2.203	1.997	1.686	1.390	1.139	0.763	0.514	0.349	0.164	0.079	0.039	0.014
1.00	0.369	1.308	1.519	1.456	1.294	1.108	0.776	0.532	0.364	0.172	0.083	0.041	0.014

Table 17 Values of f_s for b/D = 0.80 and d/D = 0.60

	β′												
a/D	0.05	0.10	0.15	0.20	0.25	0.30	0.40	0.50	0.60	0.80	1.00	1.20	1.50
0.00	−4.408	−3.048	−2.280	−1.763	−1.388	−1.104	−0.715	−0.471	−0.315	−0.145	−0.069	−0.034	−0.012
0.10	−4.336	−2.979	−2.216	−1.705	−1.335	−1.059	−0.681	−0.448	−0.299	−0.138	−0.066	−0.032	−0.011
0.20	−4.100	−2.755	−2.010	−1.520	−1.173	−0.919	−0.582	−0.379	−0.252	−0.116	−0.056	−0.027	−0.009
0.30	−3.636	−2.321	−1.620	−1.180	−0.884	−0.677	−0.417	−0.269	−0.178	−0.083	−0.040	−0.020	−0.007
0.40	−2.761	−1.537	−0.954	−0.633	−0.444	−0.326	−0.194	−0.126	−0.085	−0.041	−0.020	−0.010	−0.004
0.50	−0.877	−0.057	0.142	0.168	0.145	0.114	0.062	0.032	0.016	0.004	0.001	0.000	0.000
0.60	4.468	2.585	1.647	1.105	0.769	0.551	0.304	0.180	0.112	0.048	0.022	0.011	0.004
0.70	8.622	4.365	2.581	1.672	1.154	0.833	0.475	0.293	0.190	0.086	0.040	0.020	0.007
0.80	4.965	3.061	2.094	1.515	1.138	0.878	0.551	0.362	0.243	0.114	0.055	0.027	0.009
0.90	0.227	0.992	1.113	1.044	0.920	0.789	0.561	0.391	0.272	0.131	0.064	0.032	0.011
1.00	−0.736	0.341	0.725	0.829	0.808	0.736	0.556	0.399	0.280	0.137	0.067	0.033	0.012

Annex 8.1 (cont.)

Table 18 Values of f_s for $b/D = 0.80$ and $d/D = 0.50$

	β′												
a/D	0.05	0.10	0.15	0.20	0.25	0.30	0.40	0.50	0.60	0.80	1.00	1.20	1.50
0.00	−4.200	−2.848	−2.090	−1.587	−1.227	−0.961	−0.603	−0.388	−0.254	−0.114	−0.054	−0.026	−0.009
0.10	−4.108	−2.760	−2.011	−1.517	−1.167	−0.910	−0.568	−0.365	−0.239	−0.108	−0.051	−0.024	−0.008
0.20	−3.800	−2.474	−1.755	−1.295	−0.980	−0.755	−0.466	−0.298	−0.196	−0.089	−0.042	−0.021	−0.007
0.30	−3.153	−1.892	−1.259	−0.886	−0.650	−0.492	−0.301	−0.195	−0.131	−0.062	−0.030	−0.015	−0.005
0.40	−1.741	−0.762	−0.404	−0.250	−0.175	−0.135	−0.093	−0.069	−0.051	−0.028	−0.014	−0.007	−0.003
0.50	2.155	1.286	0.830	0.553	0.374	0.255	0.122	0.059	0.029	0.007	0.002	0.000	0.000
0.60	5.732	3.062	1.847	1.190	0.802	0.558	0.292	0.165	0.099	0.040	0.017	0.008	0.003
0.70	5.892	3.216	1.994	1.327	0.927	0.672	0.382	0.233	0.149	0.066	0.031	0.015	0.005
0.80	2.662	1.775	1.292	0.981	0.763	0.604	0.393	0.263	0.179	0.085	0.041	0.020	0.007
0.90	−0.786	0.150	0.445	0.524	0.516	0.475	0.366	0.268	0.192	0.096	0.048	0.024	0.008
1.00	−1.506	−0.354	0.129	0.335	0.408	0.414	0.351	0.268	0.195	0.100	0.050	0.025	0.009

Table 19 Values of f_s for $b/D = 0.80$ and $d/D = 0.40$

	β′												
a/D	0.05	0.10	0.15	0.20	0.25	0.30	0.40	0.50	0.60	0.80	1.00	1.20	1.50
0.00	−3.912	−2.572	−1.834	−1.354	−1.019	−0.778	−0.467	−0.290	−0.184	−0.079	−0.036	−0.017	−0.006
0.10	−3.784	−2.454	−1.731	−1.267	−0.948	−0.721	−0.432	−0.268	−0.171	−0.074	−0.034	−0.016	−0.006
0.20	−3.336	−2.055	−1.394	−0.993	−0.731	−0.552	−0.331	−0.208	−0.135	−0.060	−0.028	−0.014	−0.005
0.30	−2.256	−1.189	−0.739	−0.506	−0.369	−0.282	−0.177	−0.118	−0.081	−0.039	−0.019	−0.010	−0.003
0.40	0.759	0.432	0.259	0.153	0.085	0.042	−0.002	−0.018	−0.021	−0.015	−0.009	−0.005	−0.002
0.50	3.670	1.958	1.174	0.745	0.488	0.326	0.152	0.073	0.035	0.009	0.002	0.001	0.000
0.60	4.374	2.493	1.559	1.022	0.692	0.480	0.246	0.135	0.078	0.029	0.012	0.006	0.002
0.70	3.872	2.154	1.362	0.920	0.650	0.473	0.269	0.163	0.103	0.045	0.021	0.010	0.003
0.80	1.196	0.854	0.658	0.524	0.424	0.347	0.236	0.163	0.113	0.055	0.027	0.013	0.005
0.90	−1.503	−0.469	−0.067	0.107	0.180	0.203	0.189	0.151	0.114	0.060	0.031	0.015	0.006
1.00	−2.076	−0.877	−0.331	−0.057	0.079	0.142	0.168	0.145	0.114	0.062	0.032	0.016	0.006

Table 20 Values of f_s for $b/D = 0.80$ and $d/D = 0.30$

	β′												
a/D	0.05	0.10	0.15	0.20	0.25	0.30	0.40	0.50	0.60	0.80	1.00	1.20	1.50
0.00	−3.497	−2.183	−1.481	−1.042	−0.751	−0.549	−0.307	−0.179	−0.108	−0.043	−0.019	−0.009	−0.003
0.10	−3.295	−2.007	−1.339	−0.933	−0.669	−0.489	−0.274	−0.161	−0.098	−0.040	−0.018	−0.008	−0.003
0.20	−2.505	−1.385	−0.880	−0.601	−0.430	−0.317	−0.183	−0.112	−0.071	−0.031	−0.014	−0.007	−0.002
0.30	−0.104	−0.101	−0.096	−0.090	−0.083	−0.075	−0.059	−0.045	−0.034	−0.018	−0.009	−0.005	−0.002
0.40	2.278	1.167	0.674	0.411	0.257	0.162	0.063	0.022	0.005	−0.004	−0.003	−0.002	−0.001
0.50	3.005	1.732	1.087	0.707	0.470	0.317	0.149	0.072	0.035	0.009	0.002	0.001	0.000
0.60	3.053	1.780	1.132	0.750	0.510	0.353	0.178	0.094	0.052	0.018	0.007	0.003	0.001
0.70	2.431	1.315	0.815	0.543	0.379	0.273	0.151	0.089	0.055	0.023	0.010	0.005	0.002
0.80	0.178	0.171	0.161	0.148	0.134	0.119	0.091	0.068	0.049	0.025	0.013	0.006	0.002
0.90	−2.036	−0.939	−0.466	−0.227	−0.098	−0.026	0.033	0.045	0.041	0.026	0.014	0.007	0.003
1.00	−2.512	−1.282	−0.693	−0.372	−0.190	−0.085	0.009	0.036	0.038	0.026	0.014	0.008	0.003

Annex 8.1 (cont.)

Table 21 Values of f_s for b/D = 0.80 and d/D = 0.20

	β′												
a/D	0.05	0.10	0.15	0.20	0.25	0.30	0.40	0.50	0.60	0.80	1.00	1.20	1.50
0.00	−2.853	−1.601	−0.981	−0.626	−0.410	−0.273	−0.126	−0.060	−0.029	−0.007	−0.002	−0.000	−0.000
0.10	−2.447	−1.305	−0.783	−0.497	−0.325	−0.218	−0.101	−0.048	−0.024	−0.006	−0.001	−0.000	−0.000
0.20	−0.569	−0.349	−0.231	−0.157	−0.108	−0.075	−0.037	−0.018	−0.009	−0.002	−0.001	−0.000	−0.000
0.30	1.370	0.662	0.368	0.220	0.139	0.090	0.040	0.019	0.009	0.002	0.001	0.000	−0.000
0.40	1.996	1.150	0.722	0.470	0.313	0.212	0.100	0.048	0.024	0.006	0.001	0.000	−0.000
0.50	2.155	1.286	0.830	0.553	0.374	0.255	0.122	0.059	0.029	0.007	0.002	0.000	0.000
0.60	1.996	1.150	0.722	0.470	0.313	0.212	0.100	0.048	0.024	0.006	0.001	0.000	0.000
0.70	1.370	0.662	0.368	0.220	0.139	0.090	0.040	0.019	0.009	0.002	0.001	0.000	0.000
0.80	−0.569	−0.349	−0.231	−0.157	−0.108	−0.075	−0.037	−0.018	−0.009	−0.002	−0.001	−0.000	0.000
0.90	−2.447	−1.305	−0.783	−0.497	−0.325	−0.218	−0.101	−0.048	−0.024	−0.006	−0.001	−0.000	0.000
1.00	−2.853	−1.601	−0.981	−0.626	−0.410	−0.273	−0.126	−0.060	−0.029	−0.007	−0.002	−0.000	0.000

Table 22 Values of f_s for b/D = 0.70 and d/D = 0.60

	β′												
a/D	0.05	0.10	0.15	0.20	0.25	0.30	0.40	0.50	0.60	0.80	1.00	1.20	1.50
0.00	−4.256	−2.901	−2.140	−1.632	−1.266	−0.994	−0.626	−0.404	−0.264	−0.118	−0.056	−0.027	−0.009
0.10	−4.172	−2.821	−2.066	−1.565	−1.208	−0.944	−0.592	−0.380	−0.249	−0.112	−0.053	−0.025	0.009
0.20	−3.895	−2.559	−1.828	−1.355	−1.027	−0.791	−0.488	−0.311	−0.204	−0.093	−0.044	−0.021	−0.007
0.30	−3.334	−2.041	−1.371	−0.966	−0.705	−0.529	−0.318	−0.204	−0.136	−0.064	−0.031	−0.015	−0.005
0.40	−2.229	−1.075	−0.577	−0.339	−0.219	−0.156	−0.098	−0.070	−0.052	−0.028	−0.015	−0.008	−0.003
0.50	0.341	0.829	0.736	0.556	0.399	0.280	0.137	0.067	0.033	0.008	0.002	0.001	0.000
0.60	8.352	4.104	2.333	1.442	0.943	0.642	0.326	0.180	0.106	0.042	0.018	0.009	0.003
0.70	8.504	4.251	2.473	1.573	1.064	0.752	0.414	0.248	0.157	0.069	0.032	0.016	0.005
0.80	0.820	1.293	1.176	0.967	0.775	0.621	0.405	0.271	0.184	0.088	0.043	0.021	0.007
0.90	−1.339	−0.219	0.228	0.402	0.450	0.440	0.359	0.269	0.195	0.098	0.049	0.024	0.009
1.00	−1.841	−0.626	−0.069	0.203	0.323	0.363	0.335	0.265	0.197	0.102	0.051	0.026	0.009

Table 23 Values of f_s for b/D = 0.70 and d/D = 0.50

	β′												
a/D	0.05	0.10	0.15	0.20	0.25	0.30	0.40	0.50	0.60	0.80	1.00	1.20	1.50
0.00	−4.020	−2.674	−1.925	−1.434	−1.086	−0.833	−0.503	−0.312	−0.198	−0.085	−0.039	−0.018	−0.006
0.10	−3.912	−2.572	−1.834	−1.354	−1.019	−0.778	−0.467	−0.290	−0.184	−0.079	−0.036	−0.017	−0.006
0.20	−3.547	−2.235	−1.536	−1.101	−0.810	−0.609	−0.361	−0.224	−0.144	−0.064	−0.030	−0.014	−0.005
0.30	−2.761	−1.537	−0.954	−0.633	−0.444	−0.326	−0.194	−0.126	−0.085	−0.041	−0.020	−0.010	−0.004
0.40	−0.965	−0.144	0.059	0.089	0.072	0.045	0.006	−0.013	−0.018	−0.015	−0.009	−0.005	−0.002
0.50	4.280	2.401	1.471	0.939	0.615	0.410	0.189	0.090	0.044	0.011	0.003	0.001	0.000
0.60	8.306	4.060	2.290	1.401	0.905	0.607	0.297	0.158	0.089	0.032	0.013	0.006	0.002
0.70	4.468	2.585	1.647	1.105	0.769	0.551	0.304	0.180	0.112	0.048	0.022	0.011	0.004
0.80	−0.562	0.248	0.433	0.439	0.395	0.339	0.240	0.168	0.117	0.057	0.028	0.014	0.005
0.90	−2.076	−0.877	−0.331	−0.057	0.079	0.142	0.168	0.145	0.114	0.062	0.032	0.016	0.006
1.00	−2.444	−1.185	−0.566	−0.225	−0.035	0.067	0.138	0.135	0.111	0.063	0.033	0.017	0.006

Annex 8.1 (cont.)

Table 24 Values of f_s for b/D = 0.70 and d/D = 0.40

	β′												
a/D	0.05	0.10	0.15	0.20	0.25	0.30	0.40	0.50	0.60	0.80	1.00	1.20	1.50
0.00	−3.696	−2.364	−1.638	−1.173	−0.856	−0.632	−0.355	−0.206	−0.124	−0.048	−0.021	−0.010	−0.003
0.10	−3.545	−2.227	−1.519	−1.075	−0.777	−0.570	−0.319	−0.186	−0.112	−0.045	−0.019	−0.009	−0.003
0.20	−3.013	−1.756	−1.128	−0.763	−0.535	−0.387	−0.215	−0.128	−0.080	−0.034	−0.015	−0.007	−0.002
0.30	−1.696	−0.719	−0.362	−0.210	−0.138	−0.100	−0.065	−0.047	−0.034	−0.018	−0.010	−0.005	−0.002
0.40	2.110	1.243	0.788	0.513	0.337	0.221	0.093	0.037	0.012	−0.002	−0.003	−0.002	−0.001
0.50	5.592	2.925	1.716	1.067	0.687	0.453	0.206	0.098	0.048	0.012	0.003	0.001	0.000
0.60	5.637	2.969	1.758	1.107	0.724	0.487	0.235	0.120	0.065	0.021	0.008	0.003	0.001
0.70	2.250	1.379	0.919	0.637	0.452	0.327	0.179	0.104	0.063	0.026	0.011	0.005	0.002
0.80	−1.441	−0.471	−0.126	0.011	0.065	0.084	0.082	0.066	0.050	0.026	0.013	0.007	0.002
0.90	−2.601	−1.359	−0.755	−0.419	−0.224	−0.109	−0.002	0.031	0.036	0.026	0.015	0.008	0.003
1.00	−2.890	−1.605	−0.948	−0.562	−0.326	−0.180	−0.034	0.016	0.030	0.025	0.015	0.008	0.003

Table 25 Values of f_s for b/D = 0.70 and d/D = 0.30

	β′												
a/D	0.05	0.10	0.15	0.20	0.25	0.30	0.40	0.50	0.60	0.80	1.00	1.20	1.50
0.00	−3.232	−1.929	−1.246	−0.829	−0.561	−0.383	−0.182	−0.089	−0.044	−0.011	−0.003	−0.001	−0.000
0.10	−2.994	−1.725	−1.082	−0.706	−0.470	−0.318	−0.149	−0.072	−0.035	−0.009	−0.002	−0.001	−0.000
0.20	−2.055	−0.993	−0.552	−0.331	−0.208	−0.135	−0.060	−0.028	−0.014	−0.003	−0.001	−0.000	−0.000
0.30	0.854	0.524	0.347	0.236	0.163	0.113	0.055	0.027	0.013	0.003	0.001	0.000	−0.000
0.40	3.670	1.958	1.174	0.745	0.488	0.326	0.152	0.073	0.035	0.009	0.002	0.001	−0.000
0.50	4.280	2.401	1.471	0.939	0.615	0.410	0.189	0.090	0.044	0.011	0.003	0.001	0.000
0.60	3.670	1.958	1.174	0.745	0.488	0.326	0.152	0.073	0.035	0.009	0.002	0.001	0.000
0.70	0.854	0.524	0.347	0.236	0.163	0.113	0.055	0.027	0.013	0.003	0.001	0.000	0.000
0.80	−2.055	−0.993	−0.552	−0.331	−0.208	−0.135	−0.060	−0.028	−0.014	−0.003	−0.001	−0.000	0.000
0.90	−2.994	−1.725	−1.082	−0.705	−0.470	−0.318	−0.149	−0.072	−0.035	−0.009	−0.002	−0.001	0.000
1.00	−3.232	−1.929	−1.246	−0.829	−0.561	−0.383	−0.182	−0.089	−0.044	−0.011	−0.003	−0.001	0.000

Table 26 Values of f_s for b/D = 0.60 and d/D = 0.50

	β′												
a/D	0.05	0.10	0.15	0.20	0.25	0.30	0.40	0.50	0.60	0.80	1.00	1.20	1.50
0.00	−3.784	−2.446	−1.711	−1.235	−0.907	−0.673	−0.380	−0.221	−0.132	−0.051	−0.022	−0.010	−0.003
0.10	−3.651	−2.323	−1.602	−1.142	−0.830	−0.611	−0.343	−0.199	−0.120	−0.047	−0.020	−0.009	−0.003
0.20	−3.200	−1.911	−1.245	−0.846	−0.593	−0.426	−0.234	−0.137	−0.085	−0.035	−0.016	−0.008	−0.003
0.30	−2.187	−1.033	−0.537	−0.300	−0.183	−0.123	−0.070	−0.048	−0.035	−0.019	−0.010	−0.005	−0.002
0.40	0.298	0.788	0.695	0.517	0.362	0.247	0.109	0.045	0.016	−0.001	−0.003	−0.002	−0.001
0.50	8.218	3.973	2.207	1.322	0.831	0.539	0.241	0.113	0.055	0.013	0.003	0.001	0.000
0.60	8.261	4.015	2.247	1.361	0.867	0.573	0.269	0.136	0.072	0.023	0.008	0.004	0.001
0.70	0.432	0.918	0.821	0.637	0.474	0.350	0.194	0.112	0.067	0.027	0.012	0.006	0.002
0.80	−1.944	−0.797	−0.311	−0.088	0.014	0.057	0.075	0.064	0.050	0.027	0.014	0.007	0.003
0.90	−2.812	−1.536	−0.890	−0.516	−0.292	−0.155	−0.022	0.022	0.032	0.025	0.015	0.008	0.003
1.00	−3.047	−1.745	−1.063	−0.653	−0.394	−0.229	−0.059	0.005	0.025	0.024	0.015	0.008	0.003

Annex 8.1 (cont.)

Table 27 Values of f_s for $b/D = 0.60$ and $d/D = 0.40$

	β'												
a/D	0.05	0.10	0.15	0.20	0.25	0.30	0.40	0.50	0.60	0.80	1.00	1.20	1.50
0.00	–3.415	–2.095	–1.387	–0.994	–0.650	–0.451	–0.219	–0.108	–0.053	–0.013	–0.003	–0.001	–0.000
0.10	–3.232	–1.929	–1.246	–0.829	–0.561	–0.383	–0.182	–0.089	–0.044	–0.011	–0.003	–0.001	–0.000
0.20	–2.572	–1.354	–0.778	–0.467	–0.290	–0.184	–0.079	–0.036	–0.017	–0.004	–0.001	–0.000	–0.000
0.30	–0.877	–0.057	0.142	0.168	0.145	0.114	0.062	0.032	0.016	0.004	0.001	0.000	–0.000
0.40	4.280	2.401	1.471	0.939	0.615	0.410	0.189	0.090	0.044	0.011	0.003	0.001	–0.000
0.50	8.218	3.973	2.207	1.322	0.831	0.539	0.241	0.113	0.055	0.013	0.003	0.001	0.000
0.60	4.280	2.401	1.471	0.939	0.615	0.410	0.189	0.090	0.044	0.011	0.003	0.001	0.000
0.70	–0.877	–0.057	–0.142	–0.168	0.145	0.114	0.062	0.032	0.016	0.004	0.001	0.000	0.000
0.80	–2.572	–1.354	–0.778	–0.467	–0.290	–0.184	–0.079	–0.036	–0.017	–0.004	–0.001	–0.000	0.000
0.90	–3.232	–1.929	–1.246	–0.829	–0.561	–0.383	–0.182	–0.089	–0.044	–0.011	–0.003	–0.001	0.000
1.00	–3.415	–2.095	–1.387	–0.944	–0.650	–0.451	–0.219	–0.108	–0.053	–0.013	–0.003	–0.001	0.000

Annex 10.1 Values of $\varepsilon = f(P,e)$ (after Anonymous 1964)

P	e = 0	0.05	0.10	0.15	0.20	0.25	0.30	0.35	0.40	0.45
0.1	0.54	0.54	0.55	0.55	0.56	0.57	0.59	0.61	0.67	1.09
0.2	0.44	0.44	0.45	0.46	0.47	0.49	0.52	0.59	0.89	
0.3	0.37	0.37	0.38	0.39	0.41	0.43	0.50	0.74		
0.4	0.31	0.31	0.32	0.34	0.36	0.42	0.62			
0.5	0.25	0.26	0.27	0.29	0.34	0.51				
0.6	0.21	0.21	0.23	0.27	0.41					
0.7	0.16	0.17	0.20	0.32						
0.8	0.11	0.13	0.22							
0.9	0.06	0.12								

Annex 10.2 Values of Hantush's function M(u,B) for partially-penetrated confined aquifers (after Hantush 1962)

u	1/u	B = 0.1	0.2	0.3	0.4	0.5	0.6	0.7	0.8	0.9	1.0	1.2	1.4	1.6	1.8	2.0
0		0.1997	0.3974	0.5913	0.7801	0.9624	1.1376	1.3053	1.4653	1.6177	1.7627	2.0319	2.2759	2.4979	2.7009	2.8872
1(−6)	1.00(6)	0.1994	0.3969	0.5907	0.7792	0.9613	1.1363	1.3037	1.4635	1.6157	1.7605	2.0292	2.2728	2.4943	2.6968	2.8827
2(−6)	5.00(5)	0.1993	0.3967	0.5904	0.7788	0.9608	1.1357	1.3031	1.4628	1.6148	1.7595	2.0281	2.2715	2.4929	2.6951	2.8809
4(−6)	2.50(5)	0.1992	0.3965	0.5900	0.7783	0.9602	1.1349	1.3022	1.4617	1.6137	1.7582	2.0265	2.2696	2.4907	2.6927	2.8782
6(−6)	1.66(5)	0.1991	0.3963	0.5897	0.7779	0.9596	1.1343	1.3014	1.4609	1.6127	1.7572	2.0253	2.2682	2.4891	2.6909	2.8762
8(−6)	1.25(5)	0.1990	0.3961	0.5894	0.7775	0.9592	1.1338	1.3009	1.4602	1.6120	1.7563	2.0243	2.2670	2.4877	2.6894	2.8745
1(−5)	1.00(5)	0.1989	0.3959	0.5892	0.7772	0.9588	1.1334	1.3003	1.4596	1.6113	1.7556	2.0234	2.2660	2.4865	2.6880	2.8730
2(−5)	5.00(4)	0.1987	0.3954	0.5883	0.7760	0.9574	1.1316	1.2983	1.4572	1.6086	1.7526	2.0198	2.2618	2.4818	2.6827	2.8671
4(−5)	2.50(4)	0.1982	0.3945	0.5871	0.7744	0.9553	1.1291	1.2953	1.4539	1.6049	1.7485	2.0148	2.2560	2.4751	2.6752	2.8587
6(−5)	1.66(4)	0.1979	0.3939	0.5861	0.7731	0.9537	1.1271	1.2931	1.4513	1.6020	1.7452	2.0110	2.2515	2.4700	2.6694	2.8523
8(−5)	1.25(4)	0.1976	0.3933	0.5853	0.7720	0.9523	1.1255	1.2912	1.4492	1.5996	1.7425	2.0077	2.2477	2.4657	2.6645	2.8469
1(−4)	1.00(4)	0.1974	0.3929	0.5846	0.7710	0.9511	1.1241	1.2895	1.4473	1.5974	1.7402	2.0049	2.2444	2.4619	2.6603	2.8421
2(−4)	5.00(3)	0.1965	0.3910	0.5818	0.7673	0.9465	1.1185	1.2830	1.4398	1.5890	1.7308	1.9936	2.2313	2.4469	2.6434	2.8234
4(−4)	2.50(3)	0.1952	0.3883	0.5778	0.7620	0.9398	1.1106	1.2737	1.4292	1.5771	1.7176	1.9778	2.2128	2.4258	2.6197	2.7970
6(−4)	1.66(3)	0.1941	0.3863	0.5748	0.7580	0.9348	1.1045	1.2666	1.4211	1.5680	1.7075	1.9656	2.1986	2.4095	2.6014	2.7768
8(−4)	1.25(3)	0.1933	0.3846	0.5722	0.7545	0.9305	1.0994	1.2607	1.4143	1.5603	1.6989	1.9554	2.1866	2.3959	2.5860	2.7597
1(−3)	1.00(3)	0.1925	0.3831	0.5699	0.7515	0.9267	1.0948	1.2554	1.4083	1.5535	1.6914	1.9463	2.1761	2.3838	2.5725	2.7446
2(−3)	5.00(2)	0.1896	0.3772	0.5611	0.7397	0.9120	1.0771	1.2347	1.3846	1.5270	1.6619	1.9109	2.1348	2.3367	2.5195	2.6857
4(−3)	2.50(2)	0.1854	0.3689	0.5486	0.7231	0.8912	1.0521	1.2056	1.3513	1.4895	1.6203	1.8610	2.0766	2.2702	2.4447	2.6027
6(−3)	1.66(2)	0.1822	0.3625	0.5390	0.7103	0.8752	1.0330	1.1832	1.3258	1.4608	1.5884	1.8228	2.0320	2.2193	2.3875	2.5393
8(−3)	1.25(2)	0.1795	0.3571	0.5310	0.6995	0.8618	1.0169	1.1645	1.3044	1.4367	1.5616	1.7907	1.9946	2.1766	2.3395	2.4861
1(−2)	1.00(2)	0.1772	0.3524	0.5239	0.6901	0.8500	1.0027	1.1480	1.2855	1.4155	1.5381	1.7625	1.9617	2.1391	2.2975	2.4394
2(−2)	5.00(1)	0.1680	0.3340	0.4962	0.6533	0.8040	0.9476	1.0836	1.2121	1.3329	1.4464	1.6527	1.8340	1.9935	2.1342	2.2587
4(−2)	2.50(1)	0.1551	0.3083	0.4578	0.6020	0.7400	0.8708	0.9942	1.1100	1.2183	1.3193	1.5008	1.6577	1.7932	1.9103	2.0117
6(−2)	1.66(1)	0.1455	0.2890	0.4289	0.5635	0.6919	0.8132	0.9272	1.0336	1.1326	1.2243	1.3877	1.5268	1.6450	1.7454	1.8307
8(−2)	1.25(1)	0.1375	0.2731	0.4050	0.5317	0.6522	0.7658	0.8720	0.9707	1.0621	1.1464	1.2951	1.4200	1.5246	1.6120	1.6848
1(−1)	1.00(1)	0.1306	0.2993	0.3844	0.5043	0.6181	0.7249	0.8245	0.9167	1.0016	1.0795	1.2159	1.3290	1.4223	1.4991	1.5619
2(−1)	5.00	0.1051	0.2084	0.3081	0.4030	0.4920	0.5744	0.6500	0.7186	0.7806	0.8362	0.9297	1.0029	1.0595	1.1026	1.1352
4(−1)	2.50	7.39(−2)	0.1462	0.2153	0.2801	0.3397	0.3935	0.4415	0.4837	0.5203	0.5519	0.6015	0.6363	0.6602	0.6760	0.6863
6(−1)	1.66	5.44(−2)	0.1074	0.1575	0.2039	0.2458	0.2828	0.3149	0.3423	0.3652	0.3842	0.4122	0.4300	0.4408	0.4471	0.4506
8(−1)	1.25	4.10(−2)	8.06(−2)	0.1179	0.1519	0.1821	0.2082	0.2302	0.2484	0.2632	0.2750	0.2913	0.3007	0.3058	0.3084	0.3096
1	1.00	3.13(−2)	6.14(−2)	8.95(−2)	0.1148	0.1369	0.1555	0.1709	0.1833	0.1929	0.2004	0.2101	0.2151	0.2175	0.2186	0.2191
2	5.00(−1)	9.01(−3)	1.75(−2)	2.51(−2)	3.16(−2)	3.67(−2)	4.07(−2)	4.35(−2)	4.55(−2)	4.69(−2)	4.77(−2)	4.85(−2)	4.88(−2)			
4	2.50(−1)	9.20(−4)	1.76(−3)	2.44(−3)	2.96(−3)	3.31(−3)	3.53(−3)	3.66(−3)	3.72(−3)	3.76(−3)	3.77(−3)					

Annex 10.2 (cont.)

u	1/u	B=2.2	2.4	2.6	2.8	3.0	3.2	3.4	3.6	3.8	4.0	4.2	4.4	4.6	4.8	5.0	5.2	5.4	5.6	5.8	6.0
0		3.0593	3.2188	3.3675	3.5064	3.6369	3.7597	3.8757	3.9856	4.0900	4.1894	4.2842	4.3748	4.4616	4.5448	4.6248	4.7018	4.7760	4.8475	4.9167	4.9835
1(–6)	1.00(6)	3.0543	3.2134	3.3616	3.5001	3.6301	3.7525	3.8681	3.9775	4.0815	4.1804	4.9700	4.2747	4.3649	4.4512	4.5340	4.6136	4.6901	4.7638	4.8349	4.9036
2(–6)	5.00(5)	3.0523	3.2112	3.3592	3.4975	3.6273	3.7495	3.8649	3.9742	4.0779	4.1766	4.9644	4.2708	4.3608	4.4469	4.5295	4.6089	4.6852	4.7588	4.8297	4.8982
4(–6)	2.50(5)	3.0494	3.2080	3.3557	3.4938	3.6233	3.7453	3.8604	3.9694	4.0729	4.1714	4.9565	4.2653	4.3550	4.4408	4.5232	4.6023	4.6784	4.7516	4.8223	4.8905
6(–6)	1.66(5)	3.0471	3.2056	3.3531	3.4910	3.6203	3.7420	3.8569	3.9658	4.0690	4.1673	4.9504	4.2610	4.3505	4.4362	4.5183	4.5972	4.6731	4.7462	4.8166	4.8846
8(–6)	1.25(5)	3.0453	3.2035	3.3509	3.4886	3.6177	3.7393	3.8540	3.9627	4.0658	4.1639	4.9452	4.2574	4.3467	4.4323	4.5142	4.5929	4.6686	4.7415	4.8118	4.8797
1(–5)	1.00(5)	3.0436	3.2017	3.3489	3.4865	3.6155	3.7369	3.8515	3.9600	4.0629	4.1609	4.9407	4.2542	4.3434	4.4288	4.5106	4.5892	4.6647	4.7375	4.8076	4.8753
2(–5)	5.00(4)	3.0371	3.1946	3.3412	3.4782	3.6066	3.7274	3.8414	3.9493	4.0517	4.1490	4.9230	4.2418	4.3304	4.4152	4.4964	4.5744	4.6494	4.7215	4.7911	4.8582
4(–5)	2.50(4)	3.0279	3.1846	3.3304	3.4665	3.5941	3.7140	3.8272	3.9343	4.0358	4.1323	4.8979	4.2243	4.3120	4.3960	4.4764	4.5535	4.6276	4.6989	4.7677	4.8339
6(–5)	1.66(4)	3.0209	3.1769	3.3220	3.4575	3.5844	3.7038	3.8163	3.9227	4.0236	4.1195	4.8787	4.2108	4.2979	4.3812	4.4610	4.5375	4.6110	4.6816	4.7497	4.8153
8(–5)	1.25(4)	3.0149	3.1704	3.3150	3.4499	3.5763	3.6951	3.8071	3.9130	4.0133	4.1087	4.8625	4.1994	4.2860	4.3688	4.4480	4.5240	4.5969	4.6670	4.7346	4.7997
1(–4)	1.00(4)	3.0097	3.1647	3.3088	3.4433	3.5692	3.6875	3.7990	3.9044	4.0043	4.0992	4.8482	4.1894	4.2756	4.3578	4.4366	4.5121	4.5845	4.6542	4.7212	4.7859
2(–4)	5.00(3)	2.9891	3.1423	3.2845	3.4171	3.5412	3.6576	3.7673	3.8708	3.9688	4.0618	4.7922	4.1502	4.2345	4.3149	4.3918	4.4654	4.5360	4.6038	4.6690	4.7317
4(–4)	2.50(3)	2.9600	3.1106	3.2502	3.3801	3.5015	3.6154	3.7224	3.8233	3.9187	4.0090	4.7132	4.0948	4.1764	4.2542	4.3285	4.3995	4.4674	4.5326	4.5952	4.6553
6(–4)	1.66(3)	2.9378	3.0863	3.2238	3.3518	3.4712	3.5830	3.6880	3.7869	3.8802	3.9686	4.6527	4.0524	4.1320	4.2077	4.2800	4.3490	4.4150	4.4781	4.5387	4.5969
8(–4)	1.25(3)	2.9190	3.0658	3.2017	3.3279	3.4456	3.5557	3.6590	3.7562	3.8479	3.9345	4.6019	4.0166	4.0945	4.1686	4.2392	4.3065	4.3708	4.4323	4.4912	4.5477
1(–3)	1.00(3)	2.9024	3.0478	3.1821	3.3069	3.4231	3.5317	3.6335	3.7292	3.8194	3.9046	4.5572	3.9852	4.0616	4.1342	4.2033	4.2691	4.3320	4.3920	4.4494	4.5045
2(–3)	5.00(2)	2.8377	2.9771	3.1056	3.2245	3.3349	3.4377	3.5337	3.6236	3.7080	3.7874	4.3832	3.8623	3.9329	3.9998	4.0632	4.1233	4.1805	4.2349	4.22867	4.3360
4(–3)	2.50(2)	2.7464	2.8776	2.9980	3.1087	3.2110	3.3056	3.3936	3.4754	3.5518	3.6233	4.1416	3.6902	3.7530	3.8120	3.8676	3.9199	3.9694	4.0161	4.0602	4.1020
6(–3)	1.66(2)	2.6767	2.8018	2.9159	3.0205	3.1166	3.2052	3.2871	3.3629	3.4334	3.4989	3.9609	3.5599	3.6169	3.6702	3.7200	3.7667	3.8105	3.8517	3.8903	3.9267
8(–3)	1.25(2)	2.6183	2.7382	2.8472	2.9466	3.0377	3.1213	3.1982	3.2691	3.3346	3.3953	3.8123	3.4516	3.5038	3.5524	3.5977	3.6398	3.6792	3.7159	3.7502	3.7823
1(–2)	1.00(2)	2.5671	2.6825	2.7870	2.8820	2.9687	3.0480	3.1206	3.1873	3.2487	3.3052	3.6845	3.3574	3.4057	3.4503	3.4917	3.5300	3.5656	3.5987	3.6294	3.6580
2(–2)	5.00(1)	2.3692	2.4675	2.5552	2.6337	2.7041	2.7673	2.8243	2.8756	2.9218	2.9637	3.2153	3.0015	3.0357	3.0666	3.0946	3.1200	3.1430	3.1638	3.1827	3.1998
4(–2)	2.50(1)	2.0996	2.1759	2.2423	2.3000	2.3503	2.3942	2.4324	2.4658	2.4949	2.5202	2.6468	2.5423	2.5615	2.5782	2.5927	2.6052	2.6161	2.6256	2.6337	2.6408
6(–2)	1.66(1)	1.9031	1.9645	2.0167	2.0610	2.0986	2.1305	2.1574	2.1802	2.1995	2.2157	2.2846	2.2294	2.2408	2.2504	2.2584	2.2651	2.2706	2.2752	2.2790	2.2821
8(–2)	1.25(1)	1.7455	1.7959	1.8378	1.8725	1.9012	1.9249	1.9444	1.9604	1.9734	1.9841	2.0233	1.9928	1.9998	2.0055	2.0101	2.0137	2.0166	2.0189	2.0207	2.0221
1(–1)	1.00(1)	1.6133	1.6552	1.6892	1.7167	1.7389	1.7568	1.7711	1.7825	1.7915	1.7987	1.8216	1.8043	1.8087	1.8121	1.8147	1.8168	1.8183	1.8195	1.8204	1.8211
2(–1)	5.00	1.1596	1.1777	1.1909	1.2004	1.2073	1.2122	1.2156	1.2179	1.2195	1.2206		1.2213	1.2218	1.2221	1.2223	1.2224	1.2225	1.2226		
4(–1)	2.50	0.6928	0.6968	0.6992	0.7006	0.7014	0.7019	0.7021	0.7023	0.7023	0.7004										
6(–1)	1.66	0.4525	0.4535	0.4540	0.4542	0.4543	0.4543														
8(–1)	1.25	0.3102	0.3104	0.3105																	

M(u,B) = W(u): see Annex 3.1

Annex 10.2 (cont.)

u	1/u	B=6.2	6.4	6.6	6.8	7.0	7.2	7.4	7.6	7.8	8.0	8.2	8.4	8.6	8.8	9.0	9.2	9.4	9.6	9.8	10.0
0		5.0482	5.1109	5.1718	5.2308	5.2882	5.3440	5.3983	5.4511	5.5026	5.5529	5.6019	5.6497	5.6965	5.7421	5.7868	5.8305	5.8733	5.9151	5.9562	5.9964
1(−6)	1.00(6)	5.0343	5.0965	5.1569	5.2155	5.2724	5.3278	5.3816	5.4340	5.4851	5.5349	5.5834	5.6308	5.6771	5.7223	5.7666	5.8098	5.8521	5.8935	9.9341	5.9739
2(−6)	5.00(5)	5.0285	5.0905	5.1507	5.2091	5.2659	5.3210	5.3747	5.4269	5.4778	5.5274	5.5758	5.6230	5.6691	5.7141	5.7581	5.8012	5.8433	5.8846	5.9250	5.9645
4(−6)	2.50(5)	5.0203	5.0821	5.1420	5.2002	5.2566	5.3115	5.3649	5.4169	5.4675	5.5168	5.5649	5.6119	5.6577	5.7025	5.7463	5.7890	5.8309	5.8719	5.9120	5.9513
6(−6)	1.66(5)	5.0140	5.0756	5.1353	5.1933	5.2495	5.3042	5.3574	5.4092	5.4596	5.5087	5.5566	5.6034	5.6490	5.6936	5.7371	5.7797	5.8214	5.8621	5.9021	5.9412
8(−6)	1.25(5)	5.0087	5.0701	5.1297	5.1874	5.2435	5.2981	5.3511	5.4027	5.4529	5.5019	5.5496	5.5962	5.6416	5.6860	5.7294	5.7718	5.8133	5.8539	5.8937	5.9326
1(−5)	1.00(5)	5.0040	5.0653	5.1247	5.1823	5.2383	5.2926	5.3455	5.3969	5.4470	5.4958	5.5434	5.5898	5.6352	5.6794	5.7226	5.7649	5.8063	5.8467	5.8863	5.9251
2(−5)	5.00(4)	4.9857	5.0464	5.1052	5.1622	5.2176	5.2714	5.3236	5.3745	5.4240	5.4722	5.5192	5.5650	5.6097	5.6534	5.6961	5.7377	5.7785	5.8183	5.8573	5.8955
4(−5)	2.50(4)	4.9598	5.0196	5.0776	5.1338	5.1883	5.2413	5.2927	5.3427	5.3914	5.4388	5.4849	5.5299	5.5738	5.6167	5.6585	5.6993	5.7392	5.7782	5.8164	5.8538
6(−5)	1.66(4)	4.9399	4.9991	5.0565	5.1120	5.1659	5.2182	5.2690	5.3184	5.3664	5.4131	5.4587	5.5030	5.5463	5.5885	5.6296	5.6698	5.7091	5.7475	5.7850	5.8217
8(−5)	1.25(4)	4.9232	4.9818	5.0386	5.0937	5.1470	5.1988	5.2490	5.2979	5.3453	5.3915	5.4365	5.4803	5.5231	5.5647	5.6053	5.6450	5.6837	5.7216	5.7586	5.7948
1(−4)	1.00(4)	4.9084	4.9666	5.0229	5.0775	5.1303	5.1816	5.2314	5.2798	5.3268	5.3725	5.4170	5.4604	5.5026	5.5438	5.5840	5.6231	5.6614	5.6988	5.7353	5.7710
2(−4)	5.00(3)	4.8506	4.9069	4.9614	5.0141	5.0651	5.1145	5.1624	5.2089	5.2541	5.2980	5.3406	5.3821	5.4225	5.4619	5.5002	5.5375	5.5739	5.6095	5.6441	5.6780
4(−4)	2.50(3)	4.7689	4.8227	4.8745	4.9246	4.9730	5.0198	5.0652	5.1091	5.1516	5.1929	5.2330	5.2719	5.3097	5.3464	5.3822	5.4169	5.4508	5.4837	5.5158	5.5471
6(−4)	1.66(3)	4.7065	4.7582	4.8081	4.8562	4.9026	4.9475	4.9908	5.0327	5.0733	5.1127	5.1508	5.1877	5.2236	5.2583	5.2921	5.3249	5.3568	5.3879	5.4180	5.4474
8(−4)	1.25(3)	4.6540	4.7040	4.7522	4.7987	4.8435	4.8867	4.9284	4.9687	5.0076	5.0453	5.0818	5.1171	5.1513	5.1845	5.2166	5.2478	5.2781	5.3075	5.3361	5.3639
1(−3)	1.00(3)	4.6078	4.6565	4.7032	4.7482	4.7915	4.8333	4.8736	4.9124	4.9500	4.9862	5.0213	5.0552	5.0880	5.1197	5.1505	5.1803	5.2092	5.2372	5.2644	5.2908
2(−3)	5.00(2)	4.4282	4.4713	4.5125	4.5519	4.5898	4.6260	4.6609	4.6943	4.7264	4.7573	4.7870	4.8155	4.8430	4.8695	4.8950	4.9196	4.9433	4.9662	4.9882	5.0095
4(−3)	2.50(2)	4.1792	4.2148	4.2487	4.2808	4.3114	4.3405	4.3682	4.3945	4.4197	4.4436	4.4664	4.4881	4.5089	4.5287	4.5476	4.5656	4.5829	4.5993	4.6150	4.6301
6(−3)	1.66(2)	3.9932	4.0236	4.0523	4.0793	4.1048	4.1290	4.1517	4.1733	4.1936	4.2129	4.2311	4.2483	4.2646	4.2800	4.2946	4.3084	4.3214	4.3338	4.3455	4.3566
8(−3)	1.25(2)	3.8404	3.8668	3.8914	3.9146	3.9362	3.9566	3.9756	3.9935	4.0103	4.0261	4.0409	4.0548	4.0678	4.0801	4.0916	4.1024	4.1125	4.1220	4.1309	4.1393
1(−2)	1.00(2)	3.7093	3.7323	3.7537	3.7737	3.7923	3.8096	3.8258	3.8408	3.8548	3.8679	3.8801	3.8914	3.9020	3.9119	3.9210	3.9296	3.9375	3.9449	3.9518	3.9582
2(−2)	5.00(1)	3.2293	3.2419	3.2534	3.2638	3.2731	3.2816	3.2892	3.2961	3.3023	3.3079	3.3130	3.3175	3.3215	3.3252	3.3285	3.3314	3.3340	3.3364	3.3385	3.3403
4(−2)	2.50(1)	2.6520	2.6565	2.6603	2.6636	2.6664	2.6688	2.6708	2.6725	2.6740	2.6752	2.6762	2.6771	2.6778	2.6784	2.6789	2.6793	2.6797	2.6800	2.6802	2.6804
6(−2)	1.66(1)	2.2867	2.2884	2.2898	2.2909	2.2918	2.2926	2.2931	2.2936	2.2940	2.2943	2.2945	2.2947	2.2948	2.2949	2.2950	2.2951	2.2951	2.2952		
8(−2)	1.25(1)	2.0241	2.0248	2.0253	2.0257	2.0260	2.0263	2.0264	2.0266	2.0267	2.0267	2.0268	2.0268	2.0269							
1(−1)	1.00(1)	1.8219	1.8222	1.8224	1.8226	1.8227	1.8227	1.8228	1.8228	1.8229	M(u,B) = W(u): see Annex 3.1										

Annex 10.2 (cont.)

u	1/u	B=12	14	16	18	20	22	24	26	28	30	32	34	36	38	40	42	44	46	48	50
0		6.3595	6.6668	6.9333	7.1684	7.3789	7.5692	7.7431	7.9030	8.0511	8.1890	8.3180	8.4392	8.5535	8.6615	8.7641	8.8616	8.9546	9.0435	9.1286	9.2102
1(−6)	1.00(6)	6.3325	6.6353	6.8973	7.1279	7.3339	7.5197	7.6891	7.8445	7.9881	8.1215	8.2460	8.3627	8.4725	8.5761	8.6741	8.7671	8.8556	8.9400	9.0206	9.0977
2(−6)	5.00(5)	6.3213	6.6223	6.8823	7.1111	7.3152	7.4992	7.6667	7.8202	7.9620	8.0935	8.2161	8.3309	8.4388	8.5406	8.6367	8.7279	8.8145	8.8971	8.9758	9.0510
4(−6)	2.50(5)	6.3054	6.6038	6.8612	7.0873	7.2887	7.4701	7.6350	7.7859	7.9250	8.0539	8.1739	8.2861	8.3913	8.4904	8.5839	8.6725	8.7565	8.8364	8.9125	8.9851
6(−6)	1.66(5)	6.2932	6.5896	6.8450	7.0691	7.2685	7.4478	7.6106	7.7596	7.8966	8.0235	8.1415	8.2516	8.3549	8.4519	8.5435	8.6300	8.7120	8.7899	8.8640	8.9346
8(−6)	1.25(5)	6.2830	6.5775	6.8313	7.0537	7.2514	7.4290	7.5901	7.7374	7.8727	7.9979	8.1142	8.2226	8.3242	8.4196	8.5094	8.5942	8.6745	8.7507	8.8231	8.8921
1(−5)	1.00(5)	6.2739	6.5671	6.8193	7.0402	7.2363	7.4125	7.5721	7.7178	7.8517	7.9753	8.0901	8.1971	8.2972	8.3910	8.4794	8.5628	8.6416	8.7163	8.7872	8.8547
2(−5)	5.00(4)	6.2385	6.5257	6.7720	6.9870	7.1773	7.3476	7.5013	7.6412	7.7692	7.8871	7.9960	8.0972	8.1914	8.2795	8.3621	8.4397	8.5127	8.5817	8.6469	8.7087
4(−5)	2.50(4)	6.1884	6.4673	6.7053	6.9120	7.0940	7.2551	7.4016	7.5332	7.6531	7.7627	7.8636	7.9566	8.0428	8.1229	8.1975	8.2671	8.3322	8.3933	8.4507	8.5047
6(−5)	1.66(4)	6.1500	6.4225	6.6542	6.8546	7.0303	7.1861	7.3253	7.4508	7.5644	7.6679	7.7626	7.8495	7.9298	8.0038	8.0725	8.1362	8.1955	8.2508	8.3024	8.3507
8(−5)	1.25(4)	6.1177	6.3848	6.6112	6.8063	6.9767	7.1212	7.2613	7.3815	7.4901	7.5885	7.6781	7.7601	7.8353	7.9044	7.9682	8.0271	8.0817	8.1323	8.1792	8.2229
1(−4)	1.00(4)	6.0892	6.3517	6.5734	6.7638	6.9296	7.0756	7.2051	7.3208	7.4249	7.5189	7.6042	7.6818	7.7527	7.8177	7.8773	7.9321	7.9826	8.0292	8.0723	8.1122
2(−4)	5.00(3)	5.9778	6.2221	6.4257	6.5982	6.7463	6.8747	6.9869	7.0856	7.1729	7.2504	7.3194	7.3811	7.4364	7.4861	7.5307	7.5709	7.6072	7.6399	7.6695	7.6962
4(−4)	2.50(3)	5.8214	6.0406	6.2194	6.3677	6.4920	6.5972	6.6868	6.7635	6.8294	6.8852	6.9353	6.9778	7.0146	7.0465	7.0742	7.0982	7.1191	7.1371	7.1528	7.1663
6(−4)	1.66(3)	5.7026	5.9031	6.0638	6.1945	6.3019	6.3908	6.4648	6.5266	6.5784	6.6218	6.6583	6.6890	6.7147	6.7363	6.7545	6.7696	6.7823	6.7929	6.8017	6.8090
8(−4)	1.25(3)	5.6034	5.7887	5.9348	6.0515	6.1456	6.2219	6.2841	6.3349	6.3763	6.4103	6.4380	6.4607	6.4791	6.4942	6.5063	6.5162	6.5241	6.5305	6.5357	6.5397
1(−3)	1.00(3)	5.5168	5.6892	5.8230	5.9281	6.0113	6.0775	6.1303	6.1724	6.2061	6.2330	6.2543	6.2713	6.2848	6.2954	6.3037	6.3102	6.3153	6.3192	6.3222	6.3246
2(−3)	5.00(2)	5.1861	5.3123	5.4037	5.4701	5.5184	5.5534	5.5788	5.5970	5.6101	5.6193	5.6257	5.6302	5.6333	5.6354	5.6368	5.6377	5.6383	5.6387	5.6390	5.6391
4(−3)	2.50(2)	4.7481	4.8235	4.8714	4.9017	4.9205	4.9320	4.9390	4.9430	4.9454	4.9467	4.9474	4.9487	4.9480	4.9481						
6(−3)	1.66(2)	4.4396	4.4872	4.5140	4.5288	4.5367	4.5409	4.5429	4.5439	4.5444	4.5446	4.5447									
8(−3)	1.25(2)	4.1991	4.2300	4.2455	4.2530	4.2565	4.2565	4.2580	4.2537	4.2589	4.2590	4.2591									
1(−2)	1.00(2)	4.0020	4.0224	4.0316	4.0355	4.0370	4.0376	4.0378													
2(−2)	5.00(1)	3.3507	3.3537	3.3545	3.3547																
4(−2)	2.50(1)	2.6812	2.6812																		

M(u,B) = W(u): see Annex 3.1

Annex 10.2 (cont.)

u	1/u	B = 52	54	56	58	60	62	64	66	68	70
0		9.2886	9.3641	9.4368	9.506 9	9.5747	9.6403	9.7037	9.7653	9.8249	9.8829
1(−6)	1.00(6)	9.1716	9.2426	9.3108	9.3765	9.4398	9.5008	9.5598	9.6168	9.6720	9.7255
2(−6)	5.00(5)	9.1231	9.1922	9.2585	9.3223	9.3838	9.4430	9.5001	9.5553	9.6086	9.6602
4(−6)	2.50(5)	9.0545	9.1210	9.1847	9.2459	9.3047	9.3613	9.4158	9.4684	9.5191	9.5681
6(−6)	1.66(5)	9.0020	9.0665	9.1282	9.1874	9.2442	9.2988	9.3513	9.4019	9.4507	9.4977
8(−6)	1.25(5)	8.9578	9.0206	9.0807	9.1382	9.1933	9.2413	9.2971	9.3460	9.3931	9.4385
1(−5)	1.00(5)	8.9190	8.9803	9.0389	9.0949	9.1486	9.2001	9.2495	9.2970	9.3426	9.3865
2(−5)	5.00(4)	8.7673	8.8229	8.8759	8.9263	8.9743	9.0202	9.0640	9.1059	9.1461	9.1845
4(−5)	2.50(4)	8.5555	8.6035	8.6488	8.6916	8.7321	8.7705	8.8069	8.8414	8.8742	8.9053
6(−5)	1.66(4)	8.3959	8.4383	8.4780	8.5154	8.5505	8.5836	8.6147	8.6440	8.6716	8.6977
8(−5)	1.25(4)	8.2636	8.3016	8.3370	8.3700	8.4009	8.4297	8.4568	8.4821	8.5057	8.5279
1(−4)	1.00(4)	8.1491	8.1833	8.2151	8.2446	8.2720	8.2974	8.3211	8.3431	8.3636	8.3827
2(−4)	5.00(3)	7.7203	7.7421	7.7618	7.7797	7.7958	7.8104	7.8236	7.8355	7.8463	7.8560
4(−4)	2.50(3)	7.1780	7.1881	7.1968	7.2043	7.2108	7.2163	7.2211	7.2251	7.2286	7.2315
6(−4)	1.66(3)	6.8151	6.8201	6.8242	6.8276	6.8304	6.8327	6.8345	6.8360	6.8372	6.8382
8(−4)	1.25(3)	6.5430	6.5456	6.5476	6.5492	6.5504	6.5514	6.5521	6.5527	6.5531	6.5535
1(−3)	1.00(3)	6.3262	6.3277	6.3287	6.3294	6.3300	6.3304	6.3307	6.3310	6.3311	6.3312
2(−3)	5.00(2)	5.6392	5.6393					M(u,B) = W(u): see Annex 3.1			

u	1/u	B = 72	74	76	78	80	82	84	86	88	90
0		9.9392	9.9940	10.0473	10.0992	10.1498	10.1992	10.2474	10.2944	10.3404	10.3853
1(−6)	1.00(6)	9.7773	9.8276	9.8764	9.9236	9.9700	10.0148	10.0585	10.1011	10.1425	10.1830
2(−6)	5.00(5)	9.7102	9.7586	9.8056	9.8512	9.8955	9.9385	9.9803	10.0210	10.0606	10.0992
4(−6)	2.50(5)	9.6155	9.6613	9.7057	9.7487	9.7904	9.8308	9.8700	9.9081	9.9452	9.9812
6(−6)	1.66(5)	9.5431	9.5869	9.6293	9.6703	9.7101	9.7485	9.7858	9.8220	9.8571	9.8911
8(−6)	1.25(5)	9.4822	9.5244	9.5652	9.6046	9.6426	9.6795	9.7151	9.7497	9.7831	9.8156
1(−5)	1.00(5)	9.4288	9.4696	9.5089	9.5469	9.5835	9.6189	9.6532	9.6863	9.7183	9.7494
2(−5)	5.00(4)	9.2213	9.2566	9.2905	9.3230	9.3542	9.3843	9.4132	9.4410	9.4677	9.4935
4(−5)	2.50(4)	8.9349	8.9630	8.9898	9.0153	9.0396	9.0628	9.0848	9.1059	9.1260	9.1451
6(−5)	1.66(4)	8.7223	8.7455	8.7675	8.7882	8.8076	8.8263	8.8438	8.8603	8.8760	8.8908
8(−5)	1.25(4)	8.5487	8.5682	8.5865	8.6036	8.6197	8.6348	8.6490	8.6623	8.6747	8.6864
1(−4)	1.00(4)	8.4005	8.4170	8.4324	8.4468	8.4601	8.4726	8.4842	8.4949	8.5050	8.5143
2(−4)	5.00(3)	7.8648	7.8727	7.8798	7.8862	7.8920	7.8972	7.9019	7.9061	7.9098	7.9132
4(−4)	2.50(3)	7.2341	7.2362	7.2380	7.2395	7.2408	7.2419	7.2428	7.2436	7.2442	7.2447
6(−4)	1.66(3)	6.8390	6.8396	6.8401	6.8405	6.8408	6.8411	6.8413	6.8414	6.8416	6.8417
8(−4)	1.25(3)	6.5537	6.5539	6.5541	6.5542	6.5543	6.5543	6.5544	6.5544	6.5544	6.5544
1(−3)	1.00(3)	6.3313	6.3314	6.3314	6.3315						

Annex 10.3 Values of Streltsova's function $W(u_A,\beta,b_1/D,b_2/D)$ for partially-penetrated unconfined aquifers (after Streltsova 1974)

Table 1 Values of $W(u_A,\beta,b_1/D,b_2/D)$ for $b_1/D = 0.1$ and $b_2/D = 0.1$

	$\sqrt{\beta}$						
$1/u_A$	0.05	0.1	0.2	0.3	0.5	0.75	1.0
0.2	5.7×10^{-4}	5.7×10^{-4}	4.7×10^{-4}	2.9×10^{-4}	1.1×10^{-4}	3.8×10^{-5}	1.7×10^{-5}
0.4	0.0125	0.0121	0.0075	0.0037	0.0011	3.7×10^{-4}	1.6×10^{-4}
0.6	0.0392	0.0363	0.0184	0.0083	0.0023	7.6×10^{-4}	3.3×10^{-4}
0.8	0.0731	0.0642	0.0285	0.0122	0.0033	0.0011	4.6×10^{-4}
1.0	0.1094	0.0908	0.0367	0.0152	0.0040	0.0013	5.5×10^{-4}
2.0	0.2723	0.1824	0.0586	0.0227	0.0058	0.0018	7.8×10^{-4}
4.0	0.4674	0.2530	0.0714	0.0268	0.0067	0.0021	9.0×10^{-4}
6.0	0.5676	0.2788	0.0755	0.0281	0.0070	0.0022	9.3×10^{-4}
8.0	0.6257	0.2914	0.0773	0.0286	0.0071	0.0022	9.5×10^{-4}
10	0.6626	0.2986	0.0783	0.0289	0.0072		
20	0.7375	0.3113	0.0800	0.0295	0.0073		
40	0.7711	0.3162	0.0807	0.0296	0.0073		
60	0.7805	0.3175	0.0809	0.0297			
80	0.7846	0.3181	0.0809	0.0297			
100	0.7868	0.3184	0.0810				
200	0.7905	0.3188	0.0810				
400	0.7918	0.3190					
1000	0.7922	0.3191					

Table 2 Values of $W(u_A,\beta,b_1/D,b_2/D)$ for $b_1/D = 0.2$ and $b_2/D = 0.2$

	$\sqrt{\beta}$						
$1/u_A$	0.05	0.1	0.2	0.3	0.5	0.75	1.0
0.2	5.7×10^{-4}	5.7×10^{-4}	5.7×10^{-4}	5.5×10^{-4}	3.8×10^{-4}	2.0×10^{-4}	1.1×10^{-4}
0.4	0.0125	0.0125	0.0121	0.0101	0.0053	0.0023	0.0011
0.6	0.0392	0.0392	0.0363	0.0272	0.0123	0.0049	0.0023
0.8	0.0732	0.0731	0.0642	0.0444	0.0184	0.0071	0.0033
1.0	0.1097	0.1094	0.0908	0.0592	0.0232	0.0087	0.0040
2.0	0.2799	0.2723	0.1824	0.1024	0.0355	0.0127	0.0058
4.0	0.5215	0.4674	0.2530	0.1298	0.0424	0.0149	0.0067
6.0	0.6828	0.5676	0.2788	0.1388	0.0445	0.0155	0.0070
8.0	0.7992	0.6257	0.2914	0.1430	0.0455	0.0158	0.0071
10	0.8873	0.6626	0.2986	0.1453	0.0460	0.0160	0.0072
20	1.1236	0.7375	0.3113	0.1493	0.0469	0.0162	0.0073
40	1.2774	0.7711	0.3162	0.1508	0.0472	0.0163	0.0073
60	1.3310	0.7805	0.3175	0.1512	0.0473	0.0164	
80	1.3567	0.7846	0.3181	0.1514	0.0473	0.0164	
100	1.3713	0.7868	0.3184	0.1515			
200	1.3971	0.7905	0.3188	0.1516			
400	1.4072	0.7918	0.3190	0.1517			
1000	1.4098	0.7922	0.3191	0.1517			

Annex 10.3 (cont.)

Table 3 Values of $W(u_A,\beta,b_1/D,b_2/D)$ for $b_1/D = 0.4$ and $b_2/D = 0.2$

	$\sqrt{\beta}$							
$1/u_A$	0.05	0.1	0.2	0.3	0.5	0.75	1.0	1.5
0.2	0.0011	0.0011	0.0011	0.0011	8.5×10^{-4}	5.3×10^{-4}	3.2×10^{-4}	1.3×10^{-4}
0.4	0.0249	0.0249	0.0244	0.0214	0.0133	0.0069	0.0037	0.0014
0.6	0.0783	0.0783	0.0740	0.0600	0.0332	0.0156	0.0081	0.0028
0.8	0.1464	0.1463	0.1328	0.1016	0.0518	0.0232	0.0117	0.0040
1.0	0.2194	0.2190	0.1909	0.1397	0.0672	0.0290	0.0144	0.0048
2.0	0.5598	0.5483	0.4079	0.2624	0.1094	0.0439	0.0211	0.0070
4.0	1.0433	0.9617	0.6030	0.3516	0.1347	0.0522	0.0247	0.0084
6.0	1.3679	0.1917	0.6835	0.3833	0.1428	0.0547	0.0259	0.0089
8.0	1.6047	1.3352	0.7251	0.3985	0.1465	0.0559	0.0266	0.0092
10	1.7866	1.4321	0.7498	0.4072	0.1486	0.0566	0.0270	0.0093
20	2.3005	1.6489	0.7957	0.4225	0.1522	0.0581	0.0278	0.0095
40	2.6808	1.7599	0.8145	0.4284	0.1539	0.0588	0.0280	0.0095
60	2.8317	1.7934	0.8195	0.4301	0.1545	0.0590	0.0281	
80	2.9100	1.8085	0.8216	0.4309	0.1548	0.0590	0.0281	
100	2.9569	1.8168	0.8228	0.4314	0.1549			
200	3.0452	1.8309	0.8250	0.4324	0.1550			
400	3.0819	1.8362	0.8262	0.4327	0.1550			
1000	3.0919	1.8377	0.8265	0.4327				

Table 4 Values of $W(u_A,\beta,b_1/D,b_2/D)$ for $b_1/D = 0.4$ and $b_2/D = 0.4$

	$\sqrt{\beta}$							
$1/u_A$	0.05	0.1	0.2	0.3	0.5	0.75	1.0	1.5
0.2	5.7×10^{-4}	5.7×10^{-4}	5.7×10^{-4}	5.7×10^{-4}	5.7×10^{-4}	4.9×10^{-4}	3.8×10^{-4}	2.0×10^{-4}
0.4	0.0125	0.0125	0.0125	0.0124	0.0113	0.0081	0.0053	0.0023
0.6	0.0392	0.0392	0.0392	0.0387	0.0321	0.0204	0.0123	0.0049
0.8	0.0732	0.0732	0.0731	0.0710	0.0544	0.0319	0.0184	0.0071
1.0	0.1097	0.1097	0.1094	0.1041	0.0745	0.0413	0.0232	0.0087
2.0	0.2799	0.2799	0.2723	0.2337	0.1371	0.0671	0.0355	0.0128
4.0	0.5221	0.5215	0.4674	0.3540	0.1800	0.0823	0.0425	0.0155
6.0	0.6873	0.6828	0.5676	0.4036	0.1947	0.0872	0.0448	0.0167
8.0	0.8117	0.7992	0.6257	0.4291	0.2016	0.0895	0.0461	0.0172
10	0.9114	0.8873	0.6626	0.4441	0.2055	0.0909	0.0470	0.0175
20	1.2315	1.1236	0.7375	0.4719	0.2125	0.0939	0.0486	0.0178
40	1.5414	1.2774	0.7711	0.4832	0.2160	0.0953	0.0490	0.0178
60	1.6978	1.3310	0.7805	0.4864	0.2172	0.0956	0.0491	
80	1.7910	1.3567	0.7846	0.4880	0.2177	0.0956	0.0491	
100	1.8521	1.3713	0.7869	0.4890	0.2179	0.0957		
200	1.9821	1.3971	0.7913	0.4910	0.2182	0.0957		
400	2.0444	1.4073	0.7936	0.4916	0.2182			
1000	2.0624	1.4102	0.7941	0.4917				

Annex 10.3 (cont.)

Table 5 Values of $W(u_A,\beta,b_1/D,b_2/D)$ for $b_1/D = 0.6$ and $b_2/D = 0.3$

	$\sqrt{\beta}$							
$1/u_A$	0.05	0.1	0.2	0.3	0.5	0.75	1.0	1.5
0.2	0.0011	0.0011	0.0011	0.0011	0.0011	8.5×10^{-4}	6.2×10^{-4}	3.2×10^{-4}
0.4	0.0249	0.0249	0.0249	0.0244	0.0200	0.0133	0.0086	0.0037
0.6	0.0783	0.0783	0.0781	0.0740	0.0548	0.0332	0.0199	0.0081
0.8	0.1464	0.1464	0.1450	0.1328	0.0913	0.0518	0.0299	0.0118
1.0	0.2194	0.2194	0.2151	0.1909	0.1240	0.0672	0.0378	0.0149
2.0	0.5598	0.5594	0.5138	0.4079	0.2254	0.1095	0.0589	0.0243
4.0	1.0443	1.0334	0.8446	0.6030	0.2957	0.1360	0.0739	0.0307
6.0	1.3744	1.3355	1.0079	0.6835	0.3202	0.1465	0.0800	0.0323
8.0	1.6228	1.5442	1.1019	0.7251	0.3324	0.1523	0.0830	0.0328
10	1.8209	2.6969	1.1616	0.7498	0.3400	0.1559	0.0845	0.0329
20	2.4397	2.0902	1.2836	0.7965	0.3565	0.1621	0.0862	0.0330
40	2.9944	2.3395	1.3396	0.8195	0.3643	0.1634	0.0863	0.0330
60	3.2571	2.4265	1.3568	0.8278	0.3658	0.1635	0.0863	
80	3.4100	2.4686	1.3657	0.8317	0.3662	0.1635		
100	3.5091	2.4926	1.3713	0.8338	0.3663			
200	3.7196	2.5367	1.3822	0.8363	0.3663			
400	3.8215	2.5583	1.3855	0.8365				
1000	3.8516	2.5658	1.3859	0.8365				

Table 6 Values of $W(u_A,\beta,b_1/D,b_2/D)$ for $b_1/D - 0.6$ and $b_2/D = 0.6$

	$\sqrt{\beta}$							
$1/u_A$	0.05	0.1	0.2	0.3	0.5	0.75	1.0	1.5
0.2	5.7×10^{-4}	5.7×10^{-4}	5.7×10^{-4}	5.7×10^{-4}	5.7×10^{-4}	5.7×10^{-4}	5.3×10^{-4}	3.8×10^{-4}
0.4	0.0125	0.0125	0.0125	0.0125	0.0124	0.0113	0.0092	0.0053
0.6	0.0392	0.0392	0.0392	0.0392	0.0381	0.0321	0.0240	0.0124
0.8	0.0732	0.0732	0.0732	0.0731	0.0693	0.0544	0.0384	0.0189
1.0	0.1097	0.1097	0.1097	0.1094	0.1004	0.0745	0.0506	0.0245
2.0	0.2799	0.2799	0.2796	0.2723	0.2167	0.1373	0.0864	0.0423
4.0	0.5221	0.5221	0.5149	0.4674	0.3175	0.1835	0.1147	0.0547
6.0	0.6873	0.6872	0.6613	0.5676	0.3580	0.2032	0.1265	0.0578
8.0	0.8117	0.8113	0.7591	0.6257	0.3798	0.2144	0.1322	0.0587
10	0.9115	0.9101	0.8281	0.6627	0.3938	0.2213	0.1351	0.0589
20	1.2339	1.2151	0.9914	0.7399	0.4253	0.2333	0.1384	0.0590
40	1.5668	1.4742	1.0814	0.7828	0.4403	0.2357	0.1386	0.0590
60	1.7590	1.5863	1.1127	0.7987	0.4432	0.2358	0.1386	
80	1.8888	1.6468	1.1296	0.8062	0.4439	0.2358		
100	1.9831	1.6838	1.1402	0.8102	0.4440			
200	2.2219	1.7583	1.1612	0.8149	0.4441			
400	2.3674	1.7990	1.1676	0.8153	0.4441			
1000	2.4172	1.8135	1.1682	0.8153				

Annex 10.3 (cont.)

Table 7 Values of $W(u_A,\beta,b_1/D,b_2/D)$ for $b_1/D = 0.8$ and $b_2/D = 0.4$

	$\sqrt{\beta}$							
$1/u_A$	0.05	0.1	0.2	0.3	0.5	0.75	1.0	1.5
0.2	0.0011	0.0011	0.0011	0.0011	0.0011	0.0010	8.5×10^{-4}	1.3×10^{-4}
0.4	0.0249	0.0249	0.0249	0.0249	0.0232	0.0182	0.0134	0.0072
0.6	0.0783	0.0783	0.0783	0.0776	0.0677	0.0486	0.0335	0.0178
0.8	0.1464	0.1464	0.1463	0.1432	0.1177	0.0796	0.0530	0.0285
1.0	0.2194	0.2194	0.2190	0.2110	0.1652	0.1069	0.0701	0.0379
2.0	0.5598	0.5598	0.5483	0.4898	0.3296	0.1942	0.1268	0.0647
4.0	1.0443	1.0443	0.9617	0.7813	0.4676	0.2689	0.1714	0.0763
6.0	1.3745	1.367	1.1917	0.9190	0.5293	0.3005	0.1851	0.0776
8.0	1.6234	1.6047	1.3353	0.9975	0.5655	0.3157	0.1898	0.0778
10	1.8229	1.7866	1.4324	1.0485	0.5890	0.3235	0.1945	0.0778
20	2.4642	2.3005	1.6545	1.1666	0.6349	0.3325	0.1983	
40	3.0961	2.6811	1.7881	1.2364	0.6526	0.3410	0.1995	
60	3.4296	2.8339	1.8406	1.2564	0.6616	0.3461		
80	3.6390	2.9160	1.8677	1.2634	0.6697	0.3502		
100	3.7830	2.9679	1.8830	1.2721	0.6754			
200	4.1218	3.0849	1.9151	1.2779	0.6781			
400	4.3209	3.1495	1.9260	1.2815				
1000	4.3966	3.1664	1.9342	1.2856				

Table 8 Values of $W(u_A,\beta,b_1/D,b_2/D)$ for $b_1/D = 0.8$ and $b_2/D = 0.8$

	$\sqrt{\beta}$							
$1/u_A$	0.05	0.1	0.2	0.3	0.5	0.75	1.0	1.5
0.2	5.7×10^{-4}	5.7×10^{-4}	5.7×10^{-4}	5.7×10^{-4}	5.7×10^{-4}	5.7×10^{-4}	5.7×10^{-4}	5.2×10^{-4}
0.4	0.0125	0.0125	0.0125	0.0125	0.0125	0.0123	0.0116	0.0096
0.6	0.0392	0.0392	0.0392	0.0392	0.0391	0.0376	0.0343	0.0267
0.8	0.0732	0.0732	0.0732	0.0732	0.0727	0.0682	0.0607	0.0450
1.0	0.1097	0.1097	0.1097	0.1097	0.1082	0.0993	0.0865	0.0614
2.0	0.2799	0.2799	0.2799	0.2791	0.2632	0.2238	0.1820	0.0984
4.0	0.5221	0.5221	0.5216	0.5109	0.4488	0.3514	0.2602	0.1187
6.0	0.6873	0.6873	0.6837	0.6557	0.5506	0.4068	0.2841	0.1210
8.0	0.8117	0.8117	0.8021	0.7555	0.6132	0.4335	0.2973	0.1213
10	0.9115	0.9114	0.8934	0.8292	0.6544	0.4572	0.3053	0.1214
20	1.2340	1.2319	1.1587	1.0236	0.7346	0.4738	0.3121	0.1214
40	1.5682	1.5482	1.3725	1.1456	0.7670	0.4839	0.3171	
60	1.7663	1.7182	1.4639	1.1907	0.7888	0.4939	0.3198	
80	1.9065	1.8283	1.5113	1.2429	0.8000	0.5044		
100	2.0139	1.9066	1.5581	1.2775	0.8190			
200	2.3286	2.1028	1.5969	1.3007	0.8215			
400	2.5944	2.2659	1.6271	1.3268				
1000	2.7293	2.3455	1.6322	1.3508				

Annex 10.4 Values of the function $W(u_B,\beta,b_1/D,b_2/D)$ for partially-penetrated unconfined aquifers (after Streltsova 1974)

Table 1 Values of $W(u_B,\beta,b_1/D,b_2/D)$ for $b_1/D = 0.1$ and $b_2/D = 0.1$

	$\sqrt{\beta}$						
$1/u_B$	0.05	0.1	0.2	0.3	0.5	0.75	1.0
0.001	0.7930	0.3196	0.0815	0.0302	0.0077	0.0026	0.0012
0.002	0.7930	0.3196	0.0815	0.0302	0.0077	0.0026	0.0012
0.005	0.7931	0.3196	0.0816	0.0303	0.0078	0.0026	0.0012
0.010	0.7931	0.3197	0.0817	0.0304	0.0079	0.0027	0.0013
0.020	0.7931	0.3198	0.0819	0.0305	0.0080	0.0028	0.0014
0.050	0.7933	0.3202	0.0824	0.0311	0.0084	0.0031	0.0016
0.100	0.7935	0.3209	0.0834	0.0320	0.0091	0.0037	0.0021
0.200	0.7941	0.3222	0.0853	0.0338	0.0105	0.0048	0.0032
0.500	0.7956	0.3260	0.0909	0.0392	0.0151	0.0088	0.0071
1.0	0.7981	0.3325	0.1003	0.0484	0.0233	0.0169	0.0156
2.0	0.8032	0.3452	0.1190	0.0674	0.0417	0.0364	0.0373
5.0	0.8182	0.3820	0.1736	0.1251	0.1017	0.0998	0.1033
10	0.8425	0.4392	0.2565	0.2122	0.1856	0.1764	0.1745
20	0.8885	0.5398	0.3873	0.3358	0.2839	0.2572	0.2479
50	1.0088	0.7569	0.5975	0.4995	0.4006	0.3572	0.3426
100	1.1649	0.9612	0.7342	0.5980	0.4782	0.4293	0.4130
200	1.3743	1.1498	0.8392	0.6808	0.5514	0.5000	0.4829
500	1.6696	1.3326	0.9500	0.7798	0.6453	0.5925	0.5748
1000	1.8513	1.4309	1.0250	0.8514	0.7154	0.6621	0.6443
2000	1.9837	1.5129	1.0970	0.9219	0.7851	0.7315	0.7136
5000	2.1099	1.6114	1.1902	1.0142	0.8769	0.8232	0.8053
10000	2.1891	1.6829	1.2600	1.0837	0.9463	0.8926	0.8746

Table 2 Values of $W(u_B,\beta,b_1/D,b_2/D)$ for $b_1/D = 0.2$ and $b_2/D = 0.2$

	$\sqrt{\beta}$						
$1/u_B$	0.05	0.1	0.2	0.3	0.5	0.75	1.0
0.001	1.4167	0.7963	0.3227	0.1552	0.0505	0.0188	0.0091
0.002	1.4167	0.7963	0.3228	0.1552	0.0505	0.0188	0.0091
0.005	1.4167	0.7963	0.3228	0.1553	0.0506	0.0189	0.0092
0.010	1.4167	0.7964	0.3230	0.1555	0.0508	0.0191	0.0094
0.020	1.4167	0.7965	0.3232	0.1558	0.0512	0.0195	0.0097
0.050	1.4168	0.7968	0.3240	0.1569	0.0524	0.0205	0.0107
0.100	1.4170	0.7973	0.3254	0.1587	0.0544	0.0224	0.0123
0.200	1.4173	0.7983	0.3280	0.1623	0.0583	0.0261	0.0158
0.500	1.4182	0.8014	0.3358	0.1729	0.0703	0.0377	0.0271
1.0	1.4197	0.8065	0.3487	0.1903	0.0904	0.0584	0.0484
2.0	1.4226	0.8166	0.3739	0.2434	0.1307	0.1019	0.0953
5.0	1.4315	0.8462	0.4444	0.3818	0.2427	0.2230	0.2218
10	1.4459	0.8926	0.5470	0.4749	0.3847	0.3617	0.3554
20	1.4740	0.9765	0.7074	0.6775	0.5531	0.5120	0.4967
50	1.5516	1.1732	0.9864	0.8779	0.7677	0.7055	0.6828
100	1.6624	1.3881	1.1977	1.0637	0.9176	0.8477	0.8225
200	1.8343	1.6266	1.3815	1.2211	1.0617	0.9881	0.9617
500	1.1397	1.9105	1.5915	1.4150	1.2482	1.1724	1.1453
1000	1.3810	2.0874	1.7384	1.5570	1.3879	1.3114	1.2840
2000	1.5921	2.2442	1.8811	1.6973	1.5271	1.4502	1.4227
5000	1.8216	2.4375	2.0667	1.8816	1.7107	1.6336	1.6060
10000	1.9744	2.5794	2.2061	2.0205	1.8494	1.7723	1.7446

Annex 10.4 (cont.)

Table 3 Values of $W(u_B,\beta,b_1/D,b_2/D)$ for $b_1/D = 0.4$ and $b_2/D = 0.2$

	$\sqrt{\beta}$							
$1/u_B$	0.05	0.1	0.2	0.3	0.5	0.75	1.0	1.5
0.001	3.1197	1.8544	0.8406	0.4461	0.1666	0.0678	0.0341	0.0115
0.002	3.1197	1.8544	0.8406	0.4462	0.1667	0.0679	0.0342	0.0116
0.005	3.1198	1.8545	0.8408	0.4464	0.1669	0.0681	0.0344	0.0118
0.010	3.1198	1.8546	0.8410	0.4467	0.1673	0.0685	0.0348	0.0121
0.020	3.1198	1.8547	0.8414	0.4473	0.1681	0.0693	0.0356	0.0127
0.050	3.1199	1.8551	0.8426	0.4491	0.1705	0.0718	0.0379	0.0147
0.100	3.1201	1.8559	0.8446	0.4520	0.1744	0.0758	0.0418	0.0181
0.200	3.1206	1.8573	0.8486	0.4580	0.1822	0.0840	0.0498	0.0254
0.500	3.1218	1.8616	0.8605	0.4758	0.2055	0.1091	0.0752	0.0506
1.0	3.1238	1.8687	0.8801	0.5050	0.2442	0.1521	0.1206	0.1003
2.0	3.1278	1.8829	0.9184	0.5616	0.3198	0.2387	0.2152	0.2080
5.0	3.1398	1.9242	1.0261	0.7168	0.5235	0.4688	0.4596	0.4661
10	3.1595	1.9897	1.1841	0.9308	0.7785	0.7298	0.7168	0.7158
20	3.1979	2.1092	1.4355	1.2347	1.0872	0.0182	0.9929	0.9819
50	3.3049	2.3967	1.8930	1.6955	1.4962	1.3947	1.3611	1.3425
100	3.4606	2.7255	2.2637	2.0258	1.7900	1.6794	1.6392	1.6179
200	3.7090	3.1127	2.6056	2.3312	2.0755	1.9591	1.9169	1.8943
500	4.1765	3.6128	3.0127	2.7142	2.4469	2.3271	2.2838	2.2603
1000	4.5758	3.9466	3.3029	2.9967	2.7258	2.6048	2.5611	2.5373
2000	4.9517	4.2517	3.5864	3.2766	3.0038	2.8823	2.8384	2.8145
5000	5.3869	4.6341	3.9566	3.6447	3.3708	3.2490	3.2050	3.1810
10000	5.6863	4.9164	4.2351	3.9225	3.6482	3.5263	3.4822	3.4582

Table 4 Values of $W(u_B,\beta,b_1/D,b_2/D)$ for $b_1/D = 0.4$ and $b_2/D = 0.4$

	$\sqrt{\beta}$							
$1/u_B$	0.05	0.1	0.2	0.3	0.5	0.75	1.0	1.5
0.001	2.1193	1.4457	0.8244	0.5202	0.2425	0.1138	0.0612	0.0217
0.002	2.1193	1.4458	0.8244	0.5202	0.2426	0.1139	0.0613	0.0217
0.005	2.1193	1.4458	0.8245	0.5204	0.2428	0.1141	0.0616	0.0220
0.010	2.1193	1.4458	0.8246	0.5206	0.2432	0.1146	0.0621	0.0224
0.020	2.1193	1.4459	0.8248	0.5210	0.2439	0.1155	0.0630	0.0233
0.050	2.1193	1.4461	0.8255	0.5222	0.2460	0.1182	0.0659	0.0259
0.100	2.1194	1.4464	0.8266	0.5242	0.2496	0.1227	0.0707	0.0304
0.200	2.1196	1.4470	0.8288	0.5283	0.2566	0.1317	0.0803	0.0396
0.500	2.1201	1.4489	0.8354	0.5403	0.2775	0.1584	0.1095	0.0697
1.0	2.1209	1.4520	0.8462	0.5600	0.3115	0.2023	0.1584	0.1238
2.0	2.1225	1.4583	0.8673	0.5979	0.3758	0.2855	0.2528	0.2319
5.0	2.1274	1.4768	0.9275	0.7018	0.5421	0.4925	0.4812	0.4803
10	2.1355	1.5065	1.0180	0.8468	0.7488	0.7254	0.7211	0.7222
20	2.1514	1.5627	1.1695	1.0632	1.0109	0.9927	0.9861	0.9837
50	2.1972	1.7085	1.4778	1.4297	1.3860	1.3588	1.3474	1.3414
100	2.2676	1.8964	1.7669	1.7242	1.6693	1.6366	1.6232	1.6158
200	2.3900	2.1524	2.0662	2.0138	1.9499	1.9142	1.8998	1.8917
500	2.6586	2.5464	2.4513	2.3883	2.3186	2.2810	2.2660	2.2574
1000	2.9343	2.8464	2.7351	2.6682	2.5965	2.5583	2.5431	2.5344
2000	3.2369	3.1372	3.0157	2.9468	2.8741	2.8356	2.8203	2.8115
5000	3.6322	3.5121	3.3842	3.3141	3.2408	3.2022	3.1868	3.1779
10000	3.9206	3.7921	3.6621	3.5917	3.5182	3.4794	3.4640	3.4551

Annex 10.4 (cont.)

Table 5 Values of $W(u_B,\beta,b_1/D,b_2/D)$ for $b_1/D = 0.6$ and $b_2/D = 0.3$

	$\sqrt{\beta}$							
$1/u_B$	0.05	0.1	0.2	0.3	0.5	0.75	1.0	1.5
0.001	3.9479	2.6184	1.4298	0.8779	0.4008	0.1879	0.1015	0.0360
0.002	3.9480	2.6184	1.4298	0.8779	0.4009	0.1880	0.1016	0.0362
0.005	3.9480	2.6184	1.4299	0.8781	0.4012	0.1884	0.1020	0.0365
0.010	3.9480	2.6185	1.4301	0.8784	0.4017	0.1890	0.1027	0.0372
0.020	3.9480	2.6186	1.4304	0.8791	0.4028	0.1904	0.1041	0.0384
0.050	3.9481	2.6189	1.4315	0.8810	0.4060	0.1943	0.1083	0.0423
0.100	3.9482	2.6194	1.4333	0.8841	0.4112	0.2008	0.1152	0.0489
0.200	3.9485	2.6205	1.4369	0.8904	0.4216	0.2139	0.1293	0.0626
0.500	3.9494	2.6238	1.4476	0.9092	0.4527	0.2530	0.1721	0.1070
1.0	3.9508	2.6292	1.4652	0.9399	0.5031	0.3174	0.2441	0.1875
2.0	3.9537	2.6399	1.4998	0.9992	0.5995	0.4407	0.3842	0.3490
5.0	3.9622	2.6714	1.5976	1.1615	0.8507	0.7510	0.7263	0.7213
10	3.9763	2.7221	1.7438	1.3878	1.1644	1.1014	1.0866	1.0842
20	4.0039	2.8169	1.9857	1.7226	1.5607	1.5031	1.4842	1.4765
50	4.0828	3.0586	2.4677	2.2812	2.1248	2.0522	2.0261	2.0130
100	4.2025	3.3620	2.9097	2.7251	2.5498	2.4689	2.4398	2.4246
200	4.4066	3.7637	3.3614	3.1598	2.9707	2.8853	2.8547	2.8383
500	4.8398	4.3653	3.9395	3.7215	3.5236	3.4355	3.4039	3.3868
1000	5.2692	4.8172	4.3653	4.1414	3.9405	3.8515	3.8169	3.8023
2000	5.7300	5.2537	4.7861	4.5593	4.3569	4.2674	4.2354	4.2180
5000	6.3248	5.8159	5.3388	5.1102	4.9070	4.8173	4.7851	4.7676
10000	6.7576	6.2359	5.7556	5.5265	5.3230	5.2331	5.2010	5.1835

Table 6 Values of $W(u_B,\beta,b_1/D,b_2/D)$ for $b_1/D = 0.6$ and $b_2/D = 0.6$

	$\sqrt{\beta}$							
$1/u_B$	0.05	0.1	0.2	0.3	0.5	0.75	1.0	1.5
0.001	2.6242	1.9387	1.2748	0.9142	0.5229	0.2878	0.1684	0.0632
0.002	2.6242	1.9387	1.2748	0.9134	0.5229	0.2879	0.1685	0.0634
0.005	2.6242	1.9387	1.2749	0.9144	0.5232	0.2883	0.1690	0.0638
0.010	2.6242	1.9387	1.2750	0.9146	0.5236	0.2889	0.1697	0.0646
0.020	2.6242	1.9388	1.2752	0.9150	0.5244	0.2902	0.1713	0.0662
0.050	2.6243	1.9389	1.2757	0.9161	0.5269	0.2940	0.1759	0.0711
0.100	2.6243	1.9392	1.2767	0.9180	0.5309	0.3003	0.1835	0.0792
0.200	2.6244	1.9397	1.2785	0.9218	0.5391	0.3128	0.1987	0.0956
0.500	2.6248	1.9411	1.2840	0.9330	0.5630	0.3498	0.2436	0.1462
1.0	2.6254	1.9436	1.2932	0.9515	0.6018	0.4089	0.3160	0.2312
2.0	2.6267	1.9485	1.3112	0.9874	0.6750	0.5184	0.4491	0.3907
5.0	2.6304	1.9630	1.3632	1.0873	0.8654	0.7855	0.7608	0.7454
10	2.6366	1.9867	1.4437	1.2323	1.1097	1.0918	1.0932	1.0953
20	2.6489	2.0322	1.5833	1.4626	1.4390	1.4600	1.4721	1.4795
50	2.6847	2.1563	1.9033	1.8981	1.9505	1.9870	2.0022	2.0110
100	2.7414	2.3295	2.2419	2.2886	2.3580	2.3965	2.4119	2.4208
200	2.8448	2.5919	2.6317	2.6974	2.7705	2.8093	2.8248	2.8337
500	3.0947	3.0588	3.1743	3.2448	3.3185	3.3573	3.3728	3.3817
1000	3.3875	3.4601	3.5892	3.6601	3.7338	3.7726	3.7881	3.7970
2000	3.7526	3.8730	4.0048	4.0757	4.1494	4.1882	4.2037	4.2126
5000	4.2869	4.4223	4.5545	4.6253	4.6990	4.7378	4.7533	4.7622
10000	4.7013	4.8382	4.9703	5.0412	5.1149	5.1537	5.1691	5.1780

Annex 10.4 (cont.)

Table 7 Values of $W(u_B,\beta,b_1/D,b_2/D)$ for $b_1/D = 0.8$ and $b_2/D = 0.4$

	$\sqrt{\beta}$							
$1/u_B$	0.05	0.1	0.2	0.3	0.5	0.75	1.0	1.5
0.001	4.6137	3.2588	1.9878	1.3410	0.7047	0.3671	0.2094	0.0773
0.002	4.6137	3.2588	1.9879	1.3410	0.7048	0.3672	0.2096	0.0775
0.005	4.6137	3.2588	1.9880	1.3412	0.7052	0.3677	0.2102	0.0781
0.010	4.6137	3.2588	1.9881	1.3415	0.7058	0.3686	0.2112	0.0791
0.020	4.6137	3.2589	1.9884	1.3422	0.7070	0.3703	0.2131	0.0811
0.050	4.6138	3.2592	1.9894	1.3440	0.7106	0.3754	0.2190	0.0871
0.100	4.6139	3.2596	1.9910	1.3471	0.7165	0.3839	0.2288	0.0972
0.200	4.6141	3.2605	1.9942	1.3532	0.7284	0.4007	0.2485	0.1179
0.500	4.6148	3.2632	2.0037	1.3715	0.7634	0.4507	0.3071	0.1825
1.0	4.6160	3.2677	2.0195	1.4014	0.8202	0.5315	0.4029	0.2936
2.0	4.6183	3.2765	2.0504	1.4594	0.9279	0.6830	0.5828	0.5065
5.0	4.6251	3.3027	2.1389	1.6195	1.2077	1.0566	1.0102	0.9862
10	4.6365	3.3451	2.2737	1.8473	1.5624	1.4828	1.4644	1.4573
20	4.6589	3.4257	2.5046	2.1989	2.0291	1.9870	1.9769	1.9732
50	4.7237	3.6390	2.9974	2.8305	2.7321	2.6983	2.6836	2.6832
100	4.8244	3.9238	3.4919	3.3728	3.2823	3.2470	3.2360	3.2308
200	5.0037	4.3319	4.0370	3.9281	3.8355	3.7989	3.7873	3.7817
500	5.4120	5.0105	4.7744	4.6634	4.5680	4.5184	4.518	4.5125
1000	5.8595	5.5660	5.3318	5.2189	5.1224	5.0843	5.0723	5.0663
2000	6.3860	6.1258	5.8879	5.7739	5.6768	5.6386	5.6265	5.6204
5000	7.1228	6.8631	6.6220	6.5027	6.4098	6.3715	6.3533	6.3523
10000	7.6825	7.4192	7.1768	7.0619	6.9643	6.9259	6.9138	6.9077

Table 8 Values of $W(u_B,\beta,b_1/D,b_2/D)$ for $b_1/D = 0.8$ and $b_2/D = 0.8$

	$\sqrt{\beta}$							
$1/u_B$	0.05	0.1	0.2	0.3	0.5	0.75	1.0	1.5
0.001	3.2447	2.5476	1.8414	1.4220	0.9015	0.5324	0.3222	0.1237
0.002	3.2447	2.5476	1.8414	1.4220	0.9016	0.5326	0.3224	0.1239
0.005	3.2447	2.5476	1.8415	1.4222	0.9019	0.5330	0.3230	0.1246
0.010	3.2447	2.5476	1.8416	1.4223	0.9023	0.5338	0.3240	0.1258
0.020	3.2447	2.5476	1.8417	1.4227	0.9032	0.5353	0.3260	0.1281
0.050	3.2447	2.5478	1.8423	1.4239	0.9059	0.5399	0.3320	0.1351
0.100	3.2448	2.5480	1.8432	1.4257	0.9103	0.5476	0.3420	0.1467
0.200	3.2449	2.5485	1.8449	1.4295	0.9192	0.5627	0.3618	0.1701
0.500	3.2453	2.5499	1.8502	1.4407	0.9453	0.6072	0.4198	0.2403
1.0	3.2458	2.5521	1.8590	1.4592	0.9877	0.6781	0.5119	0.2545
2.0	3.2470	2.5567	1.8764	1.4953	1.0682	0.8089	0.6789	0.5625
5.0	3.2505	2.5703	1.9270	1.5972	1.2803	1.1288	1.0670	1.0195
10	3.2562	2.5926	2.0065	1.7486	1.5600	1.5034	1.4873	1.4746
20	3.2676	2.6360	2.1503	1.9986	1.9528	1.9676	1.9768	1.9799
50	3.3012	2.7568	2.4911	2.5026	2.5941	2.6516	2.6731	2.6859
100	3.3551	2.9330	2.8800	2.9843	3.1225	3.1910	3.2158	3.2288
200	3.4556	3.2131	3.3558	3.5081	3.6651	3.7383	3.7646	3.7785
500	3.7105	3.7538	4.0503	4.2252	4.3914	4.4670	4.4941	4.5087
1000	4.0298	4.2512	4.5940	4.7749	4.9437	5.0201	5.0475	5.0623
2000	4.4565	4.7823	5.1435	5.3270	5.4971	5.5739	5.6014	5.6163
5000	5.1234	5.5033	5.8736	6.0586	6.2295	6.3065	6.3341	6.3490
10000	5.6604	6.0541	6.4272	6.6127	6.7838	6.8609	6.8885	6.9035

Annex 11.1 Values of Papadopulos's function $F(u,\alpha,r/r_{ew})$ for large-diameter wells in confined aquifers (after Papadopulos 1967)

Table 1 Values of $F(u,\alpha,r/r_{ew})$ for $\alpha = 10^{-1}$

1/u	$r/r_{ew} = 1$	2	5	10	20	50	100	200
5(−1)	4.88(−2)	1.96(−2)	1.75(−2)	2.41(−2)	3.48(−2)	4.24(−2)	4.48(−2)	4.50(−2)
1(0)	9.19(−2)	7.01(−2)	9.55(−2)	1.41(−1)	1.85(−1)	2.09(−1)	2.14(−1)	2.15(−1)
2(0)	1.77(−1)	1.95(−1)	3.21(−1)	4.44(−1)	5.20(−1)	5.49(−1)	5.55(−1)	5.59(−1)
5(0)	4.06(−1)	5.78(−1)	9.42(−1)	1.13 (0)	1.19 (0)	1.22 (0)	1.22 (0)	1.22 (0)
1(1)	7.34(−1)	1.11 (0)	1.60 (0)	1.76 (0)	1.80 (0)	1.80 (0)	1.80 (0)	1.80 (0)
2(1)	1.26 (0)	1.84 (0)	2.33 (0)	2.43 (0)	2.46 (0)	2.46 (0)	2.46 (0)	2.46 (0)
5(1)	2.30 (0)	2.97 (0)	3.28 (0)	3.34 (0)	3.35 (0)	3.35 (0)	3.35 (0)	3.35 (0)
1(2)	3.28 (0)	3.81 (0)	4.00 (0)	4.03 (0)	4.03 (0)	4.03 (0)	4.03 (0)	4.03 (0)
2(2)	4.26 (0)	4.60 (0)	4.70 (0)	4.72 (0)	4.72 (0)	4.72 (0)	4.72 (0)	4.72 (0)
5(2)	5.42 (0)	5.58 (0)	5.63 (0)	5.64 (0)	5.64 (0)	5.64 (0)	5.64 (0)	5.64 (0)
1(3)	6.21 (0)	6.30 (0)	6.33 (0)	6.33 (0)	6.33 (0)	6.33 (0)	6.33 (0)	6.33 (0)
2(3)	6.96 (0)	7.01 (0)	7.01 (0)	7.01 (0)	7.01 (0)	7.01 (0)	7.01 (0)	7.01 (0)
5(3)	7.87 (0)	7.93 (0)	7.93 (0)	7.93 (0)	7.93 (0)	7.93 (0)	7.93 (0)	7.93 (0)
1(4)	8.57 (0)	8.63 (0)	8.63 (0)	8.63 (0)	8.63 (0)	8.63 (0)	8.63 (0)	8.63 (0)
2(4)	9.32 (0)	9.32 (0)	9.32 (0)	9.32 (0)	9.32 (0)	9.32 (0)	9.32 (0)	9.32 (0)
5(4)	1.02 (1)	1.02 (1)	1.02 (1)	1.02 (1)	1.02 (1)	1.02 (1)	1.02 (1)	1.02 (1)

Table 2 Values of $F(u,\alpha,r/r_{ew})$ for $\alpha = 10^{-2}$

1/u	$r/r_{ew} = 1$	2	5	10	20	50	100	200
5(−1)	4.99(−3)	2.13(−3)	2.11(−3)	3.52(−3)	7.47(−3)	2.03(−2)	3.44(−2)	4.35(−2)
1(0)	9.91(−3)	7.99(−3)	1.32(−2)	2.69(−2)	6.12(−2)	1.42(−1)	1.91(−1)	2.11(−1)
2(0)	1.97(−2)	2.40(−2)	5.40(−2)	1.21(−1)	2.63(−1)	4.65(−1)	5.31(−1)	5.51(−1)
5(0)	4.89(−2)	8.34(−2)	2.33(−1)	5.12(−1)	9.15(−1)	1.16 (0)	1.20 (0)	1.22 (0)
1(1)	9.67(−2)	1.93(−1)	5.67(−1)	1.12 (0)	1.58 (0)	1.78 (0)	1.81 (0)	1.82 (0)
2(1)	1.90(−1)	4.16(−1)	1.18 (0)	1.95 (0)	2.32 (0)	2.44 (0)	2.46 (0)	2.47 (0)
5(1)	4.53(−1)	1.03 (0)	2.42 (0)	3.11 (0)	3.29 (0)	3.34 (0)	3.35 (0)	3.35 (0)
1(2)	8.52(−1)	1.87 (0)	3.48 (0)	3.90 (0)	4.00 (0)	4.03 (0)	4.03 (0)	4.03 (0)
2(2)	1.54 (0)	3.05 (0)	4.43 (0)	4.65 (0)	4.71 (0)	4.72 (0)	4.73 (0)	4.73 (0)
5(2)	3.04 (0)	4.78 (0)	5.52 (0)	5.61 (0)	5.63 (0)	5.64 (0)	5.64 (0)	5.64 (0)
1(3)	4.55 (0)	5.90 (0)	6.27 (0)	6.31 (0)	6.33 (0)	6.33 (0)	6.33 (0)	6.33 (0)
2(3)	6.03 (0)	6.81 (0)	6.99 (0)	7.01 (0)	7.02 (0)	7.02 (0)	7.02 (0)	7.02 (0)
5(3)	7.56 (0)	7.85 (0)	7.92 (0)	7.94 (0)	7.94 (0)	7.94 (0)	7.94 (0)	7.94 (0)
1(4)	8.44 (0)	8.59 (0)	8.63 (0)	8.63 (0)	8.63 (0)	8.63 (0)	8.63 (0)	8.63 (0)
2(4)	9.23 (0)	9.30 (0)	9.33 (0)	9.33 (0)	9.33 (0)	9.33 (0)	9.33 (0)	9.33 (0)
5(4)	1.02 (1)	1.02 (1)	1.02 (1)	1.02 (1)	1.02 (1)	1.02 (1)	1.02 (1)	1.02 (1)
1(5)	1.09 (1)	1.09 (1)	1.09 (1)	1.09 (1)	1.09 (1)	1.09 (1)	1.09 (1)	1.09 (1)
2(5)	1.16 (1)	1.16 (1)	1.16 (1)	1.16 (1)	1.16 (1)	1.16 (1)	1.16 (1)	1.16 (1)
5(5)	1.25 (1)	1.25 (1)	1.25 (1)	1.25 (1)	1.25 (1)	1.25 (1)	1.25 (1)	1.25 (1)
1(6)	1.32 (1)	1.32 (1)	1.32 (1)	1.32 (1)	1.32 (1)	1.32 (1)	1.32 (1)	1.32 (1)

Annex 11.1 (cont.)

Table 3 Values of F $(u,\alpha,r/r_{ew})$ for $\alpha = 10^{-3}$

1/u	$r/r_{ew} = 1$	2	5	10	20	50	100	200
5(–1)	5.00(–4)	2.15(–4)	2.15(–4)	3.70(–4)	8.35(–4)	3.05(–3)	8.38(–3)	1.50(–2)
1(0)	9.99(–4)	8.11(–4)	1.37(–3)	2.95(–3)	7.58(–3)	2.81(–2)	7.56(–2)	1.47(–1)
2(0)	2.00(–3)	2.45(–3)	5.77(–3)	1.42(–2)	3.90(–2)	1.54(–1)	3.23(–1)	4.78(–1)
5(0)	4.99(–3)	8.71(–3)	2.67(–2)	7.24(–2)	2.03(–1)	6.59(–1)	1.02 (0)	1.17 (0)
1(1)	9.97(–3)	2.07(–2)	7.16(–2)	2.01(–1)	5.41(–1)	1.38 (0)	1.70 (0)	1.79 (0)
2(1)	1.99(–2)	4.66(–2)	1.74(–1)	4.87(–1)	1.19 (0)	2.27 (0)	2.40 (0)	2.45 (0)
5(1)	4.95(–2)	1.29(–1)	5.05(–1)	1.31 (0)	2.52 (0)	3.22 (0)	3.32 (0)	3.35 (0)
1(2)	9.83(–2)	2.70(–1)	1.04 (0)	2.38 (0)	3.59 (0)	3.96 (0)	4.02 (0)	4.02 (0)
2(2)	1.95(–1)	5.47(–1)	1.96 (0)	3.68 (0)	4.50 (0)	4.69 (0)	4.72 (0)	4.72 (0)
5(2)	4.73(–1)	1.31 (0)	3.81 (0)	5.23 (0)	5.55 (0)	5.63 (0)	5.64 (0)	5.64 (0)
1(3)	9.07(–1)	2.39 (0)	5.34 (0)	6.13 (0)	6.28 (0)	6.32 (0)	6.32 (0)	6.32 (0)
2(3)	1.69 (0)	3.98 (0)	6.57 (0)	6.92 (0)	7.00 (0)	7.02 (0)	7.02 (0)	7.02 (0)
5(3)	3.52 (0)	6.44 (0)	7.77 (0)	7.90 (0)	7.93 (0)	7.93 (0)	7.93 (0)	7.93 (0)
1(4)	5.53 (0)	7.95 (0)	8.55 (0)	8.61 (0)	8.63 (0)	8.63 (0)	8.63 (0)	8.63 (0)
2(4)	7.63 (0)	9.02 (0)	9.28 (0)	9.31 (0)	9.31 (0)	9.31 (0)	9.31 (0)	9.31 (0)
5(4)	9.68 (0)	1.01 (1)	1.02 (1)	1.02 (1)	1.02 (1)	1.02 (1)	1.02 (1)	1.02 (1)
1(5)	1.07 (1)	1.09 (1)	1.09 (1)	1.09 (1)	1.09 (1)	1.09 (1)	1.09 (1)	1.09 (1)
2(5)	1.15 (1)	1.16 (1)	1.16 (1)	1.16 (1)	1.16 (1)	1.16 (1)	1.16 (1)	1.16 (1)
5(5)	1.25 (1)	1.25 (1)	1.25 (1)	1.25 (1)	1.25 (1)	1.25 (1)	1.25 (1)	1.25 (1)
1(6)	1.32 (1)	1.32 (1)	1.32 (1)	1.32 (1)	1.32 (1)	1.32 (1)	1.32 (1)	1.32 (1)
2(6)	1.39 (1)	1.39 (1)	1.39 (1)	1.39 (1)	1.39 (1)	1.39 (1)	1.39 (1)	1.39 (1)
5(6)	1.48 (1)	1.48 (1)	1.48 (1)	1.48 (1)	1.48 (1)	1.48 (1)	1.48 (1)	1.48 (1)
1(7)	1.55 (1)	1.55 (1)	1.55 (1)	1.55 (1)	1.55 (1)	1.55 (1)	1.55 (1)	1.55 (1)

Table 4 Values of F $(u,\alpha,r/r_{ew})$ for $\alpha = 10^{-4}$

1/u r/r_{ew} = 1		2	5	10	20	50	100	200
5(–1)	5.00(–5)	2.17(–5)	2.18(–5)	3.73(–5)	8.46(–5)	3.16(–4)	9.56(–4)	3.83(–3)
1(0)	1.00(–4)	8.15(–5)	1.38(–4)	2.98(–4)	7.77(–4)	3.23(–3)	1.01(–2)	3.42(–2)
2(0)	2.00(–4)	2.47(–4)	5.81(–4)	1.45(–3)	4.10(–3)	1.80(–2)	5.62(–2)	1.75(–1)
5(0)	5.00(–4)	8.76(–4)	2.71(–3)	7.54(–3)	2.27(–2)	1.03(–1)	3.04(–1)	7.10(–1)
1(1)	1.00(–3)	2.09(–3)	7.34(–3)	2.16(–2)	6.69(–2)	2.97(–1)	7.92(–1)	1.43 (0)
2(1)	2.00(–3)	4.72(–3)	1.82(–2)	5.55(–2)	1.74(–1)	7.30(–1)	1.62 (0)	2.24 (0)
5(1)	5.00(–3)	1.32(–2)	5.56(–2)	1.74(–1)	5.36(–1)	1.87 (0)	2.95 (0)	3.28 (0)
1(2)	9.98(–3)	2.81(–2)	1.23(–1)	3.86(–1)	1.14 (0)	3.08 (0)	3.84 (0)	4.02 (0)
2(2)	1.99(–2)	5.88(–2)	2.64(–1)	8.13(–1)	2.17 (0)	4.25 (0)	4.63 (0)	4.71 (0)
5(2)	4.97(–2)	1.53(–1)	6.89(–1)	1.97 (0)	4.14 (0)	5.47 (0)	5.60 (0)	5.63 (0)
1(3)	9.90(–2)	3.10(–1)	1.36 (0)	3.44 (0)	5.61 (0)	6.24 (0)	6.31 (0)	6.33 (0)
2(3)	1.97(–1)	6.18(–1)	2.53 (0)	5.26 (0)	6.71 (0)	6.98 (0)	7.01 (0)	7.02 (0)
5(3)	4.81(–1)	1.48 (0)	4.95 (0)	7.33 (0)	7.82 (0)	7.92 (0)	7.94 (0)	7.94 (0)
1(4)	9.34(–1)	2.72 (0)	7.03 (0)	8.37 (0)	8.57 (0)	8.62 (0)	8.63 (0)	8.63 (0)
2(4)	1.77 (0)	4.65 (0)	8.65 (0)	9.20 (0)	9.29 (0)	9.32 (0)	9.33 (0)	9.33 (0)
5(4)	3.83 (0)	7.87 (0)	1.00 (1)	1.02 (1)	1.02 (1)	1.02 (1)	1.02 (1)	1.02 (1)
1(5)	6.25 (0)	9.92 (0)	1.08 (1)	1.09 (1)	1.09 (1)	1.09 (1)	1.09 (1)	1.09 (1)
2(5)	8.99 (0)	1.12 (1)	1.16 (1)	1.16 (1)	1.16 (1)	1.16 (1)	1.16 (1)	1.16 (1)
5(5)	1.17 (1)	1.24 (1)	1.25 (1)	1.25 (1)	1.25 (1)	1.25 (1)	1.25 (1)	1.25 (1)
1(6)	1.29 (1)	1.32 (1)	1.32 (1)	1.32 (1)	1.32 (1)	1.32 (1)	1.32 (1)	1.32 (1)
2(6)	1.38 (1)	1.39 (1)	1.39 (1)	1.39 (1)	1.39 (1)	1.39 (1)	1.39 (1)	1.39 (1)
5(6)	1.48 (1)	1.48 (1)	1.48 (1)	1.48 (1)	1.48 (1)	1.48 (1)	1.48 (1)	1.48 (1)
1(7)	1.55 (1)	1.55 (1)	1.55 (1)	1.55 (1)	1.55 (1)	1.55 (1)	1.55 (1)	1.55 (1)
2(7)	1.62 (1)	1.62 (1)	1.62 (1)	1.62 (1)	1.62 (1)	1.62 (1)	1.62 (1)	1.62 (1)
5(7)	1.71 (1)	1.72 (1)	1.72 (1)	1.72 (1)	1.72 (1)	1.72 (1)	1.72 (1)	1.72 (1)
1(8)	1.78 (1)	1.78 (1)	1.78 (1)	1.78 (1)	1.78 (1)	1.78 (1)	1.78 (1)	1.78 (1)

Table 5 Values of F $(u,\alpha,r/r_{ew})$ for $\alpha = 10^{-5}$

1/u r/r_{ew} = 1		2	5	10	20	50	100	200
5(–1)	5.00(–6)	2.27(–6)	2.48(–6)	4.19(–6)	9.00(–6)	3.21(–5)	9.77(–5)	3.15(–4)
1(0)	1.00(–5)	8.36(–6)	1.44(–5)	3.07(–5)	7.89(–5)	3.27(–4)	1.04(–3)	3.44(–3)
2(0)	2.00(–5)	2.51(–5)	5.94(–5)	1.47(–4)	4.14(–4)	1.84(–3)	6.02(–3)	2.00(–2)
5(0)	5.00(–5)	8.87(–5)	2.74(–4)	7.61(–4)	2.31(–3)	1.08(–2)	3.61(–2)	1.19(–1)
1(1)	1.00(–4)	2.11(–4)	7.42(–4)	2.18(–3)	6.85(–3)	3.30(–2)	1.10(–1)	3.50(–1)
2(1)	2.00(–4)	4.77(–4)	1.84(–3)	5.65(–3)	1.82(–2)	8.90(–2)	2.92(–1)	8.57(–1)
5(1)	5.00(–4)	1.34(–3)	5.64(–3)	1.80(–2)	5.92(2)	2.89(–1)	8.91(–1)	2.12 (0)
1(2)	1.00(–3)	2.84(–3)	1.26(–2)	4.09(–2)	1.36(–1)	6.49(–1)	1.80 (0)	3.34 (0)
2(2)	2.00(–3)	5.96(–3)	2.74(–2)	9.03(–2)	3.01(–1)	1.35 (0)	3.14 (0)	4.40 (0)
5(2)	5.00(–3)	1.56(–2)	7.43(–2)	2.47(–1)	8.06(–1)	3.03 (0)	5.01 (0)	5.52 (0)
1(3)	9.99(–3)	3.20(–2)	1.55(–1)	5.15(–1)	1.60 (0)	4.75 (0)	6.06 (0)	6.27 (0)
2(3)	2.00(–2)	6.54(–2)	3.20(–1)	1.04 (0)	2.96 (0)	6.31 (0)	6.90 (0)	6.99 (0)
5(3)	4.98(–2)	1.66(–1)	8.08(–1)	2.45 (0)	5.58 (0)	7.71 (0)	7.89 (0)	7.93 (0)
1(4)	9.93(–2)	3.34(–1)	1.58 (0)	4.28 (0)	7.54 (0)	8.52 (0)	8.61 (0)	8.63 (0)
2(4)	1.98(–1)	6.62(–1)	2.93 (0)	6.63 (0)	8.90 (0)	9.21 (0)	9.31 (0)	9.31 (0)
5(4)	4.86(–1)	1.59 (0)	5.86 (0)	9.36 (0)	1.01 (1)	1.02 (1)	1.02 (1)	1.02 (1)
1(5)	9.49(–1)	2.95 (0)	8.53 (0)	1.06 (1)	1.09 (1)	1.09 (1)	1.09 (1)	1.09 (1)
2(5)	1.82 (0)	5.15 (0)	1.07 (1)	1.15 (1)	1.16 (1)	1.16 (1)	1.16 (1)	1.16 (1)
5(5)	4.03 (0)	9.08 (0)	1.23 (1)	1.25 (1)	1.25 (1)	1.25 (1)	1.25 (1)	1.25 (1)
1(6)	6.78 (0)	1.18 (1)	1.31 (1)	1.32 (1)	1.32 (1)	1.32 (1)	1.32 (1)	1.32 (1)
2(6)	1.01 (1)	1.34 (1)	1.39 (1)	1.39 (1)	1.39 (1)	1.39 (1)	1.39 (1)	1.39 (1)
5(6)	1.37 (1)	1.47 (1)	1.48 (1)	1.49 (1)	1.49 (1)	1.49 (1)	1.49 (1)	1.49 (1)
1(7)	1.51 (1)	1.55 (1)	1.55 (1)	1.55 (1)	1.55 (1)	1.55 (1)	1.55 (1)	1.55 (1)
2(7)	1.61 (1)	1.62 (1)	1.62 (1)	1.62 (1)	1.62 (1)	1.62 (1)	1.62 (1)	1.62 (1)
5(7)	1.71 (1)	1.71 (1)	1.71 (1)	1.71 (1)	1.71 (1)	1.71 (1)	1.71 (1)	1.71 (1)
1(8)	1.78 (1)	1.78 (1)	1.78 (1)	1.78 (1)	1.78 (1)	1.78 (1)	1.78 (1)	1.78 (1)
2(8)	1.85 (1)	1.85 (1)	1.85 (1)	1.85 (1)	1.85 (1)	1.85 (1)	1.85 (1)	1.85 (1)
5(8)	1.94 (1)	1.94 (1)	1.94 (1)	1.94 (1)	1.94 (1)	1.94 (1)	1.94 (1)	1.94 (1)
1(9)	2.02 (1)	2.02 (1)	2.02 (1)	2.02 (1)	2.02 (1)	2.02 (1)	2.02 (1)	2.02 (1)

Annex 11.2 Values of Boulton-Streltsova's function $W(u_A,S_A,\beta,r/r_{ew},b_1/D,d/D,b_2/D)$ for large-diameter wells in unconfined aquifers (after Boulton and Streltsova 1976)

Tabel 1 Values of $W(u_A,S_A,\beta,r/r_{ew},b_1/D,d/D,b_2/D)$ for $b_1/D = 1.0$, $d/D = 0.0$, $b_2/D = 0.4$, $S_A = 10^{-3}$

$1/u_A$	$r/r_{ew} = 1.0$				$r/r_{ew} = 2.0$				$r/r_{ew} = 5.0$			
$\sqrt{\beta}$:	0.001	0.1	0.5	1.0	0.001	0.1	0.5	1.0	0.001	0.1	0.5	1.0
1.0	0.0010	0.0010	0.0010	0.0010	0.0008	0.0008	0.0006	0.0005	0.0013	0.0013	0.0012	0.0008
2.0	0.0020	0.0020	0.0020	0.0020	0.0024	0.0024	0.0022	0.0019	0.0058	0.0051	0.0048	0.0038
5.0	0.0050	0.0050	0.0050	0.0049	0.0087	0.0087	0.0073	0.0057	0.0266	0.0251	0.0197	0.0131
10.0	0.0100	0.0099	0.0099	0.0098	0.0207	0.0207	0.0182	0.0104	0.0715	0.0683	0.0602	0.0300
20.0	0.0199	0.0197	0.0195	0.0192	0.0463	0.0467	0.0375	0.0211	0.1736	0.1657	0.1346	0.0568
50.0	0.0436	0.0492	0.0489	0.0484	0.1293	0.1285	0.0867	0.0517	0.5009	0.4735	0.3226	0.1193
100.0	0.0923	0.0972	0.0968	0.0960	0.2700	0.2493	0.1702	0.0982	1.0011	0.9430	0.5036	0.1910
200.0	0.1973	0.1967	0.1959	0.1948	0.5468	0.5138	0.3015	0.1728	1.9542	1.6365	0.6839	0.2452
500.0	0.4735	0.4665	0.4523	0.4002	1.3107	1.1730	0.5543	0.2731	3.7839	2.6654	0.8612	0.2739
1 000.0	0.9068	0.8631	0.7219	0.5841	2.3995	2.0799	0.7750	0.3017	5.2538	3.4979	0.9235	0.2821
2 000.0	1.6938	1.5367	1.0572	0.7468	3.9852	2.8912	0.8998	0.3232	6.4339	3.5602	0.9391	0.2903
5 000.0	3.5244	2.7517	1.3977	0.8554	6.4437	3.5919	1.0537	0.3397	7.6825	3.6281	0.9568	0.3052
10 000.0	5.5332	3.4835	1.4672	0.8660	7.9585	3.6723	1.0962	0.3397	8.4690	3.6503	0.9620	0.3097
100 000.0	10.6505	3.7684	1.4703	0.8661	10.8851	3.6744	1.0962	0.3397	10.9787	3.6523	0.9626	0.3099

$1/u_A$	$r/r_{ew} = 10.0$			$r/r_{ew} = 20.0$			$r/r_{ew} = 05.0$			$r/r_{ew} = 100.0$		
$\sqrt{\beta}$:	0.1	0.5	1.0	0.1	0.5	1.0	0.1	0.5	1.0	0.1	0.5	1.0
0.5				0.0009	0.0007	0.0001	0.0019	0.0010	0.0005	0.0083	0.0072	0.0038
1.0	0.0028	0.0026	0.0018	0.0076	0.0068	0.0054	0.0279	0.0268	0.0152	0.0753	0.0692	0.0423
2.0	0.0139	0.0116	0.0082	0.0395	0.0305	0.0215	0.1534	0.1332	0.0585	0.3218	0.2578	0.1329
5.0	0.0661	0.0562	0.0282	0.2036	0.1350	0.0705	0.6547	0.4354	0.1872	0.9211	0.5632	0.2735
10.0	0.1896	0.1551	0.0615	0.5087	0.3333	0.1402	1.2157	0.6605	0.2663	1.5933	0.8003	0.2858
20.0	0.4787	0.3130	0.1127	1.0849	0.6018	0.2225	1.9395	0.8007	0.2877	2.2071	0.8882	0.2899
50.0	0.1210	0.5512	0.1789	2.1003	0.8251	0.2806	2.8573	0.9116	0.2936	2.8357	0.9125	0.2947
100.0	1.9747	0.6886	0.2235	2.8085	0.9250	0.2880	3.0318	0.9197	0.2961	3.2891	0.9183	0.2958
1 000.0	3.5122	0.9271	0.2858	3.5217	0.9356	0.2982	3.5252	0.9253	0.2970	3.6049	0.9215	0.2960
10 000.0	3.6321	0.9372	0.2897	3.6301	0.9365	0.2996	3.6293	0.9256	0.2972	3.6256	0.9240	0.2961

Annex 11.2 (cont.)

Tabel 2 Values of $W(u_A,S_A,\beta,r/r_{ew},b_1/D,d/D,b_2/D)$ for $b_1/D = 0.6$, $d/D = 0.0$, $b_2/D = 0.3$, $S_A = 10^{-3}$

$1/u_A$	$r/r_{ew} = 1.0$			$r/r_{ew} = 5.0$			$r/r_{ew} = 20.0$			$r/r_{ew} = 50.0$			$r/r_{ew} = 100.0$		
	$\sqrt{\beta}$: 0.001	0.1	0.5	0.001	0.1	0.5	0.001	0.1	0.5	0.001	0.1	0.5	0.001	0.1	0.5
1.0	0.0010	0.0010	0.0010	0.0013	0.0013	0.0010	0.0074	0.0074	0.0050	0.0301	0.0292	0.0175	0.0790	0.0761	0.0503
2.0	0.0020	0.0020	0.0020	0.0058	0.0057	0.0044	0.0409	0.0409	0.0218	0.1606	0.1503	0.0779	0.3438	0.3085	0.2010
5.0	0.0050	0.0050	0.0050	0.0273	0.0265	0.0173	0.2159	0.2159	0.0835	0.7802	0.6291	0.2734	1.1237	0.9253	0.4525
10.0	0.0100	0.0100	0.0100	0.0743	0.0675	0.0400	0.5669	0.5563	0.1830	1.8194	1.2137	0.4225	2.1240	1.5437	0.6095
20.0	0.0200	0.0200	0.0198	0.1814	0.1685	0.0815	1.3944	1.3281	0.3367	3.1661	2.0453	0.5429	3.3009	2.2569	0.6120
50.0	0.0500	0.0494	0.0482	0.5347	0.4680	0.1863	3.0645	2.7865	0.5640	5.2069	3.0028	0.6110	5.2213	3.1886	0.6129
100.0	0.0999	0.0988	0.0971	1.1338	0.9311	0.2600	5.6002	3.3972	0.6125	6.5213	3.6015	0.6120	6.7220	3.6021	0.6136
1 000.0	0.9845	0.9049	0.7514	6.1993	4.0172	0.5930	10.4528	4.3683	0.6147	10.5503	4.3015	0.6126	10.5525	4.2841	0.6141
10 000.0	6.7033	4.1975	1.6330	13.5249	4.3977	0.7112	14.3775	4.3720	0.6150	14.3822	4.3032	0.6129	14.3887	4.2849	0.6143
100 000.0	17.4883	4.8257	1.6510	18.2229	4.3988	0.7112	18.2247	4.3721	0.6150	18.2262	4.3036	0.6131	18.2262	4.2851	0.6144

Tabel 3 Values of $W(u_A,S_A,\beta,r/r_{ew},b_1/D,d/D,b_2/D)$ for $b_1/D = 0.4$, $d/D = 0.0$, $b_2/D = 0.2$, $S_A = 10^{-3}$

$1/u_A$	$r/r_{ew} = 1.0$			$r/r_{ew} = 5.0$			$r/r_{ew} = 20.0$			$r/r_{ew} = 50.0$			$r/r_{ew} = 100.0$		
	$\sqrt{\beta}$: 0.001	0.1	0.5	0.001	0.1	0.5	0.001	0.1	0.5	0.001	0.1	0.5	0.001	0.1	0.5
1.0	0.0010	0.0010	0.0010	0.0013	0.0013	0.0008	0.0071	0.0071	0.0031	0.0271	0.0203	0.0076	0.0522	0.0437	0.0154
2.0	0.0020	0.0020	0.0020	0.0048	0.0048	0.0020	0.0408	0.0408	0.0105	0.1905	0.1327	0.0409	0.3426	0.2045	0.0700
5.0	0.0050	0.0050	0.0049	0.0270	0.0260	0.0067	0.2256	0.2205	0.0418	0.9569	0.6759	0.1392	1.3972	1.1012	0.1619
10.0	0.0100	0.0100	0.0096	0.0740	0.0641	0.0147	0.6000	0.5486	0.0910	2.3804	1.7921	0.2541	3.0825	2.3541	0.2934
20.0	0.0200	0.0200	0.0191	0.1815	0.1552	0.0307	1.5147	1.1791	0.1725	4.6397	2.9895	0.3377	5.0573	3.4989	0.3431
50.0	0.0500	0.0500	0.0483	0.5353	0.4292	0.0750	3.9961	2.6179	0.3132	7.9908	4.1512	0.3861	8.0937	4.4236	0.3893
100.0	0.1000	0.1000	0.0947	1.1685	0.8295	0.1393	6.8764	3.8867	0.3605	10.0945	4.4939	0.3915	10.0948	4.5671	0.3910
1 000.0	1.0000	0.9745	0.7220	10.1713	4.0503	0.4267	15.7928	4.7150	0.3782	15.8287	4.6010	0.3929	15.8288	4.5883	0.3920
10 000.0	8.7036	4.7560	1.4672	21.5181	5.1752	0.5110	21.5829	4.7152	0.3887	21.5830	4.6011	0.3932	21.5830	4.5889	0.3925
100 000.0	27.2770	5.7269	1.4703	27.3204	5.1923	0.5121	27.3393	4.7152	0.3959	27.3393	4.6011	0.3933	27.3393	4.5991	0.3926

Annex 12.1 Values of Hantush's function $A(u_w, r/r_{ew})$ for free-flowing wells in confined aquifers (after Hantush 1964; Reed 1980)

	r/r_{ew}													r/r_{ew}													
$1/u_w$	1.0	1.1	1.2	1.3	1.4	1.5	1.6	1.7	1.8	1.9	2.0	3.0	4.0	5.0	6.0	7.0	8.0	9.0	10	20	30	40	50	60	70	80	90	100
4(−3)	1.000	0.024	0.000	0.000	0.000	0.000	0.000	0.000	0.000	0.000	0.000	0.000	0.000	0.000	0.000	0.000	0.000	0.000	0.000	0.000	0.000	0.000	0.000	0.000	0.000	0.000	0.000	0.000
8 (−3)	1.000	0.109	0.001																									
1.2 (−2)	1.000	0.188	0.009	0.000																								
1.6 (−2)	1.000	0.251	0.023	0.001																								
2.0 (−2)	1.000	0.303	0.042	0.002																								
2.4 (−2)	1.000	0.345	0.062	0.005	0.000																							
2.8 (−2)	1.000	0.380	0.083	0.010	0.001																							
3.2 (−2)	1.000	0.410	0.104	0.016	0.001																							
3.6 (−2)	1.000	0.435	0.124	0.022	0.002																							
4 (−2)	1.000	0.458	0.144	0.030	0.004	0.000	0.000																					
8 (−2)	1.000	0.589	0.290	0.117	0.039	0.010	0.002	0.000	0.000																			
1.2 (−1)	1.000	0.652	0.379	0.194	0.087	0.034	0.011	0.003	0.001	0.000																		
1.6 (−1)	1.000	0.691	0.439	0.254	0.133	0.063	0.027	0.010	0.004	0.001	0.000																	
2.0 (−1)	1.000	0.718	0.483	0.302	0.175	0.093	0.046	0.021	0.009	0.003	0.001																	
2.4 (−1)	1.000	0.739	0.517	0.341	0.211	0.122	0.066	0.033	0.016	0.007	0.003																	
2.8 (−1)	1.000	0.754	0.544	0.373	0.242	0.149	0.087	0.047	0.024	0.012	0.005																	
3.2 (−1)	1.000	0.767	0.566	0.400	0.270	0.174	0.106	0.062	0.034	0.018	0.009																	
3.6 (−1)	1.000	0.778	0.585	0.423	0.294	0.196	0.125	0.077	0.045	0.025	0.013																	
4 (−1)	1.000	0.787	0.601	0.443	0.316	0.217	0.143	0.091	0.055	0.032	0.018	0.000																
8 (−1)	1.000	0.837	0.691	0.562	0.450	0.355	0.275	0.209	0.156	0.114	0.082	0.001																
1.2 (0)	1.000	0.860	0.733	0.620	0.519	0.430	0.352	0.286	0.229	0.181	0.142	0.006																
1.6 (o)	1.000	0.873	0.758	0.655	0.562	0.479	0.405	0.339	0.282	0.233	0.191	0.015	0.000															
2.0 (0)	1.000	0.883	0.776	0.680	0.592	0.514	0.443	0.380	0.323	0.274	0.230	0.027	0.001															
2.4 (0)	1.000	0.890	0.789	0.698	0.615	0.540	0.472	0.411	0.356	0.307	0.263	0.040	0.003															
2.8 (0)	1.000	0.895	0.800	0.713	0.634	0.562	0.496	0.436	0.382	0.334	0.290	0.054	0.006	0.000														
3.2 (0)	1.000	0.899	0.808	0.725	0.649	0.579	0.515	0.457	0.405	0.357	0.313	0.068	0.009	0.001														
3.6 (0)	1.000	0.903	0.815	0.735	0.661	0.594	0.532	0.475	0.424	0.377	0.334	0.082	0.013	0.001														
4 (0)	1.000	0.906	0.821	0.743	0.672	0.606	0.546	0.491	0.440	0.394	0.351	0.095	0.018	0.002	0.000	0.000												
8 (0)	1.000	0.924	0.854	0.790	0.732	0.677	0.627	0.580	0.536	0.496	0.458	0.194	0.071	0.022	0.005	0.001	0.000											
1.2 (1)	1.000	0.932	0.870	0.812	0.760	0.711	0.665	0.623	0.583	0.546	0.511	0.256	0.119	0.049	0.018	0.006	0.002	0.000										
1.6 (1)	1.000	0.937	0.879	0.826	0.777	0.731	0.689	0.649	0.612	0.577	0.544	0.300	0.157	0.076	0.034	0.014	0.005	0.002	0.000									
2.0 (1)	1.000	0.940	0.886	0.835	0.789	0.746	0.706	0.668	0.633	0.599	0.568	0.332	0.188	0.101	0.051	0.024	0.010	0.004	0.002									

Annex 12.1 (cont.)

	r/r_{ew}												
$1/u_w$	1.0	1.1	1.2	1.3	1.4	1.5	1.6	1.7	1.8	1.9	2.0	3.0	4.0
2.4 (1)	1.000	0.943	0.890	0.842	0.798	0.757	0.718	0.682	0.648	0.616	0.586	0.357	0.214
2.8 (1)	1.000	0.945	0.894	0.848	0.805	0.765	0.728	0.693	0.661	0.630	0.601	0.378	0.235
3.2 (1)	1.000	0.946	0.898	0.853	0.811	0.773	0.737	0.703	0.671	0.641	0.613	0.395	0.254
3.6 (1)	1.000	0.948	0.900	0.857	0.816	0.779	0.743	0.711	0.680	0.650	0.623	0.409	0.270
4 (1)	1.000	0.949	0.903	0.860	0.820	0.784	0.749	0.717	0.687	0.658	0.631	0.422	0.284
8 (1)	1.000	0.956	0.916	0.879	0.845	0.813	0.783	0.756	0.729	0.704	0.681	0.497	0.370
1.2 (2)	1.000	0.959	0.922	0.888	0.856	0.827	0.800	0.774	0.750	0.726	0.705	0.534	0.415
1.6 (2)	1.000	0.962	0.926	0.894	0.864	0.836	0.810	0.786	0.762	0.741	0.720	0.558	0.444
2.0 (2)	1.000	0.963	0.929	0.898	0.869	0.843	0.818	0.794	0.772	0.751	0.731	0.574	0.464
2.4 (2)	1.000	0.964	0.931	0.901	0.874	0.848	0.823	0.800	0.779	0.759	0.739	0.588	0.481
2.8 (2)	1.000	0.965	0.933	0.904	0.877	0.851	0.828	0.806	0.785	0.765	0.746	0.598	0.494
3.2 (2)	1.000	0.966	0.935	0.906	0.879	0.855	0.832	0.810	0.789	0.770	0.752	0.607	0.505
3.6 (2)	1.000	0.966	0.936	0.908	0.882	0.857	0.835	0.813	0.793	0.774	0.756	0.614	0.514
4 (2)	1.000	0.967	0.937	0.909	0.884	0.860	0.838	0.817	0.797	0.778	0.760	0.621	0.522
8 (2)	1.000	0.970	0.943	0.918	0.895	0.874	0.854	0.835	0.817	0.800	0.784	0.658	0.569
1.2 (3)	1.000	0.972	0.946	0.923	0.901	0.881	0.862	0.844	0.827	0.811	0.796	0.677	0.593
1.6 (3)	1.000	0.973	0.948	0.926	0.905	0.886	0.867	0.850	0.834	0.819	0.804	0.690	0.609
2.0 (3)	1.000	0.974	0.950	0.928	0.908	0.889	0.871	0.855	0.839	0.824	0.810	0.699	0.620
2.4 (3)	1.000	0.974	0.951	0.930	0.910	0.891	0.874	0.858	0.842	0.828	0.814	0.706	0.629
2.8 (3)	1.000	0.975	0.952	0.931	0.912	0.893	0.877	0.861	0.846	0.831	0.818	0.712	0.636
3.2 (3)	1.000	0.975	0.953	0.932	0.913	0.895	0.879	0.863	0.848	0.834	0.821	9.716	0.642
3.6 (3)	1.000	0.976	0.954	0.933	0.914	0.897	0.880	0.865	0.850	0.837	0.824	0.720	0.647
4.0 (3)	1.000	0.976	0.954	0.934	0.915	0.898	0.882	0.867	0.852	0.839	0.826	0.724	0.652

	r/r_{ew}														
$1/u_w$	5.0	6.0	7.0	8.0	9.0	10	20	30	40	50	60	70	80	90	100
2.4 (1)	0.123	0.067	0.035	0.017	0.008	0.003									
2.8 (1)	0.142	0.082	0.046	0.024	0.012	0.006									
3.2 (1)	0.159	0.097	0.056	0.032	0.017	0.009									
3.6 (1)	0.175	0.110	0.067	0.039	0.022	0.012									
4 (1)	0.188	0.122	0.077	0.047	0.028	0.016	0.000								
8 (1)	0.277	0.207	0.153	0.112	0.081	0.057	0.001								
1.2 (2)	0.325	0.256	0.201	0.157	0.122	0.094	0.004								
1.6 (2)	0.358	0.290	0.235	0.190	0.153	0.123	0.009	0.000							
2.0 (2)	0.381	0.314	0.260	0.215	0.177	0.146	0.016	0.001							
2.4 (2)	0.399	0.334	0.281	0.236	0.199	0.167	0.023	0.002							
2.8 (2)	0.414	0.350	0.297	0.253	0.216	0.184	0.031	0.003							
3.2 (2)	0.427	0.364	0.312	0.268	0.230	0.198	0.038	0.005	0.000						
3.6 (2)	0.437	0.374	0.323	0.280	0.242	0.210	0.046	0.007	0.001						
4 (2)	0.446	0.385	0.334	0.291	0.254	0.222	0.053	0.010	0.001	0.000	0.000				
8 (2)	0.500	0.444	0.397	0.357	0.322	0.291	0.110	0.038	0.011	0.001	0.001	0.000			
1.2 (3)	0.528	0.475	0.430	0.392	0.358	0.328	0.146	0.064	0.026	0.009	0.003	0.001	0.000		
1.6 (3)	0.546	0.495	0.451	0.414	0.382	0.353	0.173	0.086	0.040	0.018	0.007	0.003	0.001	0.000	
2.0 (3)	0.559	0.510	0.468	0.432	0.400	0.372	0.194	0.104	0.054	0.026	0.012	0.005	0.002	0.001	0.000
2.4 (3)	0.569	0.520	0.479	0.444	0.413	0.385	0.210	0.119	0.066	0.035	0.018	0.008	0.004	0.002	0.001
2.8 (3)	0.578	0.530	0.490	0.455	0.424	0.397	0.223	0.132	0.077	0.044	0.024	0.012	0.006	0.003	0.001
3.2 (3)	0.585	0.538	0.498	0.464	0.434	0.407	0.235	0.143	0.087	0.052	0.030	0.016	0.009	0.004	0.002
3.6 (3)	0.591	0.544	0.505	0.471	0.44[illegible]	0.415	0.245	0.153	0.096	0.059	0.035	0.020	0.011	0.006	0.003
4.0 (3)	0.596	0.550	0.511	0.478	0.449	0.422	0.254	0.162	0.104	0.066	0.041	0.024	0.014	0.008	0.004

Annex 12.1 (cont.)

$1/u_w$	r/r_{ew}							
	5	10	20	50	100	200	500	1000
4 (3)	0.596	0.422	0.254	0.066	0.004			
6 (3)	.615	.450	.287	.094	.012			
8 (3)	.627	.467	.309	.116	.021	0.000		
1.2 (4)	.644	.490	.338	.147	.039	.001		
2 (4)	.662	.517	.372	.186	.068	.006		
2.8 (4)	.673	.533	.392	.211	.089	.014		
4 (4)	.685	.549	.413	.237	.114	.025		
6 (4)	.696	.566	.435	.264	.142	.043	0.000	
8 (4)	.704	.577	.450	.283	.161	.058	.001	
1.2 (5)	.715	.592	.469	.308	.188	.081	.005	
2 (5)	.727	.609	.492	.337	.221	.113	.014	0.000
2.8 (5)	.734	.620	.506	.355	.242	.134	.025	.001
4 (5)	.742	.631	.520	.373	.263	.156	.039	.002
6 (5)	.750	.642	.532	.392	.285	.180	.058	.007
8 (5)	.755	.650	.544	.405	.300	.197	.072	.013
1.2 (6)	.762	.660	.558	.423	.321	.220	.094	.024
2 (6)	.771	.672	.574	.443	.345	.247	.122	.044
2.8 (6)	.776	.680	.584	.456	.360	.264	.141	.059
4 (6)	.782	.688	.594	.470	.376	.282	.160	.076
6 (6)	.788	.696	.604	.484	.392	.301	.181	.096
8 (6)	.792	.702	.612	.493	.403	.314	.196	.111
1.2 (7)	.797	.709	.622	.506	.418	.331	.216	.132
2 (7)	.803	.718	.633	.521	.436	.352	.240	.157
2.8 (7)	.807	.724	.641	.531	.448	.365	.255	.173
4 (7)	.811	.730	.648	.541	.459	.378	.270	.190
6 (7)	.815	.736	.656	.551	.472	.392	.287	.208
8 (7)	.818	.740	.662	.558	.480	.402	.299	.221
1.2 (8)	.822	.746	.669	.568	.492	.415	.314	.238
2 (8)	.827	.753	.678	.580	.506	.431	.333	.258
2.8 (8)	.830	.757	.684	.587	.514	.441	.344	.271
4 (8)	.833	.762	.690	.595	.523	.452	.357	.285
6 (8)	.837	.766	.696	.603	.533	.463	.370	.300
8 (8)	.839	.770	.701	.609	.540	.470	.379	.310
1.2 (9)	.842	.774	.706	.617	.549	.481	.391	.323
2 (9)	.846	.780	.714	.626	.560	.494	.406	.340
2.8 (9)	.849	.783	.718	.632	.567	.502	.415	.350
4 (9)	.851	.787	.723	.638	.574	.510	.425	.361
6 (9)	.854	.791	.728	.645	.582	.519	.435	.372
8 (9)	.856	.794	.731	.649	.587	.525	.443	.380
1.2 (10)	.858	.797	.736	.655	.594	.533	.452	.392
2 (10)	.861	.802	.742	.663	.603	.544	.464	.405
2.8 (10)	.863	.804	.746	.668	.609	.550	.472	.413
4 (10)	.865	.807	.749	.673	.615	.557	.480	.422
6 (10)	.867	.810	.753	.678	.621	.564	.488	.431
8 (10)	.869	.813	.756	.682	.625	.569	.494	.438
1.2 (11)	.871	.816	.760	.687	.631	.576	.502	.447
2 (11)	.874	.819	.765	.693	.638	.584	.512	.457
2.8 (11)	.875	.821	.768	.696	.643	.589	.518	.464

Annex 14.1 Values of $s_{w(n)}/Q_n$ corresponding to values of Q_n and P for B = 1, C = 1, P > 1, $Q_n < Q_i$ and $Q_i^{P-1} = 100$ for well performance tests

Q_n	P = 1.7 Q_i = 719.7	1.8 316.2	1.9 166.8	2.0 100.0	2.1 65.8	2.2 46.4	2.3 34.6
0.1	1.20	1.16	1.13	1.10	1.08	1.06	1.05
0.15	1.27	1.22	1.18	1.15	1.12	1.10	1.08
0.2	1.32	1.28	1.23	1.20	1.17	1.14	1.12
0.3	1.43	1.36	1.34	1.30	1.27	1.24	1.21
0.4	1.53	1.48	1.44	1.40	1.36	1.33	1.30
0.5	1.62	1.57	1.54	1.50	1.47	1.44	1.41
0.6	1.70	1.66	1.63	1.60	1.57	1.54	1.51
0.8	1.86	1.84	1.82	1.80	1.78	1.77	1.75
1.0	2.00	2.00	2.00	2.00	2.00	2.00	2.00
1.5	2.33	2.38	2.44	2.50	2.56	2.63	2.69
2.0	2.62	2.74	2.87	3.00	3.14	3.30	3.46
3.0	3.16	3.41	3.69	4.00	4.35	4.74	5.17
4.0	3.64	4.03	4.48	5.00	5.55	6.28	7.06
5.0	4.09	4.62	5.26	6.00	6.87	7.90	9.10
6.0	4.51	5.19	6.02	7.00	8.18	9.59	11.27
8.0	5.29	6.28	7.50	9.00	10.85	13.13	15.93
10	6.01	7.31	8.94	11.00	13.59	16.85	20.95
15	7.66	9.73	12.44	16.00	20.67	26.78	34.80
20	9.14	11.99	15.82	21.00	27.99	37.41	50.13
30	11.81	16.19	22.35	31.00	43.15	60.23	84.23
40	14.23	20.13	28.66	41.00	58.85	84.65	
50	16.46	23.87	34.81	51.00	74.94		
60	18.57	27.46	40.84	61.00	91.36		
80	22.49	34.30	52.62	81.00			
100	26.12	40.81	64.10				

Annex 14.1 (continued)

Q_n	P = 2.4 Q_i = 26.8	2.5 21.5	2.6 17.8	2.8 12.9	3.0 10.0	3.2 8.1	3.4 6.8	3.6 5.9	4.0 4.6
0.1	1.04	1.03	1.03	1.02	1.01	1.01	1.00	1.00	1.00
0.15	1.07	1.06	1.05	1.03	1.02	1.02	1.01	1.01	1.00
0.2	1.11	1.09	1.08	1.06	1.04	1.03	1.02	1.02	1.01
0.3	1.19	1.16	1.15	1.11	1.09	1.07	1.06	1.04	1.03
0.4	1.28	1.25	1.23	1.19	1.16	1.13	1.11	1.09	1.06
0.5	1.38	1.35	1.33	1.29	1.25	1.22	1.19	1.16	1.13
0.6	1.49	1.46	1.44	1.40	1.36	1.33	1.29	1.26	1.22
0.8	1.73	1.72	1.70	1.67	1.64	1.61	1.59	1.56	1.51
1.0	2.00	2.00	2.00	2.00	2.00	2.00	2.00	2.00	2.00
1.5	2.76	2.84	2.91	3.07	3.25	3.44	3.65	3.87	4.38
2.0	3.64	3.83	4.03	4.48	5.00	5.59	6.28	7.06	9.00
3.0	5.66	6.20	6.80	8.22	10.00	12.21	14.97	18.40	28.00
4.0	7.96	9.00	10.19	13.13	17.00	22.11	28.86	37.76	65.00
5.0	10.52	12.18	14.13	19.12	26.00	35.49	48.59	66.66	
6.0	13.29	15.70	18.58	26.16	37.00	52.51	74.72		
8.0	19.38	23.63	28.86	43.22	65.00	98.01			
10	26.12	32.62	40.81	64.10					
15	45.31	59.09	77.16						
20	67.29	90.44							

Annex 15.1 Values of Papadopulos-Cooper's function $F(u_w,\alpha)$ for single-well constant-discharge tests in confined aquifers (after Papadopulos and Cooper 1967)

$1/u_w$	$\alpha=10^{-1}$	$\alpha=10^{-2}$	$\alpha=10^{-3}$	$\alpha=10^{-4}$	$\alpha=10^{-5}$
1(−1)	9.75(−3)	9.98(−4)	1.00(−4)	1.00(−5)	1.00(−6)
1(0)	9.19(−2)	9.91(−3)	9.99(−4)	1.00(−4)	1.00(−5)
2(0)	1.77(−1)	1.97(−2)	2.00(−3)	2.00(−4)	2.00(−5)
5(0)	4.06(−1)	4.89(−2)	4.99(−3)	5.00(−4)	5.00(−5)
1(1)	7.34(−1)	9.66(−2)	9.97(−3)	1.00(−3)	1.00(−4)
2(1)	1.26	1.90(−1)	1.99(−2)	2.00(−3)	2.00(−4)
5(1)	2.30	4.53(−1)	4.95(−2)	4.99(−3)	5.00(−4)
1(2)	3.28	8.52(−1)	9.83(−2)	9.98(−3)	1.00(−3)
2(2)	4.25	1.54	1.94(−1)	1.99(−2)	2.00(−3)
5(2)	5.42	3.04	4.72(−1)	4.97(−2)	5.00(−3)
1(3)	6.21	4.54	9.07(−1)	9.90(−2)	9.99(−3)
2(3)	6.96	6.03	1.69	1.96(−1)	2.00(−2)
5(3)	7.87	7.56	3.52	4.81(−1)	4.98(−2)
1(4)	8.57	8.44	5.53	9.34(−1)	9.93(−2)
2(4)	9.32	9.23	7.63	1.77	1.97(−1)
5(4)	1.02(1)	1.02(1)	9.68	3.83	4.86(−1)
1(5)	1.09(1)	1.09(1)	1.07(1)	6.24	9.49(−1)
2(5)	1.16(1)	1.16(1)	1.15(1)	8.99	1.82
5(5)	1.25(1)	1.25(1)	1.25(1)	1.17(1)	4.03
1(6)	1.32(1)	1.32(1)	1.32(1)	1.29(1)	6.78
2(6)	1.39(1)	1.39(1)	1.39(1)	1.38(1)	1.01(1)
5(6)	1.48(1)	1.48(1)	1.48(1)	1.48(1)	1.37(1)
1(7)	1.55(1)	1.55(1)	1.55(1)	1.55(1)	1.51(1)
2(7)	1.62(1)	1.62(1)	1.62(1)	1.62(1)	1.60(1)
5(7)	1.70(1)	1.70(1)	1.70(1)	1.71(1)	1.71(1)
1(8)	1.78(1)	1.78(1)	1.78(1)	1.78(1)	1.78(1)
2(8)	1.85(1)	1.85(1)	1.85(1)	1.85(1)	1.85(1)
5(8)	1.94(1)	1.94(1)	1.94(1)	1.94(1)	1.94(1)
1(9)	2.01(1)	2.01(1)	2.01(1)	2.01(1)	2.01(1)

Annex 15.2 Values of $s_t/s_{0.4t}$ for single-well constant-discharge tests in confined aquifers (after Rushton and Singh 1983)

	S					
$4KDt/r_{ew}^2$	10^{-1}	10^{-2}	10^{-3}	10^{-4}	10^{-5}	10^{-6}
1.0 (−2)	2.49	2.49	2.50	2.50	2.50	2.50
1.78 (−2)	2.48	2.49	2.49	2.50	2.50	2.50
3.16 (−2)	2.47	2.48	2.49	2.50	2.50	2.50
5.62 (−2)	2.45	2.47	2.49	2.49	2.49	2.50
1.0 (−1)	2.43	2.46	2.48	2.49	2.49	2.49
1.78 (−1)	2.39	2.44	2.47	2.48	2.48	2.49
3.16 (−1)	2.34	2.42	2.45	2.46	2.47	2.48
5.62 (−1)	2.28	2.38	2.42	2.44	2.46	2.46
1.0	2.19	2.31	2.37	2.41	2.43	2.44
1.78	2.08	2.22	2.30	2.35	2.38	2.40
3.16	1.94	2.10	2.19	2.26	2.30	2.33
5.62	1.78	1.93	2.04	2.12	2.18	2.22
10	1.62	1.73	1.84	1.94	2.01	2.07
17.8	1.47	1.53	1.62	1.71	1.79	1.86
31.6	1.35	1.36	1.41	1.47	1.54	1.60
56.2	1.26	1.24	1.25	1.28	1.32	1.36
100	1.21	1.17	1.15	1.16	1.17	1.19

Annex 15.3 Values of $u_w W(u_w)$ for single-well constant-discharge tests

u_w	$u_w W(u_w)$	u_w	$u_w W(u_w)$
8	3.014(–4)	8(–6)	8.928(–5)
6	2.161(–3)	6(–6)	6.870(–5)
4	1.512(–2)	4(–6)	4.740(–5)
2	9.780(–1)	2(–6)	2.510(–5)
1	2.194(–1)	1(–6)	1.324(–5)
8(–1)	2.485(–1)	8(–7)	1.077(–5)
6(–1)	2.726(–1)	6(–7)	8.250(–6)
4(–1)	2.810(–1)	4(–7)	5.660(–6)
2(–1)	2.446(–1)	2(–7)	2.970(–6)
1(–1)	1.823(–1)	1(–7)	1.554(–6)
8(–2)	1.622(–1)	8(–8)	1.261(–6)
6(–2)	1.377(–1)	6(–8)	9.630(–7)
4(–2)	1.072(–1)	4(–8)	6.584(–7)
2(–2)	6.710(–2)	2(–8)	3.430(–7)
1(–2)	4.038(–2)	1(–8)	1.784(–7)
8(–3)	3.407(–2)	8(–9)	1.446(–7)
6(–3)	2.727(–2)	6(–9)	1.101(–7)
4(–3)	1.979(–2)	4(–9)	7.504(–8)
2(–3)	1.128(–2)	2(–9)	3.890(–8)
1(–3)	6.332(–3)	1(–9)	2.015(–8)
8(–4)	5.244(–3)	8(–10)	1.630(–8)
6(–4)	4.105(–3)	6(–10)	1.240(–8)
4(–4)	2.899(–3)	4(–10)	8.424(–9)
2(–4)	1.588(–3)	2(10)	4.352(–9)
1(–4)	8.633(–4)	1(–10)	2.245(–9)
8(–5)	7.085(–4)	8(–11)	1.824(–9)
6(–5)	5.486(–4)	6(–11)	1.378(–9)
4(–5)	3.820(–4)	4(–11)	9.344(–10)
2(–5)	2.048(–4)	2(–11)	4.812(–10)
1(–5)	1.094(–4)	1(–11)	2.475(–10)

Annex 15.4 Values of $s_t/s_{0.4t}$ for single-well tests with decreasing discharge rates in confined aquifers (after Rushton and Singh 1983)

Values of $s_t/s_{0.4t}$ for S = 0.001

$\frac{4KDT}{r_{ew}^2}$	Discharge Reduction Factor (F)									
	1.0	0.7	0.4	0.2	0.1	0.07	0.04	0.02	0.01	0.0
1.0×10^{-2}	2.48	2.49	2.49	2.50	2.50	2.50	2.50	2.50	2.50	2.50
1.78×10^{-2}	2.47	2.48	2.49	2.49	2.49	2.49	2.49	2.49	2.49	2.49
3.16×10^{-2}	2.46	2.47	2.48	2.49	2.49	2.49	2.49	2.49	2.49	2.49
5.62×10^{-2}	2.44	2.45	2.47	2.48	2.48	2.48	2.49	2.49	2.49	2.49
1.0×10^{-1}	2.39	2.41	2.45	2.46	2.47	2.47	2.48	2.48	2.48	2.48
1.78×10^{-1}	2.32	2.35	2.41	2.44	2.45	2.45	2.46	2.47	2.47	2.47
3.16×10^{-1}	2.21	2.27	2.36	2.40	2.43	2.43	2.44	2.44	2.45	2.45
5.62×10^{-1}	2.04	2.13	2.26	2.34	2.38	2.39	2.40	2.41	2.42	2.42
1.0	1.81	1.94	2.12	2.24	2.30	2.32	2.35	2.36	2.37	2.37
1.78	1.55	1.69	1.92	2.09	2.19	2.22	2.26	2.28	2.29	2.30
3.16	1.30	1.43	1.66	1.88	2.02	2.07	2.12	2.16	2.18	2.19
5.62	1.13	1.21	1.39	1.63	1.80	1.87	1.94	1.99	2.02	2.04
10.0	1.04	1.09	1.18	1.37	1.55	1.63	1.71	1.77	1.81	1.84
17.8	1.02	1.04	1.07	1.18	1.32	1.39	1.47	1.54	1.58	1.62
31.6	1.01	1.02	1.03	1.08	1.16	1.21	1.27	1.33	1.37	1.41
56.2	1.00	1.01	1.02	1.04	1.08	1.12	1.15	1.19	1.22	1.25
100	1.00	1.00	1.01	1.03	1.05	1.07	1.09	1.12	1.13	1.15

Values of $s_t/s_{0.4t}$ for S = 0.01

$\frac{4KDT}{r_{ew}^2}$	Discharge Reduction Factor (F)									
	1.0	0.7	0.4	0.2	0.1	0.07	0.04	0.02	0.01	0.0
1.0×10^{-2}	2.49	2.49	2.49	2.49	2.49	2.49	2.49	2.49	2.49	2.49
1.78×10^{-2}	2.48	2.48	2.48	2.49	2.49	2.49	2.49	2.49	2.49	2.49
3.16×10^{-2}	2.47	2.47	2.48	2.48	2.48	2.48	2.48	2.48	2.48	2.48
5.62×10^{-2}	2.44	2.46	2.47	2.47	2.47	2.47	2.47	2.47	2.47	2.47
1.0×10^{-1}	2.40	2.43	2.44	2.45	2.46	2.46	2.46	2.46	2.46	2.46
1.78×10^{-1}	2.34	2.38	2.40	2.42	2.43	2.44	2.44	2.44	2.44	2.44
3.16×10^{-1}	2.23	2.29	2.33	2.37	2.39	2.40	2.41	2.41	2.41	2.42
5.62×10^{-1}	2.07	2.16	2.23	2.30	2.33	2.35	2.36	2.37	2.37	2.38
1.0	1.84	1.97	2.08	2.19	2.25	2.27	2.29	2.30	2.31	2.31
1.78	1.56	1.72	1.88	2.03	2.12	2.15	2.18	2.20	2.21	2.22
3.16	1.30	1.44	1.64	1.82	1.95	1.99	2.03	2.06	2.08	2.10
5.62	1.12	1.21	1.39	1.58	1.73	1.79	1.84	1.89	1.91	1.93
10.0	1.03	1.07	1.20	1.36	1.50	1.56	1.63	1.68	1.70	1.73
17.8	1.01	1.02	1.09	1.19	1.31	1.36	1.42	1.47	1.50	1.53
31.6	1.00	1.01	1.05	1.10	1.18	1.21	1.26	1.30	1.33	1.36
56.2	1.00	1.00	1.04	1.06	1.11	1.13	1.16	1.19	1.21	1.24
100	1.00	1.00	1.03	1.04	1.08	1.09	1.11	1.14	1.15	1.17

Annex 15.4 (cont.)

Values of $s_t/s_{0.4t}$ for S = 0.1

$\frac{4\,KDT}{r_{ew}^2}$	Discharge Reduction Factor (F)									
	1.0	0.7	0.4	0.2	0.1	0.07	0.04	0.02	0.01	0.0
1.0×10^{-2}	2.48	2.48	2.49	2.49	2.49	2.49	2.49	2.49	2.49	2.49
1.78×10^{-2}	2.47	2.47	2.48	2.48	2.48	2.48	2.48	2.48	2.48	2.48
3.16×10^{-2}	2.45	2.46	2.46	2.47	2.47	2.47	2.47	2.47	2.47	2.47
5.62×10^{-2}	2.41	2.43	2.44	2.45	2.45	2.45	2.45	2.45	2.45	2.45
1.0×10^{-1}	2.36	2.38	2.40	2.42	2.42	2.42	2.43	2.43	2.43	2.43
1.78×10^{-1}	2.28	2.31	2.35	2.37	2.38	2.39	2.39	2.39	2.39	2.39
3.16×10^{-1}	2.16	2.21	2.27	2.31	2.33	2.33	2.34	2.34	2.34	2.34
5.62×10^{-1}	1.99	2.07	2.16	2.22	2.25	2.26	2.27	2.28	2.28	2.28
1.0	1.77	1.88	2.00	2.09	2.15	2.16	2.18	2.19	2.19	2.19
1.78	1.53	1.65	1.81	1.93	2.01	2.03	2.05	2.07	2.08	2.08
3.16	1.31	1.42	1.59	1.74	1.84	1.87	1.90	1.93	1.94	1.94
5.62	1.16	1.24	1.38	1.54	1.65	1.69	1.73	1.76	1.77	1.78
10.0	1.07	1.12	1.22	1.36	1.47	1.51	1.55	1.59	1.60	1.62
17.8	1.04	1.07	1.13	1.22	1.31	1.35	1.40	1.43	1.45	1.47
31.6	1.03	1.04	1.08	1.14	1.21	1.24	1.28	1.31	1.33	1.35
56.2	1.02	1.03	1.05	1.10	1.15	1.17	1.20	1.23	1.24	1.26
100	1.02	1.02	1.04	1.08	1.11	1.13	1.16	1.18	1.19	1.21

Annex 15.5 Values of the Hantush function $G(u_w, r_{ew}/L)$ for free-flowing single-well tests in leaky aquifers (after Hantush 1964)

	r_{ew}/L																
$1/u_w$	0	1×10^{-5}	2×10^{-5}	4×10^{-5}	6×10^{-5}	8×10^{-5}	10^{-4}	2×10^{-4}	4×10^{-4}	6×10^{-4}	8×10^{-4}	10^{-3}	2×10^{-3}	4×10^{-3}	6×10^{-3}	8×10^{-3}	10^{-2}
1×10^2	0.346																0.346
2	0.311												0.311	0.311	0.311	0.312	0.312
3	0.294												0.294	0.294	0.294	0.295	0.295
4	0.283												0.283	0.283	0.283	0.284	0.285
5	0.274												0.274	0.274	0.275	0.275	0.276
6	0.268												0.268	0.268	0.268	0.269	0.271
7	0.263												0.263	0.263	0.263	0.264	0.266
8	0.258												0.258	0.258	0.259	0.260	0.261
9	0.254												0.254	0.255	0.256	0.257	0.258
1×10^3	0.251												0.251	0.252	0.252	0.254	0.255
2	0.232												0.232	0.233	0.234	0.236	0.239
3	0.222												0.222	0.223	0.225	0.227	0.231
4	0.215												0.215	0.216	0.219	0.222	0.226
5	0.210												0.210	0.212	0.215	0.218	0.222
6	0.206												0.206	0.208	0.211	0.215	0.220
7	0.203												0.203	0.205	0.209	0.213	0.219
8	0.201												0.201	0.203	0.207	0.212	0.218
9	0.198												0.198	0.201	0.205	0.210	0.217
1×10^4	0.196											0.196	0.197	0.200	0.204	0.209	0.216
2	0.185											0.185	0.185	0.190	0.197	0.205	0.213
3	0.178											0.178	0.179	0.186	0.194	0.203	0.212
4	0.173											0.173	0.176	0.183	0.193	0.202	
5	0.170											0.170	0.173	0.181	0.192		
6	0.168											0.168	0.171	0.180	0.192		
7	0.166										0.166	0.167	0.170	0.179	0.191		
8	0.164										0.164	0.165	0.169	0.179			
9	0.163										0.163	0.164	0.168	0.179			

Annex 15.5 (cont.)

	r_{ew}/L																
$1/u_w$	0	1×10^{-5}	2×10^{-5}	4×10^{-5}	6×10^{-5}	8×10^{-5}	10^{-4}	2×10^{-4}	4×10^{-4}	6×10^{-4}	8×10^{-4}	10^{-3}	2×10^{-3}	4×10^{-3}	6×10^{-3}	8×10^{-3}	10^{-2}
1×10^{5}	0.161								0.161	0.162	0.162	0.162	0.167	0.178			
2	0.152							0.152	0.153	0.153	0.154	0.155	0.163	0.177			
3	0.148							0.148	0.148	0.149	0.150	0.152	0.162				
4	0.145							0.145	0.145	0.146	0.147	0.150	0.162				
5	0.143							0.143	0.143	0.144	0.145	0.148	0.161				
6	0.141							0.141	0.142	0.143	0.144	0.147	0.160				
7	0.140							0.140	0.140	0.141	0.143	0.146	0.160				
8	0.138							0.138	0.139	0.141	0.143	0.145	0.160				
9	0.137							0.137	0.138	0.140	0.142	0.144	0.160				
1×10^{6}	0.130						0.136	0.137	0.138	0.139	0.141	0.144	0.159				
2	0.137						0.130	0.131	0.133	0.135	0.139	0.143	0.159				
3	0.127						0.127	0.127	0.130	0.134	0.138	0.142	0.158				
4	0.124						0.124	0.125	0.129	0.134							
5	0.123						0.123	0.124	0.128	0.133							
6	0.121						0.121	0.123	0.128								
7	0.120						0.120	0.122	0.127								
8	0.119						0.119	0.121	0.127								
9	0.118						0.118	0.121	0.127								
1×10^{7}	0.118						0.118	0.120	0.127								
2	0.114						0.114	0.116	0.126								
3	0.111					0.111	0.112										
4	0.109				0.109	0.110	0.111										
5	0.108				0.108	0.109	0.110										
6	0.107			0.107	0.108	0.109	0.110										
7	0.106			0.106	0.107	0.108	0.109										
8	0.105			0.105	0.106	0.108	0.109										
9	0.104		0.104	0.105	0.106	0.107	0.108										

Annex 15.5 (cont.)

$1/u_w$	r_{ew}/L																
	0	1×10^{-5}	2×10^{-5}	4×10^{-5}	6×10^{-5}	8×10^{-5}	10^{-4}	2×10^{-4}	4×10^{-4}	6×10^{-4}	8×10^{-4}	10^{-3}	2×10^{-3}	4×10^{-3}	6×10^{-3}	8×10^{-3}	10^{-2}
1×10^{8}	0.104		0.104	0.104	0.105	0.106	0.108										
2	0.100	0.100	0.101	0.102	0.103	0.105	0.107										
3	0.0982	0.0982	0.0986	0.100	0.103												
4	0.0968	0.0968	0.0974	0.0994	0.102												
5	0.0958	0.0958	0.0966	0.0989													
6	0.0950	0.0951	0.0959	0.0986													
7	0.0943	0.0944	0.0954	0.0984													
8	0.0937	0.0939	0.0949	0.0982													
9	0.0932	0.0934	0.0946	0.0981													
1×10^{9}	0.0927	0.0930	0.0943	0.0980													
2	0.0899	0.0906	0.0927	0.0977													
3	0.0883	0.0893	0.0920	0.0976													
4	0.0872	0.0885	0.0917														
5	0.0864	0.0880	0.0916														
6	0.0857	0.0876	0.0915														
7	0.0851	0.0873	0.0915														
8	0.0846	0.0870	0.0915														
9	0.0842	0.0869	0.0914														
1×10^{10}	0.0838	0.0867	0.0914														
2	0.0814	0.0862															
3	0.0861	0.0860															
4	0.0792																
5	0.0785																
6	0.0779																
7	0.0774																
8	0.0770																
9	0.0767																
10	0.0764	0.0860	0.0914	0.0976	0.102	0.105	0.107	0.116	0.126	0.133	0.138	0.142	0.158	0.177	0.191	0.202	0.212

Annex 16.1 Values of the function f(α,β) for slug tests in confined aquifers (after Cooper et al. 1967; Papadopulos et al. 1973; Bredehoeft and Papadopulos 1980)

Table 1. $10^{-10} \le \alpha \le 10^{-6}$

β	$\alpha = 10^{-6}$	$\alpha = 10^{-7}$	$\alpha = 10^{-8}$	$\alpha = 10^{-9}$	$\alpha = 10^{-10}$
0.001	0.9994	0.9996	0.9996	0.9997	0.9997
0.002	0.9989	0.9992	0.9993	0.9994	0.9995
0.004	0.9980	0.9985	0.9987	0.9989	0.9991
0.006	0.9972	0.9978	0.9982	0.9984	0.9986
0.008	0.9964	0.9971	0.9976	0.9980	0.9982
0.01	0.9956	0.9965	0.9971	0.9975	0.9978
0.02	0.9919	0.9934	0.9944	0.9952	0.9958
0.04	0.9848	0.9875	0.9894	0.9908	0.9919
0.06	0.9782	0.9819	0.9846	0.9866	0.9881
0.08	0.9718	0.9765	0.9799	0.9824	0.9844
0.1	0.9655	0.9712	0.9753	0.9784	0.9807
0.2	0.9361	0.9459	0.9532	0.9587	0.9631
0.4	0.8828	0.8995	0.9122	0.9220	0.9298
0.6	0.8345	0.8569	0.8741	0.8875	0.8984
0.8	0.7901	0.8173	0.8383	0.8550	0.8686
1.0	0.7489	0.7801	0.8045	0.8240	0.8401
2.0	0.5800	0.6235	0.6591	0.6889	0.7139
3.0	0.4554	0.5033	0.5442	0.5792	0.6096
4.0	0.3613	0.4093	0.4517	0.4891	0.5222
5.0	0.2893	0.3351	0.3768	0.4146	0.4487
6.0	0.2337	0.2759	0.3157	0.3525	0.3865
7.0	0.1903	0.2285	0.2655	0.3007	0.3337
8.0	0.1562	0.1903	0.2243	0.2573	0.2888
9.0	0.1292	0.1594	0.1902	0.2208	0.2505
10.0	0.1078	0.1343	0.1620	0.1900	0.2178
20.0	0.02720	0.03343	0.04129	0.05071	0.06149
30.0	0.01286	0.01448	0.01667	0.01956	0.02320
40.0	0.008337	0.008898	0.009637	0.01062	0.01190
50.0	0.006209	0.006470	0.006789	0.007192	0.007709
60.0	0.004961	0.005111	0.005283	0.005487	0.005735
80.0	0.003547	0.003617	0.003691	0.003773	0.003863
100.0	0.002763	0.002803	0.002845	0.002890	0.002938
200.0	0.001313	0.001322	0.001330	0.001339	0.001348

Annex 16.1 (cont.)

Table 2. $10^{-5} \leq \alpha \leq 10^{-1}$

β	$\alpha = 10^{-1}$	$\alpha = 10^{-2}$	$\alpha = 10^{-3}$	$\alpha = 10^{-4}$	$\alpha = 10^{-5}$
1.00×10^{-3}	0.9771	0.9920	0.9969	0.9985	0.9992
2.15×10^{-3}	0.9658	0.9876	0.9949	0.9974	0.9985
4.64×10^{-3}	0.9490	0.9807	0.9914	0.9954	0.9970
1.00×10^{-2}	0.9238	0.9693	0.9853	0.9915	0.9942
2.15×10^{-2}	0.8860	0.9505	0.9744	0.9841	0.9888
4.64×10^{-2}	0.8293	0.9187	0.9545	0.9701	0.9781
1.00×10^{-1}	0.7460	0.8655	0.9183	0.9434	0.9572
2.15×10^{-1}	0.6289	0.7782	0.8538	0.8935	0.9167
4.64×10^{-1}	0.4782	0.6436	0.7436	0.8031	0.8410
1.00×10^{0}	0.3117	0.4598	0.5729	0.6520	0.7080
2.15×10^{0}	0.1665	0.2597	0.3543	0.4364	0.5038
4.64×10^{0}	0.07415	0.1086	0.1554	0.2082	0.2620
7.00×10^{0}	0.04625	0.06204	0.08519	0.1161	0.1521
1.00×10^{1}	0.03065	0.03780	0.04821	0.06355	0.08378
1.40×10^{1}	0.02092	0.02414	0.02844	0.03492	0.04426
2.15×10^{1}	0.01297	0.01414	0.01545	0.01723	0.01999
3.00×10^{1}	0.009070	0.009615	0.01016	0.01083	0.01169
4.64×10^{1}	0.005711	0.005919	0.006111	0.006319	0.006554
7.00×10^{1}	0.003722	0.003809	0.003884	0.003962	0.004046
1.00×10^{2}	0.002577	0.002618	0.002653	0.002688	0.002725
2.15×10^{2}	0.001179	0.001187	0.001194	0.001201	0.001208

Annex 16.1 (cont.)

Table 3. $10^{-1} \leq \alpha \leq 10$

β	α = 0.1	α = 0.2	α = 0.5	α = 1	α = 2	α = 5	α = 10
0.000001	0.9993	0.9990	0.9984	0.9977	0.9968	0.9948	0.9923
0.000002	0.9990	0.9986	0.9977	0.9968	0.9955	0.9927	0.9894
0.000004	0.9986	0.9980	0.9968	0.9955	0.9936	0.9898	0.9853
0.000006	0.9982	0.9975	0.9961	0.9945	0.9922	0.9876	0.9822
0.000008	0.9980	0.9971	0.9955	0.9936	0.9910	0.9857	0.9796
0.00001	0.9977	0.9968	0.9949	0.9929	0.9900	0.9841	0.9773
0.00002	0.9968	0.9955	0.9929	0.9900	0.9858	0.9776	0.9683
0.00004	0.9955	0.9936	0.9899	0.9858	0.9801	0.9687	0.9558
0.00006	0.9944	0.9922	0.9877	0.9827	0.9757	0.9619	0.9464
0.00008	0.9936	0.9909	0.9858	0.9800	0.9720	0.9562	0.9387
0.0001	0.9928	0.9899	0.9841	0.9777	0.9688	0.9512	0.9318
0.0002	0.9898	0.9857	0.9776	0.9687	0.9562	0.9321	0.9059
0.0004	0.9855	0.9797	0.9685	0.9560	0.9389	0.9061	0.8711
0.0006	0.9822	0.9752	0.9615	0.9465	0.9258	0.8869	0.8458
0.0008	0.9794	0.9713	0.9557	0.9385	0.9151	0.8711	0.8253
0.001	0.9769	0.9679	0.9505	0.9315	0.9057	0.8576	0.8079
0.002	0.9670	0.9546	0.9307	0.9048	0.8702	0.8075	0.7450
0.004	0.9528	0.9357	0.9031	0.8686	0.8232	0.7439	0.6684
0.006	0.9417	0.9211	0.8825	0.8419	0.7896	0.7001	0.6178
0.008	0.9322	0.9089	0.8654	0.8202	0.7626	0.6662	0.5797
0.01	0.9238	0.8982	0.8505	0.8017	0.7400	0.6384	0.5492
0.02	0.8904	0.8562	0.7947	0.7336	0.6595	0.5450	0.4517
0.04	0.8421	0.7980	0.7214	0.6489	0.5654	0.4454	0.3556
0.06	0.8048	0.7546	0.6697	0.5919	0.5055	0.3872	0.3030
0.08	0.7734	0.7190	0.6289	0.5486	0.4618	0.3469	0.2682
0.1	0.7459	0.6885	0.5951	0.5137	0.4276	0.3168	0.2428
0.2	0.6418	0.5774	0.4799	0.4010	0.3234	0.2313	0.1740
0.4	0.5095	0.4458	0.3566	0.2902	0.2292	0.1612	0.1207
0.6	0.4227	0.3642	0.2864	0.2311	0.1817	0.1280	0.09616
0.8	0.3598	0.3072	0.2397	0.1931	0.1521	0.1077	0.08134
1	0.3117	0.2648	0.2061	0.1663	0.1315	0.09375	0.07120
2	0.1786	0.1519	0.1202	0.09912	0.08044	0.05940	0.04620
4	0.08761	0.07698	0.06420	0.05521	0.04668	0.03621	0.02908
6	0.05527	0.04999	0.04331	0.03830	0.03326	0.02663	0.02185
8	0.03963	0.03658	0.03254	0.02933	0.02594	0.02125	0.01771
10	0.03065	0.02870	0.02600	0.02376	0.02130	0.01776	0.01499
20	0.01408	0.01361	0.01288	0.01219	0.01133	0.009943	0.008716
40	0.006680	0.006568	0.006374	0.006171	0.005897	0.005395	0.004898
60	0.004367	0.004318	0.004229	0.004132	0.003994	0.003726	0.003445
80	0.003242	0.003214	0.003163	0.003105	0.003022	0.002853	0.002668
100	0.002577	0.002559	0.002526	0.002487	0.002431	0.002313	0.002181
200	0.001271	0.001266	0.001258	0.001247	0.001230	0.001194	0.001149
400	0.0006307	0.0006295	0.0006272	0.0006242	0.0006195	0.0006085	0.0005944
600	0.0004193	0.0004188	0.0004177	0.0004163	0.0004141	0.0004087	0.0004016
800	0.0003140	0.0003137	0.0003131	0.0003123	0.0003110	0.0003078	0.0003035
1000	0.0002510	0.0002508	0.0002504	0.0002499	0.0002490	0.0002469	0.0002440

Annex 18.1 Values of the function $F(u_{vf},r')$ for different values of u_{vf}/r' and r' (after Merton 1987)

Table 1 For a vertical fracture with an observation well located on the x-axis

	r′											
u_{vf}/r'	0.2	0.4	0.6	0.8	0.9	1.02	1.05	1.1	1.2	1.5	3.0	5.0
1.0 (–3)	0.05013	0.07090	0.08683	0.10027	0.10623	0.01883	0.00434	0.00025	0.00000	0.00000	0.00000	0.00000
1.5 (–3)	0.06140	0.08683	0.10635	0.12280	0.12972	0.02760	0.00854	0.00099	0.00000	0.00000	0.00000	0.00000
2.0 (–3)	0.07090	0.10027	0.12280	0.14179	0.14912	0.03551	0.01298	0.00218	0.00003	0.00000	0.00000	0.00000
3.0 (–3)	0.08683	0.12280	0.15040	0.17364	0.18065	0.04949	0.02187	0.00545	0.00022	0.00000	0.00000	0.00000
4.0 (–3)	0.10027	0.14180	0.17366	0.20040	0.20621	0.06182	0.03045	0.00943	0.00070	0.00000	0.00000	0.00000
6.0 (–3)	0.12280	0.17366	0.21269	0.24490	0.24711	0.08323	0.04653	0.01834	0.00255	0.00000	0.00000	0.00000
8.0 (–3)	0.14180	0.20053	0.24560	0.28178	0.27993	0.10182	0.06130	0.02766	0.00539	0.00002	0.00000	0.00000
1.0 (–2)	0.15853	0.22420	0.27459	0.31366	0.30779	0.11849	0.07502	0.03701	0.00894	0.00007	0.00000	0.00000
1.5 (–2)	0.19416	0.27459	0.33626	0.37939	0.36441	0.15468	0.10586	0.05971	0.01964	0.00056	0.00000	0.00000
2.0 (–2)	0.22420	0.31707	0.38813	0.43254	0.40992	0.18572	0.13316	0.08115	0.03168	0.00176	0.00000	0.00000
3.0 (–2)	0.27459	0.38832	0.47449	0.51749	0.48290	0.23857	0.18083	0.12057	0.05703	0.00639	0.00000	0.00000
4.0 (–2)	0.31707	0.44839	0.54620	0.58562	0.54196	0.28366	0.22235	0.15630	0.08244	0.01348	0.00000	0.00000
6.0 (–2)	0.38833	0.54902	0.66339	0.69421	0.63744	0.36006	0.29388	0.21989	0.13136	0.03243	0.00004	0.00000
8.0 (–2)	0.44840	0.63349	0.75878	0.78141	0.71536	0.42502	0.35553	0.27611	0.17728	0.05494	0.00029	0.00000
1.0 (–1)	0.50132	0.70740	0.84024	0.85566	0.78254	0.48255	0.41059	0.32711	0.22045	0.07920	0.00096	0.00000
1.5 (–1)	0.61397	0.86218	1.00627	1.00753	0.92215	0.60571	0.52955	0.43906	0.31879	0.14256	0.00554	0.00012
2.0 (–1)	0.70883	0.98924	1.13941	1.13054	1.03714	0.71004	0.63110	0.53603	0.40671	0.20601	0.01485	0.00073
3.0 (–1)	0.86724	1.19438	1.35117	1.32902	1.22553	0.88508	0.80262	0.70175	0.56084	0.32742	0.04568	0.00528
4.0 (–1)	0.99929	1.35932	1.51996	1.48961	1.37998	1.03135	0.94672	0.84226	0.69414	0.43950	0.08743	0.01570
6.0 (–1)	1.21559	1.61965	1.78513	1.74526	1.62851	1.27030	1.18312	1.07440	0.91757	0.63686	0.18529	0.05246
8.0 (–1)	1.39103	1.82357	1.99218	1.94705	1.82625	1.46257	1.37392	1.26273	1.10086	0.80463	0.28708	0.10295

Annex 18.1 (cont.)

	r′											
u_{vf}/r'	0.2	0.4	0.6	0.8	0.9	1.02	1.05	1.1	1.2	1.5	3.0	5.0
1.0 (0)	1.53940	1.99199	2.16283	2.11442	1.99107	1.62368	1.53410	1.42124	1.25609	0.94961	0.38544	0.15998
1.5 (0)	1.83451	2.31834	2.49276	2.43972	2.31269	1.94021	1.84935	1.73400	1.56406	1.24251	0.60594	0.30814
2.0 (0)	2.06164	2.56348	2.74001	2.68459	2.55560	2.18039	2.08887	1.97218	1.79964	1.46988	0.79194	0.44842
3.0 (0)	2.40238	2.92418	3.10308	3.04526	2.91413	2.53614	2.44393	2.32593	2.15055	1.81206	1.08872	0.69144
4.0 (0)	2.65597	3.18860	3.36880	3.30973	3.17756	2.79809	2.70552	2.58684	2.40995	2.06691	1.31920	0.89154
6.0 (0)	3.02611	3.57019	3.75180	3.69151	3.55799	3.17722	3.08428	2.96512	2.78646	2.43868	1.66541	1.20474
8.0 (0)	3.29564	3.84569	4.02807	3.96718	3.83291	3.45151	3.35846	3.23898	3.05936	2.70918	1.92265	1.44442
1.0 (1)	3.50775	4.06148	4.24433	4.18303	4.04831	3.66656	3.57342	3.45379	3.27379	2.92189	2.12726	1.63813
1.5 (1)	3.89834	4.45709	4.64060	4.57860	4.44340	4.06118	3.96789	3.84808	3.66747	3.31338	2.50776	2.00359
2.0 (1)	4.17853	4.73986	4.92367	4.86141	4.72590	4.34344	4.25008	4.13016	3.94956	3.59396	2.78278	2.27092
3.0 (1)	4.57648	5.14048	5.32468	5.26151	5.12621	4.74353	4.65009	4.53009	4.34951	3.99237	3.17547	2.65595
4.0 (1)	4.86045	5.42583	5.61018	5.54644	5.41128	5.02850	4.93503	4.81515	4.63445	4.27634	3.45672	2.93326
6.0 (1)	5.26228	5.82895	6.01361	5.94928	5.81412	5.43141	5.33791	5.21801	5.03730	4.67891	3.85551	3.32829
8.0 (1)	5.54817	6.11564	6.29997	6.23573	6.10066	5.71791	5.62440	5.50450	5.32384	4.96559	4.13987	3.61046
1.0 (2)	5.77033	6.33829	6.52259	6.45819	6.32318	5.94046	5.84689	5.72698	5.54636	5.18860	4.36129	3.83039
1.5 (2)	6.17463	6.74334	6.92657	6.86290	6.72801	6.34522	6.25167	6.13178	5.95151	5.59415	4.76406	4.23151
2.0 (2)	6.46189	7.03089	7.21356	7.15038	7.01552	6.63279	6.53920	6.41931	6.23907	5.88219	5.05035	4.51697
3.0 (2)	6.86706	7.43585	7.61809	7.55529	7.42086	7.03812	6.94452	6.82461	6.64448	6.28790	5.45488	4.92081
4.0 (2)	7.15488	7.72313	7.90564	7.84318	7.70883	7.32610	7.23250	7.11261	6.93250	6.57631	5.74377	5.20739
6.0 (2)	7.56052	8.12756	8.31136	8.24925	8.11502	7.73232	7.63884	7.51886	7.33884	6.98295	6.15148	5.61169
8.0 (2)	7.84877	8.41485	8.59898	8.53720	8.40296	8.02030	7.92683	7.80685	7.62680	7.27204	6.44392	5.89874
1.0 (3)	8.07147	8.63753	8.82263	8.76092	8.62679	8.24415	8.15057	8.03070	7.85073	7.49606	6.66932	6.12665
1.5 (3)	8.47653	9.04151	9.22584	9.16486	9.03065	8.64791	8.55441	8.43439	8.25429	7.89972	7.07649	6.53763
2.0 (3)	8.76193	9.32881	9.51199	9.45030	9.31586	8.93295	8.83941	8.71931	8.53926	8.18476	7.36643	6.82816
3.0 (3)	9.16578	9.73037	9.91361	9.85175	9.71687	9.33391	9.24030	9.12029	8.93978	8.58471	7.76451	7.24116
4.0 (3)	9.45036	10.01467	10.19906	10.13716	10.00216	9.61883	9.52517	9.40488	9.22463	8.86906	8.04828	7.52138
6.0 (3)	9.85226	10.41766	10.60245	10.53972	10.40471	10.02141	9.92772	9.80757	9.62655	9.27127	8.45020	7.91890
8.0 (3)	10.13825	10.70410	10.88851	10.82617	10.69110	10.30775	10.21403	10.09384	9.91271	9.55646	8.73502	8.20156

Annex 18.1 (cont.)

Table 2 For a vertical fracture with an observation well located on the y-axis

u_{vf}/r'	r′										
	0.05	0.07	0.10	0.20	0.30	0.40	0.50	1	2	5	10
1.00 (−3)	0.00000	0.00000	0.00000	0.00000	0.00000	0.00000	0.00000	0.00000	0.00000	0.00000	0.00000
1.50 (−3)	0.00000	0.00000	0.00000	0.00000	0.00000	0.00000	0.00000	0.00000	0.00000	0.00000	0.00000
2.00 (−3)	0.00000	0.00000	0.00000	0.00000	0.00000	0.00000	0.00000	0.00000	0.00000	0.00000	0.00000
3.00 (−3)	0.00006	0.00001	0.00000	0.00000	0.00000	0.00000	0.00000	0.00000	0.00000	0.00000	0.00000
4.00 (−3)	0.00025	0.00007	0.00001	0.00000	0.00000	0.00000	0.00000	0.00000	0.00000	0.00000	0.00000
6.00 (−3)	0.00117	0.00047	0.00012	0.00000	0.00000	0.00000	0.00000	0.00000	0.00000	0.00000	0.00000
8.00 (−3)	0.00275	0.00139	0.00050	0.00002	0.00000	0.00000	0.00000	0.00000	0.00000	0.00000	0.00000
1.00 (−2)	0.00483	0.00280	0.00124	0.00008	0.00001	0.00000	0.00000	0.00000	0.00000	0.00000	0.00000
1.50 (−2)	0.01129	0.00792	0.00460	0.00075	0.00013	0.00002	0.00000	0.00000	0.00000	0.00000	0.00000
2.00 (−2)	0.01860	0.01442	0.00965	0.00248	0.00064	0.00017	0.00004	0.00000	0.00000	0.00000	0.00000
3.00 (−2)	0.03376	0.02906	0.02259	0.00921	0.00372	0.00151	0.00062	0.00001	0.00000	0.00000	0.00000
4.00 (−2)	0.04857	0.04423	0.03721	0.01931	0.00978	0.00495	0.00252	0.00009	0.00000	0.00000	0.00000
6.00 (−2)	0.07619	0.07375	0.06751	0.04517	0.02896	0.01841	0.01169	0.00123	0.00001	0.00000	0.00000
8.00 (−2)	0.10127	0.10137	0.09713	0.07442	0.05399	0.03862	0.02750	0.00500	0.00017	0.00000	0.00000
1.00 (−1)	0.12428	0.12712	0.12544	0.10470	0.08205	0.06315	0.04825	0.01221	0.00077	0.00000	0.00000
1.50 (−1)	0.17515	0.18491	0.19042	0.17969	0.15702	0.13376	0.11257	0.04425	0.00636	0.00002	0.00000
2.00 (−1)	0.21934	0.23570	0.24856	0.25086	0.23236	0.20868	0.18443	0.09011	0.01954	0.00025	0.00000
3.00 (−1)	0.29505	0.32347	0.35030	0.38060	0.37467	0.35480	0.32899	0.19899	0.06527	0.00290	0.00002
4.00 (−1)	0.35983	0.39903	0.43866	0.49599	0.50331	0.48882	0.46367	0.31134	0.12603	0.01059	0.00026
6.00 (−1)	0.46966	0.52764	0.58982	0.69430	0.72420	0.71941	0.69679	0.51869	0.26073	0.04203	0.00305
8.00 (−1)	0.56293	0.63706	0.71840	0.86071	0.90750	0.91001	0.88970	0.69738	0.39250	0.08854	0.01103
1.00 (0)	0.64538	0.73369	0.83136	1.00390	1.06340	1.07133	1.05286	0.85155	0.51432	0.14276	0.02472
1.50 (0)	0.82086	0.93796	1.06706	1.29282	1.37310	1.38976	1.37437	1.16057	0.77457	0.28699	0.07785
2.00 (0)	0.96745	1.10628	1.25749	1.51735	1.61002	1.63180	1.61825	1.39799	0.98489	0.42538	0.14565
3.00 (0)	1.20679	1.37566	1.55542	1.85598	1.96262	1.99008	1.97862	1.75176	1.30959	0.66679	0.29140
4.00 (0)	1.39929	1.58768	1.78487	2.10875	2.22304	2.25358	2.24326	2.01304	1.55555	0.86624	0.43082
6.00 (0)	1.69952	1.91147	2.12874	2.47823	2.60066	2.63444	2.62533	2.39172	1.91836	1.17899	0.67347
8.00 (0)	1.93022	2.15581	2.38427	2.74751	2.87419	2.90967	2.90118	2.66587	2.18437	1.41849	0.87370

Annex 18.1 (cont.)

u_{vf}/r'	r′										
	0.05	0.07	0.10	0.20	0.30	0.40	0.50	1	2	5	10
1.00 (1)	2.11762	2.35208	2.58766	2.95948	3.08878	3.12529	3.11719	2.88086	2.39439	1.61218	1.04225
1.50 (1)	2.47347	2.72064	2.96622	3.34987	3.48274	3.52066	3.51309	3.27539	2.78222	1.97763	1.37243
2.00 (1)	2.73545	2.98939	3.24022	3.62996	3.76464	3.80327	3.79597	3.55757	3.06104	2.24499	1.62126
3.00 (1)	3.11460	3.37560	3.63182	4.02777	4.16430	4.20366	4.19662	3.95755	3.45759	2.62997	1.98717
4.00 (1)	3.38889	3.65353	3.91251	4.31162	4.44907	4.48879	4.48189	4.24247	3.74083	2.90737	2.25484
6.00 (1)	3.78079	4.04915	4.31094	4.71323	4.85162	4.89171	4.88495	4.64519	4.14185	3.30245	2.64014
8.00 (1)	4.06164	4.33189	4.59509	4.99899	5.13785	5.17812	5.17143	4.93151	4.42730	3.58496	2.91767
1.00 (2)	4.28066	4.55206	4.81611	5.22097	5.36013	5.40050	5.39385	5.15383	4.64911	3.80504	3.13461
1.50 (2)	4.68060	4.95353	5.21874	5.62490	5.76443	5.80496	5.79837	5.55821	5.05282	4.20641	3.53206
2.00 (2)	4.96552	5.23922	5.50501	5.91182	6.05154	6.09213	6.08557	5.84536	5.33962	4.49208	3.81567
3.00 (2)	5.36821	5.64269	5.90906	6.31653	6.45642	6.49711	6.49057	6.25029	5.74424	4.89555	4.21724
4.00 (2)	5.65451	5.92938	6.19604	6.60382	6.74383	6.78455	6.77802	6.53771	6.03151	5.18223	4.50306
6.00 (2)	6.05859	6.33386	6.60081	7.00892	7.14902	7.18977	7.18326	6.94294	6.43657	5.58679	4.90672
8.00 (2)	6.34559	6.62105	6.88814	7.29642	7.43657	7.47734	7.47085	7.23049	6.72408	5.87407	5.19347
1.00 (3)	6.56832	6.84390	7.11108	7.51945	7.65963	7.70042	7.69391	7.44869	6.94709	6.09691	5.41611
1.50 (3)	6.97323	7.24897	7.51627	7.92479	8.06458	8.10190	8.09275	7.84665	7.33935	6.50213	5.82106
2.00 (3)	7.26065	7.53647	7.80392	8.21240	8.34770	8.38526	8.37639	8.12959	7.61873	6.78967	6.10851
3.00 (3)	7.66586	7.94174	8.20917	8.61230	8.74778	8.78576	8.77714	8.52989	8.01469	7.16713	6.51390
4.00 (3)	7.95341	8.22934	8.49678	8.89628	9.03236	9.07036	9.06194	8.81432	8.29670	7.43627	6.80153
6.00 (3)	8.35876	8.63414	8.89639	9.29732	9.43398	9.47241	9.46439	9.21665	8.69575	7.81941	7.14822
8.00 (3)	8.64637	8.91758	9.18051	9.58259	9.71925	9.75780	9.75002	9.50161	8.97966	8.09528	7.40036

Annex 18.1 (cont.)

Table 3 For a vertical fracture with an observation well located at 45° from the fracture

	r′									
u_{vf}/r'	0.05	0.07	0.10	0.20	0.30	0.40	0.50	1	2	5
1.00 (−3)	0.00000	0.00000	0.00000	0.00000	−0.00000	−0.00000	0.00000	0.00000	0.00000	0.00000
1.50 (−3)	0.00000	0.00000	0.00000	0.00000	0.00000	−0.00000	−0.00000	0.00000	0.00000	0.00000
2.00 (−3)	0.00003	0.00000	0.00000	0.00000	0.00000	0.00000	−0.00000	0.00000	0.00000	0.00000
3.00 (−3)	0.00023	0.00006	0.00001	0.00000	0.00000	0.00000	0.00000	−0.00000	0.00000	0.00000
4.00 (−3)	0.00068	0.00026	0.00006	0.00000	0.00000	0.00000	0.00000	−0.00000	0.00000	0.00000
6.00 (−3)	0.00227	0.00120	0.00046	0.00002	0.00000	0.00000	0.00000	0.00000	0.00000	0.00000
8.00 (−3)	0.00452	0.00280	0.00136	0.00013	0.00001	0.00000	0.00000	0.00000	0.00000	0.00000
1.00 (−2)	0.00714	0.00489	0.00275	0.00040	0.00006	0.00001	0.00000	0.00000	0.00000	0.00000
1.50 (−2)	0.01445	0.01139	0.00782	0.00217	0.00061	0.00017	0.00005	0.00000	0.00000	0.00000
2.00 (−2)	0.02203	0.01872	0.01428	0.00550	0.00210	0.00081	0.00031	0.00000	0.00000	0.00000
3.00 (−2)	0.03682	0.03387	0.02889	0.01564	0.00825	0.00434	0.00229	0.00009	0.00000	0.00000
4.00 (−2)	0.05066	0.04866	0.04407	0.02856	0.01779	0.01099	0.00679	0.00057	0.00000	0.00000
6.00 (−2)	0.07568	0.07621	0.07363	0.05778	0.04284	0.03129	0.02273	0.00397	0.00003	0.00000
8.00 (−2)	0.09791	0.10121	0.10131	0.08814	0.07168	0.05712	0.04511	0.01138	0.00022	0.00000
1.00 (−1)	0.11807	0.12414	0.12714	0.11812	0.10183	0.08569	0.07122	0.02242	0.00079	0.00000
1.50 (−1)	0.16215	0.17481	0.18514	0.18915	0.17715	0.16080	0.14316	0.06080	0.00513	0.00000
2.00 (−1)	0.20010	0.21880	0.23614	0.25428	0.24906	0.23507	0.21653	0.10743	0.01438	0.00004
3.00 (−1)	0.26472	0.29417	0.32429	0.37019	0.38019	0.37307	0.35511	0.20784	0.04573	0.00063
4.00 (−1)	0.31977	0.35865	0.40019	0.47185	0.49638	0.49614	0.47972	0.30710	0.08844	0.00300
6.00 (−1)	0.41282	0.46796	0.52939	0.64594	0.69476	0.70590	0.69271	0.48912	0.18833	0.01599
8.00 (−1)	0.49168	0.56077	0.63928	0.79289	0.86028	0.87986	0.86938	0.64781	0.29177	0.03984
1.00 (0)	0.56135	0.64281	0.73627	0.92059	1.00232	1.02829	1.01995	0.78669	0.39143	0.07164
1.50 (0)	0.70994	0.81739	0.94111	1.18238	1.28850	1.32504	1.32039	1.07049	0.61412	0.16855
2.00 (0)	0.83499	0.96322	1.10970	1.38965	1.51093	1.55387	1.55151	1.29282	0.80142	0.27224
3.00 (0)	1.04230	1.20137	1.37927	1.70781	1.84677	1.89706	1.89737	1.62959	1.09967	0.46852
4.00 (0)	1.21247	1.39302	1.59131	1.94890	2.09785	2.15221	2.15398	1.88154	1.33096	0.64050
6.00 (0)	1.48403	1.69211	1.91505	2.30563	2.46541	2.52411	2.52753	2.25037	1.67807	0.92176
8.00 (0)	1.69728	1.92212	2.15932	2.56807	2.73362	2.79464	2.79891	2.51937	1.93578	1.14401

Annex 18.1 (cont.)

u_{vf}/r'	r'									
	0.05	0.07	0.10	0.20	0.30	0.40	0.50	1	2	5
1.00 (1)	1.87298	2.10905	2.35552	2.77579	2.94492	3.00737	3.01217	2.73118	2.14068	1.32669
1.50 (1)	2.21162	2.46420	2.72398	3.16033	3.33441	3.39877	3.40430	3.12138	2.52152	1.67678
2.00 (1)	2.46421	2.72580	2.99269	3.43742	3.61398	3.67937	3.68527	3.40142	2.79679	1.93613
3.00 (1)	2.83340	3.10456	3.37883	3.83218	4.01133	4.07773	4.08404	3.79917	3.18964	2.31285
4.00 (1)	3.10249	3.37865	3.65672	4.11448	4.29496	4.36183	4.36831	4.08302	3.47106	2.58612
6.00 (1)	3.48906	3.77032	4.05235	4.51453	4.69632	4.76373	4.77040	4.48464	3.87020	2.97686
8.00 (1)	3.76713	4.05108	4.33507	4.79951	4.98195	5.04967	5.05640	4.77038	4.15467	3.25731
1.00 (2)	3.98451	4.27005	4.55522	5.02104	5.20386	5.27171	5.27855	4.99243	4.37595	3.47595
1.50 (2)	4.38225	4.66990	4.95671	5.42434	5.60773	5.67581	5.68269	5.39638	4.77888	3.87545
2.00 (2)	4.66601	4.95478	5.24241	5.71098	5.89462	5.96278	5.96974	5.68332	5.06534	4.15998
3.00 (2)	5.06759	5.35746	5.64590	6.11539	6.29932	6.36757	6.37456	6.08810	5.46955	4.56261
4.00 (2)	5.35333	5.64375	5.93261	6.40258	6.58661	6.65493	6.66195	6.37431	5.75649	4.84890
6.00 (2)	5.75687	6.04782	6.33710	6.80752	6.99170	7.06008	7.06715	6.77734	6.16133	5.25271
8.00 (2)	6.04357	6.33486	6.62430	7.09495	7.27925	7.34762	7.35322	7.06372	6.45177	5.53980
1.00 (3)	6.26617	6.55758	6.84721	7.31798	7.50228	7.56975	7.57515	7.28588	6.67692	5.76245
1.50 (3)	6.67089	6.96254	7.25230	7.72331	7.90653	7.97296	7.97894	7.69000	7.08560	6.16729
2.00 (3)	6.95821	7.24996	7.53985	8.01087	8.19263	8.25938	8.26545	7.97637	7.37522	6.45484
3.00 (3)	7.36334	7.65521	7.94513	8.41414	8.59553	8.66137	8.66654	8.37542	7.77701	6.87116
4.00 (3)	7.65085	7.94277	8.23276	8.69970	8.87950	8.94536	8.95094	8.65926	8.06077	7.16421
6.00 (3)	8.05618	8.34819	8.63676	9.09984	9.28061	9.34699	9.35235	9.06018	8.46165	7.56116
8.00 (3)	8.34381	8.63518	8.92099	9.38443	9.56550	9.63197	9.63766	9.34528	8.74655	7.84564

Annex 18.2 Values of the function $F(u_{vf})$ for different values of u_{vf} (after Gringarten, Ramey and Raghavan 1974)

u_{vf}	$F(u_{vf})$	u_{vf}	$F(u_{vf})$
1.0 (–2)	0.3544	1.0 (1)	5.1200
1.5 (–2)	0.4342	1.5 (1)	5.5226
2.0 (–2)	0.5014	2.0 (1)	5.8090
3.0 (–2)	0.6140	3.0 (1)	6.2130
4.0 (–2)	0.7090	4.0 (1)	6.5000
5.0 (–2)	0.7926	5.0 (1)	6.7228
6.0 (–2)	0.8680	6.0 (1)	6.9048
8.0 (–2)	1.0014	8.0 (1)	7.1922
1.0 (–1)	1.1174	1.0 (2)	7.4150
1.5 (–1)	1.3580	1.5 (2)	7.8202
2.0 (–1)	1.5512	2.0 (2)	8.1078
3.0 (–1)	1.8522	3.0 (2)	8.5132
4.0 (–1)	2.0834	4.0 (2)	8.8008
5.0 (–1)	2.2710	5.0 (2)	9.0238
6.0 (–1)	2.4290	6.0 (2)	9.2062
8.0 (–1)	2.6854	8.0 (2)	9.4938
1.0 (0)	2.8894	1.0 (3)	9.7168
1.5 (0)	3.2688	1.5 (3)	10.1224
2.0 (0)	3.5432	2.0 (3)	10.4100
3.0 (0)	3.9352	3.0 (3)	10.8154
4.0 (0)	4.2160	4.0 (3)	11.1032
5.0 (0)	4.4350	5.0 (3)	11.3262
6.0 (0)	4.6146	6.0 (3)	11.5086
8.0 (0)	4.8988	8.0 (3)	11.7962

Annex 18.3 Values of the function $F(u_{vf},C'_{vf})$ for different values of u_{vf} and C'_{vf} (after Ramey and Gringarten 1976)

	C'_{vf}					
u_{vf}	0.001	0.005	0.01	0.05	0.1	0.5
1 (–6)	1.1205 (–3)					
1.5 (–6)	1.6450 (–3)					
2 (–6)	2.1159 (–3)					
3 (–6)	2.9508 (–3)					
4 (–6)	3.6983 (–3)					
6 (–6)	4.9975 (–3)					
8 (–6)	6.1444 (–3)					
1 (–5)	7.1851 (–3)	2.6122 (–3)	1.5802 (–3)	3.7982 (–4)		
1.5 (–5)	9.4121 (–3)	3.9039 (–3)	2.3653 (–3)	5.6945 (–4)		
2 (–5)	1.1347 (–2)	5.0794 (–3)	3.1096 (–3)	7.5684 (–4)		
3 (–5)	1.4623 (–2)	7.2394 (–3)	4.5236 (–3)	1.1270 (–3)		
4 (–5)	1.7436 (–2)	9.2119 (–3)	5.8623 (–3)	1.4924 (–3)		
6 (–5)	2.2169 (–2)	1.2740 (–2)	8.3519 (–3)	2.2095 (–3)		
8 (–5)	2.6210 (–2)	1.5924 (–2)	1.0681 (–2)	2.9145 (–3)		
1 (–4)	2.9806 (–2)	1.8867 (–2)	1.2892 (–2)	3.6097 (–3)	1.8440 (–3)	
1.5 (–4)	3.7392 (–2)	2.5279 (–2)	1.7894 (–2)	5.2967 (–3)	2.7646 (–3)	
2 (–4)	4.3822 (–2)	3.0952 (–2)	2.2479 (–2)	6.9421 (–3)	3.6675 (–3)	
3 (–4)	5.4598 (–2)	4.0687 (–2)	3.0617 (–2)	1.0104 (–2)	5.4404 (–3)	
4 (–4)	6.3724 (–2)	4.9174 (–2)	3.7945 (–2)	1.3160 (–2)	7.1772 (–3)	
6 (–4)	7.8964 (–2)	6.3566 (–2)	5.0710 (–2)	1.8955 (–2)	1.0553 (–2)	
8 (–4)	9.1807 (–2)	7.5941 (–2)	6.1989 (–2)	2.4490 (–2)	1.3841 (–2)	
1 (–3)	1.0315 (–1)	8.7001 (–2)	7.2239 (–2)	2.9815 (–2)	1.7060 (–2)	3.7963 (–3)
1.5 (–3)	1.2716 (–1)	1.1039 (–1)	9.4200 (–2)	4.2169 (–2)	2.4757 (–2)	5.6926 (–3)
2 (–3)	1.4739 (–1)	1.3033 (–1)	1.1326 (–1)	5.3746 (–2)	3.2169 (–2)	7.5650 (–3)
3 (–3)	1.8114 (–1)	1.6367 (–1)	1.4545 (–1)	7.4811 (–2)	4.6145 (–2)	1.1265 (–2)
4 (–3)	2.0957 (–1)	1.9193 (–1)	1.7306 (–1)	9.4207 (–2)	5.9436 (–2)	1.4915 (–2)
6 (–3)	2.5719 (–1)	2.3926 (–1)	2.1950 (–1)	1.2881 (–2)	8.4083 (–2)	2.2076 (–2)
8 (–3)	2.9732 (–1)	2.7929 (–1)	2.5905 (–1)	1.6009 (–2)	1.0716 (–1)	2.9111 (–2)
1 (–2)	3.3259 (–1)	3.1459 (–1)	2.9411 (–1)	1.8894 (–1)	1.2901 (–1)	3.6045 (–2)
1.5 (–2)	4.0667 (–1)	3.8857 (–1)	3.6420 (–1)	2.5162 (–1)	1.7833 (–1)	5.2848 (–2)
2 (–2)	4.6837 (–1)	4.5039 (–1)	4.2572 (–1)	3.0676 (–1)	2.2335 (–1)	6.9215 (–2)
3 (–2)	5.6981 (–1)	5.5205 (–1)	5.2624 (–1)	4.0032 (–1)	3.0260 (–1)	1.0059 (–1)
4 (–2)	6.5351 (–1)	6.3608 (–1)	6.0913 (–1)	4.8071 (–1)	3.7319 (–1)	1.3083 (–1)
6 (–2)		7.7284 (–1)	7.4343 (–1)	6.1415 (–1)	4.9407 (–1)	1.8789 (–1)
8 (–2)		8.8453 (–1)	8.5289 (–1)	7.2609 (–1)	5.9880 (–1)	2.4209 (–1)
1 (–1)		9.8038 (–1)	9.4673 (–1)	8.2383 (–1)	6.9218 (–1)	2.9388 (–1)
1.5 (–1)			1.1367 (0)	1.0222 (0)	8.8500 (–1)	4.1248 (–1)
2 (–1)			1.2891 (0)	1.1835 (0)	1.0456 (0)	5.2188 (–1)
3 (–1)			1.5296 (0)	1.4377 (0)	1.3019 (0)	7.1547 (–1)
4 (–1)			1.7198 (0)	1.6389 (0)	1.5081 (0)	8.8837 (–1)
6 (–1)				1.9478 (0)	1.8265 (0)	1.1811 (0)
8 (–1)				2.1856 (0)	2.0738 (0)	1.4313 (0)
1 (0)				2.3803 (0)	2.2774 (0)	1.6505 (0)
1.5 (0)				2.7476 (0)	2.6594 (0)	2.0818 (0)
2 (0)				3.0212 (0)	2.9444 (0)	2.4241 (0)
3 (0)					3.3628 (0)	2.9293 (0)
4 (0)					3.6792 (0)	3.3057 (0)
6 (0)						3.8331 (0)
8 (0)						4.2077 (0)
1 (+1)						4.4985 (0)

Annex 19.1 Values of F(χ,τ) according to Equation 19.2

τ	χ=.01	χ=.025	χ=.05	χ=.1	χ=.25	χ=.5	χ=1.0	χ=2.5	χ=5.0	χ=10
0.0010	0.0261	0.0158	0.0058	–	–	–	–	–	–	–
0.0015	0.0337	0.0226	0.0104	0.0014	–	–	–	–	–	–
0.0025	0.0458	0.0338	0.0192	0.0048	–	–	–	–	–	–
0.0040	0.0599	0.0474	0.0308	0.0112	–	–	–	–	–	–
0.0065	0.0783	0.0653	0.0471	0.0224	0.0010	–	–	–	–	–
0.010	0.0985	0.0851	0.0657	0.0370	0.0039	–	–	–	–	–
0.015	0.1216	0.1079	0.0877	0.0557	0.0102	–	–	–	–	–
0.025	0.1573	0.1433	0.1221	0.0868	0.0257	0.0016	–	–	–	–
0.040	0.1980	0.1839	0.1620	0.1241	0.0497	0.0070	–	–	–	–
0.065	0.2496	0.2353	0.2129	0.1729	0.0865	0.0213	–	–	–	–
0.10	0.3046	0.2902	0.2673	0.2258	0.1305	0.0448	0.0027	–	–	–
0.15	0.3654	0.3509	0.3278	0.2851	0.1826	0.0788	0.0096	–	–	–
0.25	0.4558	0.4412	0.4178	0.3739	0.2639	0.1398	0.0310	–	–	–
0.40	0.5538	0.5392	0.5155	0.4707	0.3552	0.2151	0.0689	–	–	–
0.65	0.6710	0.6563	0.6324	0.5869	0.4666	0.3126	0.1301	0.0045	–	–
1.0	0.7888	0.7741	0.7501	0.7040	0.5802	0.4157	0.2040	0.0157	–	–
1.5	0.9120	0.8973	0.8731	0.8266	0.7001	0.5270	0.2905	0.0376	–	–
2.5	1.0843	1.0695	1.0453	0.9983	0.8687	0.6865	0.4222	0.0857	0.0034	–
4.0	1.2603	1.2455	1.2212	1.1739	1.0419	0.8523	0.5654	0.1530	0.0128	–
6.5	1.4610	1.4462	1.4218	1.3741	1.2401	1.0439	0.7360	0.2574	0.0342	–
10	1.6568	1.6420	1.6175	1.5696	1.4339	1.2324	0.9077	0.3537	0.0670	0.0016
15	1.8594	1.8435	1.8190	1.7709	1.6339	1.4279	1.0886	0.4748	0.1126	0.0049
25	2.1393	2.1244	2.0998	2.0515	1.9129	1.7020	1.3458	0.6595	0.1947	0.0149
40	2.4283	2.4134	2.3888	2.3403	2.2004	1.9854	1.6152	0.8650	0.3000	0.0334
65	2.7620	2.7471	2.7225	2.6738	2.5328	2.3139	1.9305	1.1176	0.4445	0.0674
100	3.0919	3.0771	3.0524	3.0035	2.8617	2.6397	2.2456	1.3798	0.6085	0.1152
150	3.4354	3.4204	3.3957	3.3468	3.2041	2.9795	2.5761	1.6631	0.7983	0.1808
250	3.9197	3.9037	3.8790	3.8299	3.6864	3.4590	3.0450	2.0757	1.0931	0.3001
400	4.4197	4.4049	4.3801	4.3309	4.1867	3.9569	3.5341	2.5168	1.4270	0.4559
650	5.0019	4.9870	4.9622	4.9129	4.7680	4.5359	4.1049	3.0416	1.8436	0.6748
1000	5.5809	5.5649	5.5401	5.4907	5.3453	5.1115	4.6739	3.5727	2.2816	0.9283

Annex 19.2 Values of the function $F(u_a)$ according to Equation 19.6

$1/u_a^2$	$F(u_a)$	$1/u_a^2$	$F(u_a)$	$1/u_a^2$	$F(u_a)$
0.10	0.0000	10.0	0.5379	1000	0.9449
0.15	0.0001	15.0	0.6083	1500	0.9549
0.25	0.0017	25.0	0.6852	2500	0.9650
0.40	0.0110	40.0	0.7446	4000	0.9722
0.65	0.0401	65.0	0.7955	6500	0.9782
1.00	0.0891	100.0	0.8327	10000	0.9824
1.50	0.1542	150.0	0.8619	15000	0.9856
2.50	0.2543	250.0	0.8919	25000	0.9888
4.00	0.3539	400.0	0.9139	40000	0.9912
6.5	0.4548	650.0	0.9320	65000	0.9931

Annex 19.3 Values of the function $F(\tau)$ according to Equation 19.9

τ	$F(\tau)$	τ	$F(\tau)$	τ	$F(\tau)$	τ	$F(\tau)$
0.0010	0.0352	0.065	0.259	4.0	1.27	250	3.93
0.0015	0.0430	0.10	0.315	6.5	1.47	400	4.43
0.0025	0.0552	0.15	0.375	10	1.67	650	5.01
0.0040	0.0695	0.25	0.466	15	1.87	1000	5.59
0.0065	0.0879	0.40	0.564	25	2.15	1500	6.19
0.010	0.1082	0.65	0.681	40	2.44	2500	7.04
0.015	0.1313	1.0	0.799	65	2.77	4000	7.93
0.025	0.1671	1.5	0.922	100	3.10	6500	8.96
0.040	0.2079	2.5	1.094	150	3.45		

References

Anonymous 1964. Steady flow of groundwater towards wells. Proc. Comm. Hydrol. Research TNO No. 10, 179 pp.

Abdul Khader, M.H. and M.K. Veerankutty 1975. Transient well flow in an unconfined-confined aquifer system. J. Hydrol., 26, pp. 123-140.

Aron, G. and V.H. Scott 1965. Simplified solution for decreasing flow in wells. J. Hydraul. Div., Proc. Am. Soc. Civil Engrs., 91(HY5), pp. 1-12.

Barenblatt, G.E., I.P. Zheltov, and I.N. Kochina 1960. Basic concepts in the theory of seepage of homogeneous liquids in fissured rocks. Journal of Applied Mathematics and Mechanics, 24(5), pp. 1286-1303.

Bierschenk, W.H. 1963. Determining well efficiency by multiple step-drawdown tests. Intern. Assoc. Sci. Hydrol. Publ. 64, pp. 493-507.

Birsoy, Y.K. and W.K. Summers 1980. Determination of aquifer parameters from step tests and intermittent pumping data. Ground Water, 18, pp. 137-146.

Boehmer, W.K. and J. Boonstra 1986. Flow to wells in intrusive dikes. Ph.D. Thesis, Free University, Amsterdam, 262 pp.

Boehmer, W.K. and J. Boonstra 1987. Analysis of drawdown in the country rock of composite dike aquifers. J. Hydrol., 94, pp. 199-214.

Boonstra, J. and W.K. Boehmer 1986. Analysis of data from aquifer and well tests in intrusive dikes. J. Hydrol. 88, pp. 301-317.

Boonstra, J. and W.K. Boehmer 1989. Analysis of data from well tests in dikes and fractures. In: G. Jousma et al. (eds.), Proc. Intern. Conf. on Groundwater Contamination: Use of models in decision-making, Amsterdam, 1987. Kluwer Acad. Press, Dordrecht, pp. 171-180.

Boulton, N.S. 1954. The drawdown of the watertable under non-steady conditions near a pumped well in an unconfined formation. Proc. Inst. Civil Engrs. 3, pp. 564-579.

Boulton, N.S. 1963. Analysis of data from non-equilibrium pumping tests allowing for delayed yield from storage. Proc. Inst. Civil Engrs., 26, pp. 469-482.

Boulton, N.S. and T.D. Streltsova 1976. The drawdown near an abstraction of large diameter under non-steady conditions in an unconfined aquifer. J. Hydrol., 30, pp. 29-46.

Boulton, N.S. and T.D. Streltsova 1977. Unsteady flow to a pumped well in a fissured waterbearing formation. J. Hydrol., 35, pp. 257-270.

Bourdet, D. and A.C. Gringarten 1980. Determination of fissure volume and block size in fractured reservoirs by type-curve analysis. Paper SPE 9293 presented at the 1980 SPE Annual Fall Techn. Conf. and Exhib., Dallas.

Bouwer, H. 1978. Groundwater hydrology. McGraw-Hill Book, New York, 480 pp.

Bouwer, H. and R.C. Rice 1976. A slug test for determining hydraulic conductivity of unconfined aquifers with completely or partially penetrating wells. Water Resources Res. Vol. 12, pp. 423-428.

Bouwer, H. and R.C. Rice 1978. Delayed aquifer field as a phenomenon of delayed air entry. Water Resources Res., Vol. 14, pp. 1068-1074.

Bruggeman, G.A. 1966. Analyse van de bodemconstanten in een grondpakket, bestaande uit twee of meer watervoerende lagen gescheiden door semi-permeabele lagen. Unpublished research report.

Bukhari, S.A., A. Vandenberg, and D.H. Lennox 1969. Iterative analysis: bounded leaky artesian aquifer. J. Irr. Drain. Div., Proc. Am. Soc. Civil Engrs., Vol. 95(IR1), pp. 1-14.

Butler, J.J. 1988. Pumping tests in nonuniform aquifers – The radially symmetric case. J. Hydrol., 101, pp. 15-30.

Cinco Ley, H., F. Samaniego, and N. Dominguez 1978. Transient pressure behavior for a well with a finite-conductivity vertical fracture. Soc. Petrol. Engrs. J., 18, 253-264 pp.

Clark, L. 1977. The analysis and planning of step drawdown tests. Quart.J. Engng. Geol. Vol. 10, pp. 125-143.

Cooley, R.L. and C.M. Case 1973. Effect of a watertable aquitard on drawdown in an underlying pumped aquifer. Water Resources Res., Vol. 9, pp. 434-447.

Cooper, H.H., J.D. Bredehoeft, and I.S. Papadopulos 1967. Response of a finite-diameter well to an instantaneous charge of water. Water Resources Res., Vol. 3, pp. 263-269.

Cooper, H.H. and C.E. Jacob 1946. A generalized graphical method for evaluating formation constants and summarizing well field history. Am. Geophys. Union Trans. Vol. 27, pp. 526-534.

Dagan, G. 1967. A method of determining the permeability and effective porosity of unconfined anisotropic aquifers. Water Resources Res., Vol. 3, pp. 1059-1071.

Darcy, H. 1856. Les fontaines publiques de la ville de Dijon, V. Dalmont, Paris, 647 pp.

De Glee, G.J. 1930. Over grondwaterstromingen bij wateronttrekking door middel van putten. Thesis. J. Waltman, Delft (The Netherlands), 175 pp.

De Glee, G.J. 1951. Berekeningsmethoden voor de winning van grondwater. In: Drinkwatervoorziening, 3e Vacantiecursus: 38-80 Moorman's periodieke pers, The Hague.

De Marsily, G. 1986. Quantitative hydrogeology. Academic Press, London, 440 pp.

De Ridder, N.A. 1961. The hydraulic characteristics of the Tielerwaard calculated from pumping test data (in Dutch). Inst. Land and Water Manag. Res., Wageningen, Report no. 83, 15 pp.

De Ridder, N.A. 1966. Analysis of the pumping test 'De Vennebulten' near Varsseveld (in Dutch). Inst. Land and Water Manag. Res., Wageningen, Report no. 335, 5 pp.

Dietz, D.N. 1943. De toepassing van invloedsfuncties bij het berekenen van de verlaging van het grondwater ten gevolge van wateronttrekking. Water, Vol. 27(6), pp. 51-54.

Driscoll, F.G. (ed.) 1986. Groundwater and wells. 2nd edition, Johnson Division, St. Paul, Minnesota, 1089 pp.

Dupuit, J. 1863. Etudes théoriques et pratiques sur le mouvement des eaux dans les canaux découverts et a travers les terrains permeables. 2ème edition; Dunot, Paris, 304 pp.

Earlougher, R.C. 1977. Advances in well test analysis. Monograph Vol.5, Soc. Petrol. Engrs. of Am. Inst. Mining Met. Engrs., Dallas, 264 pp.

Eden, R.N. and C.P. Hazel 1973. Computer and graphical analysis of variable discharge pumping test of wells. Inst. Engrs. Australia, Civil Engng. Trans., pp. 5-10.

Edelman, J.H. 1947. On the calculation of groundwater flows. Ph.D. Thesis, Univ. Techn. Delft, 77 pp. (in Dutch).

Edelman, J.H. 1972. Groundwater hydraulics of extensive aquifers. Int. Inst. for Land Reclam. and Improv., Wageningen, Bull. 13. 216 pp.

Ferris, J.G. 1950. Quantitative method for determining groundwater characteristics for drainage design. Agr. Engineering, Vol. 31, pp. 285-291.

Ferris, J.G., D.B. Knowless, R.H. Brown, and R.W. Stallman 1962. Theory of aquifer tests. U.S. Geological Survey, Water-Supply Paper 1536E, 174 pp.

Gambolati, G. 1976. Transient free surface flow to a well: An analysis of theoretical solutions. Water Resources Res., Vol. 12, pp. 27-39.

Genetier, B. 1984. La pratique des pompages d'essai en hydrogéologie. Bur. Rech. Géol. Min.. Manuels et méthodes, No. 9, 132 pp.

Gringarten, A.C. 1982. Flow-test evaluation of fractured reservoirs. In: Narasimhan, T.N. Recent trends in hydrogeology. Geological Soc. Am. Special Paper 189, pp. 237-263.

Gringarten, A.C. and H.J. Ramey Jr. 1974. Unsteady state pressure distributions created by a well with a single horizontal fracture, partial penetration or restricted entry. Soc. Petrol. Engrs. J., pp. 413-426.

Gringarten, A.C., H.J. Ramey Jr, and R. Raghavan 1975. Applied pressure analysis for fractured wells. J. Petrol. Techn. pp. 887-892.

Gringarten, A.C. and P.A. Witherspoon 1972. A method of analyzing pump test data from fractured aquifers. In: Int. Soc. Rock Mechanics and Int. Ass. Eng. Geol., Proc. Symp. Rock Mechanics, Stuttgart, Vol.3-B pp. 1-9.

Groundwater Manual 1981. A water resources technical publication. U.S. Department of the Interior; Water and Power Resources Service. U.S. Government Printing Office, Denver, 480 pp.

Hantush, M.S. 1956. Analysis of data from pumping tests in leaky aquifers. Trans. Amer. Geophys. Union, Vol. 37, pp. 702-714.

Hantush, M.S. 1959a. Non-steady flow to flowing wells in leaky aquifers. J. Geophys. Res., Vol. 64, pp. 1043-1052.

Hantush, M.S. 1959b. Analysis of data from pumping wells near a river. J. Geophys. Res., Vol. 94, pp. 1921-1932.

Hantush, M.S. 1960. Modification of the theory of leaky aquifers. J. Geophys. Res. Vol. 65, pp. 3713-3725.

Hantush, M.S. 1961a. Drawdown around a partially penetrating well. J. Hydraul. Div., Proc. Amer. Soc. Civil. Engrs. Vol. 87(HY4), pp. 83-98.

Hantush, M.S. 1961b. Aquifer tests on partially penetrating wells. J. Hydraul. Div., Proc. Amer. Soc. Civil Engrs., Vol. 87(HY5), pp. 171-195.
Hantush, M.S. 1962. Flow of ground water in sands of nonuniform thickness; 3. Flow to wells. J. Geophys. Res., Vol. 67, pp. 1527-1534.
Hantush, M.S. 1964. Hydraulics of wells. In: V.T. Chow (editor). Advances in hydroscience, Vol. I, pp. 281-432. Academic Press, New York and London.
Hantush, M.S. 1966. Analysis of data from pumping tests in anisotropic aquifers. J. Geophys. Res., Vol. 71, pp. 421-426.
Hantush, M.S. 1967. Flow to wells in aquifers separated by a semipervious layer. J. Geophys. Res., Vol. 72, pp. 1709-1720.
Hantush, M.S. and C.E. Jacob 1955. Non-steady radial flow in an infinite leaky aquifer. Trans. Amer. Geophys. Union Vol. 36, pp. 95-100.
Hantush, M.S. and R.G. Thomas 1966. A method for analyzing a drawdown test in anisotropic aquifers. Water Resources Res., Vol. 2, pp. 281-285.
Hemker, C.J. 1984. Steady groundwater flow in leaky multiple-aquifer systems. J.Hydrol., 72, pp. 355-374.
Hemker, C.J. 1985. Microcomputer aquifer test evaluation. Aqua-VU No. 9. Free Univ., Inst. Earth Sciences, Amsterdam, 52 pp.
Huisman, L. 1972. Groundwater recovery. MacMillan, 336 pp.
Hurr, R.T. 1966. A new approach for estimating transmissibility from specific capacity. Water Resources Res. Vol. 2, pp. 657-664.
Jacob, C.E. 1940. On the flow of water in an elastic artesian aquifer. Trans. Amer. Geophys. Union, Vol. 21, Part 2, pp. 574-586.
Jacob, C.E. 1944. Notes on determining permeability by pumping tests under watertable conditions. U.S. Geol. Surv. open. file rept.
Jacob, C.E. 1947. Drawdown test to determine effective radius of artesian well. Trans. Amer. Soc. of Civil. Engrs., Vol. 112, Paper 2321, pp. 1047-1064.
Jacob, C.E. and S.W. Lohman 1952. Non-steady flow to a well of constant drawdown in an extensive aquifer. Trans. Amer. Geophys. Union, Vol. 33, pp. 559-569.
Jahnke, E. and F. Embde 1945. Tables of functions with formulas and curves. Dover Publ., New York, 306 p.
Javandel, I. and P.A. Witherspoon 1980. A semi-analytical solution for partial penetration in two-layer aquifers. Water Resources Res., Vol. 16, pp. 1099-1106.
Javandel, I. and P.A. Witherspoon 1983. Analytical solution of a partially penetrating well in a two-layer aquifer. Water Resources Res., Vol. 19, pp. 567-578.
Jenkins, D.N. and J.K. Prentice 1982. Theory for aquifer test analysis in fractured rocks under linear (nonradial) flow conditions. Ground Water, Vol. 20, pp. 12-21.
Johnson, A.J. 1967. Specific yield. Compilation of specific yields for various materials. U.S. Geol. Survey Water Supply Paper 1662-D, 74 pp.
Kazemi, H., M.S. Seth, and G.W. Thomas 1969. The interpretation of interference tests in naturally fractured reservoirs with uniform fracture distribution. Soc. of Petrol. Engrs. J., pp. 463-472.
Kohlmeier, R., G. Strayle, and W. Giesel 1983. Determination of water levels in observation wells by measurement of transit-time of ultrasonic pulses and calculation of hydraulic parameters. In: Proceedings International Symposium on Methods and instrumentation for the investigation of groundwater systems. Noordwijkerhout, 1983, Netherlands Org. Appl. Sci. Res. TNO, pp. 489-501.
Krauss, I. 1974. Die Bestimmung der Transmissivitat von Grundwasser leitern aus dem Einschwingverhalten des Brunnen-Grundwasserleiter-Systems. Z. Geophys., Vol. 40, pp. 381-400.
Kroszynski, U.I. and G. Dagan 1975. Well pumping in unconfined aquifers: the influence of the unsaturated zone. Water Resources Res., Vol. 11, pp. 479-490.
Lennox, D.H. 1966. Analysis of step-drawdown test. J. Hydr. Div., Proc. of the Amer. Soc. Civil Engrs., Vol. 92(HY6), pp. 25-48.
Maini, F. and G. Hocking 1977. An examination of the feasibility of hydrologic isolation of a high level waste repository in crystalline rocks. Invited paper, Geologic Disposal of High Radio-active Waste Session. Ann. Meeting, Geol. Soc. Am., Seattle, Washington.
Matthess, G. 1982. The properties of groundwater. John Wiley & Sons, New York, pp. 406.
Matthews, C.S. and D.G. Russel 1967. Pressure buildup and flow tests in wells. Soc. Petrol. Engrs. of Am. Inst. Min. Met. Engrs., Monograph 1, 167 pp.

Mavor, M.J. and H. Cinco Ley 1979. Transient pressure behavior of naturally fractured reservoirs. Paper SPE 7977 presented at the 1979 California SPE Regional Meeting, Ventura, California.

Merton, J.G. 1987. Interpretations des essais de pompage en milieu fissuré. IWACO – Bureau d'Etudes en Eau et Environnement. Rotterdam, Pays-Bas, 120 p.

Moench, A.F. 1984. Double-porosity models for a fissured groundwater reservoir with fracture skin. Water Resources Res., Vol. 20, pp. 831-846.

Moench, A.F. and P.A. Hsieh 1985. Analysis of slug test data in a well with finite thickness skin. In: IAH Memoires, Vol. XVII, Part 1. Proceedings of the 17th IAH Congress on 'The Hydrology of Rocks of Low Permeability'. Tucson, Arizona, pp. 17-29.

Mulder, P.J.M. 1983. Rapportage putproef Hoogezand. Dienst Grondwaterverkenning TNO Delft, OS.83-24, 7 p.

Muskat, M. 1937. The flow of homogeneous fluids through porous media. McGraw Hill Book Co., New York, 763 pp.

Najurietta, H.L. 1980. A theory for pressure transient analysis in naturally fractured reservoirs. J. of Petrol. Techn. July 1980, pp. 1241-1250.

Nespak-Ilaco 1985. Optimal Well and Well-field Design Report. Panjnad-Abbasia Salinity Control and Reclamation project. Supplement to the Final Plan Report. Pakistan Water and Power Devel. Auth., Lahore/Arnhem, 109 pp.

Neuman, S.P. 1972. Theory of flow in unconfined aquifers considering delayed response of the watertable. Water Resources Res., Vol.8, pp. 1031-1045.

Neuman, S.P. 1973. Supplementary comments on Theory of flow in unconfined aquifers considering delayed response of the watertable. Water Resources Res., Vol. 9, pp. 1102-1103.

Neuman, S.P. 1974. Effect of partial penetration on flow in unconfined aquifers considering delayed gravity response. Water Resources Res., Vol. 10, pp. 303-312.

Neuman, S.P. 1975. Analysis of pumping test data from anisotropic unconfined aquifers considering delayed gravity response. Water Resources Res., Vol. 11, pp. 329-342.

Neuman, S.P. 1979. Perspective on 'Delayed yield'. Water Resources Res., Vol. 15, pp. 899-908.

Neuman, S.P., G.R. Walter, H.W. Bentley, J.J. Ward, and D.D. Gonzalez. 1984. Determination of horizontal anisotropy with three wells. Ground Water, Vol. 22, pp. 66-72.

Neuman, S.P. and P.A. Witherspoon 1968. Theory of flow in aquicludes adjacent to slightly leaky aquifers. Water Resources Res., Vol. 4, pp. 103-112.

Neuman, S.P. and P.A. Witherspoon 1969a. Theory of flow in a confined two aquifer system. Water Resources Res., Vol. 5, pp. 803-816.

Neuman, S.P. and P.A. Witherspoon 1969b. Applicability of current theories of flow in leaky aquifers. Water Resources Res., Vol. 5, pp. 817-829.

Neuman, S.P. and P.A. Witherspoon 1972. Field determination of the hydraulic properties of leaky multiple aquifer systems. Water Resources Res., Vol. 8, pp. 1284-1298.

Papadopulos, I.S. 1965. Nonsteady flow to a well in an infinite anisotropic aquifer. Intern. Assoc. Sci. Hydrol., Proc. Dubrovnik Symposium on the Hydrology of fractured rocks, pp. 21-31.

Papadopulos, I.S. 1966. Nonsteady flow to multi-aquifer wells. J. Geophys. Res., Vol. 71, pp. 4791-4797.

Papadopulos, I.S. 1967. Drawdown distribution around a large diameter well. Symp. on Groundwater Hydrology, San Fransisco. Proc. Amer. Water Resources Assoc., No. 4, pp. 157-168.

Papadopulos, S.S., J.D. Bredehoeft, and H.H. Cooper 1973. On the analysis of 'slug test' data. Water Resources Res., Vol. 9, pp. 1087-1089.

Papadopulos, I.S. and H.H. Cooper Jr 1967. Drawdown in a well of large diameter. Water Resources Res., Vol. 3, pp. 241-244.

Ramey Jr, H.J. 1976. Practical use of modern well test analysis. Paper SPE 5878, presented at the SPE-AIME 46th Annual California Regional Meeting, Long Beach, Ca.

Ramey, H.J. 1982. Well-loss function and the skin effect: A review. In: Narasimhan, T.N. (ed.) Recent trends in hydrogeology. Geol. Soc. Am., Special Paper 189, pp. 265-271.

Ramey, H.J., R.G. Agarwal, and I. Martin 1975. Analysis of 'Slug test' or DST flow period data. J. Can. Petrol. Technology, July-September, pp. 37-47.

Ramey Jr, H.J. and A.C. Gringarten 1976. Effect of high-volume vertical fractures on geothermal steam well behavior. In: Proceedings Second U.N. Devel. and Use of Geothermal Resources, San Fransisco, 1976, U.S. Government Printing Office, Washington D.C., Vol. 3, pp. 1759-1762.

Reed, J.E. 1980. Type curves for selected problems of flow to wells in confined aquifers. In: Techniques of water resources investigations of the U.S. Geological Survey, Book 3, Chapter 3, Washington D.C., 54 p.

Robinson, T.W. 1939. Earth-tides shown by fluctuations of water-levels in wells in New Mexico and Iowa. Trans. Amer. Geophys. Union, Vol. 20, pp. 656-666.
Rorabaugh, M.J. 1953. Graphical and theoretical analysis of step-drawdown test of artesian well. Proc. Amer. Soc. Civil Engrs., Vol. 79, separate no. 362, 23 pp.
Ross, B. 1985. Theory of the oscillating slug test in deep wells. In: Interb. Assoc. Hydrogeol. Memoires, Vol. 17, Part 1. Proc. 17th IAH Congress on hydrology of rocks of low permeability, Tucson, Arizona, pp. 44-51.
Rushton, K.R. and K.W.F. Howard 1982. The unreliability of open observation boreholes in unconfined aquifer pumping tests. Ground Water, Vol. 20, pp. 546-550.
Rushton, K.R. and K.S. Rathod 1980. Overflow tests analyzed by theoretical and numerical methods. Ground Water, Vol. 18, pp. 61-69.
Rushton, K.R. and V.S. Singh 1983. Drawdowns in large-diameter wells due to decreasing abstraction rates. Ground Water, Vol. 21, pp. 671-677.
Serra, K., A.C. Reynolds, and R. Raghavan 1983. New pressure transient analysis methods for naturally fractured reservoirs. J. of Petrol. Technol., pp. 2271-2283.
Sheahan, N.T. 1971. Type-curve solution of step-drawdown test. Ground Water, Vol. 9, pp. 25-29.
Skinner, A.C. 1988. Practical experience of borehole performance evaluation. J. Inst. Water Environm. Manag., Vol. 2, pp. 332-340.
Streltsova, T.D. 1972a. Unconfined aquifer and slow drainage. J. Hydrol., Vol. 16. pp. 117-124.
Streltsova, T.D. 1972b. Unsteady radial flow in an unconfined aquifer. Water Resources Res., Vol. 8, pp. 1059-1066.
Streltsova, T.D. 1973. On the leakage assumption applied to equations of groundwater flow. J. Hydrol., Vol. 20, pp. 237-253.
Streltsova, T.D. 1974. Drawdown in compressible unconfined aquifer. J. Hydraul. Div., Proc. Amer. Soc. Civil Engrs., Vol. 100(HY11), pp. 1601-1616.
Streltsova, T.D. 1976. Progress in research on well hydraulics. Advances in groundwater hydrology. Amer. Water Resources Assoc., pp. 15-28.
Streltsova-Adams, T.D. 1978. Well hydraulics in heterogeneous aquifer formations. Adv. in Hydrosci., Vol. 11, pp. 357-423.
Theis, C.V. 1935. The relation between the lowering of the piezometric surface and the rate and duration of discharge of a well using groundwater storage. Trans. Amer. Geophys. Union, Vol. 16, pp. 519-524.
Thiem, G. 1906. Hydrologische Methoden. Gebhardt, Leipzig, 56 pp.
Thiery, D., M. Vandenbeusch, and P. Vaubourg 1983. Interprétation des pompages d'essai en milieu fissuré aquifère. Documents du BRGM, no. 57, 53 pp.
Uffink, G.J.M. 1979. De bepaling van k-waarden in het veld aan de hand van 'putproeven' en 'puntproeven'. Resultaten van het onderzoek aan de Parkweg (Den Haag). R.I.D. mededeling 79-02, Voorburg, 61 pp.
Uffink, G.J.M. 1980. De bepaling van de doorlatendheid van watervoerende pakketten. R.I.D. mededeling 80-8, Voorburg, 37 pp.
Uffink, G.J.M. 1982. Richtlijnen voor het uitvoeren van putproeven H_2O, Vol. 15, pp. 202-205.
Uffink, G.J.M. 1984. Theory of the oscillating slug test. Nat. Institute for Public Health and Environmental Hygiene, Bilthoven. Unpublished research report, 18 pp. (in Dutch).
Vandenberg, A. 1975. Determining aquifer coefficients from residual drawdown data. Water Resources Res., Vol. 11, pp. 1025-1028.
Vandenberg, A. 1976. Tables and type curves for analysis of pump tests in leaky parallel-channel aquifers. Techn. Bull. No. 96. Inland Waters Directorate, Water Resources Branch, Ottawa, 28 pp.
Vandenberg, A. 1977. Type curves for analysis of pump tests in leaky strip aquifers. J. Hydrol., Vol. 33, pp. 15-26.
Van der Kamp, G. 1976. Determining aquifer transmissivity by means of well response tests: The underdamped case. Water Resources Res., Vol. 12, pp. 71-77.
Van der Kamp, G. 1985. Brief quantitative guidelines for the design and analysis of pumping tests. In: Hydrology in the Service of Man. Mem. 18th Congress Intern. Ass. Hydrogeol., Cambridge, pp. 197-206.
Van Golf-Racht, T.D. 1982. Fundaments of fractured reservoir engineering. Developments in Petroleum Science, 12. Elsevier Sci. Publ. Co., Amsterdam-Oxford-New York, 710 p.
Walton, W.C. 1962. Selected analytical methods for well and aquifer evaluation. Illinois State Water Survey Bull., No. 49; 81 pp.
Warren, J.E. and P.J. Root 1963. The behavior of naturally fractured reservoirs. Soc. of Petrol. Engrs. J., Vol. 3, pp. 245-255.

Weeks, E.P. 1969. Determining the ratio of horizontal to vertical permeability by aquifer-test analysis. Water Resources Res., Vol. 5, pp. 196-214.

Wenzel, L.K. 1942. Methods for determining permeability of water-bearing materials, with special reference to discharging well methods. U.S. Geol. Survey, Water Supply Paper 887, 192 pp.

Wit, K.E. 1963. The hydraulic characteristics of the Oude Korendijk polder, calculated from pumping test data and laboratory measurements of core samples (in Dutch). Inst. Land and Water Manag. Res., Wageningen, Report No. 190, 24 pp.

Witherspoon, P.A., J. Javandel, S.P. Neuman, and R.A. Freeze 1967. Interpretation of aquifer gas storage conditions from water pumping tests. Amer. Gas. Assoc. New York, 273 pp.

Worthington, P.F. 1981. Estimation of the transmissivity of thin leaky-confined aquifers from single-well pumping tests. J. of Hydrol., Vol. 49, pp. 19-30.

Author's index

Currently Available ILRI Publications

No.	Publications	Author	ISBN No.
11	Reclamation of saft-affected soils in Iraq	P. J. Dieleman (ed.)	–
15	Planning of service centres in rural areas of developing countries	D. B. W. M. van Dusseldorp	–
16	Drainage principes and applications (in 4 volumes)	–	90 70260 123, –131, –62 X and –63 8
16[S]	Principos y aplicationes del drenaje (en 4 volúmenes)	–	–
17	Land evaluation for rural purposes	R. Brinkman and A. J. Smyth	90 70260 859
19	On irrigation efficiencies	M. G. Bos and J. Nugteren	90 70260 875
20	Discharge measurements structures (Third revised edition)	M. G. Bos	90 70754 150
21	Optimum use of water resources	N. A. de Ridder and A. Erez	–
24	Drainage and reclamation of salt-affected soils	J. Martinez Beltrán	–
25	Proceedings of the International Drainage Workshop	J. Wesseling (ed.)	90 70260 549
26	Framework for regional planning in developing countries	J. M. van Staveren and D. B. W. M. van Dusseldorp	90 70260 832
27	Land reclamation and water management	–	90 70260 689
29	Numerical modelling of groundwater basins: A user-oriented manual	J. Boonstra and N. A. de Ridder	90 70260 697
30	Proceedings of the Symposium on Peat Lands Below Sea Level	H. de Bakker and M. W. van den Berg	90 70260 700
31	Proceedings of the Bangkok Symposium an Acid Sulphate Soils	H. Dost and N. Breeman (eds.)	90 70260 719
32	Monitoring and evaluation of agricultural change	Josette Murphy and Leendert H. Sprey	90 70260 743
33	Introduction to farm surveys	Josette Murphy and Leendert H. Sprey	90 70260 735
34	Evaluation permanente du développement agricole	Josette Murphy and Leendert H. Sprey	90 70260 891
35	Introduction aux enquêtes agricoles en Afrique	Josette Murphy and Leendert H. Sprey	90 70260 956
36	Proceedings of the International Workshop on Land Evaluation for Extensive Grazing (LEEG)	W. Siderius (ed.)	90 70260 948
37	Proceedings of the ISSS Symposium on 'Water and solute movement in heavy clay soils'	J. Bouma, P. A. C. Raats (ed.)	90 72060 972
38	Aforadores de caudal para canales abiertoss	M. G. Bos, J. A. Replogle and A. J. Clemmens	90 70260 921
39	Acid Sulphate Soils: A baseline for research and development	D. Dent	90 70260 980
40	Land evaluation for land-use planning and conservation in sloping areas	W. Siderius (ed.)	90 70260 999
41	Research on water management of rice fields in the Nile Delta, Egypt	S. EL. Guindy & I. A. Risseeuw; H. J. Nijland (Ed.)	90 70754 08 8
43	BASCAD, a mathematical model for level basin irrigation	J. Boonstra and M. Jurriëns	90 70754 12 6
44	Selected Papers of the Dakar Symposium on Acid Sulphate Soils	H. Dost (ed.)	90 70754 13 4
45	Health and Irrigation Volume 1	J. M. V. Oomen, J. de Wolf and W. R. Jobin	90 70754 21 5
45	Health and Irrigation Volume 2	J. M. V. Oomen, J. de Wolf and W. R. Jobin	90 70754 177

46	CRIWAR: A Simulation Program for Calculating the Crop Irrigation Water Requirement of a Cropped Area	J. Vos, P. Kabat, M. G. Bos, R. A. Feddes	90 70754 22 3
47	Analysis and Evaluation of Pumping Test Data	G. P. Kruseman and N. A. de Ridder	90 70754 20 7
48	SATEM: Selected Aquifer Test Evaluation Methods	J. Boonstra	90 70754 19 3
49	Screening of hydrological data: Tests for stationarity and relative consistency	E. R. Dahmen and M. J. Hall	90 70754 23 1

No.	Bulletins	Author	ISBN No.
1	The auger hole method	W. F. van Beers	90 70260 816
4	On the calcium carbonate content of young marine sediments	B. Verhoeven	–
6	Mud transport studies in coastal water from the Western Scheldt to the Danish frontier	A. J. de Groot	–
8	Some nomographs for the calculation of drain spacings	W. F. J. van Beers	–
9	The Managil South-Western Extension: An extension to the Gezira Scheme	D. J. Shaw	–
10	A viscous fluid model for demonstration of groundwater flow to parallel drains	F. Homma	90 70260 824
11[S]	Análisis y evaluación de los datos de ensayos por bombeo	G. P. Kruseman and N. A. de Ridder	–
11[F]	Interprétation et discussion des pompages d'essai	G. P. Kruseman and N. A. de Ridder	–
12	Gypsifereous Soils	J. G. van Alphen and F. de los Rios Romero	–
13	Groundwater hydraulics of extensive aquifers	J. H. Edelman	90 70260 794

No.	Bibliographies	Author	ISBN No.
7	Agricultural extension in developing countries	C. A. de Vries	–
8	Bibliography on cotton irrigation	C. J. Brouwer and L. F. Abell	–
9	Annotated bibliography or surface irrigation methods	S. Raadsma, G. Schrale	–
10	Soil Survey interpretation	R. H. Brook	–
13	Abstract journals on irrigation, drainage and water resources engineering	L. F. Abell	–
18	Drainage: An annotated guide to books and journals	G. Naber	90 70260 93 X

Other publications

Papers International Symposium Polders of the World (3 volumes)	90 70260 751 76 X and –778
Final Report Symposium Polders of the World	–
Proceedings Symposium Lowland Development Indonesia	90 70754 07 X